·开发宝典丛书·

iOS开发范例实战宝典

（基础篇）

杨佩璐　魏彩娟　编著

清华大学出版社
北　京

内 容 简 介

《iOS 开发范例实战宝典》分为基础篇和进阶篇两个分册，其内容包含了 iOS 开发必知必会的 240 个经典实例和几百个开发模块。书中的实例紧跟技术趋势，以最新的 iOS 8 为版本编写，内容覆盖了 iOS 开发的方方面面，几乎涉及 iOS 开发的所有重要知识。书中给出了每个实例的具体实现过程，并对程序代码做了详细注释，对其中的重点和难点进行了专门分析，而且精讲了每个实例的重点代码，读者可以在这些实例的基础上做出更多更新的功能。

本书为《iOS 开发范例实战宝典（基础篇）》，共 13 章，包含了 117 个开发实例。其中包括 7 个按钮类实例、9 个滑块类实例、2 个开关类实例、7 个进度条类实例、8 个指示器类实例、6 个选择器类实例、4 个视图类实例、3 个分段控件类实例、11 个导航栏类实例、6 个标签栏类实例、11 个菜单类实例、6 个提醒对话框类实例、18 个文本处理类实例和 19 个表实例。

本书涉及面广，涉及 iOS 软件开发的各种常用应用。适合所有想全面学习 iOS 开发技术的人员阅读，也适合 iOS 专业开发人员作为案头必备的参考书。

本书封面贴有清华大学出版社防伪标签，无标签者不得销售。
版权所有，侵权必究。侵权举报电话：010-62782989 13701121933

图书在版编目（CIP）数据

iOS 开发范例实战宝典. 基础篇 / 杨佩璐，魏彩娟编著. —北京：清华大学出版社，2015
（开发宝典丛书）
ISBN 978-7-302-39576-8

Ⅰ. ①i… Ⅱ. ①杨… ②魏… Ⅲ. ①移动终端–应用程序–程序设计 Ⅳ. ①TN929.53

中国版本图书馆 CIP 数据核字（2015）第 046516 号

责任编辑：杨如林
封面设计：欧振旭
责任校对：胡伟民
责任印制：沈　露

出版发行：清华大学出版社
网　　址：http://www.tup.com.cn, http://www.wqbook.com
地　　址：北京清华大学学研大厦 A 座　　邮　编：100084
社 总 机：010-62770175　　邮　购：010-62786544
投稿与读者服务：010-62776969, c-service@tup.tsinghua.edu.cn
质 量 反 馈：010-62772015, zhiliang@tup.tsinghua.edu.cn

印 刷 者：清华大学印刷厂
装 订 者：三河市新茂装订有限公司
经　　销：全国新华书店
开　　本：185mm×260mm　　印　张：43.25　　字　数：1080 千字
版　　次：2015 年 5 月第 1 版　　印　次：2015 年 5 月第 1 次印刷
印　　数：1～3500
定　　价：99.80 元

产品编号：063389-01

前　　言

移动应用开发是当前 IT 开发的热点。由于苹果提供了完备的开发工具和成熟的软件盈利方式，苹果的 iOS 开发成为热点中的热点。苹果开发技术较为封闭，尤其是相对于开源技术的 Android 开发而言更是如此。同时，由于移动开发的发展时间较短，这使得开发资料相对匮乏，尤其是实用性比较强的开发资料更是为数不多，开发者往往缺乏应用指导资料。

笔者结合自己多年的 iOS 开发经验和心得体会，花费了一年多的时间分析了 iOS 开发中常见的几百个应用场景，并进行了精心整理，挑选了最为典型的 240 个 iOS 开发实例，编写成了《iOS 开发范例实战宝典》（分为基础篇和进阶篇两个分册）。

本书为《iOS 开发范例实战宝典（基础篇）》，包含了 117 个经典实例，涉及 iOS 开发中较为基础的 14 个界面开发专题。希望各位读者能在本书的引领下跨入 iOS 开发的大门，并成为一名开发高手。

本书特色

1．实例丰富，代码精讲

本书详细讲解了 117 个 iOS 开发经典实例，并对重点代码做了大量注释和讲解，以便于读者更加轻松地学习。通过对这些实例的演练，可以快速提高读者的开发水平。

2．内容全面，涵盖广泛

本书全面介绍了 iOS 开发中最为常见的 14 类界面模块，包括按钮、滑块、开关、进度条、指示器、选择器、视图、分段控件、导航栏、标签栏、菜单、提醒对话框、文本处理和表。这些内容是 iOS 开发必知必会的内容，需要读者重点掌握。

3．讲解详细，循序渐进

本书中的每个实例都给出了详细的分析过程和实现步骤，书中的每个实例都按照"实现原理→实现过程→重点代码"三个步骤进行分析。对于复杂的实例，还给出了完备的流程图来帮助读者理解实例的工作机制，掌握起来更加容易。

4．专注核心，举一反三

为了在有限的篇幅内讲解更多的开发实例，本书只给出了每个实例的核心代码及分析。完整的实例代码读者可以自己下载阅读，并进行测试和练习，而且还可以对这些代码进行改造，以用于实际的开发之中，从而起到举一反三的作用。

本书内容及体系结构

第 1 章 按钮类效果

本章 7 个实例，主要内容包括发光的按钮、弹出式按钮、超文本链接和抛光效果的按钮等内容。通过本章的学习，读者可以实现各种常见按钮的开发。

第 2 章 滑块类效果

本章 9 个实例，主要内容包括具有多个颜色的滑块控件、环形滑块控件、具有范围的滑块控件以及具有文字的滑块控件等内容。通过本章的学习，读者可以掌握滑块类控件的技术与应用。

第 3 章 开关类

本章 2 个实例，主要内容包括自定义开关的外观、实现滑块窗口滑动切换的效果。通过本章的学习，读者可以掌握开关的一些常见技术及应用。

第 4 章 进度条类和指示器类效果

本章 15 个实例，主要内容包括扁平带梯度效果的进度条、扇形进度条、环形进度条、具有范围的进度条、文本上传进度条、倒计时进度条、带进度条的工具栏、变色的指示器，以及仿 Facebook web 上正在加载中的效果等内容。通过本章的学习，读者可以掌握进度条类和指示器类的各种常见技术及应用。

第 5 章 选择器类效果

本章 6 个实例，主要内容包括时间设置器、闹铃、城市经纬度查询、定制多选功能选择器、转盘选择器和老虎机。通过本章的学习，读者可以掌握选择器一些常见技术及应用。

第 6 章 视图

本章 4 个实例，主要内容包括更改空白视图的背景颜色、关闭应用程序、手电筒及旋转大挑战。通过本章的学习，读者可以掌握视图的一些常见技术及应用。

第 7 章 分段控件

本章 3 个实例，主要内容包括滑块式分段控件、开关式分段控件和自定义分段控件。通过本章的学习，读者可以掌握关于分段控件的一些常见技术及应用。

第 8 章 导航栏

本章 11 个实例，主要内容包括具有阴影的导航栏、具有图片的导航栏、具有分段控件的导航栏、具有子标题的导航栏、上下滑动的导航栏和具有下拉菜单的导航栏等内容。通过本章的学习，读者可以掌握关于导航栏的一些常见技术及应用。

第 9 章　标签栏

本章 6 个实例，主要内容包括右上角带有数字的标签栏、具有渐变效果的标签栏、中间凸起的标签栏、标签栏控制器实现的视图切换效果、具有动画效果的标签栏以及滚动的标签栏。通过本章的学习，读者可以掌握关于标签栏的一些常见技术及应用。

第 10 章　菜单

本章 11 个实例，主要内容包括立方体菜单、仿 Windows 8 的 Metro 风格、下拉菜单、浮动的菜单、具有按钮的菜单、仿 Tumblr iOS App 菜单、边栏菜单和九宫格菜单等内容。通过本章的学习，读者可以掌握关于菜单的一些常见技术及应用。

第 11 章　提醒对话框

本章 6 个实例，主要内容包括具有文本框的警告视图、全屏的警告视图、具有进度条的警告视图、具有列表的警告视图、坠落的警告视图、自定义的动作表单、弹出视图，以及模糊界面背景。通过本章的学习，读者可以掌握关于提醒对话框的一些常见技术及应用。

第 12 章　文本处理

本章 18 个实例，主要内容包括具有多个颜色的标签、发光的标签、循环渐变的标签、滚动的标签、具有光晕效果的标签和标签云等内容。通过本章的学习，读者可以掌握关于文本处理的一些常见技术及应用。

第 13 章　表

本章 19 个实例，主要内容包括邮编查询、水平列表、表的自动调整、排排看、归归类、自定义索引的表、自制的列表单选控件、下拉刷新列表、背景随动和卡片插入式列表等内容。通过本章的学习，读者可以掌握关于表的一些常见技术及应用。

本书读者对象

- 想全面学习 iOS 开发技术的人员；
- iOS 专业开发人员；
- iOS 开发爱好者；
- 大中专院校的学生；
- 社会培训班学员；
- 需要一本案头必备手册的程序员。

本书配套资源获取方式

本书提供以下的配套资源：
- 本书开发环境；
- 本书实例源代码；

为了节省读者的购书开支，本书放弃以配书光盘的方式提供这些资源，而是改为采用提供下载的方式。读者可以登录清华大学出版社网站（www.tup.com.cn），搜索到本书页面，然后按照提示下载，也可以在本书服务网站（www.wanjuanchina.net）的相关版块上下载这些配套资源。

本书售后服务方式

编程学习的最佳方式是共同学习。但是由于实际环境所限，大部分读者都是独自前行。为了便于读者更好地学习iOS语言，我们构建了多样的学习环境，力图打造立体化的学习方式，除了对内容精雕细琢之外，还提供了完善的学习交流和沟通方式。主要有以下几种方式：

- 提供技术论坛 http://www.wanjuanchina.net，读者可以将学习过程中遇到的问题发布到论坛上以获得帮助。
- 提供QQ交流群336212690，读者申请加入该群后便可以和作者及广大读者交流学习心得，解决学习中遇到的各种问题。
- 提供book@wanjuanchina.net和bookservice2008@163.com服务邮箱，读者可以将自己的疑问发电子邮件以获取帮助。

本书作者

本书主要由山东中医药大学的杨佩璐和河南牧业经济学院的魏彩娟编写。其中，杨佩璐编写了第1~7章，魏彩娟编写了第8~13章。其他参与编写的人员有陈晓建、陈振东、程凯、池建、崔久、崔莎、邓凤霞、邓伟杰、董建中、耿璐、韩红轲、胡超、黄格力、黄缙华、姜晓丽、李学军、刘娣、刘刚、刘宁、刘艳梅、刘志刚、司其军、滕川、王连心、沃怀凯、闫玉宝。另外，刘媛媛负责了各款iOS硬件环境下的代码验证和调试。

虽然笔者对本书中所述内容都尽量核实，并多次进行文字校对，但因时间所限，可能还存在疏漏和不足之处，恳请读者批评指正。

编者

目　　录

第 1 章　按钮类效果 ⋯⋯⋯1
　实例 1　发光的按钮 ⋯⋯⋯1
　实例 2　弹出式按钮 ⋯⋯⋯3
　实例 3　超文本链接 ⋯⋯⋯11
　实例 4　抛光效果的按钮 ⋯⋯⋯14
　实例 5　具有进度条的按钮 ⋯⋯⋯17
　实例 6　唱碟机按钮 ⋯⋯⋯25
　实例 7　环形按钮 ⋯⋯⋯31

第 2 章　滑块类效果 ⋯⋯⋯40
　实例 8　具有多个颜色的滑块控件 ⋯⋯⋯40
　实例 9　环形滑块控件 ⋯⋯⋯43
　实例 10　具有范围的滑块控件 ⋯⋯⋯49
　实例 11　具有文字的滑块控件 ⋯⋯⋯56
　实例 12　自定义的滑块控件 ⋯⋯⋯62
　实例 13　自定义的声音调节滑块控件 ⋯⋯⋯71
　实例 14　具有步长的滑块控件 ⋯⋯⋯75
　实例 15　模拟现实音量控制条 ⋯⋯⋯80
　实例 16　iOS 视频修剪控件 ⋯⋯⋯87

第 3 章　开关类 ⋯⋯⋯101
　实例 17　自定义开关的外观 ⋯⋯⋯101
　实例 18　实现滑块窗口滑动切换的效果 ⋯⋯⋯109

第 4 章　进度条类和指示器类效果 ⋯⋯⋯116
　实例 19　扁平带梯度效果的进度条 ⋯⋯⋯116
　实例 20　扇形进度条 ⋯⋯⋯125
　实例 21　环形进度条 ⋯⋯⋯129
　实例 22　具有范围的进度条 ⋯⋯⋯135
　实例 23　文本上传进度条 ⋯⋯⋯143
　实例 24　倒计时进度条 ⋯⋯⋯151
　实例 25　带进度条的工具栏 ⋯⋯⋯158
　实例 26　变色的指示器 ⋯⋯⋯164
　实例 27　仿 Facebook web 上正在加载中的效果 ⋯⋯⋯167

实例 28	Windows Phone 风格的指示器	174
实例 29	三个方块组成的指示器	179
实例 30	三个点的指示器	182
实例 31	多个方块组成的指示器	187
实例 32	仿 Yahoo 天气应用的加载效果	190
实例 33	消息提示指示器	197

第 5 章 选择器类效果 203

实例 34	时间设置器	203
实例 35	闹铃	206
实例 36	城市经纬度查询	211
实例 37	定制多选功能选择器	216
实例 38	转盘选择器	228
实例 39	老虎机	235

第 6 章 视图 245

实例 40	更改空白视图的背景颜色	245
实例 41	关闭应用程序	247
实例 42	手电筒	252
实例 43	旋转大挑战	254

第 7 章 分段控件 263

实例 44	滑块式分段控件	263
实例 45	开关式分段控件	273
实例 46	自定义分段控件	289

第 8 章 导航栏 295

实例 47	具有阴影的导航栏	295
实例 48	具有图片的导航栏	296
实例 49	具有分段控件的导航栏	300
实例 50	具有子标题的导航栏	302
实例 51	上下滑动的导航栏	310
实例 52	具有下拉菜单的导航栏	313
实例 53	具有页面控件的导航栏	327
实例 54	包含多个按钮的导航栏	332
实例 55	导航栏的颜色调节	338
实例 56	滚动的导航栏	343
实例 57	具有导航记录的导航栏	351

第 9 章 标签栏 368

实例 58	右上角带有数字的标签栏	368
实例 59	具有渐变效果的标签栏	370

实例 60　中间凸起的标签栏 ································· 373
实例 61　标签栏控制器实现的视图切换效果 ················ 379
实例 62　具有动画效果的标签栏 ························· 384
实例 63　滚动的标签栏 ································· 388

第 10 章　菜单 ·· 396
实例 64　立方体菜单 ··································· 396
实例 65　仿 Windows 8 的 Metro 风格 ···················· 400
实例 66　下拉菜单 ····································· 403
实例 67　浮动的菜单 ··································· 409
实例 68　具有按钮的菜单 ······························· 417
实例 69　仿 Tumblr iOS App 菜单 ······················· 425
实例 70　边栏菜单 ····································· 435
实例 71　九宫格菜单 ··································· 440
实例 72　侧面弹出式菜单 ······························· 450
实例 73　分享菜单 ····································· 456
实例 74　扇形菜单 ····································· 463

第 11 章　提醒对话框 ································· 471
实例 75　具有文本框的警告视图 ························· 471
实例 76　全屏的警告视图 ······························· 472
实例 77　具有进度条的警告视图 ························· 477
实例 78　具有列表的警告视图 ··························· 482
实例 79　坠落的警告视图 ······························· 488
实例 80　弹出视图，模糊界面背景 ······················· 493

第 12 章　文本处理 ··································· 506
实例 81　具有多个颜色的标签 ··························· 506
实例 82　发光的标签 ··································· 510
实例 83　循环渐变的标签 ······························· 514
实例 84　滚动的标签 ··································· 517
实例 85　具有光晕效果的标签 ··························· 519
实例 86　标签云 ······································· 524
实例 87　自动计算文本长度 ····························· 530
实例 88　仿 QQ 登录 ··································· 533
实例 89　阅读浏览器 ··································· 540
实例 90　艺术字 ······································· 544
实例 91　网址管理器 ··································· 551
实例 92　拨号器 ······································· 555
实例 93　我的邮箱管理器 ······························· 557
实例 94　数字天才 ····································· 563

实例 95	九宫格	568
实例 96	单位换算器	574
实例 97	计算器	577
实例 98	表情键盘	580

第 13 章 表 587

实例 99	邮编查询	587
实例 100	水平列表	590
实例 101	表的自动调整	592
实例 102	排排看	595
实例 103	归归类	599
实例 104	自定义索引的表	604
实例 105	自制的列表单选控件	611
实例 106	下拉刷新列表	617
实例 107	背景随动	620
实例 108	卡片插入式列表	624
实例 109	嵌套的表	626
实例 110	仿 QQ 聊天	630
实例 111	树形展开列表	640
实例 112	圆角表视图	646
实例 113	表单元格的自定义折叠	651
实例 114	具有搜索功能的表视图	655
实例 115	自定义表单元格的动画效果	660
实例 116	两个列表的显示	666
实例 117	表单元格内容的复制	672

第 1 章　按钮类效果

按钮是用户交互的最基础控件。即使是在 iPhone 或者 iPad 中，用户使用最多的操作也是单击。而单击操作最多的控件往往是按钮控件。本章将主要讲解按钮控件相关的实例。

实例 1　发光的按钮

【实例描述】

本实例实现的功能是当用户单击按钮时，按钮会发光。以这样的方式，提示用户已经单击按钮了。运行效果如图 1.1 所示。

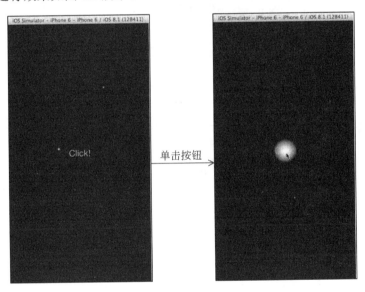

图 1.1　实例 1 运行效果

【实现过程】

当用户单击界面上的按钮时，按钮发光，当按钮没有被单击时，不会发光。具体的实现步骤如下。

（1）创建一个项目，命名为"发光按钮"。

（2）打开 ViewController.h 文件，编写代码，实现插座变量和动作的声明。程序代码如下：

```
#import <UIKit/UIKit.h>
@interface ViewController : UIViewController{
    IBOutlet UIButton *button;                    //按钮的插座变量
```

```
}
- (IBAction)click:(id)sender;                              //动作click:的声明
@end
```

（3）打开 Main.storyboard 文件，对 View Controller 视图控制器的设计界面进行设计，效果如图 1.2 所示。

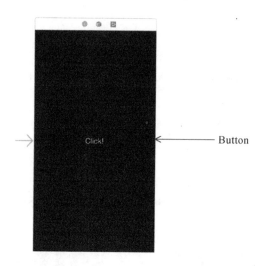

图 1.2　实例 1 View Controller 视图控制器的设计界面效果

需要添加的视图、控件以及对它们的设置如表 1-1 所示。

表 1-1　实例 1 视图、控件设置

视图、控件	属 性 设 置	其　　他
设计界面	Background：黑色	
Button	Title：Click! Font：System 20.0 Text Color：浅橘黄色	与动作 click:方法关联

（4）打开 ViewController.m 文件，编写代码，实现单击按钮时，按钮发光。程序代码如下：

```
- (IBAction)click:(id)sender {
    [button setShowsTouchWhenHighlighted:YES];             //发光
}
```

【代码解析】

本实例关键功能是单击按钮时按钮发光。下面就是这个知识点的详细讲解。

要单击按钮，使按钮发光，可以使用 UIButton 的 setShowsTouchWhenHighlighted 属性。其语法形式如下：

```
@property(nonatomic)BOOL showaTouchWhenHighlighted;
```

在代码中，使用了 setShowsTouchWhenHighlighted 属性对按钮在单击时是否发光进行了设置，代码如下：

```
[button setShowsTouchWhenHighlighted:YES];
```

此代码实现的功能，就是在单击按钮时，使按钮发光。如果不想设置在单击按钮时，使按钮发光，开发者可以不写发光的代码，因为它默认就是不发光的，或者将 YES 改为 NO。程序代码如下：

```
[button setShowsTouchWhenHighlighted:NO];
```

实例 2　弹出式按钮

【实例描述】

在 iPhone 或者 iPad 上，屏幕的空间大小非常有限，为了腾出更多的空间，开发者一般都会采用一种弹出式按钮。本实例即实现了此功能。当单击屏幕上的按钮后，就会以此按钮为中心弹出三个按钮。单击弹出的三个按钮中的任意一个按钮，就会弹出相应的警告视图；再次单击中心按钮之后，就会将这三个按钮隐藏。运行效果如图 1.3 所示。

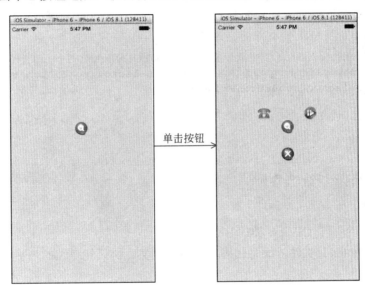

图 1.3　实例 2 运行效果

【实现过程】

当用户单击界面的按钮时，会在此按钮的周围弹出三个按钮。具体的实现步骤如下。

（1）创建一个项目，命名为"弹出式按钮"。

（2）添加图片 button1.png、button2.png、button3.png 和 button4.png 到创建项目的 Supporting Files 文件夹中。

（3）创建一个基于 UIButton 类的 but 类。

（4）打开 but.h 文件，编写代码，实现协议和属性的声明。程序代码如下：

```
#import <UIKit/UIKit.h>
@class but;
//协议
@protocol butDelegate <NSObject>
- (void)subButtonPress:(but*)button;                          //单击按钮
```

```
@end
@interface but : UIButton
@property (nonatomic, weak) id<butDelegate> delegate;         //属性的声明
@end
```

（5）打开 but.m 文件，编写代码，通过触摸实现单击的功能。程序代码如下：

```
#import "but.h"
@implementation but
//初始化
- (id)initWithFrame:(CGRect)frame
{
    self = [super initWithFrame:frame];
    if (self) {
        // Initialization code
    }
    return self;
}
//触摸开始
- (void)touchesBegan:(NSSet *)touches withEvent:(UIEvent *)event{
    self.highlighted = YES;                      //以高亮状态显示
}
//触摸结束
- (void)touchesEnded:(NSSet *)touches withEvent:(UIEvent *)event{
    if ([_delegate respondsToSelector:@selector(subButtonPress:)]) {
        [_delegate subButtonPress:self];         //调用 subButtonPress:方法
    }
    self.highlighted= NO;                        //不以高亮状态显示
}
//触摸取消
- (void)touchesCancelled:(NSSet *)touches withEvent:(UIEvent *)event{
    self.highlighted = NO;                       //不以高亮状态显示
}
@end
```

（6）创建一个基于 UIView 类的 Button 类。

（7）打开 Button.h 文件，编写代码，实现宏定义、协议和属性的声明等功能。程序代码如下：

```
#import <UIKit/UIKit.h>
#import "but.h"                                  //头文件
//宏定义
#define kDCPathButtonParentView self.parentView
#define kDCPathButtonCurrentFrameWidth kDCPathButtonParentView.frame.size.width
#define kDCPathButtonCurrentFrameHeight kDCPathButtonParentView.frame.size.height
#define kDCCovertAngelToRadian(x) ((x)*M_PI)/180
//枚举类型
typedef enum{
    kDCPathButtonRotationNormal = 0,
    kDCPathButtonRotationReverse,
}DCPathButtonRotationOrientation;
//协议的声明
@protocol ButtonDelegate <NSObject>
@optional
//单击按钮的动作
- (void)button_0_action;
```

```
- (void)button_1_action;
- (void)button_2_action;
- (void)button_3_action;
- (void)button_4_action;
- (void)button_5_action;
@end
@interface Button : UIView<butDelegate>{
    CGPoint kDCPathButtonSubButtonBirthLocation;
    CGPoint kDCPathButtonSubButtonTag_0_AppearLocation;
    CGPoint kDCPathButtonSubButtonTag_1_AppearLocation;
    CGPoint kDCPathButtonSubButtonTag_2_AppearLocation;
    CGPoint kDCPathButtonSubButtonTag_3_AppearLocation;
    CGPoint kDCPathButtonSubButtonTag_4_AppearLocation;
    CGPoint kDCPathButtonSubButtonTag_5_AppearLocation;
    CGPoint kDCPathButtonSubButtonFinalLocation;
}
//方法的声明
- (id)initDCPathButtonWithSubButtons:(NSInteger)buttonCount
                  totalRadius:(CGFloat)totalRadius
                  centerRadius:(NSInteger)centerRadius
                    subRadius:(CGFloat)subRadius
                  centerImage:(NSString *)centerImageName
              centerBackground:(NSString *)centerBackgroundName
                    subImages:(void(^)(Button *))imageBlock
              subImageBackground:(NSString *)subImageBackgroundName
                    inLocationX:(CGFloat)xAxis
                      locationY:(CGFloat)yAxis
                    toParentView:(UIView *)parentView;
//属性的声明
@property (nonatomic, weak) id<ButtonDelegate> delegate;
@property (nonatomic, getter = isExpanded) BOOL expanded;
……
- (void)subButtonImage:(NSString *)imageName withTag:(NSInteger)tag;
@end
```

（8）打开 Button.m 文件，编写代码。实现弹出式按钮的动作和位置等功能。使用的方法如表 1-2 所示。

表 1-2 Button.m 文件方法总结

方　　法	功　　能
initDCPathButtonWithSubButtons:totalRadius:centerRadius:subRadius:centerImage:centerBackground:subImages: subImageBackground: inLocationX: locationY: toParentView:	按钮的初始化
configureCenterButton:image:backgroundImage:	对中心按钮进行配置
centerButtonPress	按下中心按钮的动作
configureTheButtons:	对其他按钮的配置
subButtonImage:withTag:	设置按钮的图像
isExpanded	获取按钮是否展开
button:appearAt:withDalay:duration:	按钮出现在指定的位置上的动画
button:shrinkAt:offsetAxisXoffSEtAxisY:withDelay:rotateDirection:animationDuration:	按钮消失的动画
matchRotationOrientation:	匹配旋转的方向
offsetAxisY: withAngel:	获取 Y 的偏移轴
subButtonPress:	实现按下其他的按钮

这里需要讲解几个重要的方法（其他方法请读者参考源代码）。initDCPathButtonWithSubButtons:totalRadius:centerRadius:subRadius:centerImage:centerBackground:subImages:subImageBackground:inLocationX: locationY:toParentView:方法，实现按钮的初始化。程序代码如下：

```
- (id)initDCPathButtonWithSubButtons:(NSInteger)buttonCount totalRadius:
(CGFloat)totalRadius   centerRadius:(NSInteger)centerRadius   subRadius:
(CGFloat)subRadius   centerImage:(NSString *)centerImageName   centerBack
ground:(NSString *)centerBackgroundName   subImages:(void   (^)(Button
*))imageBlock   subImageBackground:(NSString *)subImageBackgroundName
inLocationX:(CGFloat)xAxis locationY:(CGFloat)yAxis toParentView:(UIView
*)parentView{
    //判断 parentView 是否为空
    parentView == nil? (self.parentView = parentView):(self.parentView
= parentView);
    xAxis == 0? (self.centerLocationAxisX = kDCPathButtonCurrentFrameWidth/2) :
    (self.centerLocationAxisX = xAxis);       //判断 xAxis 是否为 0
    yAxis == 0? (self.centerLocationAxisY = kDCPathButtonCurrentFrameHeight/2) :
    (self.centerLocationAxisY = yAxis);       //判断 yAxis 是否为 0
    self.buttonCount = buttonCount;           //按钮的个数
    self.totalRaiuds = totalRadius;
    self.subRadius = subRadius;
    _expanded = NO;                           //展开设置为 NO
    kDCPathButtonSubButtonBirthLocation = CGPointMake(-kDCPathButton
    Current FrameWidth/2,
    -kDCPathButtonCurrentFrameHeight/2);
    kDCPathButtonSubButtonFinalLocation = CGPointMake(self.centerLocationAxisX,
self.centerLocationAxisY);
    if (self = [super initWithFrame:self.parentView.bounds]) {
        [self configureCenterButton:centerRadius image:centerImageName backgroundImage:
centerBackgroundName];                        //配置中心按钮
        [self configureTheButtons:buttonCount];
        imageBlock(self);
        [self.parentView addSubview:self];    //添加视图
    }
    return self;
}
```

要实现单击中心按钮后，展开其他按钮或者关闭其他按钮，需要添加：configureCenterButton:image:backgroundImage:、centerButtonPress、subButtonImage:withTag:、isExpanded、button:appearAt:withDalay:duration:、button:shrinkAt:offsetAxisXoffSEtAxisY:withDelay:rotateDirection:animationDuration: 和 matchRotationOrientation: 方法。其中 configureCenterButton: image: backgroundImage:方法，对中心按钮进行配置。程序代码如下：

```
- (void)configureCenterButton:(CGFloat)centerRadius image:(NSString *)
imageName backgroundImage:
(NSString *)backgroundImageName{
    self.centerButton = [[UIButton alloc]init];
    self.centerButton.frame = CGRectMake(0, 0, centerRadius * 2, centerRadius * 2);
                                              //中心按钮框架
    //中心按钮的中心点
    self.centerButton.center = CGPointMake(self.centerLocationAxisX,
    self.centerLocationAxisY);
    //判断 imageName 是否为空
    if (imageName == nil) {
        //如果为空
```

```objc
        imageName = @"dc-center";              //设置 imageName
    }
    //判断 backgroundImageName 是否为空
    if (backgroundImageName == nil) {
        //如果为空
        backgroundImageName = @"dc-background"; //设置 backgroundImageName
    }
    //设置按钮图像
    [self.centerButton setImage:[UIImage imageNamed:imageName] forState:UIControlStateNormal];
    [self.centerButton setBackgroundImage:[UIImage imageNamed:backgroundImageName] forState:UIControlStateNormal];                //设置按钮的背景图像
    [self.centerButton addTarget:self action:@selector(centerButtonPress) forControlEvents:UIControlEventTouchUpInside]; //设置按钮的动作
    self.centerButton.layer.zPosition = 1;
    [self addSubview:self.centerButton];
}
```

centerButtonPress 方法,实现按下中心按钮的动作,如果按钮没有展开,将按钮展开;如果按钮展开了,就将按钮关闭。程序代码如下:

```objc
- (void)centerButtonPress{
    //判断按钮是否展开
    if (![self isExpanded]) {
        //如果没有展开按钮,将按钮展开
        [self button:[self.buttons objectAtIndex:0] appearAt:
kDCPathButtonSubButtonTag_0_AppearLocation withDalay:0.5 duration:0.35];
                                            //展开索引为 0 的按钮
        [self button:[self.buttons objectAtIndex:1] appearAt:
kDCPathButtonSubButtonTag_1_AppearLocation withDalay:0.55 duration:0.4];
                                            //展开索引为 1 的按钮
        [self button:[self.buttons objectAtIndex:2] appearAt:
kDCPathButtonSubButtonTag_2_AppearLocation withDalay:0.6 duration:0.45];
                                            //展开索引为 2 的按钮
        self.expanded = YES;               //将 expanded 设置为 YES
    }else{
        //如果展开按钮,将展开的按钮关闭
        [self button:[self.buttons objectAtIndex:0]
            shrinkAt:kDCPathButtonSubButtonTag_0_AppearLocation
         offsetAxisX:kDCPathButtonLeftOffSetX
         offSEtAxisY:[self offsetAxisY:kDCPathButtonLeftOffSetX withAngel:
            kDCPathButtonAngel60C]
           withDelay:0.4
     rotateDirection:kDCPathButtonRotationNormal animationDuration:1];
                                            //关闭索引为 0 的按钮
        [self button:[self.buttons objectAtIndex:1]
            shrinkAt:kDCPathButtonSubButtonTag_1_AppearLocation
         offsetAxisX:kDCPathButtonRightOffSetX
         offSEtAxisY:-[self offsetAxisY:kDCPathButtonRightOffSetX withAngel:
            kDCPathButtonAngel60C] withDelay:0.5
     rotateDirection:kDCPathButtonRotationReverse animationDuration:1.2];
                                            //关闭索引为 1 的按钮
        [self button:[self.buttons objectAtIndex:2]
            shrinkAt:kDCPathButtonSubButtonTag_2_AppearLocation
         offsetAxisX:0 offSEtAxisY:kDCPathButtonVerticalOffSetX
           withDelay:0.6
     rotateDirection:kDCPathButtonRotationNormal animationDuration:1.4];
                                            //关闭索引为 2 的按钮
```

```
    self.expanded = NO;
  }
}
```

configureTheButtons:方法实现对其他按钮的配置。程序代码如下：

```
- (void)configureTheButtons:(NSInteger)buttonCount{
  //设置显示后按钮的位置
  kDCPathButtonSubButtonTag_0_AppearLocation = CGPointMake(
                                              self.centerLocationAxisX
- self.totalRaiuds * sinf(kDCCovertAngelToRadian(kDCPathButtonAngel60C)),
                                              self.centerLocationAxisY
- self.totalRaiuds * cosf(kDCCovertAngelToRadian(kDCPathButtonAngel60C)));
                         //设置显示后索引为 0 的按钮的位置
  kDCPathButtonSubButtonTag_1_AppearLocation = CGPointMake(
                                              self.centerLocationAxisX
+ self.totalRaiuds * sinf(kDCCovertAngelToRadian(kDCPathButtonAngel60C)),
                                              self.centerLocationAxisY
- self.totalRaiuds * cosf(kDCCovertAngelToRadian(kDCPathButtonAngel60C)));
                         //设置显示后索引为 1 的按钮的位置
  kDCPathButtonSubButtonTag_2_AppearLocation = CGPointMake(
                                              self.centerLocationAxisX ,
                                              self.centerLocationAxisY
+self.totalRaiuds);
  self.buttons = [NSMutableArray array];
  //对显示后的按钮进行设置
  for (NSInteger i = 0; i<3; i++) {
    subButton = [[but alloc]init];
    subButton.delegate = self;                    //设置委托
    subButton.frame = CGRectMake(0, 0, self.subRadius * 2, self.subRadius * 2);
                                                  //设置框架
    subButton.center = kDCPathButtonSubButtonBirthLocation; //设置中心点
    NSString *imageFormat = [NSString stringWithFormat:@"dc-button_%d",i];
    [subButton setImage:[UIImage imageNamed:imageFormat] forState:
    UIControlStateNormal];
    [self addSubview:subButton];
    [self.buttons addObject:subButton];
  }
}
```

subButtonImage:方法，实现设置按钮的图像这一功能。程序代码如下：

```
- (void)subButtonImage:(NSString *)imageName withTag:(NSInteger)tag{
  //判断 tag 值是否大于按钮的个数
  if (tag > self.buttonCount) {
    tag = self.buttonCount;                       //设置 tag 值
  }
  but *currentButton = [self.buttons objectAtIndex:tag];
  //设置当前按钮的图像
  [currentButton setImage:[UIImage imageNamed:imageName] forState:
  UIControlStateNormal];
}
```

button:appearAt:withDalay:duration:方法，实现使用延迟时间将按钮出现在指定的位置上的动画。程序代码如下：

```
- (void)button:(but *)button appearAt:(CGPoint)location withDalay:
(CGFloat)delay duration:(CGFloat)duration{
button.center = location;                         //设置按钮的中心位置
```

```objc
//创建并设置缩放动画
CAKeyframeAnimation *scaleAnimation = [CAKeyframeAnimation animationWithKeyPath:
@"transform"];
    scaleAnimation.duration = duration;              //设置所需时间
    scaleAnimation.values   =   @[[NSValue   valueWithCATransform3D:
CATransform3DMakeScale(0.1,  0.1,  1)],[NSValue  valueWithCATransform3D:
CATransform3DMakeScale(1.3,  1.3,  1)],[NSValue  valueWithCATransform3D:
CATransform3DMakeScale(1, 1, 1)]];
    scaleAnimation.calculationMode = kCAAnimationLinear;
    scaleAnimation.keyTimes = @[[NSNumber numberWithFloat:0.0f],
[NSNumber numberWithFloat:delay],[NSNumber numberWithFloat:1.0f]];
    button.layer.anchorPoint = CGPointMake(0.5f, 0.5f);
                                                     //设置锚点
    [button.layer addAnimation:scaleAnimation forKey:@"buttonAppear"];
                                                     //添加动画
}
```

button:shrinkAt:offsetAxisX:offSEtAxisY:withDelay:rotateDirection:animationDuration:方法，实现按钮消失的动画。程序代码如下：

```objc
- (void)button:(but *)button shrinkAt:(CGPoint)location offsetAxisX:
(CGFloat)axisX offSEtAxisY:(CGFloat)axisY withDelay:(CGFloat)delay rotateDirection:
(DCPathButtonRotationOrientation)orientation animationDuration:
(CGFloat)duration{
    //旋转动画
    CAKeyframeAnimation *rotation = [CAKeyframeAnimation animationWithKeyPath:
@"transform.rotation.z"];
    rotation.duration = duration * delay;            //时间设置
    rotation.values = @[[NSNumber numberWithFloat:0.0f],[NSNumber numberWithFloat:
[self matchRotationOrientation:orientation]],[NSNumber numberWithFloat:0.0f]];
    rotation.keyTimes = @[[NSNumber numberWithFloat:0.0f],[NSNumber
numberWithFloat:delay],
[NSNumber numberWithFloat:1.0f]];
    //缩小的动画
    CAKeyframeAnimation *shrink = [CAKeyframeAnimation animationWithKeyPath:
@"position"];
    shrink.duration = duration * (1 - delay);        //设置时间
    CGMutablePathRef path = CGPathCreateMutable();
    CGPathMoveToPoint(path, NULL, location.x, location.y); //设置路径的初始点
    CGPathAddLineToPoint(path, NULL, location.x + axisX, location.y + axisY);
    CGPathAddLineToPoint(path, NULL, kDCPathButtonSubButtonFinalLocation.x,
kDCPathButtonSubButtonFinalLocation.y);
    shrink.path = path;                              //设置缩放动画的路径
    CGPathRelease(path);
    //组合动画
    CAAnimationGroup *totalAnimation = [CAAnimationGroup animation];
    totalAnimation.duration = 1.0f;                  //设置时间
    totalAnimation.animations = @[rotation,shrink];  //动画效果
    totalAnimation.fillMode = kCAFillModeForwards;
    totalAnimation.timingFunction = [CAMediaTimingFunction functionWithName:
kCAMediaTimingFunctionEaseOut];
    totalAnimation.delegate = self;                  //设置委托
    button.layer.anchorPoint = CGPointMake(0.5f, 0.5f);
    button.center = kDCPathButtonSubButtonBirthLocation; //设置中心位置
    [button.layer addAnimation:totalAnimation forKey:@"buttonDismiss"];
}
```

添加subButtonPress:方法，实现单击中心按钮以外的按钮。程序代码如下：

```
- (void)subButtonPress:(but *)button{
    if ([_delegate respondsToSelector:@selector(button_0_action)] &&
        button == [self.buttons objectAtIndex:0]) {
        [_delegate button_0_action];              //调用button_0_action方法
    }
    else if ([_delegate respondsToSelector:@selector(button_1_action)] &&
            button == [self.buttons objectAtIndex:1]){
        [_delegate button_1_action];              //调用button_1_action方法
    }
    else if ([_delegate respondsToSelector:@selector(button_2_action)] &&
            button == [self.buttons objectAtIndex:2]){
        [_delegate button_2_action];              //调用button_2_action方法
    }
}
```

（9）打开 ViewController.m 文件，编写代码，实现在界面创建一个弹出式的按钮。在 viewDidLoad 方法中编写代码，实现弹出式按钮对象的创建。程序代码如下：

```
- (void)viewDidLoad
{
    self.view.frame = CGRectMake(0, 0, 320, 460);
    self.view.backgroundColor = [UIColor cyanColor];
//创建按钮对象
    Button *dcPathButton = [[Button alloc]
                            initDCPathButtonWithSubButtons:3
                                                                //设置子按钮的个数
                            totalRadius:60
                            centerRadius:15
                            subRadius:15
                            centerImage:@"button2.png"//设置中心按钮的图像
                            centerBackground:nil
                            subImages:^(Button *dc){
                                //设置子按钮的图像
                                [dc subButtonImage:@"button1.png" withTag:0];
                                [dc subButtonImage:@"button3.png" withTag:1];
                                [dc subButtonImage:@"button4.png" withTag:2];
                            }
                            subImageBackground:nil
                            inLocationX:0 locationY:0 toParentView:self.view];
    dcPathButton.delegate = self;
    [super viewDidLoad];
    // Do any additional setup after loading the view, typically from a nib.
}
```

添加 button_0_action:方法，实现单击弹出的按钮，显示一个警告视图。程序代码如下：

```
- (void)button_0_action{
    UIAlertView *A=[[UIAlertView alloc]initWithTitle:@"你选择的是"电话"按钮
" message:nil delegate:nil cancelButtonTitle:@"OK" otherButtonTitles:nil];
    [A show];
}
```

> 注意：本实例的代码只是部分，如果读者想要看到完整的代码，请参考本书的源代码。

【代码解析】

由于本实例中的代码和方法非常多，为了方便读者的阅读，笔者绘制了一些执行流程图。以下就是此实例代码的执行流程，如图1.4所示。

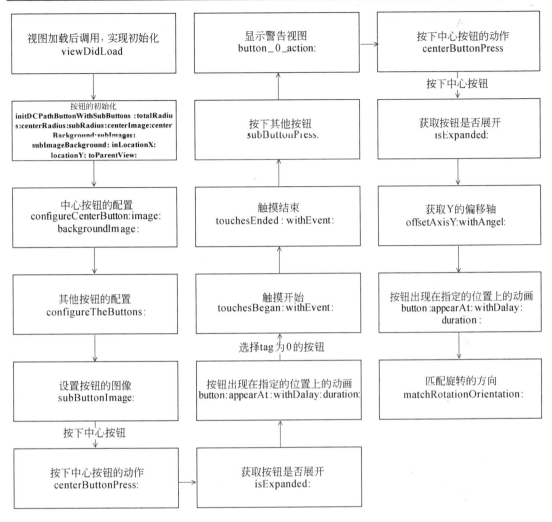

图 1.4　程序执行流程

实例 3　超文本链接

【实例描述】

本实例实现的功能是单击超文本链接，将直接在下方显示网页内容。其中，超文本链接是使用按钮实现的，出现的内容是用网页视图实现的。运行效果如图 1.5 所示。

【实现过程】

当用户单击超文本链接时，会在网页视图中显示链接所指的内容。具体的实现步骤如下。

（1）创建一个项目，命名为"超文本链接"。

（2）创建一个基于 UIButton 类的 HyperLinks 类。

（3）打开 HyperLinks.h 文件，编写代码，实现颜色对象、实例变量以及类方法 hyperlinksButton 的声明。程序代码如下：

```
#import <UIKit/UIKit.h>
@interface HyperLinks : UIButton{
    UIColor *lineColor;                              //颜色对象
    BOOL isHighlight;                                //实例变量
}
+ (HyperLinks *) hyperlinksButton;
@end
```

图 1.5　实例 3 运行效果

（4）打开 HyperLinks.m 文件，编写代码。其中，超文本链接的绘制需要使用 hyperlinksButton 和 drawRect:方法实现。hyperlinksButton 方法实现文本超链接的创建。程序代码如下：

```
+ (HyperLinks*) hyperlinksButton {
    HyperLinks* button = [[HyperLinks alloc] init];        //创建
    return button;
}
```

（5）drawRect:方法，实现超文本链接的绘制。程序代码如下：

```
- (void) drawRect:(CGRect)rect {
    CGRect textRect = self.titleLabel.frame;
    CGContextRef contextRef = UIGraphicsGetCurrentContext();
    CGFloat descender = self.titleLabel.font.descender;
    if([lineColor isKindOfClass:[UIColor class]]){
        CGContextSetStrokeColorWithColor(contextRef, lineColor.CGColor);
    }
    CGContextMoveToPoint(contextRef, textRect.origin.x, textRect.origin.y
+ textRect.size.height + descender+5);                     //设置初始点
    CGContextAddLineToPoint(contextRef, textRect.origin.x + textRect.
size.width, textRect.origin.y + textRect.size.height + descender+5);
                                                           //设置结束点
    CGContextClosePath(contextRef);
    CGContextDrawPath(contextRef, kCGPathStroke);           //绘制路径
}
```

（6）实现按钮的触摸，需要添加 touchesBegan:withEvent:、touchesMoved:withEvent:、touchesEnded:withEvent:以及 touchesCancelled:withEvent:方法。程序代码如下：

```
//触摸开始
-(void)touchesBegan:(NSSet *)touches withEvent:(UIEvent *)event{
    [super touchesBegan:touches withEvent:event];
    [self.titleLabel setTextColor:[UIColor redColor]];        //设置标题的颜色
}
//触摸移动
-(void)touchesMoved:(NSSet *)touches withEvent:(UIEvent *)event{
    [super touchesMoved:touches withEvent:event];
}
//触摸结束
-(void)touchesEnded:(NSSet *)touches withEvent:(UIEvent *)event{
    [super touchesEnded:touches withEvent:event];
}
//触摸取消
-(void)touchesCancelled:(NSSet *)touches withEvent:(UIEvent *)event{
    [super touchesCancelled:touches withEvent:event];
}
```

（7）打开 ViewController.h 文件，编写代码，实现插座变量和动作的声明。程序代码如下：

```
#import <UIKit/UIKit.h>
@interface ViewController : UIViewController{
    IBOutlet UIWebView *web;
    IBOutlet UIButton *bb;
}
- (IBAction)open:(id)sender;
@end
```

（8）打开 Main.storyboard 文件，对 View Controller 视图控制器的设计界面进行设计，效果如图 1.6 所示。

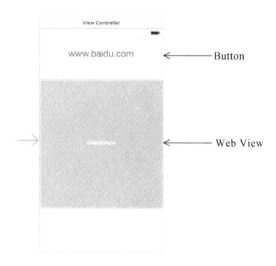

图 1.6　实例 3 View Controller 视图控制器的设计界面效果

需要添加的视图、控件以及对它们的设置如表 1-3 所示。

表 1-3 实例 3 视图、控件设置

视图、控件	属 性 设 置	其 他
Button	Title：www.baidu.com Font：System 24.0	Class：HyperLinks 与插座变量 bb 关联 与动作 open:关联
Web View		与插座变量 web 关联

（9）打开 ViewController.m 文件，编写代码，实现单击超文本链接时，在网页视图中显示内容。程序代码如下：

```
- (IBAction)open:(id)sender {
  NSString *a=[NSString stringWithFormat:@"http://%@",bb.titleLabel.text];
  NSURL *url=[NSURL URLWithString:a];
  NSURLRequest *request=[NSURLRequest requestWithURL:url];
  [web loadRequest:request];                               //加载
}
```

【代码解析】

本实例关键功能是在触摸时按钮的标题颜色变化。下面就对这个知识点做一个详细的讲解。

要在触摸时改变按钮标题的颜色，可以使用 UILabel 的 textColor 属性实现，其语法形式如下：

```
@property(nonatomic,retain)UIColor *textColor;
```

在代码中，使用了 textColor 属性改变按钮标题的颜色，代码如下：

```
[self.titleLabel setTextColor:[UIColor redColor]];
```

实例 4　抛光效果的按钮

【实例描述】

本实例实现的是按钮的抛光效果。抛光效果可以使得表面发亮，更具有质感。运行效果如图 1.7 所示。

图 1.7　实例 4 运行效果

【实现过程】

（1）创建一个项目，命名为"抛光效果的按钮"。

（2）创建一个基于 UIButton 类的 shiny 类。

（3）打开 shiny.h 文件，编写代码，实现属性和方法的声明。程序代码如下：

```
#import <UIKit/UIKit.h>
@interface shiny : UIButton
//属性的声明
@property(strong, nonatomic)UIColor *myColor;
//方法的声明
- (id)initWithFrame:(CGRect)frame withBackgroundColor:(UIColor*)backgroundColor;
                                                              //初始化
-(void)wasPressed;
-(void)endedPress;                                            //结束按钮的单击
- (void)makeButtonShiny:(shiny*)button withBackgroundColor:(UIColor*)backgroundColor;
@end
```

（4）打开 shiny.m 文件，编写代码。此代码分为两个大的部分：按钮的抛光效果；按钮的按下弹起效果。要实现按钮的抛光效果，需要添加 initWithFrame:withBackgroundColor: 和 makeButtonShiny:withBackgroundColor:方法。initWithFrame:withBackgroundColor:方法实现对抛光效果的按钮初始化。程序代码如下：

```
- (id)initWithFrame:(CGRect)frame withBackgroundColor:(UIColor*)backgroundColor
{
    self = [super initWithFrame:frame];
    if (self) {
        self.myColor = backgroundColor;
        [self makeButtonShiny:self withBackgroundColor:backgroundColor];
        [self addTarget:self action:@selector(wasPressed) forControlEvents:
UIControlEventTouchDown];                                     //添加动作
        [self addTarget:self action:@selector(endedPress) forControlEvents:
UIControlEventTouchUpInside];                                 //添加动作
    }
    return self;
}
```

makeButtonShiny:withBackgroundColor:方法，实现按钮的抛光效果。程序代码如下：

```
- (void)makeButtonShiny:(shiny*)button withBackgroundColor:(UIColor*)backgroundColor
{
    CALayer *layer = button.layer;
    layer.cornerRadius = 8.0f;                                //圆角的设置
    layer.masksToBounds = YES;
    layer.borderWidth = 2.0f;
    layer.borderColor = [UIColor colorWithWhite:0.4f alpha:0.2f].CGColor;
    CAGradientLayer *shineLayer = [CAGradientLayer layer];
    shineLayer.frame = button.layer.bounds;
    //抛光的颜色
    shineLayer.colors = [NSArray arrayWithObjects:
                    (id)[UIColor colorWithWhite:1.0f alpha:0.4f].CGColor,
                    (id)[UIColor colorWithWhite:1.0f alpha:0.2f].CGColor,
                    (id)[UIColor colorWithWhite:0.75f alpha:0.2f].CGColor,
                    (id)[UIColor colorWithWhite:0.4f alpha:0.2f].CGColor,
                    (id)[UIColor colorWithWhite:1.0f alpha:0.4f].CGColor,
                    nil];
    //抛光的位置
```

```objc
shineLayer.locations = [NSArray arrayWithObjects:
                        [NSNumber numberWithFloat:0.0f],
                        [NSNumber numberWithFloat:0.5f],
                        [NSNumber numberWithFloat:0.5f],
                        [NSNumber numberWithFloat:0.8f],
                        [NSNumber numberWithFloat:1.0f],
                        nil];
[button.layer addSublayer:shineLayer];                    //为按钮添加图层
[button setBackgroundColor:backgroundColor];
}
```

这时按钮是没有按下或弹起效果的。要实现这一效果,可以添加 wasPressed 和 endedPress 方法。wasPressed 方法,实现按钮的抛光效果。程序代码如下:

```objc
-(void)wasPressed
{
    UIColor *newColor;
    CGFloat red = 0.0, green = 0.0, blue = 0.0, alpha = 0.0, white = 0.0;
    if([self.myColor respondsToSelector:@selector(getRed:green:blue:alpha:)]) {
        [self.myColor getRed:&red green:&green blue:&blue alpha:&alpha];
                                                                //获取颜色
        [self.myColor getWhite:&white alpha:&alpha];
        }
    self.backgroundColor = newColor;
}
```

endedPress 方法实现按钮的弹起效果。程序代码如下:

```objc
-(void)endedPress
{
    self.backgroundColor = self.myColor;
}
```

(5) 打开 ViewController.m 文件,编写代码,实现抛光效果按钮的创建。程序代码如下:

```objc
- (void)viewDidLoad
{
    //实现蓝色效果的抛光按钮
    CGRect rect1 = CGRectMake(120,252,100,100);
    shiny *shinyBtn1 = [[shiny alloc] initWithFrame:rect1 withBackgroundColor:
    [UIColor blueColor]];
    CGRect rect2 = CGRectMake(100,100,150,50);
    //实现黄色效果的抛光按钮
    shiny *shinyBtn2 = [[shiny alloc] initWithFrame:rect2 withBackgroundColor:
    [UIColor yellowColor]];
    [self.view addSubview:shinyBtn1];                         //添加视图
    [self.view addSubview:shinyBtn2];
    [super viewDidLoad];
    // Do any additional setup after loading the view, typically from a nib.
}
```

【代码解析】

本实例关键功能是抛光按钮的添加。下面就对这个知识点做一个详细的讲解。

要实现抛光按钮的添加,可以使用 UIView 的 addSubView 方法,其语法形式如下:

```objc
-(void)addSubView:(UIView *)view;
```

其中，(UIView *)view 表示添加的视图。在代码中使用了 addSubView:方法实现抛光按钮的添加，代码如下：

```
[self.view addSubview:shinyBtn1];
```

其中，shinyBtn1 表示添加的视图。

实例 5　具有进度条的按钮

【实例描述】

本实例实现的是具有进度条的按钮效果。当单击按钮后，将在按钮上以扇形的方式显示进度。在加载的过程中，单击按钮，就变为暂停状态，进度条也停止。运行效果如图 1.8 所示。

图 1.8　实例 5 运行效果

【实现过程】

绘制一个圆形的进度条，在此进度条上放置一个环形的按钮控件，当用户单击此按钮控件时，圆形的进度条就会被加载。具体的实现步骤如下。

（1）创建一个项目，命名为"具有进度条的按钮"。
（2）添加图片 Pause.png 和 Play.png 到创建项目的 Supporting Files 文件夹中。
（3）创建一个 CALayer 类的 ProgressLayer 类。
（4）打开 ProgressLayer.h 文件，编写代码，实现属性的声明。程序代码如下：

```
#import <QuartzCore/QuartzCore.h>
@interface ProgressLayer : CALayer
//属性
@property (nonatomic, assign) float progress;
@property (nonatomic, assign) float startAngle;
@property (nonatomic, retain) UIColor *tintColor;
@property (nonatomic, retain) UIColor *trackColor;
@end
```

（5）打开 ProgressLayer.m 文件，编写代码。对进度条的图层进行初始化设置以及绘制。添加 initWithLayer:方法，实现进度条的图层初始化。程序代码如下：

```objc
- (id)initWithLayer:(id)layer
{
    self = [super initWithLayer:layer];
    if(self)
    {
        if([layer isKindOfClass:[ProgressLayer class]])
        {
            ProgressLayer *otherLayer = layer;
            self.progress = otherLayer.progress;        //设置进度条的进度
            self.startAngle = otherLayer.startAngle;
            self.tintColor = otherLayer.tintColor;
            self.trackColor = otherLayer.trackColor;    //设置进度条的通道颜色
        }
    }
    return self;
}
```

添加 needsDisplayForKey:方法，实现显示一些需要的关键字进行判断。程序代码如下：

```objc
+ (BOOL)needsDisplayForKey:(NSString *)key
{
    //判断 key 是否为@"progress"
    if([key isEqualToString:@"progress"])
        return YES;
    else
        return [super needsDisplayForKey:key];          //重写绘制
}
```

添加 drawInContext:方法，对进度条进行绘制。程序代码如下：

```objc
- (void)drawInContext:(CGContextRef)context
{
    CGFloat radius = MIN(self.bounds.size.width, self.bounds.size.height)/2.0;
    CGPoint center = {self.bounds.size.width/2.0, self.bounds.size.height/2.0};
    CGRect circleRect = {center.x-radius, center.y-radius, radius*2.0, radius*2.0};
    //对背景圆的绘制
    CGContextAddEllipseInRect(context, circleRect);                 //设置圆
    CGContextSetFillColorWithColor(context, trackColor.CGColor);
                                                                    //设置填充颜色
    CGContextFillPath(context);                                     //绘制路径
    //对经过弧的绘制
    CGContextAddArc(context, center.x, center.y, radius, startAngle, startAngle + progress*2.0*M_PI, 0); CGContextAddLineToPoint(context, center.x, center.y);                                 //设置直线
    CGContextClosePath(context);
    CGContextSetFillColorWithColor(context, tintColor.CGColor);
    CGContextFillPath(context);
}
```

（6）创建一个基于 UIView 类的 ProgressView 类。在此类中没有任何的绘图，它的工作就是在可视界面中放置 ProgressLayer 图层。

（7）打开 ProgressView.h 文件，编写代码，实现头文件、属性以及方法的声明。程序代码如下：

```objc
#import <UIKit/UIKit.h>
```

```
#import "ProgressLayer.h"
@interface ProgressView : UIView
//属性的声明
@property (nonatomic, assign) float progress;
@property (nonatomic, assign) float startAngle;
@property (nonatomic, retain) UIColor *tintColor ;
@property (nonatomic, retain) UIColor *trackColor;
//方法的声明
- (void) setProgress:(float)progress animated:(BOOL)animated;
@end
```

(8) 打开 ProgressView.m 文件，编写代码。实现将绘制的进度条放到可视界面中。使用的方法如表 1-4 所示。

表 1-4　ProgressView.m文件中的方法总结

方　　法	功　　能
layerClass	获取 ProgressLayer 的实例
initWithFrame:	进度条的初始化
initWithCode:	初始化实例对象
_initIVars	对进度条中的属性进行设置
progress	获取进度条
setProgress:	设置进度条的增长动画
setProgress: animated:	设置进度条中的动画
tintColor	获取进度条中加载的颜色
setTintColor:	设置进度条中加载的颜色
trackColor	获取轨道颜色
setTrackColor:	设置轨道颜色
startAngle	获取开始加载的角度
setStartAngle:	设置开始加载的角度

这里需要讲解几个重要的方法（其他方法请读者参考源代码）。initWithFrame:方法对进度条进行初始化。程序代码如下：

```
- (id)initWithFrame:(CGRect)frame
{
    self = [super initWithFrame:frame];
    if (self) {
        [self _initIVars];
    }
    return self;
}
```

_initIVars 方法，实现对进度条的属性进行设置。程序代码如下：

```
- (void) _initIVars
{
    self.backgroundColor = [UIColor clearColor];        //设置背景颜色
    self.opaque = NO;                                    //可见
    self.tintColor = [UIColor colorWithRed:0.2 green:0.45 blue:0.8 alpha:1.0];
    self.trackColor = [UIColor whiteColor];             //设置轨道颜色
}
```

setProgress:animated:方法，设置进度条的动作。程序代码如下：

```objc
- (void) setProgress:(float)progress animated:(BOOL)animated
{
    if(progress < 0.0f)
        progress = 0.0f;
    else if(progress > 1.0f)
        progress = 1.0f;
    //应用与进度条图层
    ProgressLayer *layer = (ProgressLayer *)self.layer;
    if(animated)
    {
        //核心动画
        CABasicAnimation *animation = [CABasicAnimation animationWithKeyPath:
        @"progress"];
        animation.duration = 0.25;                              //设置时间
        animation.fromValue = [NSNumber numberWithFloat:layer.progress];
                                                                //设置开始值
        animation.toValue = [NSNumber numberWithFloat:progress];
                                                                //设置结束值
        [layer addAnimation:animation forKey:@"progressAnimation"];
        layer.progress = progress;                              //设置进度
    }
    else {
        layer.progress = progress;
         //设置进度
        [layer setNeedsDisplay];
    }
    //重写绘制
}
```

setTintColor:方法，对进度条加载的颜色进行设置。程序代码如下：

```objc
- (void) setTintColor:(UIColor *)tintColor
{
    ProgressLayer *layer = (ProgressLayer *)self.layer;
    layer.tintColor = tintColor;                                //设置颜色
    [layer setNeedsDisplay];                                    //重写绘制
}
```

setTrackColor:方法，对圆形按钮中的轨道颜色进行设置。程序代码如下：

```objc
- (void) setTrackColor:(UIColor *)trackColor
{
    ProgressLayer *layer = (ProgressLayer *)self.layer;
    layer.trackColor = trackColor;
    [layer setNeedsDisplay];
}
```

setStartAngle:方法，对进度条加载的角度进行设置。程序代码如下：

```objc
- (void) setStartAngle:(float)startAngle
{
    ProgressLayer *layer = (ProgressLayer *)self.layer;
    layer.startAngle = startAngle;
    [layer setNeedsDisplay];
}
```

（9）创建一个基于 NSObject 类的 ProgressPlay 类。

（10）打开 ProgressPlay.h 文件，编写代码，实现宏定义、协议以及属性、方法和对象的声明。程序代码如下：

```objc
#import <Foundation/Foundation.h>
//宏定义
#define DURATION  20.0
#define PERIOD    0.5
//引入类
@class ProgressPlay;
//协议的声明
@protocol ProgressPlayDelegate <NSObject>
- (void) player:(ProgressPlay *)player didReachPosition:(float)position;
- (void) playerDidStop:(ProgressPlay *)player;
@end
@interface ProgressPlay : NSObject{
    NSTimer *timer;
}
//方法的声明
- (void) play;
- (void) pause;
//属性的声明
@property (assign) float position;
@property (retain) id <ProgressPlayDelegate> delegate;
@end
```

（11）打开 ProgressPlay.m 文件，编写代码，实现播放动画和暂停。play 方法实现播放动画。程序代码如下：

```objc
- (void) play
{
    if(timer)
        return;
    timer = [NSTimer scheduledTimerWithTimeInterval:PERIOD target:self selector:
@selector(timerDidFire:) userInfo:nil repeats:YES];         //创建定时器
}
```

pause 方法，实现动画的暂停。程序代码如下：

```objc
- (void) pause
{
    [timer invalidate];                                     //让定时器失效
    timer = nil;
}
```

timerDidFire:方法，实现加载的动画功能。程序代码如下：

```objc
- (void) timerDidFire:(NSTimer *)theTimer
{
    //判断进度是否大于等于1.0
    if(self.position >= 1.0)
    {
        self.position = 0.0;
        [timer invalidate];                                 //让定时器失效
        timer = nil;
        [self.delegate playerDidStop:self];                 //播放停止
    }
    else
    {
        self.position += PERIOD/DURATION;
        [self.delegate player:self didReachPosition:self.position];
    }
}
```

（12）打开 ViewController.h 文件，编写代码，实现头文件、插座变量、属性和动作的声明。程序代码如下：

```
#import <UIKit/UIKit.h>
//头文件
#import "ProgressView.h"
#import "ProgressPlay.h"
@interface ViewController : UIViewController<ProgressPlayDelegate>{
    IBOutlet UIButton *playPauseButton;           //按钮的插座变量
    IBOutlet ProgressView *progress;              //进度条视图的插座变量
}
@property (retain, nonatomic) ProgressPlay *player;
- (IBAction)playPauseButton:(UIButton *)sender;   //方法
@end
```

（13）打开 Main.storyboard 文件，对 View Controller 视图控制器的设计界面进行设计，效果如图 1.9 所示。

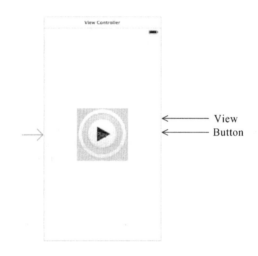

图 1.9　实例 5 View Controller 视图控制器的设计界面

需要添加的视图、控件以及对它们的设置如表 1-5 所示。

表 1-5　实例 5 视图、控件设置

视图、控件	属 性 设 置	其　　他
View	Background：CyanColor	Class：ProgressView 与插座变量 progress 关联
UIButton	Title：Play Text Color：桃红色 Background：Play.png	与插座变量 playPauseButton 关联 与动作 playPauseButton:关联

（14）打开 ViewController.m 文件，编写代码。实现单击按钮进度条加载或者暂停。程序代码如下：

```
- (void)viewDidLoad
{
    [super viewDidLoad];
    // Do any additional setup after loading the view, typically from a nib.
```

```objc
    self.player = [[ProgressPlay alloc] init] ;              //实例化
    self.player.delegate = self;                              //设置委托
    UIColor *tintColor = [UIColor greenColor];
    [[ProgressView appearance] setTintColor:tintColor];
    progress.trackColor = [UIColor colorWithWhite:0.5 alpha:1.0];
                                                              //设置轨道颜色
    progress.startAngle = (3.0*M_PI)/2.0;
}
//单击按钮实现的功能
- (IBAction)playPauseButton:(UIButton *)sender
{
    if(sender.selected)
    {
        sender.selected = NO;
        [self.player pause];                                  //进度条停止播放
        [playPauseButton setBackgroundImage:[UIImage imageNamed:@"Pause"]
forState:UIControlStateNormal];                               //设置按钮的图像
        [playPauseButton setTitle:@"Pause" forState:UIControlStateNormal];
                                                              //设置标题
    }
    else
    {
        [self.player play];
        sender.selected=YES;
        [playPauseButton setBackgroundImage:[UIImage imageNamed:@"Play"]
forState:UIControlStateNormal];                               //设置图像
        [playPauseButton setTitle:@"Play" forState:UIControlStateNormal];
                                                              //设置标题
    }
}
- (void) player:(ProgressPlay *)player didReachPosition:(float)position
{
    progress.progress = position;                             //设置进度条的进度
}
//加载结束调用
- (void) playerDidStop:(ProgressPlay *)player
{
    playPauseButton.selected = NO;
    progress.progress = 0.0;                                  //设置进度条的进度
}
```

【代码解析】

由于本实例中的代码和方法非常多,为了方便读者的阅读,笔者绘制了一些执行流程图。如图 1.10~1.12 所示。其中,单击运行按钮后一直到在模拟器上显示具有进度条按钮的界面,使用了 initWithCoder、layerClass、needsDisplayForKey:、tintColor、_initIVars、setTintColor:、setTrackColor:、viewDidLoad、setTrackColor:、setStartAngle:、setTintColor:和 drawInContext:方法。它们的执行流程如图 1.10 所示。

当用户单击具有进度条的按钮后,就会进行加载,如果在加载的过程中没有再次单击按钮,它就会一直加载,直到加载完成为止。要实现这些功能需要 playPauseButton:、play、timerDidFire:、player:didReachPosition:、setProgress:、progress、setProgress: animated:、initWithLayer:、drawInContext:和 playerDidStop:方法。它们的执行流程如图 1.11 所示。

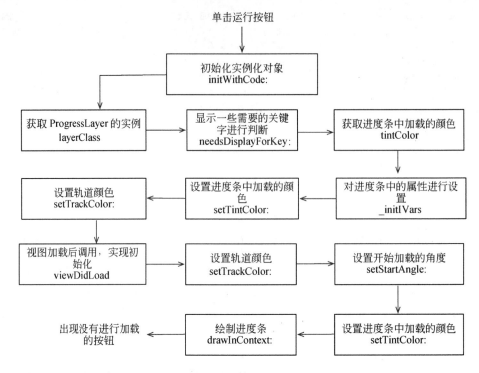

图 1.10　实例 5 程序执行流程 1

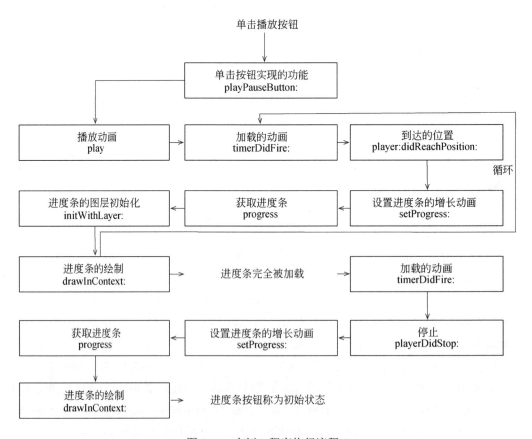

图 1.11　实例 5 程序执行流程 2

如果在进度条加载的过程中，再次单击了按钮，加载的进度条就停留在当前加载的地方。它的实现需要：timerDidFire:、player:didReachPosition:、setProgress:、progress、setProgress: animated:、playPauseButton:、pause、initWithLayer:和 drawInContext:方法共同实现。它们的执行流程如图 1.12 所示。

图 1.12　实例 5 程序执行流程 3

实例 6　唱碟机按钮

【实例描述】

本实例实现的功能是以前的唱碟机。当用户单击按钮后，唱碟机开始播放；当再次单击按钮，唱碟机就会暂停。运行效果如图 1.13 所示。

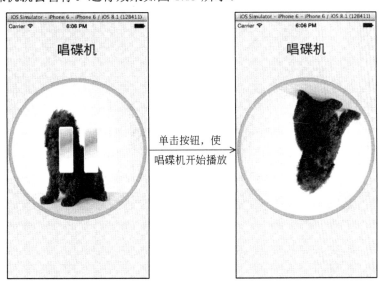

图 1.13　实例 6 运行效果

【实现过程】

当用户单击按钮后，自定义的视图就会进行旋转，当用户再次单击按钮后，自定义的视图会停止旋转。具体的实现步骤如下：

（1）创建一个项目，命名为"仿音乐播放器按钮"。

（2）添加图片button1.png、button2.png 和 dog.jpg 到创建项目的 Supporting Files 文件夹中。

（3）创建一个基于 UIView 类的 fang 类。

（4）打开 fang.h 文件，编写代码，实现属性和方法的声明。程序代码如下：

```
#import <UIKit/UIKit.h>
@interface fang : UIView
//属性的声明
@property (strong, nonatomic) UIImage *roundImage;
@property (assign, nonatomic) BOOL isPlay;
@property (assign, nonatomic) float rotationDuration;
@property (strong, nonatomic) UIImageView *roundImageView;
@property (strong, nonatomic) UIImageView *playStateView;
@property (strong, nonatomic) CABasicAnimation* rotationAnimation;
//方法的声明
-(void) play;                                    //播放
-(void) pause;                                   //暂停
@end
```

（5）打开 fang.m 文件，编写代码。在此文件中，代码分为：初始化唱碟机；绘制；设置是否播放；设置图片；触摸；旋转。添加 initJingRound 方法实现一些初始化的设置。

```
-(void) initJingRound
{
    CGPoint center = CGPointMake(self.frame.size.width / 2.0, self.frame.size.height / 2.0);
    //设置唱碟机的视图
    self.clipsToBounds = YES;
    self.userInteractionEnabled = YES;
    self.layer.cornerRadius = center.x;
    self.layer.borderWidth = 1.0;
    self.layer.borderColor = [[UIColor grayColor] CGColor];  //设置边框颜色
    self.layer.shadowColor = UIColor.blackColor.CGColor;     //设置阴影颜色
    self.layer.shadowRadius = 2;
    self.layer.shadowOpacity = 0.6;
    self.layer.shadowOffset = CGSizeMake(0, 1);              //设置阴影偏移量
    //设置唱碟机的图像视图
    UIImage *roundImage = self.roundImage;
    self.roundImageView = [[UIImageView alloc] initWithFrame:CGRectMake(0, 0, self.frame.size.width, self.frame.size.height)];  //实例化
    [self.roundImageView setCenter:center];                  //设置中心位置
    [self.roundImageView setImage:roundImage];               //设置图像
    [self addSubview:self.roundImageView];                   //添加
    //设置唱碟机播放的状态
    UIImage *stateImage;
    if (self.isPlay) {
        stateImage = [UIImage imageNamed:@"button2"];        //设置图像
    }else{
        stateImage = [UIImage imageNamed:@"button1"];        //设置图像
    }
```

```objc
self.playStateView = [[UIImageView alloc] initWithFrame:CGRectMake(0,
0, stateImage.size.width, stateImage.size.height)];      //实例化
[self.playStateView setCenter:center];
[self.playStateView setImage:stateImage];                //设置图像
[self addSubview:self.playStateView];                    //添加
//边
CGColorSpaceRef colorSpace = CGColorSpaceCreateDeviceRGB();
//创建位图图形上下文
CGContextRef context=CGBitmapContextCreate(nil, self.frame.size.width,
self.frame.size.height, 8.0, self.frame.size.width*4, colorSpace,
kCGImageAlphaPremultipliedLast|kCGBitmapByteOrder32Big);
CFRelease(colorSpace);
CGContextSetStrokeColorWithColor(context, [[UIColor colorWithRed:0.1
green:1.0 blue:0.0 alpha:0.7] CGColor]);                 //设置填充颜色
CGContextBeginPath(context);
CGContextAddArc(context, center.x, center.y, center.x , 0, 2 * M_PI, 0);
                                                         //绘制圆
CGContextClosePath(context);
CGContextSetLineWidth(context, 15.0);                    //设置线宽
CGContextStrokePath(context);
//将图形上下文转变为CGImageRef
CGImageRef image = CGBitmapContextCreateImage(context);
CGContextRelease(context);
UIImage* image2 = [UIImage imageWithCGImage:image];
UIImageView *imgv =[[UIImageView alloc] initWithFrame:CGRectMake(0,0,
self.frame.size.width , self.frame.size.height)];        //实例化
imgv.center = CGPointMake(self.frame.size.width/2, self.frame.size.
height/2);    //设置中心位置
imgv.image = image2;                                     //设置图像视图的图像
[self addSubview:imgv];
//旋转
CABasicAnimation* rotationAnimation;
rotationAnimation = [CABasicAnimation animationWithKeyPath:@
"transform.rotation.z"];
rotationAnimation.toValue = [NSNumber numberWithFloat: M_PI * 2.0 ];
                                                         //结束值
if (self.rotationDuration == 0) {
    self.rotationDuration = kRotationDuration;
}
rotationAnimation.duration = self.rotationDuration;      //设置时间
rotationAnimation.RepeatCount = FLT_MAX;                 //循环个数
rotationAnimation.cumulative = NO;
[self.roundImageView.layer addAnimation:rotationAnimation forKey:nil];
                                                         //添加动画
//暂停
if (!self.isPlay) {
    self.layer.speed = 0.0;
}
}
```

初始化唱碟机后,就可以使用 drawRect:方法绘制唱碟机。程序代码如下:

```objc
- (void)drawRect:(CGRect)rect
{
    [self initJingRound];
}
```

添加 setIsPlay:方法，用来设置唱碟机是否开始播放。程序代码如下：

```objc
-(void)setIsPlay:(BOOL)aIsPlay
{
    _isPlay = aIsPlay;
    //判断唱碟机是否播放
    if (self.isPlay) {
        [self startRotation];
    }else{
        [self pauseRotation];
    }
}
```

添加 setRoundImage:方法，对唱碟机的图像进行设置。程序代码如下：

```objc
-(void)setRoundImage:(UIImage *)aRoundImage
{
    _roundImage = aRoundImage;
    self.roundImageView.image = self.roundImage;        //设置唱碟机的图像
}
```

添加 touchesBegan:withEvent 方法，实现触摸。程序代码如下：

```objc
-(void)touchesBegan:(NSSet *)touches withEvent:(UIEvent *)event
{
    self.isPlay = !self.isPlay;
}
```

添加 startRotation 方法，实现唱碟机的触摸播放按钮开始旋转。程序代码如下：

```objc
-(void) startRotation
{
    //开始动画
    CFTimeInterval pausedTime = [self.layer timeOffset];
    self.layer.speed = 1.0;
    self.layer.timeOffset = 0.0;
    self.layer.beginTime = 0.0;
    CFTimeInterval timeSincePause = [self.layer convertTime:
    CACurrentMediaTime() fromLayer:nil] - pausedTime;
    self.layer.beginTime = timeSincePause;              //设置开始时间
    self.playStateView.image = [UIImage imageNamed:@"button2"];//设置图像
    self.playStateView.alpha = 0;                       //设置透明度
    //动画效果
    [UIView animateWithDuration:1.0 delay:0 options:UIViewAnimationOptionCurveEaseInOut
                animations:^{
                    self.playStateView.alpha = 1;       //设置透明度
                }
                completion:^(BOOL finished){
                    if (finished){
                        [UIView animateWithDuration:1.0 animations:^
                        {self.playStateView.alpha = 0;}];
                    }
                }
     ];
}
```

添加 pauseRotation 方法，实现触摸暂停按钮后，暂停动画的播放。程序代码如下：

```
-(void) pauseRotation
{
    //设置图像
    self.playStateView.image = [UIImage imageNamed:@"button1"];//设置图像
    self.playStateView.alpha = 0;                              //设置透明度
    //动画效果
    [UIView animateWithDuration:1.0 delay:0 options:UIViewAnimationOption
    CurveEaseInOut
                    animations:^{
                        self.playStateView.alpha = 1;         //设置透明度
                    }
                    completion:^(BOOL finished){
                        if (finished){
                            [UIView animateWithDuration:1.0 animations:^{
                                self.playStateView.alpha = 0;
                                //暂停
                                CFTimeInterval pausedTime = [self.layer convertTime:
                                CACurrentMediaTime() fromLayer:nil];
                                self.layer.speed = 0.0;       //设置速度
                                self.layer.timeOffset = pausedTime;
                            }];
                        }
                    }
    ];
}
```

添加 play 和 pause 方法，对播放和暂停进行设置。程序代码如下：

```
-(void)play
{
    self.isPlay = YES;
}
-(void)pause
{
    self.isPlay = NO;
}
```

（6）打开 ViewController.h 文件，编写代码，实现插座变量和头文件的声明。程序代码如下：

```
#import <UIKit/UIKit.h>
#import "fang.h"
@interface ViewController : UIViewController{
    IBOutlet fang *roundView;
}
@end
```

（7）打开 Main.storyboard 文件，对 View Controller 视图控制器的设计界面进行设计，效果如图 1.14 所示。

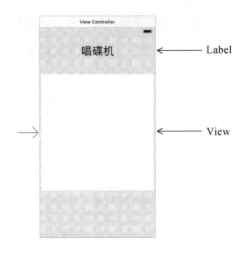

图 1.14 实例 6 View Controller 视图控制器的设计界面

需要添加的视图、控件以及对它们的设置如表 1-6 所示。

表 1-6 实例 6 视图、控件设置

视图、控件	属 性 设 置	其 他
设计界面		
Label	Text：唱碟机 Font：System 30.0 Alignment：居中对齐	
View	Background：浅粉色	Class：font 与插座变量 roundView 关联

（8）打开 ViewController.m 文件，编写代码，实现唱碟机对象的创建以及设置。程序代码如下：

```
- (void)viewDidLoad
{
    roundView.roundImage = [UIImage imageNamed:@"dog.jpg"];      //设置图片
    roundView.rotationDuration = 8.0;                             //设置时间
    roundView.isPlay = NO;                                        //设置是否进行播放
    [super viewDidLoad];
    // Do any additional setup after loading the view, typically from a nib.
}
```

【代码解析】

本实例关键功能是视图图层的旋转动画效果。下面就对这个知识点做一个详细的讲解。

要实现视图动画的旋转效果可以使用很多方法。本实例采用的是 CAPropertyAnimation 动画。要实现此动画需要使用 animationWithKeyPath:方法，创建指定的关键路径。它的语法形式如下：

```
+ (id)animationWithKeyPath:(NSString *)keyPath;
```

其中，(NSString *)keyPath 表示关键路径的属性动画。这些动画很多，如表 1-7 所示。

表 1-7 关键路径的属性动画

动画	功能
transform.scale	比例转换
transform.scale.x	宽的比例转换
transform.scale.y	高的比例转换
transform.rotation.z	平面图的旋转
opacity	透明度
zPosition	横轴上的位置
backgroundColor	背景的颜色
cornerRadius	绘制圆角时使用半径绘制
borderWidth	边框的宽度
bounds	框架的大小
contents	一个提供了该层内容的对象
frame	框架的大小位置
hidden	隐藏
mask	一个可选层的 alpha 通道，用来遮蔽层的内容
position	位置
shadowColor	阴影的颜色
shadowOffset	阴影的偏移
shadowOpacity	阴影的不透明度
shadowRadius	阴影渲染的模糊半径

在代码中，使用了 animationWithKeyPath:方法实现了图层的旋转，代码如下：

```
rotationAnimation = [CABasicAnimation animationWithKeyPath:@"transform.rotation.z"];
```

其中，@"transform.rotation.z"表示关键路径的属性动画。

实例 7 环形按钮

【实例描述】

本实例实现的功能是在界面显示一个四分之一的环形。在环形上放置一些按钮。当用户单击环形上的某一按钮后，会出现一个警告视图。将用户顺着环形拖动，环形按钮会跟着旋转。运行效果如图 1.15 所示。

【实现过程】

在界面上的每一个按钮成圆弧式分布，并且可以单击任意按钮。用户的手指放在指定的位置可以滑动这些按钮。具体的实现步骤如下。

（1）创建一个项目，命名为"环形按钮"。

（2）添加图片 button1.png、button2.png、button3.png、button4.png、button5.png、button6.png 和 yuan.jpg 到创建项目的 Supporting Files 文件夹中。

（3）创建一个基于 UIView 类的 ButtonWheel 类。

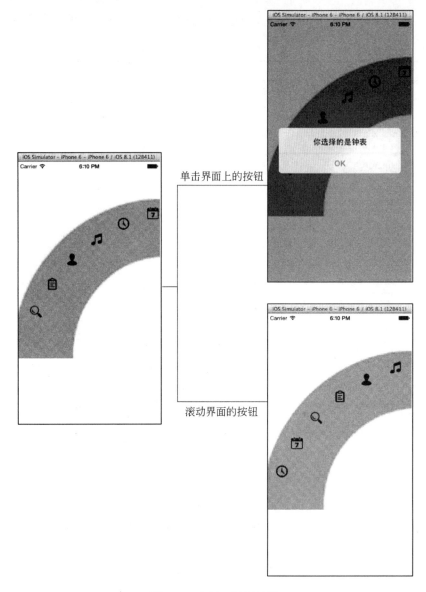

图 1.15 实例 7 运行效果

（4）打开 ButtonWheel.h 文件，编写代码，实现宏定义、属性和方法的声明。程序代码如下：

```
#import <UIKit/UIKit.h>
//宏定义
#define kTOUCHBAND 25
@interface ButtonWheel : UIView
//属性声明
@property (nonatomic,retain) UIView *controlView;
@property (nonatomic,retain) NSMutableArray *buttons;
@property (assign) CGPoint radialCenter;
@property (assign) double radius;
……
@property (assign) double touchVel;
@property (assign) BOOL spinningWheel;
```

```
@property (nonatomic,retain) NSTimer *spinTimer;
//方法声明
-(id)initWithFrame:(CGRect)frame buttonArray:(NSArray*)buttonArray arcCenter:
(CGPoint)theCenter radius:(double)theRadius;
@end
```

(5) 打开 ButtonWheel.m 文件,编写代码。实现环形按钮的绘制。添加 initWithFrame: buttonArray:arcCenter:radius:方法实现对环形按钮的初始化。程序代码如下:

```
-(id)initWithFrame:(CGRect)frame         buttonArray:(NSArray*)buttonArray
arcCenter:(CGPoint)theCenter radius:(double)theRadius {
    self = [super initWithFrame:frame];
    if (self) {
        self.buttons=[NSMutableArray arrayWithArray:buttonArray];
                                                       //创建可变数组
        self.controlView = [[UIView alloc] initWithFrame:frame];
                                                       //创建视图对象
        [self configureWithArcCenter:theCenter radius:theRadius];
        [self assembleViews];
    }
    return self;
}
```

添加 configureWithArcCenter:radius:方法实现对圆弧的设置。程序代码如下:

```
-(void)configureWithArcCenter:(CGPoint)theCenter radius:(double)theRadius {
    self.radialCenter=theCenter;
    self.radius=theRadius;                             //设置半径
    CGRect viewRect = self.frame;                      //实例化对象
    self.theta1=asin((320-radialCenter.x)/radius);
    self.theta2=-M_PI+acos((viewRect.size.height-radialCenter.y)/radius);
    self.visibleArc=fabs(theta2-theta1);
    self.angle=0;
    self.firstAngle=theta1-DEGREES_TO_RADIANS(kFirstImageOffset);
    self.delta=DEGREES_TO_RADIANS(kImageSeparation);
    self.numVisible=trunc(fabs(theta2-firstAngle)/delta);
    self.fullSweep=delta*[buttons count];
}
```

添加 assembleViews 方法,实现装备视图。程序代码如下:

```
-(void)assembleViews {
    [self positionImages];
//遍历
    for (UIButton *imv in buttons) {
        [controlView addSubview:imv];
    }
    [self addSubview:controlView];
}
```

添加 positionImages 方法,对图像的位置进行设置。程序代码如下:

```
-(void)positionImages {
    //遍历
    for (int i=0;i<[buttons count];i++) {
        int actualIndex = i;
        UIButton *btnView = [buttons objectAtIndex:actualIndex];
        double x,y,a;
        double offset=0;
        offset=angle-(delta*i);
```

```
        if (offset>0) offset-=fullSweep;
        else if (offset<(-fullSweep)) offset+=fullSweep;
        a=firstAngle+offset;
        x = radialCenter.x + sin(a)*radius;                    //设置x
        y = radialCenter.y - cos(a)*radius;                    //设置y
        btnView.center=CGPointMake(x, y);                      //设置中心
        btnView.alpha=(CGRectContainsRect(controlView.bounds, btnView.
        frame))?1.0:0.0;                                       //设置透明度
    }
}
```

这时，运行结果，会看到一个环形按钮。但是此环形按钮是不会旋转的，也不会在触摸时有任何反应。要实现旋转需要添加 clearSpin、spinWheel 和 repositionImagesWithDelta:方法。这些方法可以使环形按钮实现旋转的功能。clearSpin 方法实现对旋转的一些设置。程序代码如下：

```
-(void)clearSpin {
    self.touchVel=0;
    self.spinningWheel=FALSE;
}
```

spinWheel 方法实现旋转的功能。程序代码如下：

```
-(void)spinWheel {
    [self performSelectorOnMainThread:@selector(repositionImagesWithDelta:)withObject:
[NSNumber numberWithDouble:self.touchVel] waitUntilDone:false];
                                                              //在主线程上执行方法
    self.touchVel*=0.9;
    //判断
    if (fabs(touchVel)<0.005) {
        [self clearSpin];
        [spinTimer invalidate];                               //让定时器失效
        self.spinTimer=nil;
    }
}
```

repositionImagesWithDelta:方法实现旋转中的重新定位。程序代码如下：

```
-(void)repositionImagesWithDelta:(NSNumber*)increment {
    self.angle+=[increment doubleValue];
    if (angle<(delta-fullSweep)) angle+=2*fullSweep;
    else if (angle>(fullSweep-delta)) angle-=2*fullSweep;
    [self positionImages];
}
```

这时的环形按钮已经具有了旋转的功能，但是现在还不可以旋转，因为没有方法去触发旋转。如果想要实现触摸旋转功能，需要添加 ptInCircle:radius:point:、ptInTouchBand:、touchesBegan:withEvent:、touchesMoved:withEvent:、touchesEnded:withEvent: 和 touchesCancelled:withEvent:方法共同实现。ptInCircle:radius:point:方法实现编写圆。程序代码如下：

```
-(BOOL)ptInCircle:(CGPoint) ctr radius:(double)cradius point:(CGPoint)pt
{
    return pow(pt.x-ctr.x, 2)+pow(pt.y-ctr.y, 2)<=pow(cradius, 2);
}
```

ptInTouchBand:方法用来编写触摸带，用户可以通过这个地方触摸环形按钮。程序代

码如下：

```objc
-(BOOL)ptInTouchBand:(CGPoint)pt {
    return [self ptInCircle:radialCenter radius:radius+kTOUCHBAND point:pt]
    &&! [self ptInCircle:
radialCenter radius:radius-kTOUCHBAND point:pt];
}
```

touchesBegan:withEvent:方法，实现开始触摸。程序代码如下：

```objc
-(void)touchesBegan:(NSSet *)touches withEvent:(UIEvent *)event {
    //判断定时器是否存在
    if (spinTimer) {
        [spinTimer invalidate];                                //让定时器失效
        self.spinTimer=nil;
    }
    [self clearSpin];
    //判断触摸次数
    if ([touches count]==1) {
        UITouch *touch=(UITouch*)[[touches objectEnumerator] nextObject];
        CGPoint pt=[touch locationInView:controlView];        //获取位置
        if ([self ptInTouchBand:pt]) {
            self.spinningWheel=true;
        } else {
            [super touchesBegan:touches withEvent:event];     //开始触摸
        }
    } else {
        [super touchesBegan:touches withEvent:event];
    }
}
```

touchesMoved:withEvent:方法，实现触摸移动。程序代码如下：

```objc
-(void)touchesMoved:(NSSet *)touches withEvent:(UIEvent *)event {
    if ( [touches count]>=1) {
        UITouch *touch=(UITouch*)[[touches objectEnumerator] nextObject];
        CGPoint pt=[touch locationInView:controlView];        //获取位置
        if (self.spinningWheel) {
            CGPoint lastPt=[touch previousLocationInView:controlView];
                                                              //获取触摸位置
            double x2=asin((pt.x-radialCenter.x)/(2*radius));
            double x1=asin((lastPt.x-radialCenter.x)/(2*radius));
            double y2=M_PI-acos((pt.y-radialCenter.y)/(2*radius));
            double y1=M_PI-acos((lastPt.y-radialCenter.y)/(2*radius));
            y2=-y2;
            y1=-y1;
            double approxVel=2*MAX(x2-x1, y2-y1);
            self.touchVel=approxVel;
            [self performSelectorOnMainThread:@selector(repositionImagesWithDelta:)
withObject:[NSNumber numberWithDouble:approxVel] waitUntilDone:false];
                                                              //在主线程上执行方法
        } else {
            [super touchesMoved:touches withEvent:event];     //移动触摸
        }
    } else {
        [super touchesMoved:touches withEvent:event];
    }
}
```

touchesEnded:withEvent:方法实现触摸结束。程序代码如下：

```objc
-(void)touchesEnded:(NSSet *)touches withEvent:(UIEvent *)event {
    //判断触摸次数
    if ([touches count]>=1) {
        if (spinningWheel) {
            if (fabs(touchVel)>0.0262) {
                self.spinTimer = [NSTimer timerWithTimeInterval:0.03 target:self selector:@selector(spinWheel) userInfo:nil repeats:YES];
                //创建定时器
                [[NSRunLoop currentRunLoop] addTimer:spinTimer forMode:NSDefaultRunLoopMode];
            } else {
                [self clearSpin];
            }
        } else {
            [super touchesEnded:touches withEvent:event];    //结束触摸
        }
    }
    else {
        [super touchesEnded:touches withEvent:event];
    }
}
```

touchesCancelled:withEvent:方法实现触摸取消。程序代码如下：

```objc
-(void)touchesCancelled:(NSSet *)touches withEvent:(UIEvent *)event {
    [self clearSpin];
    [super touchesCancelled:touches withEvent:event];    //取消触摸
}
```

（6）打开 Main.storyboard 文件，在 View Controller 视图控制器的设计界面中放入一个 Image View 图像视图，将 Image 属性设置为"yuan.jpg"。

（7）打开 ViewController.h 文件，编写代码，实现头文件以及属性的声明。程序代码如下：

```objc
#import <UIKit/UIKit.h>
//头文件
#import "ButtonWheel.h"
@interface ViewController : UIViewController
//属性声明
@property (nonatomic,retain) ButtonWheel *controlView;
@property (nonatomic,retain) NSMutableArray *buttons;
@end
```

（8）打开 ViewController.m 文件，编写代码。在此文件中，代码包括环形按钮对象的创建，以及单击每一个按钮后出现的动作。首先添加 makeButton:tag 方法，实现小按钮对象的创建，以及单击按钮的动作声明。程序代码如下：

```objc
-(UIButton *)makeButton:(NSString *)imageName tag:(NSInteger)tag {
    UIButton *btn = [UIButton buttonWithType:UIButtonTypeCustom];
    UIImage *im = [UIImage imageNamed:imageName];
    [btn setImage:im forState:UIControlStateNormal];    //设置图像
    [btn setShowsTouchWhenHighlighted:TRUE];
    [btn setTag:tag];    //设置tag值
    CGRect r = CGRectMake(0,0, im.size.width, im.size.height);
```

```
    btn.frame=r;                                              //设置框架
[btn addTarget:self action:@selector(buttonPressed:) forControlEvents:
UIControlEventTouchUpInside];                                 //添加动作
    return btn;
}
```

在 viewDidLoad 方法中，实现环形按钮的显示。程序代码如下：

```
- (void)viewDidLoad
{
    [super viewDidLoad];
    // Do any additional setup after loading the view, typically from a nib.
    self.buttons = [NSMutableArray arrayWithCapacity:6];
    //为按钮添加图片
    UIButton *btn = [self makeButton:@"button1.png" tag:0];
    [buttons addObject:btn];                                  //添加对象
    [buttons addObject:[self makeButton:@"button2.png" tag:1]];
    [buttons addObject:[self makeButton:@"button3.png" tag:2]];
    [buttons addObject:[self makeButton:@"button4.png" tag:3]];
    [buttons addObject:[self makeButton:@"button5.png" tag:4]];
    [buttons addObject:[self makeButton:@"button6.png" tag:5]];
    CGRect screenRect = [[UIScreen mainScreen] bounds]; //获取主屏幕的大小
    self.controlView = [ButtonWheel alloc];
    //判断状态栏是否隐藏
    if (![[UIApplication sharedApplication] isStatusBarHidden]) {
        CGSize sz = screenRect.size;                          //获取尺寸
        sz.height -= [[UIApplication sharedApplication] statusBarFrame].size.height;
        screenRect.size=sz;                                   //设置尺寸
    }
    double radius=400;
    CGPoint radialCenter=CGPointMake(400, 500);
[controlView initWithFrame:screenRect buttonArray:buttons arcCenter:
radialCenter radius:
radius ];                                                     //初始化
    [self.view addSubview:controlView];                       //添加视图
}
```

添加 buttonPressed:方法，实现单击按钮后触发的动作。程序代码如下：

```
-(void)buttonPressed:(id)btn {
    UIButton *b = (UIButton*)btn;
    //判断 tag 是否为 0
    if (b.tag==0) {
        UIAlertView *a=[[UIAlertView alloc]initWithTitle:@"你选择的是日历"
message:nil delegate:nil cancelButtonTitle:@"OK" otherButtonTitles: nil];
                                                              //创建警告视图
        [a show];
    //判断 tag 是否为 1
    }else if (b.tag==1){
        UIAlertView *a=[[UIAlertView alloc]initWithTitle:@"你选择的是钟表"
message:nil delegate:nil cancelButtonTitle:@"OK" otherButtonTitles: nil];
        [a show];
    //判断 tag 是否为 2
```

```
        }else if(b.tag==2){
            UIAlertView *a=[[UIAlertView alloc]initWithTitle:@"你选择的是音乐"
message:nil delegate:nil cancelButtonTitle:@"OK" otherButtonTitles: nil];
            [a show];
        //判断tag是否为3
        }else if (b.tag==3){
            UIAlertView *a=[[UIAlertView alloc]initWithTitle:@"你选择的是头像"
message:nil delegate:nil cancelButtonTitle:@"OK" otherButtonTitles: nil];
            [a show];
        //判断tag是否为4
        }else if (b.tag==4){
            UIAlertView *a=[[UIAlertView alloc]initWithTitle:@"你选择的是日记"
message:nil delegate:nil cancelButtonTitle:@"OK" otherButtonTitles: nil];
            [a show];
        //判断tag是否为5
        }else if(b.tag==5){
            UIAlertView *a=[[UIAlertView alloc]initWithTitle:@"你选择的是查找"
message:nil delegate:nil cancelButtonTitle:@"OK" otherButtonTitles: nil];
            [a show];
        }
}
```

【代码解析】

由于本实例中的代码和方法非常多，为了方便读者的阅读，笔者绘制了一些执行流程图，如图1.16和图1.17所示。其中，环形按钮的显示以及单击某一按钮，使用了ViewDidLoad、makeButton:tag:、initWithFrame: configureWithArcCenter:radius:、assembleViews、positionImages 和 buttonPressed:方法。它们的执行流程如图1.16所示。

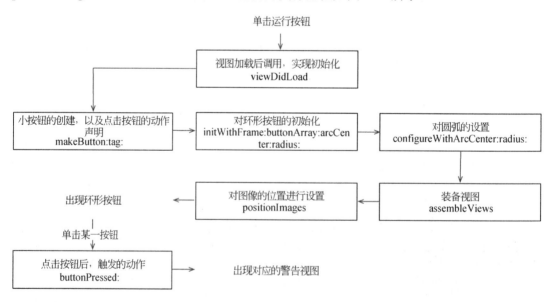

图1.16 实例7程序执行流程1

在触摸带上触摸并且滑动时环形按钮旋转，使用了touchesBegan:withEvent:、clearSpin、ptInTouchBand:、ptInCircle:、repositionImagesWithDelta:、touchesEnded::withEvent:、 touchesMoved:withEvent:、positionImages 和 spinWheel 方法。它们的实现流程如图 1.17 所示。

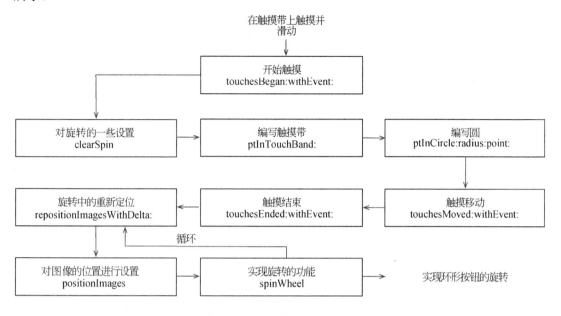

图 1.17　实例 7 程序执行流程 2

第 2 章　滑块类效果

在 iPhone 或者 iPad 中，像视频进度的调节、声音进度的调节都是使用滑块的方式实现的。它也是用户经常使用的操作之一。本章将主要讲解拖动滑块一类的实例。

实例 8　具有多个颜色的滑块控件

【实例描述】

本实例实现的功能是在界面显示一个具有多种颜色的滑块控件。运行效果如图 2.1 所示。

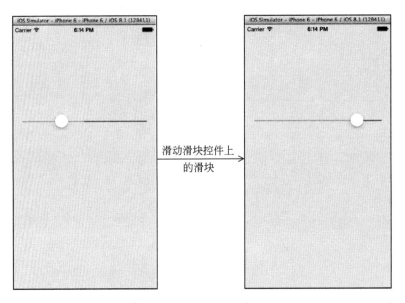

图 2.1　实例 8 运行效果

【实现过程】

（1）创建一个项目，命名为"具有多个颜色的滑块"。
（2）创建一个基于 UIControl 类的 Slider 类。
（3）打开 Slider.h 文件，编写代码。实现头文件、宏定义、对象、实例变量以及属性的声明。程序代码如下：

```
#import <UIKit/UIKit.h>                                //头文件
#import <objc/message.h>                              //头文件
#define POINT_OFFSET    (2)                           //宏定义
@interface Slider : UIControl{
```

```
    //对象
    UISlider*        _slider;
    UIProgressView*  _progressView;
    //实例变量
    BOOL             _loaded;
    id               _target;
    SEL              _action;
}
//属性
@property (nonatomic, assign) CGFloat value;
@property (nonatomic, assign) CGFloat middleValue;
@property (nonatomic, strong) UIColor* thumbTintColor;
@property (nonatomic, strong) UIColor* minimumTrackTintColor;
@property (nonatomic, strong) UIColor* middleTrackTintColor;
@property (nonatomic, strong) UIColor* maximumTrackTintColor;
@end
```

（4）打开 Slider.m 文件，编写代码。实现一个具有多个颜色的滑块的创建。使用的方法如表 2-1 所示。

表 2-1　Slider.m文件中的方法总结

方　　法	功　　能
loadSubView	创建
awakeFromNib	调用创建方法
initWithFrame:	初始化滑块控件
value	获取当前值
setValue:	设置当前值
middleValue	获取中间的值
setMiddleValue:	设置中间的值
minimumTrackTintColor	获取滑块左边的轨道填充颜色
setMinimumTrackTintColor:	设置滑块左边的轨道填充颜色
middleTrackTintColor	获取中间的轨道填充颜色
setMiddleTrackTintColor:	设置中间的轨道填充颜色
maximumTrackTintColor	获取滑块右边的轨道填充颜色
setMaximumTrackTintColor:	设置滑块右边的轨道填充颜色

这里需要讲解几个重要的方法（其他方法请读者参考源代码）。loadSubView 方法，用于创建滑块控件。程序代码如下：

```
- (void)loadSubView {
    if (_loaded) return;
    _loaded = YES;
    self.backgroundColor = [UIColor clearColor];            //设置背景
    //创建并设置滑块控件对象
    _slider = [[UISlider alloc] initWithFrame:self.bounds];//实例化对象
    _slider.autoresizingMask = UIViewAutoresizingFlexibleWidth|UIView
    AutoresizingFlexibleHeight;
    [self addSubview:_slider];                              //添加
    CGRect rect = _slider.bounds;
    rect.origin.x += POINT_OFFSET;                          //设置 x 的位置
    rect.size.width -= POINT_OFFSET*2;                      //设置宽度
    //创建并设置进度条控件对象
    _progressView = [[UIProgressView alloc] initWithFrame:rect];
```

```
    _progressView.autoresizingMask = UIViewAutoresizingFlexibleWidth|
    UIViewAutoresizingFlexibleHeight;
    _progressView.center = _slider.center;              //设置中心位置
    _progressView.userInteractionEnabled = NO;
    [_slider addSubview:_progressView];                  //添加视图
    [_slider sendSubviewToBack:_progressView];
    _slider.maximumTrackTintColor = [UIColor clearColor];
}
```

(5) 打开 ViewController.h 文件，编写代码。实现头文件以及插座变量的声明。程序代码如下：

```
#import <UIKit/UIKit.h>
#import "Slider.h"                                       //头文件
@interface ViewController : UIViewController{
    IBOutlet Slider *slider;                             //插座变量
}
@end
```

(6) 打开 Main.storyboard 文件，将设计界面的 Background 设置为浅灰色。再添加一个 View 空白视图到设计界面，将 Class 设置为 Slider。将插座变量与此视图关联。

(7) 打开 ViewController.m 文件，对创建的滑块控件对象进行设置。程序代码如下：

```
- (void)viewDidLoad
{
    slider.value = 0.3;                                  //设置值
    slider.middleValue = 0.5;                            //设置中间值
    slider.minimumTrackTintColor = [UIColor orangeColor];
    slider.middleTrackTintColor = [UIColor greenColor];
                                           //设置slider最右边的颜色
    slider.maximumTrackTintColor = [UIColor blueColor];
    [super viewDidLoad];
    // Do any additional setup after loading the view, typically from a nib.
}
```

【代码解析】

本实例关键功能是滑块控件轨道的多颜色实现。下面将详细讲解这个知识点。

对于滑块控件轨道的多颜色实现，本实例使用了进度条来实现。首先创建了一个滑块控件，将它的轨道颜色设置为了透明色，这样，在滑块控件中的滑块未滑动时，只会显示一个圆形的滑块作为具有滑块控件的滑块。然后，创建了一个进度条，将此进度条放在滑块控件的背后。这时，进度条就可以作轨道。程序代码如下：

```
_slider = [[UISlider alloc] initWithFrame:self.bounds];
_slider.autoresizingMask = UIViewAutoresizingFlexibleWidth|UIViewAutoresizing
FlexibleHeight;
[self addSubview:_slider];
CGRect rect = _slider.bounds;
rect.origin.x += POINT_OFFSET;
rect.size.width -= POINT_OFFSET*2;
//创建并设置进度条控件
_progressView = [[UIProgressView alloc] initWithFrame:rect];
_progressView.autoresizingMask = UIViewAutoresizingFlexibleWidth|UIView
AutoresizingFlexibleHeight;
_progressView.center = _slider.center;
_progressView.userInteractionEnabled = NO;
[_slider addSubview:_progressView];
```

```
[_slider sendSubviewToBack:_progressView];
_slider.maximumTrackTintColor = [UIColor clearColor];
```

实例 9　环形滑块控件

【实现描述】

本实例实现的功能是在界面显示一个环形的滑块控件。当用户滑动滑块控件中的滑块时，会将当前的值进行显示。其中，用来显示当前值的为标签。运行效果如图 2.2 所示。

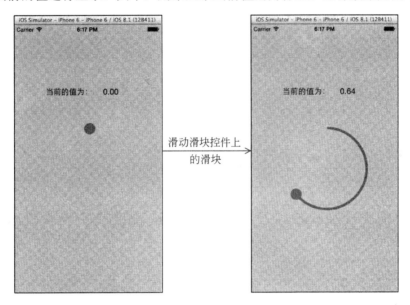

图 2.2　实例 9 运行效果

【实现过程】

（1）创建一个项目，命名为"环形滑块"。
（2）创建一个基于 UIControl 类的 CircularSlider 类。
（3）打开 CircularSlider.h 文件，编写代码。实现宏定义、属性和方法的声明。程序代码如下：

```
#import <UIKit/UIKit.h>
//宏定义
#define kLineWidth 5.0
#define kThumbRadius 12.0
@interface CircularSlider : UIControl
//属性
@property (nonatomic) float value;
@property (nonatomic) float minimumValue;
@property (nonatomic) float maximumValue;
@property(nonatomic, retain) UIColor *minimumTrackTintColor;
@property(nonatomic, retain) UIColor *maximumTrackTintColor;
@property(nonatomic, retain) UIColor *thumbTintColor;
@property(nonatomic, getter=isContinuous) BOOL continuous;
@property(nonatomic,strong)UITextField *textField;
@property (nonatomic) CGPoint thumbCenterPoint;
```

```
//方法
- (void)setup;                                   //对滑块控件进行设置
- (BOOL)isPointInThumb:(CGPoint)point;
- (CGFloat)sliderRadius;
- (void)drawThumbAtPoint:(CGPoint)sliderButtonCenterPoint inContext:
(CGContextRef)context;
- (CGPoint)drawCircularTrack:(float)track atPoint:(CGPoint)point withRadius:
(CGFloat)radius inContext:(CGContextRef)context;
@end
```

（4）打开 CircularSlider.m 文件，编写代码。实现一个环形效果的滑块控件。使用的方法如表 2-2 所示。

表 2-2　CircularSlider.m文件中的方法总结

方　　法	功　　能
translateValueFromSourceIntervalToDestinationInterval	获取目标值
angleBetweenThreePoints	获取角度
setValue:	设置当前值
setMinimumValue:	设置最小值
setMaximumValue:	设置最大值
setMinimumTrackTintColor:	设置滑块左边的轨道填充颜色
setMaximumTrackTintColor:	设置滑块右边的轨道填充颜色
setThumbTintColor:	设置滑块控件上的滑块
initWithFrame:	初始化环形滑块控件
awakeFromNib	当.nib 文件加载时调用
setup	对滑块控件进行设置
sliderRadius	获取半径
drawThumbAtPoint: inContext:	绘制滑块控件上的滑块
drawCircularTrack: atPoint: withRadius: inContext:	绘制环形的轨道
drawRect:	绘制环形的滑块控件
isPointInThumb:	判断滑块控件在滑块上的点
panGestureHappened:	拖动的手势实现

要实现一个环形效果的滑块控件，首先要对环形的滑块控件进行绘制，然后再实现滑块控件的设置，最后实现滑块控件的功能。其中，绘制滑块控件在此文件中分为了绘制环形的轨道、绘制滑块控件中的滑块以及绘制一个完整的环形滑块控件。这里需要讲解几个重要的方法（其他方法请读者参考源代码）。其中，要实现轨道的绘制需要使用 translateValueFromSourceIntervalToDestinationInterval 和 drawCircularTrack: atPoint: withRadius: inContext:方法共同实现。其中 drawCircularTrack: atPoint: withRadius: inContext:方法实现的就是绘制环形的轨道。程序代码如下：

```
- (CGPoint)drawCircularTrack:(float)track atPoint:(CGPoint)center withRadius:
(CGFloat)radius inContext:
(CGContextRef)context {
    UIGraphicsPushContext(context);                  //保存当前上下文的状态
    CGContextBeginPath(context);
    float angleFromTrack = translateValueFromSourceIntervalToDestina
tionInterval(track, self.minimumValue, self.maximumValue, 0, 2*M_PI);
    CGFloat startAngle = -M_PI_2;                    //设置开始的角度
```

```
    CGFloat endAngle = startAngle + angleFromTrack;    //设置结束的角度
    CGContextAddArc(context, center.x, center.y, radius, startAngle, endAngle, NO);
                                                       //绘制圆弧
    CGPoint arcEndPoint = CGContextGetPathCurrentPoint(context);
    CGContextStrokePath(context);                      //绘制路径
    UIGraphicsPopContext();
    return arcEndPoint;
}
```

要实现滑块控件中滑块的绘制,需要使用 drawThumbAtPoint: inContext:方法实现。程序代码如下:

```
- (void)drawThumbAtPoint:(CGPoint)sliderButtonCenterPoint inContext:
(CGContextRef)context {
    UIGraphicsPushContext(context);
    CGContextBeginPath(context);
    CGContextMoveToPoint(context, sliderButtonCenterPoint.x, sliderButton
CenterPoint.y);                                        //设置开始点
    CGContextAddArc(context, sliderButtonCenterPoint.x, sliderButtonCenterPoint.y,
    kThumbRadius, 0.0, 2*M_PI, NO);                    //绘制圆弧
    CGContextFillPath(context);
    UIGraphicsPopContext();
}
```

环形和滑块控件中的滑块都绘制好之后,就可以绘制滑块控件了,需要使用 sliderRadius 和 drawRect:方法共同实现。其中 sliderRadius 方法用来获取滑块的半径。程序代码如下:

```
- (CGFloat)sliderRadius {
    CGFloat radius = MIN(self.bounds.size.width/2, self.bounds.size.height/2);
    radius -= MAX(kLineWidth, kThumbRadius);
    return radius;
}
```

drawRect:方法,实现滑块控件的绘制。程序代码如下:

```
- (void)drawRect:(CGRect)rect {
    CGContextRef context = UIGraphicsGetCurrentContext();
    CGPoint middlePoint;
    middlePoint.x = self.bounds.origin.x + self.bounds.size.width/2;
    middlePoint.y = self.bounds.origin.y + self.bounds.size.height/2;
    CGContextSetLineWidth(context, kLineWidth);        //绘制线宽
    CGFloat radius = [self sliderRadius];
    [self.maximumTrackTintColor setStroke];            //设置滑块右边轨道的颜色
    [self drawCircularTrack:self.maximumValue atPoint:middlePoint withRadius:
radius inContext:context];                             //绘制环形
    [self.minimumTrackTintColor setStroke];            //设置滑块左边轨道的颜色
    self.thumbCenterPoint = [self drawCircularTrack:self.value atPoint:
middlePoint withRadius:radius inContext:context];
    [self.thumbTintColor setFill];
    [self drawThumbAtPoint:self.thumbCenterPoint inContext:context];
                                                       //绘制滑块控件中的滑块
}
```

绘制好环形滑块控件后,就可以对此控件进行设置了。需要使用 setValue:、setMinimumValue: 、setMaximumValue:、setMinimumTrackTintColor:、setMaximumTrackTintColor:、setThumbTintColor:和 setup 方法共同实现。其中 setup 方法实现的就是对滑块控件

的设置。程序代码如下：

```
- (void)setup {
    self.value = 0.0;
    self.minimumValue = 0.0;                              //设置最小值
    self.maximumValue = 1.0;                              //设置最大值
    self.minimumTrackTintColor = [UIColor blueColor];
    self.maximumTrackTintColor = [UIColor greenColor];
    self.thumbTintColor = [UIColor redColor];             //设置滑块的颜色
    self.continuous = YES;
    self.thumbCenterPoint = CGPointZero;
    UIPanGestureRecognizer *panGestureRecognizer = [[UIPanGestureRecognizer alloc] initWithTarget:self action:@selector(panGestureHappened:)];
                                                          //创建拖曳的手势识别器
    panGestureRecognizer.maximumNumberOfTouches = panGestureRecognizer.minimumNumberOfTouches;
    [self addGestureRecognizer:panGestureRecognizer];     //添加手势识别器
}
```

这时的环形滑块控件就绘制并设置好了，但是它还没有起到滑动的效果，要想实现此功能，需要使用 translateValueFromSourceIntervalToDestinationInterval、angleBetweenThreePoints、sliderRadius 和 panGestureHappened:方法。其中，panGestureHappened:方法实现的就是通过拖曳实现滑块滑动的功能。程序代码如下：

```
- (void)panGestureHappened:(UIPanGestureRecognizer *)panGestureRecognizer {
    CGPoint tapLocation = [panGestureRecognizer locationInView:self];
                                                          //获取手势的位置
    switch (panGestureRecognizer.state) {
        case UIGestureRecognizerStateChanged: {
            CGFloat radius = [self sliderRadius];         //获取半径
            CGPoint sliderCenter = CGPointMake(self.bounds.size.width/2, self.bounds.size.height/2);
            CGPoint sliderStartPoint = CGPointMake(sliderCenter.x, sliderCenter.y - radius);
            CGFloat angle = angleBetweenThreePoints(sliderCenter, sliderStartPoint, tapLocation);
            //判断角度是否小于0
            if (angle < 0) {
                angle = -angle;                           //设置角度
            }
            else {
                angle = 2*M_PI - angle;                   //设置角度
            }
            self.value = translateValueFromSourceIntervalToDestinationInterval(angle, 0, 2*M_PI, self.minimumValue, self.maximumValue); //设置值
            break;
        }
        default:
            break;
    }
}
```

（5）打开 ViewController.h 文件，编写代码。实现头文件以及插座变量的声明。程序代码如下：

```
#import <UIKit/UIKit.h>
#import "CircularSlider.h"
```

```
@interface ViewController : UIViewController{
    IBOutlet CircularSlider *circular;
    IBOutlet UILabel *label;                        //声明关于标签的插座变量
}
@end
```

(6)打开 Main.storyboard 文件,对 View Controller 视图控制器的设计界面进行设计,效果如图 2.3 所示。

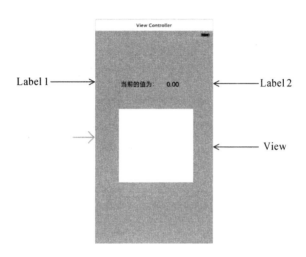

图 2.3　实例 9 View Controller 视图控制器的设计界面

需要添加的视图、控件以及对它们的设置如表 2-3 所示。

表 2-3　实例 9 视图、控件设置

视图、控件	属 性 设 置	其　　他
设计界面	Background:浅蓝色	
View		Class:CircularSlider 与插座变量 circular 关联
Label1	Text:当前的值为:	
Label2	Text:0.00	与插座变量 label 关联

(7)打开 ViewController.m 文件,编写代码,实现环形滑块控件对象的最小值、最大值的设置以及更新进度的值。程序代码如下:

```
- (void)viewDidLoad
{
    circular.backgroundColor=[UIColor clearColor];
    [circular addTarget:self action:@selector(updateProgress:) forControlEvents:
UIControlEventValueChanged];
    [circular setMinimumValue:0];                   //设置最大值
    [circular setMaximumValue:1];                   //设置最小值
    [super viewDidLoad];
    // Do any additional setup after loading the view, typically from a nib.
}
//更新进度的值
- (IBAction)updateProgress:(UISlider *)sender {
```

```
        [circular setValue:sender.value];                              //设置值
        label.text=[NSString stringWithFormat:@"%0.2f",circular.value]; //设置文本
}
```

【代码解析】

由于本实例中的代码和方法非常多，为了方便读者的阅读，笔者绘制了一些执行流程图，如图2.4和图2.5所示。其中，环形滑块控件的显示，需要使用awakeFromNib、setup、setValue:、setMinimumValue:、setMaximumValue:、setMinimumTrackTintColor:、setMaximumTrackTintColor:、setThumbTintColor:、viewDidLoad、drawRect:、sliderRadius、drawCirularTrack:atPoint:withRadius:inContext:、translateValueFromSourceIntervalToDestinationInterval和drawThumbAtPoint:inContext:方法共同实现。它们的程序执行流程如图2.4所示。

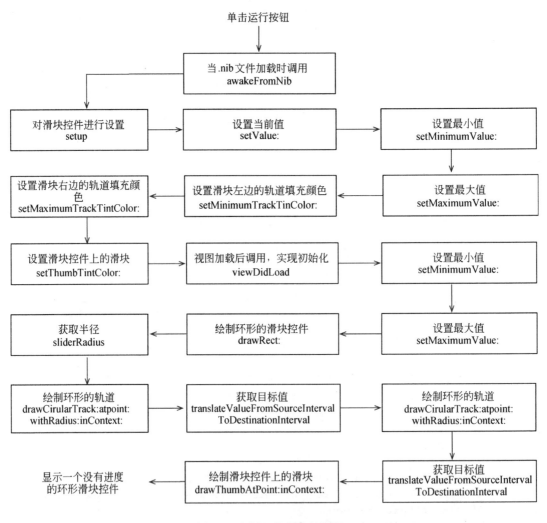

图2.4 实例9 程序执行流程1

实现环形滑块控件后，拖动滑块控件中的滑块到达一个位置需要使用panGestureHappened:、sliderRadius、angleBetweenThreePoints、translateValueFromSourceIntervalToDestinationInterval、setValue、updateProgress:、angleBetweenThreePoints、drawRect:、 drawCircularTrack:atPoint:withRadius:inContext:和 drawThumbAtPoint:inContext:方法共同实现。它们的执行流程如图 2.5 所示。

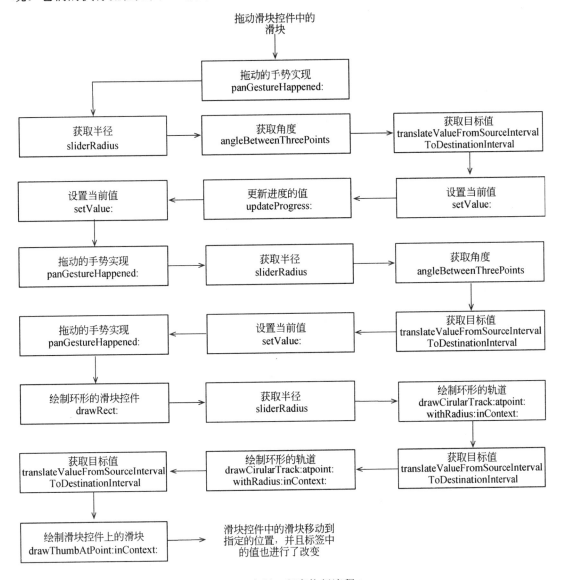

图 2.5 实例 9 程序执行流程 2

实例 10 具有范围的滑块控件

【实例描述】

本实例实现的功能是在界面显示一个具有范围的滑块控件。当用户滑动滑块控件中的两个滑块时,会将它们当前的值进行显示,并且还会显示它们之间的范围值。其中,用不同的标签显示对应滑块的值。运行效果如图 2.6 所示。

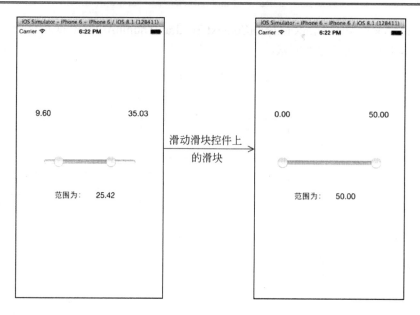

图 2.6　实例 10 运行效果

【实现过程】

（1）创建一个项目，命名为"具有范围的滑块"。
（2）添加图片 1.png、2.png 和 3.png 到创建项目的 Supporting Files 文件夹中。
（3）创建一个基于 UIControl 类的 Slider 类。
（4）打开 Slider.h 文件，编写代码。实现属性和方法的声明。程序代码如下：

```
#import <UIKit/UIKit.h>
@interface Slider : UIControl<UIGestureRecognizerDelegate>
//属性
@property (assign, nonatomic) CGFloat minValue;
@property (assign, nonatomic) CGFloat maxValue;
……
@property (readonly, nonatomic) CGFloat trackWidth;
@property (assign, nonatomic) BOOL didSetup;
//方法
- (void)setup;
- (void)leftHandlePanEngadged:(UIGestureRecognizer *)gesture;
- (void)rightHandlePanEngadged:(UIGestureRecognizer *)gesture;
@end
```

（5）打开 Slider.m 文件，编写代码，实现一个具有范围的滑块。使用到的方法如表 2-4 所示。

表 2-4　Slider.m 文件中方法总结

方　　法	功　　能
setLeftValue:	设置左边的值
setRightValue	设置右边的值
setMaxValue:	设置最大值
setMinValue:	设置最小值
layoutSubviews	为滑块控件布局

续表

方　　法	功　　能
setup	创建滑块控件
handleImage	设置拖动的图片
leftHandleImage	获取左边拖动的图片
rightHandleImag	获取右边拖动的图片
trackBackgroundImage	获取滑块轨道的背景颜色
trackFillImage	获取滑块轨道的填充颜色
hitTest: withEvent:	确定发生触摸的子视图
roundValueToStepValue:	获取增量
leftHandlePanEngadged:	左边拖动的手势
rightHandlePanEngadged:	右边拖动的手势
setFrame:	设置框架
trackWidth	设置轨道的宽度
gestureRecognizer:shouldRecognizeSimultaneouslyWithGestureRecognizer:	获取是否同时接收两个事件

在此文件中，可以将这些方法分为创建具有范围的滑块控件和拖动滑块控件上的滑块两个部分。其中创建滑块分为：对滑块控件的创建；设置；布局三个部分。这里需要讲解几个重要的方法（其他方法请读者参考源代码）。要创建滑块控件，需要使用 setup 方法实现。程序代码如下：

```
- (void)setup {
    //判断最大值是否等于0
    if (self.maxValue == 0) {
        self.maxValue = 50.0;                                      //设置最大值
    }
    if (self.rightValue == 0) {
        self.rightValue = self.maxValue;                           //设置右边的值
    }
    CGRect paddedFrame = self.frame;                               //获取框架
    paddedFrame.size.height = kREDHandleTapTargetRadius*2;
    self.frame = paddedFrame;                                      //设置框架
    //创建并设置滑块控件轨道的背景图片
    UIImage *emptySliderImage = self.trackBackgroundImage;
    self.sliderBackground = [[UIImageView alloc] initWithImage:emptySliderImage];
    [self addSubview:self.sliderBackground];                       //添加视图
    //创建并设置滑块控件轨道的填充图片
    UIImage *sliderFillImage = self.trackFillImage;
    self.sliderFillBackground = [[UIImageView alloc] initWithImage:sliderFillImage];
    [self addSubview:self.sliderFillBackground];
    //创建并设置左边用于拖动的图片
    self.leftHandle = [[UIImageView alloc] init];                  //实例化对象
    self.leftHandle.image = self.leftHandleImage;                  //设置图像
    self.leftHandle.frame = CGRectMake(0, 0, self.rightHandleImage.size.width+
kREDHandleTapTargetRadius, self.rightHandleImage.size.height+kREDHandle
TapTargetRadius);
    self.leftHandle.contentMode = UIViewContentModeCenter;
    self.leftHandle.userInteractionEnabled = YES;
    UIPanGestureRecognizer *leftPanGesture = [[UIPanGestureRecognizer alloc]
initWithTarget:self action:@selector(leftHandlePanEngadged:)];
                                                                   //实例化手势
    leftPanGesture.delegate = self;
```

```
    [self.leftHandle addGestureRecognizer:leftPanGesture];
    [self addSubview:self.leftHandle];
    //创建并设置右边用于拖动的图片
    self.rightHandle = [[UIImageView alloc] init];
    self.rightHandle.image = self.rightHandleImage;                    //设置图像
    self.rightHandle.frame = CGRectMake(0, 0, self.rightHandleImage.size.
width+kREDHandleTapTargetRadius,     self.rightHandleImage.size.height+
kREDHandleTapTargetRadius);                                             //设置框架
    self.rightHandle.contentMode = UIViewContentModeCenter;
    self.rightHandle.userInteractionEnabled = YES;
    UIPanGestureRecognizer *rightPanGesture = [[UIPanGestureRecognizer
alloc] initWithTarget:self action:@selector(rightHandlePanEngadged:)];
    rightPanGesture.delegate = self;
    [self.rightHandle addGestureRecognizer:rightPanGesture];    //添加手势
    [self addSubview:self.rightHandle];                          //添加视图
}
```

创建好滑块控件以后,需要对此控件进行重新布局。这时,需要使用 layoutSubviews 方法。程序代码如下:

```
- (void)layoutSubviews {
    if (!self.didSetup) {
        [self setup];
        self.didSetup = YES;
    }
    self.sliderBackground.frame = CGRectMake(0, 0, CGRectGetWidth(self.bounds),
CGRectGetHeight(self.sliderBackground.frame));      //设置滑块控件的背景框架
    self.sliderBackground.center = CGPointMake(floorf(CGRectGetWidth(self.
bounds)/2),floorf(CGRectGetHeight(self.bounds)/2));  //设置滑块控件背景的中心点
    CGFloat oneHundredPercent = self.maxValue - self.minValue;
    CGFloat leftValuePercentage = self.leftValue/oneHundredPercent;
                                                     //获取左值的百分比
    CGFloat leftXCoor = floorf((self.trackWidth-self.handleImage.size.
    width) * leftValuePercentage);
    self.leftHandle.frame     =    CGRectMake(0,   0,   CGRectGetWidth(self.
leftHandle.frame), CGRectGetHeight(self.leftHandle.frame));    //设置框架
    self.leftHandle.center = CGPointMake(leftXCoor, self.sliderBackground.
    center.y);                                       //设置中心位置
    CGFloat rightValuePercentage = self.rightValue/oneHundredPercent;
    CGFloat    rightXCoor   =    floorf((self.trackWidth-self.handleImage.
size.width) * rightValuePercentage) + self.handleImage.size.width;
    self.rightHandle.frame    =    CGRectMake(0,   0,   CGRectGetWidth(self.
rightHandle.frame), CGRectGetHeight(self.rightHandle.frame));  //设置框架
    self.rightHandle.center = CGPointMake(rightXCoor, self.sliderBackground.
    center.y);                                                 //设置框架
    CGFloat fillBackgroundWidth = self.rightHandle.center.x-self.
    leftHandle.center.x;
    self.sliderFillBackground.frame = CGRectMake(self.leftHandle.center.x,
0, fillBackgroundWidth, CGRectGetHeight(self.sliderFillBackground.frame));
    self.sliderFillBackground.center = CGPointMake(self.sliderFillBackground.
center.x, self.sliderBackground.center.y);
}
```

要实现拖动,需要使用 leftHandlePanEngadged:和 rightHandlePanEngadged:方法共同实现。其中 leftHandlePanEngadged:方法,实现左边滑块(左边拖动图片)的拖动。程序代码如下:

```
- (void)leftHandlePanEngadged:(UIGestureRecognizer *)gesture {
```

```objc
  UIPanGestureRecognizer *panGesture = (UIPanGestureRecognizer *)gesture;
  //判断拖动手势的状态
  if (panGesture.state == UIGestureRecognizerStateChanged) {
    CGPoint pointInView = [panGesture translationInView:self]; //获取位置
    CGFloat oneHundredPercentOfValues = self.maxValue - self.minValue;
                                                         //两个值的间隔
    CGFloat trackOneHundredPercent = self.trackWidth-self.handleImage.
    size.width;
    CGFloat trackPercentageChange = (pointInView.x / trackOneHundred
    Percent)*100;
    self.leftValue += (trackPercentageChange/100.0) * oneHundred
    PercentOfValues;                                     //设置左边的值
    [panGesture setTranslation:CGPointZero inView:self];
    [self sendActionsForControlEvents:UIControlEventValueChanged];
                                                    //发生动作给指定的控件
  }
  //判断手势的状态
  else if (panGesture.state == UIGestureRecognizerStateCancelled ||
       panGesture.state == UIGestureRecognizerStateEnded ||
       panGesture.state == UIGestureRecognizerStateCancelled) {
    self.leftValue = [self roundValueToStepValue:self.leftValue];
                                                         //设置左边的值
    [self sendActionsForControlEvents:UIControlEventValueChanged];
  }
}
```

其中 leftHandlePanEngadged:方法，实现右边滑块（右边拖动图片）的拖动。程序代码如下：

```objc
- (void)rightHandlePanEngadged:(UIGestureRecognizer *)gesture {
  UIPanGestureRecognizer *panGesture = (UIPanGestureRecognizer *)gesture;
  if (panGesture.state == UIGestureRecognizerStateChanged) {
    CGPoint pointInView = [panGesture translationInView:self]; //获取位置
    CGFloat oneHundredPercentOfValues = self.maxValue - self.minValue;
                                                         //两个值的间隔
    CGFloat trackOneHundredPercent = self.trackWidth-self.handleImage.
    size.width;
    CGFloat trackPercentageChange = (pointInView.x / trackOneHundred
    Percent)*100;
    self.rightValue += (trackPercentageChange/100.0) * oneHundred
    PercentOfValues;                                     //设置右边的值
    [panGesture setTranslation:CGPointZero inView:self];
    [self sendActionsForControlEvents:UIControlEventValueChanged];
  }
  //判断手势的状态
  else if (panGesture.state == UIGestureRecognizerStateCancelled ||
       panGesture.state == UIGestureRecognizerStateEnded ||
       panGesture.state == UIGestureRecognizerStateCancelled) {
    self.rightValue = [self roundValueToStepValue:self.rightValue];
                                                         //设置右边的值
    [self sendActionsForControlEvents:UIControlEventValueChanged];
  }
}
```

（6）打开 ViewController.h 文件，编写代码。实现头文件、插座变量、对象和方法的声明。程序代码如下：

```objc
#import <UIKit/UIKit.h>
```

```
#import "Slider.h"
@interface ViewController : UIViewController{
    //声明插座变量
    IBOutlet UILabel *label1;                    //声明关于标签的插座变量
    IBOutlet UILabel *label2;
    IBOutlet UILabel *label3;
    Slider *rangeSlider;
}
- (void)updateSliderLabels;
- (void)rangeSliderValueChanged:(id)sender;
@end
```

（7）打开 Main.storyboard 文件，对 View Controller 视图控制器的设计界面进行设计，效果如图 2.7 所示。

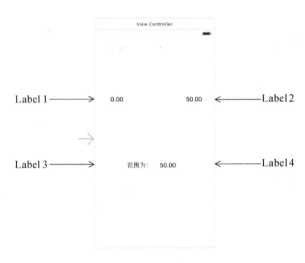

图 2.7　实例 10 View Controller 视图控制器的设计界面

需要添加的视图、控件以及对它们的设置如表 2-5 所示。

表 2-5　实例 10 视图、控件设置

视图、控件	属 性 设 置	其　　他
Label1	Text：0.00	与插座变量 label1 关联
Label2	Text：50.00	与插座变量 label2 关联
Label3	Text：范围为：	
Label4	Text：50.00	与插座变量 label3 关联

（8）打开 ViewController.m 文件，实现将创建的具有范围的滑块控件对象显示在当前的视图中，并实现当滑块控件的值发生改变时，将值显示在标签中。程序代码如下：

```
- (void)viewDidLoad
{
    rangeSlider = [[Slider alloc] initWithFrame:CGRectMake(0, 0, 200, 20)];
    rangeSlider.center = self.view.center;                  //设置中心位置
    [self.view addSubview:rangeSlider];
    [rangeSlider addTarget:self action:@selector(rangeSliderValueChanged:)
forControlEvents:UIControlEventValueChanged];               //添加动作
    [super viewDidLoad];
    // Do any additional setup after loading the view, typically from a nib.
```

```
}
- (void)rangeSliderValueChanged:(id)sender {
    [self updateSliderLabels];                     //标签中内容的更新
}
- (void)updateSliderLabels {
    label1.text = [NSString stringWithFormat:@"%.2f", rangeSlider.leftValue];
    label2.text = [NSString stringWithFormat:@"%.2f", rangeSlider.rightValue];
    label3.text=[NSString stringWithFormat:@"%.2f",rangeSlider.rightValue-
rangeSlider.leftValue];
}
```

【代码解析】

本实例关键功能是标签数据的更新。下面将详细讲解这个知识点。

数据的更新是需要使用 UIControl 类的 sendActionsForControlEvents:方法实现。它的功能是对于给定的控件事件发送消息，其语法形式如下：

```
- (void)sendActionsForControlEvents:(UIControlEvents)controlEvents;
```

其中，(UIControlEvents)controlEvents 表示要触发的控件事件。这些控件事件如表 2-6 所示。

表 2-6 控件事件

控件事件	功　　能
UIControlEventTouchDown	单击触摸按下的控件事件
UIControlEventTouchDownRepeat	多点触摸按下的控件事件
UIControlEventTouchDragInside	触摸，并在控件内拖动的控件事件
UIControlEventTouchDragOutside	触摸，并在控件边界范围之外拖动的控件事件
UIControlEventTouchDragEnter	拖动动作中，从控件边界外到内时产生的事件
UIControlEventTouchDragExit	拖动动作中，从控件边界内到外时产生的事件
UIControlEventTouchUpInsid	所有在控件之内触摸抬起事件，前提是先得按下
UIControlEventTouchUpOutside	所有在控件之外触摸抬起事件，前提是先得按下，然后拖动到控件外
UIControlEventTouchCancel	所有触摸取消事件
UIControlEventValueChanged	控件的值发生改变时的事件。用于滑块、分段控件以及其他取值的控件
UIControlEventEditingDidBegin	文本控件中开始编辑时的事件
UIControlEventEditingChanged	文本控件中的文本被改变时的事件
UIControlEventEditingDidEnd	文本控件中编辑结束时的事件
UIControlEventEditingDidOnExit	文本控件内通过按下回车键结束编辑时的事件
UIControlEventAllTouchEvents	所有的触摸事件
UIControlEventAllEditingEvents	所有关于文本编辑的事件
UIControlEventApplicationReserved	应用程序可以使用的控件事件
UIControlEventSystemReserved	内部框架可以使用的控件事件
UIControlEventAllEvents	所有的事件

在代码中，使用了 sendActionsForControlEvents:方法触发了值发生改变时的事件，从而进行了显示值的更新，代码如下：

```
[self sendActionsForControlEvents:UIControlEventValueChanged];
```

其中，UIControlEventValueChanged 表示要触发的控件事件。

实例 11　具有文字的滑块控件

【实例描述】

本实例实现的功能是在界面显示一个具有文字的滑块控件。当用户触摸滑块控件中的滑块时，具有文本内容的视图会显示出来，并且，在拖动过程中，滑块的当前值会进行显示。当触摸结束时，此视图会消失。其中，用于显示当前的值是标签和自制的视图。运行效果如图 2.8 所示。

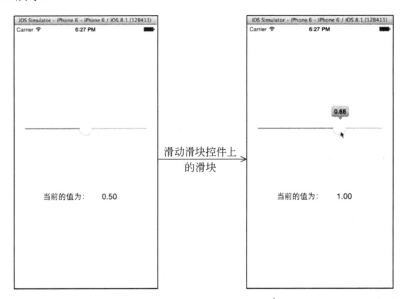

图 2.8　实例 11 运行效果

【实现过程】

（1）创建一个项目，命名为"具有文字的滑块控件"。
（2）创建一个基于 UIView 类的 ShowTextView 类。
（3）打开 ShowTextView.h 文件，编写代码。实现属性的声明。程序代码如下：

```
#import <UIKit/UIKit.h>
@interface ShowTextView : UIView
@property (nonatomic, strong) UILabel *textLabel;                    //属性
@end
```

（4）打开 ShowTextView.m 文件，编写代码。实现绘制一个具有文本内容的气泡视图，需要实现三个部分：初始化具有文本内容的气泡视图；设置框架；绘制。其中，initWithFrame:方法，实现对具有文本内容的气泡视图初始化。程序代码如下：

```
- (id)initWithFrame:(CGRect)frame
{
    self = [super initWithFrame:frame];
    if (self) {
```

```objc
    self.textLabel = [[UILabel alloc] initWithFrame:CGRectZero];
                                                            //创建标签对象
    self.textLabel.backgroundColor = [UIColor clearColor];  //设置背景颜色
    self.textLabel.textColor = [UIColor whiteColor];        //设置文本颜色
    self.textLabel.font = [UIFont boldSystemFontOfSize:13]; //设置字体
    self.textLabel.textAlignment = NSTextAlignmentCenter;   //设置对齐方式
    self.textLabel.adjustsFontSizeToFitWidth = YES;
    self.opaque = NO;
    [self addSubview:self.textLabel];
    // Initialization code
  }
  return self;
}
```

setFrame:方法，对框架进行设置。程序代码如下：

```objc
- (void)setFrame:(CGRect)frame
{
    [super setFrame:frame];
    CGFloat y = (frame.size.height - 26) / 3;
    //判断高度是否小于38
    if (frame.size.height < 38)
        y = 0;
    self.textLabel.frame = CGRectMake(0, y, frame.size.width, 26);
                                                            //设置文本标签的框架
}
```

drawRect:方法，实现绘制气泡图。程序代码如下：

```objc
- (void)drawRect:(CGRect)rect
{
    CGColorSpaceRef colorSpace = CGColorSpaceCreateDeviceRGB(); //创建颜色空间
    CGContextRef context = UIGraphicsGetCurrentContext();       //获取上下文
    //渐变颜色
    UIColor* gradientColor = [UIColor cyanColor];
    UIColor* gradientColor2 = [UIColor orangeColor];
    NSArray* gradientColors = [NSArray arrayWithObjects:(id)gradientColor.CGColor,
(id)gradientColor2.CGColor, nil];                               //实例化数组
    CGFloat gradientLocations[] = {0, 1};
    CGGradientRef gradient = CGGradientCreateWithColors(colorSpace,
(__bridge CFArrayRef)gradientColors, gradientLocations);
    CGRect frame = self.bounds;                                 //获取大小
    CGRect frame2 = CGRectMake(CGRectGetMinX(frame) + floor((CGRectGetWidth
(frame) - 11) * 0.51724 + 0.5), CGRectGetMinY(frame) + CGRectGetHeight(frame)
 - 9, 11, 9);
    //使用贝塞尔曲线绘制气泡
    UIBezierPath* bezierPath = [UIBezierPath bezierPath];
    [bezierPath moveToPoint: CGPointMake(CGRectGetMaxX(frame) - 0.5,
    CGRectGetMinY(frame) + 4.5)];                               //开始点
    [bezierPath addLineToPoint: CGPointMake(CGRectGetMaxX(frame) - 0.5,
    CGRectGetMaxY(frame) - 11.5)];
    //以3个点画一段曲线
    [bezierPath addCurveToPoint: CGPointMake(CGRectGetMaxX(frame) - 4.5,
CGRectGetMaxY(frame) - 7.5) controlPoint1: CGPointMake(CGRectGetMaxX(frame)
 - 0.5, CGRectGetMaxY(frame) - 9.29) controlPoint2: CGPointMake(CGRectGet
MaxX(frame) - 2.29, CGRectGetMaxY(frame) - 7.5)];
    [bezierPath addLineToPoint: CGPointMake(CGRectGetMinX(frame2) + 10.64,
CGRectGetMinY(frame2) + 1.5)];                                  //添加点
    [bezierPath addLineToPoint: CGPointMake(CGRectGetMinX(frame2) + 5.5,
```

```
    CGRectGetMinY(frame2) + 8)];
    [bezierPath addLineToPoint: CGPointMake(CGRectGetMinX(frame2) + 0.36,
CGRectGetMinY(frame2) + 1.5)];
    [bezierPath addLineToPoint: CGPointMake(CGRectGetMinX(frame) + 4.5,
CGRectGetMaxY(frame) - 7.5)];
    //以3个点画一段曲线
    [bezierPath addCurveToPoint: CGPointMake(CGRectGetMinX(frame) + 0.5,
CGRectGetMaxY(frame) - 11.5) controlPoint1: CGPointMake(CGRectGetMinX
(frame) + 2.29, CGRectGetMaxY(frame) - 7.5) controlPoint2: CGPointMake
(CGRectGetMinX(frame) + 0.5, CGRectGetMaxY(frame) - 9.29)];
    [bezierPath addLineToPoint: CGPointMake(CGRectGetMinX(frame) + 0.5,
CGRectGetMinY(frame) + 4.5)];
    [bezierPath addCurveToPoint: CGPointMake(CGRectGetMinX(frame) + 4.5,
CGRectGetMinY(frame) + 0.5) controlPoint1: CGPointMake(CGRectGetMinX(frame)
+ 0.5, CGRectGetMinY(frame) + 2.29) controlPoint2: CGPointMake(CGRectGet
MinX(frame) + 2.29, CGRectGetMinY(frame) + 0.5)];
    [bezierPath addLineToPoint: CGPointMake(CGRectGetMaxX(frame) - 4.5,
CGRectGetMinY(frame) + 0.5)];
    //以3个点画一条曲线
    [bezierPath addCurveToPoint: CGPointMake(CGRectGetMaxX(frame) - 0.5,
CGRectGetMinY(frame) + 4.5) controlPoint1: CGPointMake(CGRectGetMaxX(frame)
- 2.29, CGRectGetMinY(frame) + 0.5) controlPoint2: CGPointMake(CGRectGet
MaxX(frame) - 0.5, CGRectGetMinY(frame) + 2.29)];
    [bezierPath closePath];
    CGContextSaveGState(context);                        //保存状态
    [bezierPath addClip];
    CGRect bezierBounds = bezierPath.bounds;
    CGContextDrawLinearGradient(context,  gradient,CGPointMake(CGRectGet
MidX(bezierBounds),   CGRectGetMinY(bezierBounds)),CGPointMake(CGRectGet
MidX(bezierBounds),CGRectGetMaxY(bezierBounds)),0);   //绘制渐变填充
    CGContextRestoreGState(context);
    CGGradientRelease(gradient);
    CGColorSpaceRelease(colorSpace);
}
```

（5）创建一个基于 UISlider 类的 Slider 类。

（6）打开 Slider.h 文件，编写代码。实现头文件、属性以及方法的声明。程序代码如下：

```
#import <UIKit/UIKit.h>
#import "ShowTextView.h"                              //头文件
@interface Slider : UISlider
@property (nonatomic, strong) ShowTextView *popover;  //属性
//方法
- (void)showPopover;                                  //显示具有文本内容的气泡视图
- (void)showPopoverAnimated:(BOOL)animated;
- (void)hidePopover;
- (void)hidePopoverAnimated:(BOOL)animated;
@end
```

（7）打开 Slider.m 文件，编写代码。实现滑动滑块控件中的滑块，就出现一个显示文本内容的气泡。使用的方法如表 2-7 所示。

表 2-7 Slider.m文件中方法总结

方　　法	功　　能
popover	获取具有文本内容的气泡视图
setValue:	设置值

续表

方　　法	功　　能
touchesBegan:withEvent:	开始触摸
touchesEnded:withEvent:	结束触摸
touchesCancelled:withEvent:	取消触摸
updatePopoverFrame	更新
showPopover	显示具有文本内容的气泡视图
showPopoverAnimated:	以动画的形式显示具有文本内容的气泡视图
hidePopover	隐藏具有文本内容的气泡视图
hidePopoverAnimated:	以动画的形式隐藏具有文本内容的气泡视图

这里需要讲解几个重要的方法（其他方法请读者参考源代码）。popover 方法，获取具有文本内容的气泡视图。程序代码如下：

```objc
- (ShowTextView *)popover
{
    if (_popover == nil) {
        [self addTarget:self action:@selector(updatePopoverFrame) forControlEvents:UIControlEventValueChanged];        //添加动作
        _popover = [[ShowTextView alloc] initWithFrame:CGRectMake(self.frame.origin.x, self.frame.origin.y - 32, 40, 32)];        //创建
        [self updatePopoverFrame];
        _popover.alpha = 0;                             //设置透明度
        [self.superview addSubview:_popover];           //添加视图
    }
    return _popover;
}
```

updatePopoverFrame 方法用于对具有文本内容的气泡视图的框架进行更新。程序代码如下：

```objc
- (void)updatePopoverFrame
{
    CGFloat minimum = self.minimumValue;              //获取最小值
    CGFloat maximum = self.maximumValue;              //获取最大值
    CGFloat value = self.value;                       //获取当前的值
    //判断
    if (minimum < 0.0) {
        value = self.value - minimum;
        maximum = maximum - minimum;                  //获取最大值与最小值的间隔
        minimum = 0.0;
    }
    CGFloat x = self.frame.origin.x;                  //获取框架 x 的位置
    CGFloat maxMin = (maximum + minimum) / 2.0;
    x += (((value - minimum) / (maximum - minimum)) * self.frame.size.width)
- (self.popover.frame.size.width / 2.0);
    //判断 value 是否大于 maxMin
    if (value > maxMin) {
        value = (value - maxMin) + (minimum * 1.0);
        value = value / maxMin;
        value = value * 11.0;
        x = x - value;
    } else {
```

```
        value = (maxMin - value) + (minimum * 1.0);
        value = value / maxMin;
        value = value * 11.0;
        x = x + value;
    }
    CGRect popoverRect = self.popover.frame;              //获取框架
    popoverRect.origin.x = x;                             //获取 x 的位置
    popoverRect.origin.y = self.frame.origin.y - popoverRect.size.height - 1;
                                                          //获取 y 的位置
    self.popover.frame = popoverRect;                     //设置框架
}
```

touchesBegan:withEvent:、touchesEnded:withEvent:和 touchesCancelled:withEvent:方法共同实现触摸的功能。程序代码如下：

```
//触摸开始
- (void)touchesBegan:(NSSet *)touches withEvent:(UIEvent *)event
{
    [self updatePopoverFrame];                            //更新气泡的框架
    [self showPopoverAnimated:YES];                       //显示动画
    [super touchesBegan:touches withEvent:event];
}
//触摸结束
- (void)touchesEnded:(NSSet *)touches withEvent:(UIEvent *)event
{
    [self hidePopoverAnimated:YES];                       //隐藏动画
    [super touchesEnded:touches withEvent:event];
}
//触摸取消
- (void)touchesCancelled:(NSSet *)touches withEvent:(UIEvent *)event
{
    [self hidePopoverAnimated:YES];                       //隐藏动画
    [super touchesCancelled:touches withEvent:event];
}
```

showPopoverAnimated:和 hidePopoverAnimated:方法，实现具有文本内容的气泡视图的动画显示与隐藏。程序代码如下：

```
//以动画的形式显示
- (void)showPopoverAnimated:(BOOL)animated
{
    if (animated) {
        //实现动画效果
        [UIView animateWithDuration:0.25 animations:^{
            self.popover.alpha = 1.0;                     //设置透明度
        }];
    } else {
        self.popover.alpha = 1.0;
    }
}
//以动画的形式隐藏
- (void)hidePopoverAnimated:(BOOL)animated
{
    if (animated) {
```

```
    //实现动画效果
    [UIView animateWithDuration:0.25 animations:^{
        self.popover.alpha = 0;
    }];
} else {
    self.popover.alpha = 0;
}
}
```

(8) 打开 ViewController.h 文件，编写代码。实现头文件、插座变量和动作的声明。程序代码如下：

```
#import <UIKit/UIKit.h>
#import "Slider.h"
@interface ViewController : UIViewController{
    IBOutlet Slider *slider;
    IBOutlet UILabel *la;                    //声明关于标签的插座变量
}
- (IBAction)valuechange:(id)sender;
@end
```

(9) 打开 Main.storyboard 文件，对 View Controller 视图控制器的设计界面进行设计，效果如图 2.9 所示。

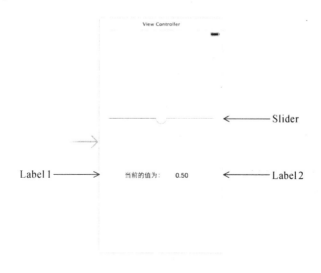

图 2.9　实例 11 View Controller 视图控制器的设计界面

需要添加的视图、控件以及对它们的设置如表 2-8 所示。

表 2-8　实例 11 视图、控件设置

视图、控件	属 性 设 置	其 他
Label1	Text：当前的值为：	
Label2	Text：0.50	与插座变量 la 关联
Slider		Class：Slider 与插座变量 slider 关联 与动作 valuechange:关联

（10）打开 ViewController.m 文件，编写代码。拖动滑块出现当前的值。程序代码如下：

```
- (void)viewDidLoad
{
  slider.popover.textLabel.textColor=[UIColor blackColor];  //设置文本的颜色
  [super viewDidLoad];
   // Do any additional setup after loading the view, typically from a nib.
}
- (IBAction)valuechange:(id)sender {
  [self updateSliderPopoverText];
}
//更新文本内容
- (void)updateSliderPopoverText
{
  slider.popover.textLabel.text = [NSString stringWithFormat:@"%.2f",
  slider.value];                                                //显示值
  la.text = [NSString stringWithFormat:@"%.2f", slider.value];  //显示值
}
```

【代码解析】

本实例关键的功能是用于显示文本内容视图的移动。下面将详细讲解这个知识点。

对于显示文本内容视图的跟随手势的位置移动，它使用 UIContril 的 addTarget: action: forControlEvents:方法，其语法形式如下：

```
-    (void)addTarget:(id)target    action:(SEL)action    forControlEvents:
(UIControlEvents)controlEvents;
```

其中，(id)target 表示要把消息发送到的位置；(SEL)action 表示该事件响应的方法；(UIControlEvents)controlEvents 表示事件类型。在代码中，使用了 addTarget: action: forControlEvents 方法，实现了当控件的值发生改变时，此滑块控件进行自动调用更改框架的方法，代码如下：

```
[self addTarget:self action:@selector(updatePopoverFrame) forControlEvents:
UIControlEventValueChanged];
```

其中，self 表示把消息发送到自身的控件上；@selector(updatePopoverFrame)表示该事件响应的方法；UIControlEventValueChanged 表示事件类型。

实例 12　自定义的滑块控件

【实例描述】

本实例实现的功能是显示一个自定义的滑块控件。滑块的移动方法除了可以拖动外，还可以进行轻拍。运行效果如图 2.10 所示。

【实现过程】

（1）创建一个项目，命名为"自定义的滑块控件"。
（2）创建一个基于 UIView 类的 drawView 类。

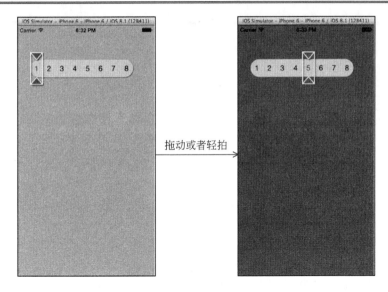

图 2.10 实例 12 运行效果

（3）打开 drawView.h 文件，编写代码。实现对象、实例变量以及方法的声明。程序代码如下：

```
#import <UIKit/UIKit.h>
@interface drawView : UIView{
   UIColor *arrowCr;
   UIColor *strokeCr;                               //声明对象
   BOOL arrow;
}
-(id)initWithFrame:(CGRect)theFrame   arrowColor:(UIColor  *)arrowColor
strokeColor:(UIColor *)strokeColor isUpArrow:(BOOL)arrowType ;
@end
```

（4）打开 drawView.m 文件，编写代码。实现初始化以及三角的绘制。其中，initWithFrame: arrowColor: strokeColor: isUpArrow:方法的功能就是实现视图的初始化。程序代码如下：

```
-  (id)initWithFrame:(CGRect)theFrame   arrowColor:(UIColor  *)arrowColor
strokeColor:(UIColor *)strokeColor isUpArrow:(BOOL)arrowType ;
{
    self = [super initWithFrame:theFrame];
    if (self) {
       self.backgroundColor = [UIColor clearColor];        //设置背影颜色
       arrowCr =arrowColor;
       strokeCr = strokeColor;
       arrow = arrowType;
    }
    return self;
}
```

drawRect:方法，实现对三角的绘制。程序代码如下：

```
- (void)drawRect:(CGRect)rect
{
    CGContextRef context = UIGraphicsGetCurrentContext();
    CGContextBeginPath(context);
    if (arrow) {
       CGContextMoveToPoint(context, rect.size.width/2,0);//设置开始的点
```

```
        CGContextAddLineToPoint(context, rect.size.width, rect.size.height);
                                                            //设置结束的点
        CGContextAddLineToPoint(context, 0, rect.size.height);
    }
    else
    {
        CGContextMoveToPoint(context, 0,0);                 //设置开始的点
        CGContextAddLineToPoint(context, rect.size.width, 0);//设置结束的点
        CGContextAddLineToPoint(context, rect.size.width/2, rect.size.height);
    }
    CGContextClosePath(context);
    CGContextSetLineWidth(context, 2.0);                    //设置线宽
    CGContextSetFillColorWithColor(context, arrowCr.CGColor);
    CGContextSetStrokeColorWithColor(context, strokeCr.CGColor);
    CGContextDrawPath(context, kCGPathFillStroke);          //绘制
}
```

（5）创建一个基于 UIView 类的 Slider 类。

（6）打开 Slider.h 文件，编写代码。实现头文件、宏定义、协议、实例变量、对象、属性以及方法的声明。程序代码如下：

```
#import <UIKit/UIKit.h>
#import "drawView.h"                                        //头文件
#define spaceBetweenSliderandRatingView 0                   //宏定义
//协议
@protocol SliderDelegate <NSObject>
@required
- (void) selectedRating:(NSString *)scale;
@end
@interface Slider : UIView<UIGestureRecognizerDelegate>{
    //实例变量
    NSUInteger maximumRating;
    ……
    NSMutableArray *itemsAry;
    NSMutableArray *itemsXPositionAry;
    //对象
    UIView *containerView;
    ……
    UIColor *sliderBorderColor;
    id<SliderDelegate>delegate;
}
//属性
@property(nonatomic,assign)NSUInteger maximumRating;
……
@property(nonatomic,strong)UIColor *sliderBorderColor;
//方法
-(void)drawRatingControlWithX:(float)x Y:(float)y;          //绘制滑块控件
-(void)drawRatingView;
-(void)createContainerView;
-(void)createSliderView;
-(void)calculateAppropriateSelectorXposition:(UIView *)view;
@end
```

（7）打开 Slider.m 文件，编写代码。实现一个自定义的滑块控件效果。使用的方法如表 2-9 所示。

表 2-9 Slider.m 文件中方法总结

方　　法	功　　能
initWithFrame:	滑块控件的初始化
drawRatingControlWithX:	绘制滑块控件
createContainerView	创建放置滑块控件的视图
createSliderView	创建滑块控件中的滑块
drawRatingView	绘制滑块控件中的内容
changeTextColor:	改变文本的颜色
handleTap:	实现轻拍
handlePan:	实现拖动
calculateAppropriateSelectorXposition:	计算滑块控件中滑块的位置

在此文件中，可以将这些方法分为三个部分：绘制一个完整的滑块控件；初始化；实现拖动和轻拍的手势。这里需要讲解几个重要的方法（其他方法请读者参考源代码）。要创建一个完整的滑块控件，需要三个步骤：绘制滑块控件；绘制滑块控件中的内容；绘制放置滑块控件的视图以及绘制滑块控件中的滑块。其中 drawRatingControlWithX:方法实现绘制滑块控件。程序代码如下：

```
-(void)drawRatingControlWithX:(float)x Y:(float)y;
{
    totalNumberOfRatingViews = 1 + ((self.maximumRating - self.
minimumRating)/self.difference); float width = totalNumberOfRatingViews
*self.widthOfEachNo + (totalNumberOfRatingViews +1)*self.spaceBetweenEachNo;
                                        //获取等级控件中内容的个数
    float height = self.heightOfEachNo + (self.sliderHeight *2);
    self.frame = CGRectMake(x, y, width, height);        //设置框架
    [self createContainerView];
}
```

drawRatingView 方法实现绘制滑块控件中的内容。程序代码如下：

```
-(void)drawRatingView
{
    float itemX = self.spaceBetweenEachNo;
    float itemY = 0;
    int differ = self.minimumRating;                //获取最小的值（等级）
    itemsAry = [NSMutableArray new];
    itemsXPositionAry = [NSMutableArray new];
    //创建标签对象，并对标签对象进行设置
    for (int i =self.minimumRating; i<self.maximumRating+1; i = i+self.
difference) {
        UILabel *lblMyLabel = [[UILabel alloc] initWithFrame:CGRectMake
(itemX, itemY, self.widthOfEachNo, self.heightOfEachNo)]; //实例化标签对象
        lblMyLabel.numberOfLines = 0;                //设置标签的行数
        lblMyLabel.tag=i;                            //设置 tag 值
        lblMyLabel.backgroundColor = [UIColor clearColor];
        lblMyLabel.textAlignment = NSTextAlignmentCenter;  //设置对齐方式
        lblMyLabel.text = [NSString stringWithFormat:@"%d",differ];
        differ = differ + self.difference;
        lblMyLabel.textColor = self.disableStateTextColor;  //设置文本的颜色
        lblMyLabel.layer.shadowColor = [lblMyLabel.textColor CGColor];
                                                    //设置阴影的颜色
        lblMyLabel.layer.shadowOffset = CGSizeMake(0.0, 0.0);  //设置阴影的偏移量
```

```objc
        lblMyLabel.layer.shadowRadius = 2.0;
        lblMyLabel.layer.shadowOpacity = 0.3;
        lblMyLabel.layer.masksToBounds = NO;
        lblMyLabel.userInteractionEnabled = YES;
        [containerView addSubview:lblMyLabel];                    //添加视图
        itemX = lblMyLabel.frame.origin.x + self.widthOfEachNo + self.space
"BetweenEachNo;
        [itemsAry addObject:lblMyLabel];                          //添加对象
        [itemsXPositionAry addObject:[NSString stringWithFormat:@"%f",
        lblMyLabel.frame.origin.x]];
        //对手势的创建以及设置
        UITapGestureRecognizer *singleTap = [[UITapGestureRecognizer alloc]
initWithTarget:self action:@selector(handleTap:)];
        [singleTap setNumberOfTapsRequired:1];
        [singleTap setNumberOfTouchesRequired:1];
        [lblMyLabel addGestureRecognizer:singleTap];
    }
    UILabel *firstLbl = [itemsAry objectAtIndex:0];
    firstLbl.textColor = self.selectedStateTextColor;             //设置文本的颜色
}
```

createContainerView 方法的功能是,绘制一个用来放置滑块控件的视图。程序代码如下:

```objc
-(void)createContainerView
{
    containerView = [[UIView alloc]initWithFrame:CGRectMake(0, self.sliderHeight, self.frame.size.width, self.heightOfEachNo)];
    //对此视图进行设置
    containerView.backgroundColor = self.scaleBgColor;            //设置背景颜色
    containerView.layer.shadowColor = [[UIColor redColor] CGColor];
                                                                  //设置阴影的颜色
    containerView.layer.shadowOffset = CGSizeMake(1.0f, 1.0f);//设置偏移量
    containerView.layer.shadowOpacity = 1.0f;
    containerView.layer.shadowRadius = 1.0f;
    containerView.layer.cornerRadius = self.heightOfEachNo/2;
    [self addSubview:containerView];                              //添加视图
    [self createSliderView];
}
```

createSliderView 方法的功能是创建滑块控件中的滑块。程序代码如下:

```objc
-(void)createSliderView
{
    float y = (self.heightOfEachNo + self.sliderHeight) + spaceBetween
    SliderandRatingView;
    float height = self.sliderHeight - (2*spaceBetweenSliderandRatingView);
    sliderView = [[UIView alloc]initWithFrame:CGRectMake(self.
    spaceBetweenEachNo, 0, self.widthOfEachNo, self.frame.size.height)];
                                                                  //实例化视图对象
    sliderView.layer.shadowColor = [[UIColor redColor] CGColor];
                                                                  //设置阴影的颜色
    sliderView.layer.shadowOffset = CGSizeMake(2.0f, 2.0f);
    sliderView.layer.shadowOpacity = 1.0f;
    sliderView.layer.shadowRadius = 2.0f;
    sliderView.layer.cornerRadius = 2;
    sliderView.layer.borderColor = [self.sliderBorderColor CGColor];
                                                                  //设置边框的颜色
    sliderView.layer.borderWidth = 2.0f;                          //设置边框的宽度
```

```
    [self insertSubview:sliderView aboveSubview:containerView];
    //创建上三角
    UIView *upArrow = [[UIView alloc]initWithFrame:CGRectMake(0, y,
    self.widthOfEachNo, height)];
    upArrow.backgroundColor = [UIColor clearColor];
    [sliderView addSubview:upArrow];                          //添加视图
    drawView *triangleUp = [[drawView alloc]initWithFrame:CGRectMake(0, 0,
 upArrow.frame.size.width,  upArrow.frame.size.height)  arrowColor:self.
 arrowColor strokeColor:self.sliderBorderColor isUpArrow:YES];//实例化对象
    [upArrow addSubview:triangleUp];
    //创建下三角
    UIView *downArrow = [[UIView alloc]initWithFrame:CGRectMake(0, 0,
    self.widthOfEachNo, height)];
    downArrow.backgroundColor = [UIColor clearColor];        //设置背景颜色
    [sliderView addSubview:downArrow];
    drawView *triangleDown = [[drawView alloc]initWithFrame:CGRectMake(0,
 0,  upArrow.frame.size.width,  upArrow.frame.size.height)  arrowColor:
 self.arrowColor strokeColor:self.sliderBorderColor isUpArrow:NO];
    [sliderView addSubview:triangleDown];
    //创建手势以及对手势进行设置
    UIPanGestureRecognizer* panRecognizer = [[UIPanGestureRecognizer alloc]
    initWithTarget:self action:@selector(handlePan:)];
    panRecognizer.minimumNumberOfTouches = 1;
    panRecognizer.delegate = self;                           //设置委托
    [sliderView addGestureRecognizer:panRecognizer];         //添加手势
    [self drawRatingView];
}
```

initWithFrame:方法的功能是滑块控件的初始化。程序代码如下:

```
- (id)initWithFrame:(CGRect)frame
{
    self = [super initWithFrame:frame];
    if (self) {
        self.maximumRating = 8;
        self.minimumRating = 1;
        self.spaceBetweenEachNo = 0;
        self.widthOfEachNo = 30;
        self.heightOfEachNo = 40;
        self.sliderHeight = 17;                              //高度
        self.difference = 1;
        self.scaleBgColor = [UIColor cyanColor];             //背景
        self.arrowColor = [UIColor redColor];                //三角
        self.disableStateTextColor = [UIColor blackColor];//未选中的字体颜色
        self.selectedStateTextColor = [UIColor redColor];  //选中的字体颜色
        self.sliderBorderColor = [UIColor whiteColor];
    }
    return self;
}
```

要实现轻拍或者拖动的手势，需要使用 handleTap:、handlePan: 和 calculateAppropriateSelectorXposition:方法共同实现。其中，handleTap:方法实现轻拍手势。程序代码如下:

```
- (void)handleTap:(UIPanGestureRecognizer *)recognizer {
    float tappedViewX = recognizer.view.frame.origin.x;
    //实现动画
    [UIView beginAnimations:@"MoveView" context:nil];
```

```objc
[UIView setAnimationCurve:UIViewAnimationCurveEaseIn];
[UIView setAnimationDuration:0.5f];                    //设置动画的持续时间
CGRect sliderFrame = sliderView.frame;                 //获取框架
sliderFrame.origin.x = tappedViewX;
sliderView.frame = sliderFrame;
[UIView commitAnimations];
//对itemsAry数组中的所有标签进行文本颜色的判断
for(UILabel *mylbl in itemsAry)
{
    if (mylbl.textColor == self.selectedStateTextColor) {
        mylbl.textColor = self.disableStateTextColor; //设置文本的颜色
    }
}
float selectedViewX =sliderView.frame.origin.x;
NSUInteger index = [itemsXPositionAry indexOfObject:[NSString stringWithFormat:@"%f",selectedViewX]];
UILabel *myLabel = [itemsAry objectAtIndex:index];     //实例化标签对象
[self performSelector:@selector(changeTextColor:) withObject:myLabel afterDelay:0.5];
[delegate selectedRating:myLabel.text];
}
```

handlePan:方法实现拖动的手势。程序代码如下：

```objc
- (void)handlePan:(UIPanGestureRecognizer *)recognizer {
    CGPoint translation = [recognizer translationInView:self]; //获取位置
    CGFloat newX = MIN(recognizer.view.frame.origin.x + translation.x, self.frame.size.width - recognizer.view.frame.size.width);
    CGRect newFrame = CGRectMake( newX,recognizer.view.frame.origin.y, recognizer.view.frame.size.width, recognizer.view.frame.size.height);
    recognizer.view.frame = newFrame;                  //设置框架
    [recognizer setTranslation:CGPointZero inView:self];
    //判断对象是否在数组中
    if ([itemsXPositionAry containsObject:[NSString stringWithFormat:@"%f",recognizer.view.frame.origin.x]]) {
        //对itemsAry数组中的所有标签进行文本颜色的判断
        for(UILabel *mylbl in itemsAry)
        {
            if (mylbl.textColor == self.selectedStateTextColor) {
                mylbl.textColor = self.disableStateTextColor;//设置文本的颜色
            }
        }
        NSUInteger index = [itemsXPositionAry indexOfObject:[NSString stringWithFormat:@"%f",recognizer.view.frame.origin.x]];
        UILabel * uil = [itemsAry objectAtIndex:index];  //实例化对象
        uil.textColor = self.selectedStateTextColor;     //设置文本内容
    }
    //判断手势是否为结束状态
    if (recognizer.state == UIGestureRecognizerStateEnded) {
        [self calculateAppropriateSelectorXposition:recognizer.view];
    }
}
```

calculateAppropriateSelectorXposition:方法的功能是用来计算滑块控件中滑块的位置，它是相对于拖动手势存在的。程序代码如下：

```objc
-(void)calculateAppropriateSelectorXposition:(UIView *)view
```

```
{
    float selectorViewX = view.frame.origin.x;              //获取 x 的位置
    float itemXposition = 0;
    float itempreviousXpostion = 0;
    //遍历
    for (int i =0; i<[itemsXPositionAry count]; i++) {
        if (i !=0) {
            itemXposition = [[itemsXPositionAry objectAtIndex:i]floatValue];
            itempreviousXpostion = [[itemsXPositionAry objectAtIndex:i-1]floatValue];
        }
        else{
            itemXposition = [[itemsXPositionAry objectAtIndex:i]floatValue];
        }
       //判断滑块控件中滑块的 x 的位置是否小于 0
        if (selectorViewX < itemXposition)
            break;
    }
    //判断滑块的 x 的位置是否大于 self.spaceBetweenEachNo
    if (selectorViewX > self.spaceBetweenEachNo) {
        float nextValue = itemXposition - selectorViewX;
        float previousValue = selectorViewX -itempreviousXpostion;
        //判断
        if (nextValue > previousValue) {
            CGRect viewFrame = view.frame;
            viewFrame.origin.x = itempreviousXpostion;      //设置 x 的位置
            view.frame = viewFrame;                         //设置框架
        }
        else {
            CGRect viewFrame = view.frame;
            viewFrame.origin.x = itemXposition;             //设置 x 的位置
            view.frame = viewFrame;                         //设置框架
        }
    }
    else{
        CGRect viewFrame = view.frame;
        viewFrame.origin.x = self.spaceBetweenEachNo;
        view.frame = viewFrame;
    }
    float selectedViewX =view.frame.origin.x;
    //对 itemsAry 数组中的所有标签进行文本颜色的判断
    for(UILabel *mylbl in itemsAry)
    {
        if (mylbl.textColor == self.selectedStateTextColor) {
            mylbl.textColor = self.disableStateTextColor;   //设置文本的颜色
        }
    }
    NSUInteger index = [itemsXPositionAry indexOfObject:[NSString stringWithFormat:
@"%f",selectedViewX]];
    UILabel *myLabel = [itemsAry objectAtIndex:index];      //实例化标签对象
    myLabel.textColor = self.selectedStateTextColor;
    [delegate selectedRating:myLabel.text];
}
```

（8）打开 ViewController.h 文件，编写代码。实现头文件以及对象的声明。程序代码如下：

```
#import <UIKit/UIKit.h>
#import "Slider.h"
```

```
@interface ViewController : UIViewController<SliderDelegate>{
    Slider *slider1;
    Slider *slider2;
}
@end
```

(9) 打开 ViewController.m 文件,编写代码。实现创建一个支持范围选择的滑块控件对象,并将其添加到当前的视图中,然后实现选择某一个值后,颜色将发生改变。程序代码如下:

```
- (void)viewDidLoad
{
    slider1 = [[Slider alloc]init];                          //创建
    slider1.delegate = self;
    [slider1 drawRatingControlWithX:30 Y:60];
    [self.view addSubview:slider1];                          //添加
    self.view.backgroundColor=[UIColor greenColor];          //设置背景颜色
    [super viewDidLoad];
     // Do any additional setup after loading the view, typically from a nib.
}
//属性选择某一个值后,调用的方法
- (void) selectedRating:(NSString *)scale;
{
    //判断 scale 是否为指定的字符串
    if ([scale isEqualToString:@"1"]) {
        self.view.backgroundColor=[UIColor greenColor];      //设置背景
    }else if ([scale isEqualToString:@"2"]){
        self.view.backgroundColor=[UIColor grayColor];
    }else if ([scale isEqualToString:@"3"]){
        self.view.backgroundColor=[UIColor yellowColor];
    }else if ([scale isEqualToString:@"4"]){
        self.view.backgroundColor=[UIColor blackColor];
    }else if ([scale isEqualToString:@"5"]){
        self.view.backgroundColor=[UIColor brownColor];
    }else if ([scale isEqualToString:@"6"]){
        self.view.backgroundColor=[UIColor orangeColor];
    }else if ([scale isEqualToString:@"7"]){
        self.view.backgroundColor=[UIColor whiteColor];
    }else{
        self.view.backgroundColor=[UIColor magentaColor];
    }
}
```

【代码解析】

由于本实例中的代码和方法非常多,为了方便读者的阅读,笔者绘制了一些执行流程图,如图 2.11 和图 2.12 所示。其中,自定义滑块控件的显示,需要使用 viewDidLoad、initWithFrame:、drawRatingControlWithX:、createContainerView、createSliderView、initWithFrame: arrowColor: strokeColor: isUpArrow:、drawRatingView 和 drawRect:方法共同实现。它们的程序执行流程如图 2.11 所示。

如果想要让自定义滑块控件中的滑块到达指定的位置,可以使用轻拍或者使用拖动手势将滑块进行拖动。轻拍的实现,需要使用 handleTap:、selectedRating:和 changeTextColor:

方法；拖动的实现可以使用 handlePan:、calculateAppropriateSelectorXposition: 和 selectedRating 方法。它们的执行流程如图 2.12 所示。

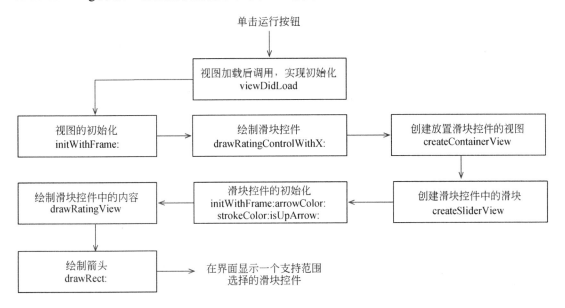

图 2.11　实例 12 程序执行流程 1

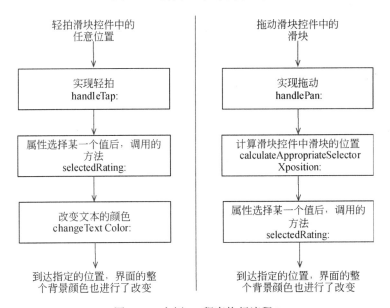

图 2.12　实例 12 程序执行流程 2

实例 13　自定义的声音调节滑块控件

【实例描述】

本实例实现的功能是在界面显示一个自定义的声音调节滑块控件。当滑动此控件的滑块时，会显示当前的值。其中，用于显示当前值的是标签。运行效果如图 2.13 所示。

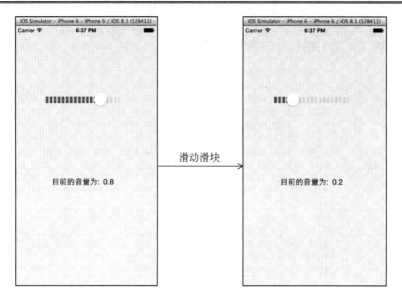

图 2.13　实例 13 运行效果

【实现过程】

（1）创建一个项目，命名为"自定义的声音调节滑块控件"。

（2）添加图片 1.png 和 2.png 到创建项目的 Supporting Files 文件夹中。

（3）创建一个基于 UIView 类的 Slider 类。

（4）打开 Slider.h 文件，编写代码。实现协议、对象、实例变量、属性以及方法的声明。程序代码如下：

```
#import <UIKit/UIKit.h>
@class Slider;
//协议
@protocol SlideDelegate <NSObject>
- (void) slideValueChange:(CGFloat) value;
@end
@interface Slider : UIView{
    //对象
    UISlider    *_slideView;
    UIView      *_processView;
    //实例变量
    CGFloat     width;
    CGFloat     height;
}
//属性
@property (assign, nonatomic) id<SlideDelegate>delegate;
@property (retain, nonatomic) UISlider *slideView;
@property (retain, nonatomic) UIView   *processView;
//方法
- (void) setSlideValue:(CGFloat) value;
- (void) slideValueChanged;
- (id)initWithFrame:(CGRect)frame;
@end
```

（5）打开 Slider.m 文件，编写代码。在此文件中，代码分为三个部分：初始化；设置值；当值发生改变时实现的效果。其中，initWithFrame:方法实现对声音调节滑块控件进行初始化。程序代码如下：

```objc
- (id)initWithFrame:(CGRect)frame
{
    self = [super initWithFrame:frame];
    if (self) {
        self.backgroundColor = [UIColor clearColor];
        width = frame.size.width;
        height= frame.size.height;
        //创建并设置空白视图对象
        UIView *view = [[UIView alloc]initWithFrame:self.bounds];
                                                                //实例化对象
        view.backgroundColor = [UIColor colorWithPatternImage:[UIImage
        imageNamed:@"1.png"]];
        [self addSubview:view];                                 //添加视图
        _processView = [[UIView alloc]init];
        _processView.backgroundColor = [UIColor colorWithPatternImage:
        [UIImage imageNamed:@"2.png"]];
        _processView.frame = CGRectMake(0, 0, width * .8, height);
                                                                //设置框架
        [self addSubview:_processView];
        //创建并设置滑块控件对象
        _slideView = [[UISlider alloc]initWithFrame:self.bounds];
        _slideView.value = 0.8;                    //设置滑块控件的值
        _slideView.maximumValue = 1.0;
        _slideView.minimumValue = 0.0;             //设置最小值
        [_slideView setMaximumTrackTintColor:[UIColor clearColor]];
        [_slideView setMinimumTrackTintColor:[UIColor clearColor]];
        [_slideView addTarget:self action:@selector(slideValueChanged)
        forControlEvents:
UIControlEventValueChanged];                                    //添加动作
        [self addSubview:_slideView];
    }
    return self;
}
```

setSlideValue:方法，用于对滑块控件中当前的值进行设置。程序代码如下：

```objc
- (void) setSlideValue:(CGFloat) value{
    _slideView.value = value;                                   //设置值
    [self slideValueChanged];
}
```

slideValueChanged 方法，当值发生改变时调用。程序代码如下：

```objc
- (void) slideValueChanged{
    CGFloat value = _slideView.value;
    _processView.frame = CGRectMake(0, 0, width * value, height); //设置框架
    [delegate slideValueChange:_slideView.value];
}
```

（6）打开 ViewController.h 文件，编写代码。实现头文件、对象以及插座变量的声明。程序代码如下：

```objc
#import <UIKit/UIKit.h>
#import "Slider.h"
@interface ViewController : UIViewController<SlideDelegate>{
    Slider *cumslider;
    IBOutlet UILabel *label;
```

}
@end

（7）打开 Main.storyboard 文件，对 View Controller 视图控制器的设计界面进行设计，效果如图 2.14 所示。

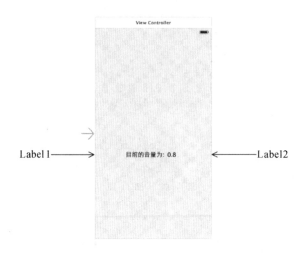

图 2.14　实例 13 View Controller 视图控制器的设计界面

需要添加的视图、控件以及对它们的设置如表 2-10 所示。

表 2-10　实例 13 视图、控件设置

视图、控件	属 性 设 置	其　　他
设计界面	Background：浅绿色	
Label1	Text：目前的音量为：	
Label2	Text：0.8 Alignment：居中对齐	与插座变量 label 关联

（8）打开 ViewController.m 文件，编写代码。实现创建一个音量调节滑块控件对象，并在滑动滑块时，显示当前的值。程序代码如下：

```
- (void)viewDidLoad
{
    cumslider = [[Slider alloc]initWithFrame:CGRectMake(70, 150, 165, 23)];
    cumslider.delegate = self;
    [self.view addSubview:cumslider];
    [super viewDidLoad];
    // Do any additional setup after loading the view, typically from a nib.
}
//当值发生变化时调用
- (void)slideValueChange:(CGFloat)value {
    NSString *s=[NSString stringWithFormat:@"%0.1f",cumslider.slideView.value];
    label.text=s;                                              //设置文本的内容
}
```

【代码解析】

本实例的关键功能是拖动滑块控件中的滑块，实现音量的增加。下面将详细讲解这个知识点。

首先在 initWithFrame:方法中，创建了两个视图和一个滑块控件，然后将滑块控件中滑块左边的颜色设置为透明，将滑块右边的颜色也设置为透明。接着获取滑块控件的当前值，然后，使用当前的值影响创建的一个视图的宽度，这样就实现了滑动滑块增加或减少音量的效果。程序代码如下：

```
CGFloat value = _slideView.value;
_proccosView.frame = CGRectMake(0, 0, width * value, height);
[delegate slideValueChange:_slideView.value];
```

实例 14　具有步长的滑块控件

【实例描述】

本实例实现的功能是在界面显示一个具有步长的滑块控件。所谓步长，就是滑块每一次滑动的长度必须是固定的。运行效果如图 2.15 所示。

图 2.15　实例 14 运行效果

【实现过程】

（1）创建一个项目，命名为"具有步长的滑块控件"。
（2）添加图片 1.png、2.png 和 3.png 到创建项目的 Supporting Files 文件夹中。
（3）创建一个基于 UIControl 类的 StepSlider 类。
（4）打开 StepSlider.h 文件，编写代码。实现实例变量、对象、属性以及方法的声明。程序代码如下：

```
#import <UIKit/UIKit.h>
@interface StepSlider : UIControl{
    //实例变量
    BOOL _thumbOn;
    //对象
    UIImageView *_thumbImageView;
```

```objc
    UIImageView *_trackImageViewNormal;
    UIImageView *_trackImageViewHighlighted;
}
//属性
@property(nonatomic) float value;
@property(nonatomic) float minimumValue;
......
@property(nonatomic, getter = isStepped) BOOL stepped;
//方法
- (void)commonInit;
- (float)xForValue:(float)value;
- (float)valueForX:(float)x;
- (float)stepMarkerXCloseToX:(float)x;
- (void)updateTrackHighlight;                           //更新轨道
- (NSString *)valueStringFormat;
@end
```

（5）打开 StepSlider.m 文件，编写代码。实现具有步长效果的滑块控件。使用的方法如表 2-11 所示。

表 2-11　StepSlider.m文件中方法总结

方　　法	功　　能
initWithFrame:	初始化具有步长的滑块控件
initWithCoder:	初始化实例化对象
layoutSubviews	布局
drawRect:	绘制
setValue:	设置值
commonInit	初始化
xForValue:	获取值的 x
valueForX:	获取 x 的值
stepMarkerXCloseToX:	获取标记接近的 x
updateTrackHighlight	更新轨道
valueStringFormat	获取字符串的格式
beginTrackingWithTouch:withEvent:	开始触摸
endTrackingWithTouch:withEvent:	结束触摸
continueTrackingWithTouch:withEvent:	继续触摸

要实现具有步长的滑块控件，首先要绘制一个滑块控件，其次是对步长的设置，最后是实现拖动功能。其中，initWithFrame:、initWithCoder:、layoutSubviews、drawRect:、setValue:、commonInit、xForValue:、valueStringFormat 和 updateTrackHighlight 方法共同实现滑块控件的绘制。这里需要讲解几个重要的方法（其他方法请读者参考源代码）。commonInit 方法，是对一些常用属性的初始化。程序代码如下：

```objc
- (void)commonInit
{
    _value = 1.f;
    _minimumValue = 1.f;                                //设置最小值
    _maximumValue = 6.f;
    _continuous = YES;
    _thumbOn = NO;
    _stepped = YES;
```

```
    _decimalPlaces = 0;
    self.backgroundColor = [UIColor clearColor];          //设置背景颜色
    //创建并设置滑块控件对象的轨道
    _trackImageViewNormal = [[UIImageView alloc] initWithImage:[UIImage
imageNamed:@"2.png"]];
    [self addSubview:_trackImageViewNormal];              //添加视图
    _trackImageViewHighlighted = [[UIImageView alloc] initWithImage:
[UIImage imageNamed:@"1.png"]];
    [self addSubview:_trackImageViewHighlighted];
    //创建滑块控件对象中的滑块
    _thumbImageView = [[UIImageView alloc] initWithImage:[UIImage
imageNamed:@"3.png"]];
    [self addSubview:_thumbImageView];
    //创建并设置滑块控件对象中滑块中的标签
    _labelOnThumb = [[UILabel alloc] init];
    _labelOnThumb.backgroundColor = [UIColor clearColor]; //设置标签的背景颜色
    _labelOnThumb.textAlignment = NSTextAlignmentCenter;  //设置对齐方式
    _labelOnThumb.text = [NSString stringWithFormat:[self valueString
Format], _value];
    _labelOnThumb.textColor = [UIColor whiteColor];
    [self addSubview:_labelOnThumb];
    //创建并设置滑块控件对象中滑块上方的标签
    _labelAboveThumb = [[UILabel alloc] init];
    _labelAboveThumb.backgroundColor = [UIColor clearColor];
    _labelAboveThumb.textAlignment = NSTextAlignmentCenter;
        //设置对齐方式
    _labelAboveThumb.text = [NSString stringWithFormat:[self valueString
Format], _value];
    _labelAboveThumb.textColor = [UIColor colorWithRed:0/255.f green:151.
f/255.f blue:79.f/255.f alpha:1.f];
    [self addSubview:_labelAboveThumb];
}
```

layoutSubviews 方法的功能是，对创建的子视图进行重新布局。程序代码如下：

```
- (void)layoutSubviews
{
    //设置滑块控件中轨道的框架
    _trackImageViewNormal.frame = self.bounds;
    _trackImageViewHighlighted.frame = self.bounds;
    //设置滑块控件中滑块的框架以及中心
    CGFloat thumbHeight = 98.f * _trackImageViewNormal.bounds.size.height / 64.f;
    CGFloat thumbWidth = 91.f * thumbHeight / 98.f;
    _thumbImageView.frame = CGRectMake(0, 0, thumbWidth, thumbHeight);
                                                          //设置框架
    _thumbImageView.center = CGPointMake([self xForValue:_value],
CGRectGetMidY(_trackImageViewNormal.frame));              //设置中心位置
    //设置滑块控件中滑块内的标签的框架
    _labelOnThumb.frame = _thumbImageView.frame;
    //设置滑块控件中滑块上方的标签的框架
    _labelAboveThumb.frame = CGRectMake(_labelOnThumb.frame.origin.x,
_labelOnThumb.frame.origin.y - _labelOnThumb.frame.size.height * 0.6f,
_labelOnThumb.frame.size.width, _labelOnThumb.frame.size.height); //设置框架
    [self updateTrackHighlight];
}
```

updateTrackHighlight 方法，实现当滑块控件的滑块经过轨道的某一位置时，将经过的轨道改变。程序代码如下：

```objc
- (void)updateTrackHighlight
{
    CAShapeLayer *maskLayer = [[CAShapeLayer alloc] init];
    CGFloat thumbMidXInHighlightTrack = CGRectGetMidX([self convertRect:
    _thumbImageView.frame toView:_trackImageViewNormal]);
    CGRect maskRect = CGRectMake(0, 0, thumbMidXInHighlightTrack, _track
    ImageViewNormal.frame.size.height);
    CGMutablePathRef path = CGPathCreateMutable();
    CGPathAddRect(path, nil, maskRect);                    //绘制矩形
    [maskLayer setPath:path];                              //设置路径
    CGPathRelease(path);
    _trackImageViewHighlighted.layer.mask = maskLayer;
}
```

drawRect:方法，实现对标签的设置以及调用更新轨道方法。程序代码如下：

```objc
- (void)drawRect:(CGRect)rect
{
    _labelOnThumb.center = _thumbImageView.center;
    _labelAboveThumb.center = CGPointMake(_thumbImageView.center.x, _thumb
ImageView.center.y - _labelAboveThumb.frame.size.height * 0.6f);
                                                          //设置中心位置
    [self updateTrackHighlight];
}
```

setValue:方法，对值进行设置。程序代码如下：

```objc
- (void)setValue:(float)value
{
    if (value < _minimumValue || value > _maximumValue) {
        return;
    }
    _value = value;
    _thumbImageView.center = CGPointMake([self xForValue:value],
    _thumbImageView.center.y);
    _labelOnThumb.text = [NSString stringWithFormat:[self valueString
Format], _value];    //设置文本内容
    _labelAboveThumb.text = [NSString stringWithFormat:[self valueS
tringFormat], _value];
    [self setNeedsDisplay];
}
```

设置步长需要使用stepMarkerXCloseToX:方法实现。程序代码如下：

```objc
- (float)stepMarkerXCloseToX:(float)x
{
    float xPercent = MIN(MAX(x / self.frame.size.width, 0), 1);
    float stepPercent = 1.f / 5.f;
    float midStepPercent = stepPercent / 2.f;
    int stepIndex = 0;
    //判断
    while (xPercent > midStepPercent) {
        stepIndex++;
        midStepPercent += stepPercent;
    }
    return stepPercent * (float)stepIndex * self.frame.size.width;
}
```

要实现触摸，需要使用 beginTrackingWithTouch:withEvent:、endTrackingWithTouch:

withEvent: 和 continueTrackingWithTouch:withEvent: 方法共同实现。其中，beginTrackingWithTouch:withEvent:方法实现触摸开始的功能。程序代码如下：

```objc
-(BOOL) beginTrackingWithTouch:(UITouch *)touch withEvent:(UIEvent *)event{
    CGPoint touchPoint = [touch locationInView:self];           //获取位置
    //判断给定的点是否在 CGRect 中
    if(CGRectContainsPoint(_thumbImageView.frame, touchPoint)){
        _thumbOn = YES;
    }else {
        _thumbOn = NO;
    }
    return YES;
}
```

endTrackingWithTouch:withEvent:方法的功能是实现触摸结束。程序代码如下：

```objc
-(void)endTrackingWithTouch:(UITouch *)touch withEvent:(UIEvent *)event{
    if (_thumbOn) {
        if (_stepped) {
            _thumbImageView.center = CGPointMake( [self stepMarkerXCloseToX:[touch locationInView:self].x], _thumbImageView.center.y);
                                                            //设置滑块控件中滑块的中心
            [self setNeedsDisplay];
        }
        //设置标签的值
        _value = [self valueForX:_thumbImageView.center.x];
        _labelOnThumb.text = [NSString stringWithFormat:[self valueStringFormat], _value];      //设置文本内容
        _labelAboveThumb.text = [NSString stringWithFormat:[self valueStringFormat], _value];
        [self sendActionsForControlEvents:UIControlEventValueChanged];
    }
    _thumbOn = NO;
}
```

continueTrackingWithTouch:withEvent:方法的功能是，实现触摸的移动。程序代码如下：

```objc
-(BOOL)continueTrackingWithTouch:(UITouch *)touch withEvent:(UIEvent *)event{
    if(!_thumbOn)
        return YES;
    CGPoint touchPoint = [touch locationInView:self];
    _thumbImageView.center = CGPointMake( MIN( MAX( [self xForValue:_minimumValue], touchPoint.x), [self xForValue:_maximumValue]), _thumbImageView.center.y);                                    //设置中心位置
    if (_continuous && !_stepped) {
        _value = [self valueForX:_thumbImageView.center.x];
        _labelOnThumb.text = [NSString stringWithFormat:[self valueStringFormat], _value];      //设置文本内容
        _labelAboveThumb.text = [NSString stringWithFormat:[self valueStringFormat], _value];
        [self sendActionsForControlEvents:UIControlEventValueChanged];
    }
    [self setNeedsDisplay];
    return YES;
}
```

（6）打开 Main.storyboard 文件，在 View Controller 视图控制器的设计界面中，添加一个 View 空白的视图。将此视图的 Class 设置为 StepSlider。

【代码解析】

本实例的关键功能是滑块滑动到的位置是固定的。下面将详细讲解这个知识点。

要实现滑块滑动到指定的位置，本实例使用的方法是，获取当前滑块在 x 位置的百分比，然后再获取滑块到达的指定位置和前一个指定位置中间的百分比，最后将两个百分比进行判断。当前滑块在 x 位置的百分比，比在中间的百分比大时，滑块就会到指定的位置。当前滑块在 x 位置的百分比，比在中间的百分比小时，滑块会到达指定位置的前一个位置。如果将指定位置设为 n，那么它的前一个位置就为 n-1。程序代码如下：

```
float xPercent = MIN(MAX(x / self.frame.size.width, 0), 1);
float stepPercent = 1.f / 5.f;
float midStepPercent = stepPercent / 2.f;
int stepIndex = 0;
while (xPercent > midStepPercent) {
    stepIndex++;
    midStepPercent += stepPercent;
}
return stepPercent * (float)stepIndex * self.frame.size.width;
```

实例 15　模拟现实音量控制条

【实例描述】

本实例实现的功能是在界面显示一个模拟现实音量控制条的滑块控件。当用户滑动此控件中的滑块时，会用其他图片填充当前的音量，并将当前的值进行显示。其中，用于显示当前值的是标签。运行效果如图 2.16 所示。

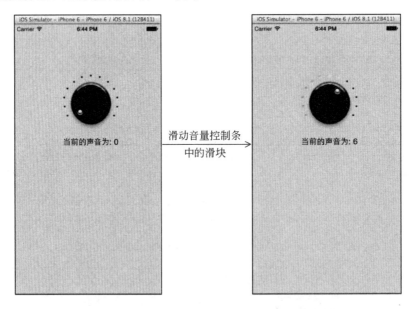

图 2.16　实例 15 运行效果

【实现过程】

音量控制条可以当作一个大的滑块控件，音量控制条中的调节按钮是滑块控件的滑块，当用户滑动滑块时，音量也会发生改变。具体的实现过程如下。

（1）创建一个项目，命名为"模拟现实音量控制条"。
（2）添加图片 1.png、2.png 和 3.png 到创建项目的 Supporting Files 文件夹中。
（3）创建一个基于 UIImageView 类的 circleSlider 类。
（4）打开 circleSlider.h 文件，编写代码。实现协议、实例变量、对象以及方法的声明。程序代码如下：

```
#import <UIKit/UIKit.h>
@class Circle;
//协议
@protocol CircleDelegate <NSObject>
@optional
- (void)circleSlide:(Circle *)circleSlide withProgress:(float)progress;
@end
@interface Circle : UIImageView<CircleDelegate>{
    //实例变量
    CGPoint      _rotatePoint;
    float        _radius;
    float        _startAngle;
    float        _endAngle;
    float        _progress;
}
//属性
@property (nonatomic, assign) id delegate;
@property (nonatomic, readonly) float progress;
//方法
- (id)initWithImage:(UIImage *)image  rotatePoint:(CGPoint)rotatePoint
radius:(float)radius startAngle:
(float)startAngle endAngle:(float)endAngle;
- (void)setProgress:(float)progress;
@end
```

（5）打开 circleSlider.m 文件，编写代码。实现环形滑块控件的效果。使用的方法如表 2-12 所示。

表 2-12　circleSlider.m 文件中方法总结

方　　法	功　　能
initWithImage: rotatePoint: radius: startAngle: endAngle:	初始化环形滑块控件
setProgress:	设置进度
progressOfAngle:	获取角度的进度
angleOfProgress:	获取进度的角度
samesign:	获取符号
progressOfPosition:	获取位置的进度
positionOfProgress:	获取进度的位置
touchesMoved:	触摸移动

这里需要讲解几个重要的方法（其他方法请读者参考源代码）。initWithImage: rotatePoint: radius: startAngle: endAngle:方法实现对环形滑块控件的初始化。程序代码如下：

```
- (id)initWithImage:(UIImage *)image  rotatePoint:(CGPoint)rotatePoint
radius:(float)radius
startAngle:(float)startAngle endAngle:(float)endAngle
```

```objc
{
    self = [super initWithImage:image];
    if (self) {
        [self setUserInteractionEnabled:YES];
        _rotatePoint = rotatePoint;
        _radius = radius;
        _startAngle = startAngle;                      //设置开始的角度
        _endAngle = endAngle;                          //设置结束的角度
        _progress = 1;
        self.center = [self positionOfProgress:_progress];
    }
    return self;
}
```

touchesMoved: withEvent:方法,实现触摸移动的功能。程序代码如下:

```objc
- (void)touchesMoved:(NSSet *)touches withEvent:(UIEvent *)event
{
    CGPoint ptCurr=[[touches anyObject] locationInView:self.delegate];
    float newProgress = [self progressOfPosition:ptCurr];
    //判断是否发生跳跃
    if ((newProgress > _progress && (newProgress - _progress) < 0.5)
        || (newProgress < _progress && (_progress - newProgress) < 0.5)) {
        _progress = [self progressOfPosition:ptCurr];
        if (delegate && [delegate respondsToSelector:@selector(circleSlide:
        withProgress:)]) {
            [delegate circleSlide:self withProgress:_progress];
                                                       //调用circleSlide:方法
        }
    }
}
```

(6)创建一个基于UIControl类的Slider类。

(7)打开Slider.h文件,编写代码。实现头文件、宏定义、对象、实例变量、属性和方法的声明。程序代码如下:

```objc
#import <UIKit/UIKit.h>
#import "Circle.h"                          //头文件
//宏定义
#define CIRCLE_X                 (67.0f)
#define CIRCLE_Y                 (66.0f)
......
#define RADIANS_TO_DEGREES(_radians)    ((_radians)*180)/M_PI
@interface Slider : UIControl{
 @private                                   //将以下的对象、变量声明为私有的
    //实例变量
    NSInteger _minimumVolume;
    NSInteger _maximumVolume;
    float _progress;
    NSInteger _currentVolume;
    //对象
    UIImageView *_backgroundView;
    Circle *_circleSlide;
    UIImageView *_contentView;
    UIImage *_contentImage;
}
@property (nonatomic, assign) NSInteger currentVolume;
- (id)initWithFrame:(CGRect)frame minimumVolume:(NSInteger)minimumVolume
```

```
maximumVolume:
(NSInteger)maximumVolume;
@end
```

（8）打开 Slider.m 文件，编写代码。实现一个音量控制条控制音量的效果。使用的方法如表 2-13 所示。

表 2-13 Slider.m文件中方法总结

方 法	功 能
initWithFrame:	使用框架初始化音量控制条
initWithFrame: minimumVolume: maximumVolume:	使用框架、最小和最大音量初始化音量控制条
drawRect:	绘制
currentVolume	获取当前的音量
setCurrentVolume:	设置当前的音量
circleSlide:withProgress:	获取滑块的进度

可以将这些方法分为两部分：设置音量控制条；实现音量的显示。其中，设置音量控制条需要使用 initWithFrame:、initWithFrame: minimumVolume: maximumVolume:、currentVolume 和 setCurrentVolume:方法。这里需要讲解几个重要的方法（其他方法请读者参考源代码）。initWithFrame:方法实现使用框架初始化音量控制条。程序代码如下：

```
- (id)initWithFrame:(CGRect)frame
{
    self = [super initWithFrame:frame];
    if (self) {
        [self setUserInteractionEnabled:YES];
        //背景视图对象的创建与设置
        _backgroundView = [[UIImageView alloc] initWithImage:[UIImage 
        imageNamed:@"2.png"]];
        [_backgroundView setFrame:_backgroundView.bounds];        //设置框架
        frame.size.width = _backgroundView.bounds.size.width;     //设置宽度
        frame.size.height = _backgroundView.bounds.size.height;   //设置高度
        [self setFrame:frame];
        [self setBackgroundColor:[UIColor clearColor]];
        [self addSubview:_backgroundView];
        //环形控件对象的创建与设置
        _circleSlide =[[Circle alloc] initWithImage:[UIImage imageNamed:
        @"3.png"] rotatePoint:CGPointMake(CIRCLE_X, CIRCLE_Y) radius:CONTROL_
        CIRCLE_RADIUS    startAngle:DEGREES_TO_RADIANS(START_ANGLE)    endAngle:
        DEGREES_TO_RADIANS (END_ANGLE)];
        _circleSlide.delegate = self;                              //设置委托
        [self addSubview:_circleSlide];
        //音量填充视图的创建与设置
        _contentImage = [UIImage imageNamed:@"1.png"];
        _contentView = [[UIImageView alloc] initWithFrame:_backgroundView.
        bounds];
        [self addSubview:_contentView];
        _progress = 1;
    }
    return self;
}
```

setCurrentVolume:方法，对当前的音量进行设置。程序代码如下：

```
- (void)setCurrentVolume:(NSInteger)currentVolume
```

```
{
    if (currentVolume >= _minimumVolume && currentVolume <= _maximumVolume) {
        _progress = 1.0f - (float)(currentVolume - _minimumVolume)/(_maximumVolume - _minimumVolume);
        _currentVolume = currentVolume;                    //设置当前音量
        [self setNeedsDisplay];                            //重新绘制
    }
}
```

drawRect:方法，实现音量的绘制以及滑块进度的绘制。程序代码如下：

```
- (void)drawRect:(CGRect)rect
{
    CGContextRef context = CGBitmapContextCreate(NULL, self.bounds.size.width, self.bounds.size.height, 8, 4 * self.bounds.size.width, CGColorSpaceCreateDeviceRGB(), kCGImageAlphaPremultipliedFirst);      //创建位图上下文
    float endAngle = (END_ANGLE-START_ANGLE)*_progress+START_ANGLE;
    endAngle = (_progress == 0)?(endAngle+0.1):endAngle;
    //绘制弧
    CGContextAddArc(context, CIRCLE_X, CIRCLE_Y, VOLUME_CIRCLE_RADIUS, DEGREES_TO_RADIANS(START_ANGLE), DEGREES_TO_RADIANS(endAngle), YES);
    CGContextAddArc(context, CIRCLE_X, CIRCLE_Y, 0, DEGREES_TO_RADIANS(0), DEGREES_TO_RADIANS(0), YES);
    CGContextClosePath(context);
    CGContextClip(context);                                //修改当前的裁剪路径
    CGContextDrawImage(context, self.bounds, _contentImage.CGImage);
                                                           //绘制位图
    CGImageRef imageMasked = CGBitmapContextCreateImage(context);
    CGContextRelease(context);
    UIImage *newImage = [UIImage imageWithCGImage:imageMasked];
    //实例化对象
    CGImageRelease(imageMasked);
    [_contentView setImage:newImage];                      //设置图像
    [_circleSlide setProgress:_progress];                  //设置进度
}
```

circleSlide: withProgress:方法，获取环形滑块进度的设置。程序代码如下：

```
- (void)circleSlide:(Circle *)circleSlide withProgress:(float)progress
{
    [self sendActionsForControlEvents:UIControlEventValueChanged];
    _progress = progress;
    [self setNeedsDisplay];
    NSInteger volume = (_maximumVolume - _minimumVolume)*(1-_progress);
    if (_currentVolume != volume) {
        _currentVolume = volume;                           //设置当前音量
        [self sendActionsForControlEvents:UIControlEventValueChanged];
    }
}
```

（9）打开 Main.storyboard 文件，将 View Controller 视图控制器的设计界面设置为浅灰色。

（10）打开 ViewController.h 文件，编写代码。实现头文件和对象的声明。程序代码如下：

```
#import <UIKit/UIKit.h>
#import "Slider.h"
@interface ViewController : UIViewController{
    UILabel *label;
}
@end
```

(11)打开 ViewController.m 文件，编写代码。实现将创建的音量控制条对象显示在当前的视图中，并且将当前的音量显示在标签中。程序代码如下：

```
- (void)viewDidLoad
{
[super viewDidLoad];
//创建并设置音量控制条
    CGRect frame = CGRectMake(100, 100, 0, 0);
    Slider *bar = [[Slider alloc] initWithFrame:frame minimumVolume:0
    maximumVolume:10];
    [bar addTarget:self action:@selector(onVolumeBarChange:) forControlEvents:
UIControlEventValueChanged];                    //添加动作
    [self.view addSubview:bar];
    //创建并设置用来显示音量数据的标签
    bar.currentVolume = 0;                       //设置当前音量
    frame = bar.frame;                           //获取框架
    frame.origin.y += frame.size.height+10;
    frame.size.height = 30;                      //设置高度
    label = [[UILabel alloc] initWithFrame:frame];
    [label setTextAlignment:NSTextAlignmentCenter]; //设置对齐方式
    [label setBackgroundColor:[UIColor clearColor]];
    [label setTextColor:[UIColor blackColor]];
    [label setText:[NSString stringWithFormat:@"当前的声音为: %d", bar.
    currentVolume]];
    [self.view addSubview:label];
}
//在标签中显示当前的音量
- (void)onVolumeBarChange:(id)sender
{
    Slider *bar = sender;
    [label setText:[NSString stringWithFormat:@"当前的声音为: %d", bar.
    currentVolume]];                             //设置文本
}
```

【代码解析】

由于本实例中的代码和方法非常多，为了方便读者的阅读，笔者绘制了一些执行流程图，如图 2.17 和图 2.18 所示。其中，音量控制条的显示，需要使用 viewDidLoad、initWithFrame: minimumVolume: maximumVolume: 、initWithFrame: 、initWithImage:rotatePoint:radius:startAngle:endAngle: 、angleOfProgress: 、positionOfProgress: 、setCurrentVolume: 、currentVolume、drawRect:和 setProgress:方法共同实现。它们的程序执行流程如图 2.17 所示。

实现拖动音量控制条上的滑块到指定的位置，并且使用图片将滑块经过的音量填充，需要使用 progressOfPosition: 、progressOfAngle: 、touchesMoved:withEvent: 、circleSlide:withProgress: 、onVolumeBarChange: 、currentVolume、drawRect: 、setProgress: 、angleOfProgress:和 positionOfProgress:方法共同实现。它们的执行流程如图 2.18 所示。

iOS 开发范例实战宝典（基础篇）

图 2.17　实例 15 程序执行流程 1

图 2.18　实例 15 程序执行流程 2

实例 16 iOS 视频修剪控件

【实例描述】

本实例实现的功能是在界面显示一个 iOS 视频修剪控件。当用户单击"播放"按钮，视频会进行播放，当用户单击"播放截取部分"按钮后，视频只播放用户使用两个滑块进行选择的部分。运行效果如图 2.19 所示。

【实现过程】

实现视频修剪控件的效果，首先需要创建一个用于视频修剪的控件，然后实现拖动功能，最后实现修剪。具体的实现步骤如下。

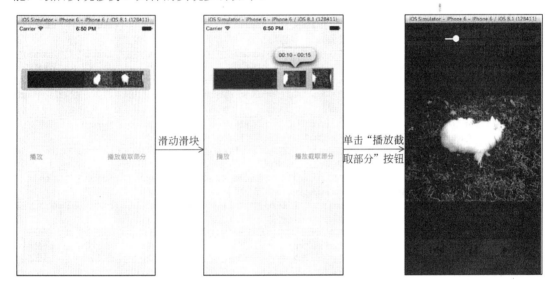

图 2.19 实例 16 运行效果

（1）创建一个项目，命名为"iOS 视频修剪控件"。
（2）添加一个 1.m4v 的视频到创建项目的 Supporting Files 文件夹中。
（3）添加一个 AVFoundation.framework 和 MediaPlayer.framework 框架到创建的项目中。
（4）创建一个 UIView 类的 Bubble 类。
（5）打开 Bubble.h 文件，编写代码。使用 drawRect:方法实现绘制一个用于显示时间的气泡。程序代码如下：

```
- (void)drawRect:(CGRect)rect
{
    CGColorSpaceRef colorSpace = CGColorSpaceCreateDeviceRGB();
    CGContextRef context = UIGraphicsGetCurrentContext();
    //气泡颜色
    UIColor* bubbleGradientTop = [UIColor colorWithRed: 1 green: 0.939 blue: 0.743 alpha: 1];
    UIColor* bubbleGradientBottom = [UIColor colorWithRed: 1 green: 0.817 blue: 0.053 alpha: 1];
    NSArray* bubbleGradientColors = [NSArray arrayWithObjects:
                        (id)bubbleGradientTop.CGColor,
```

```objc
                                    (id)bubbleGradientBottom.CGColor, nil];
                                                    //实例化数组对象
    CGFloat bubbleGradientLocations[] = {0, 1};
    CGGradientRef bubbleGradient = CGGradientCreateWithColors(colorSpace,
(__bridge CFArrayRef)bubbleGradientColors, bubbleGradientLocations);
    //阴影
    UIColor* outerShadow = [UIColor blackColor];         //设置外阴影的颜色
    CGSize outerShadowOffset = CGSizeMake(0.1, 6.1);
    CGFloat outerShadowBlurRadius = 13;
    CGRect bubbleFrame = self.bounds;
    CGRect arrowFrame = CGRectMake(CGRectGetMinX(bubbleFrame) + floor
((CGRectGetWidth(bubbleFrame) - 59) * 0.50462 + 0.5), CGRectGetMinY
(bubbleFrame) + CGRectGetHeight(bubbleFrame) - 46, 59, 46);     //框架
    //绘制气泡
    UIBezierPath* bubblePath = [UIBezierPath bezierPath];
    [bubblePath moveToPoint: CGPointMake(CGRectGetMaxX(bubbleFrame) - 12,
CGRectGetMinY(bubbleFrame) + 28.5)];                      //设置开始点
    [bubblePath addLineToPoint: CGPointMake(CGRectGetMaxX(bubbleFrame) -
12, CGRectGetMaxY(bubbleFrame) - 27.5)];                  //设置结束点
    [bubblePath addCurveToPoint: CGPointMake(CGRectGetMaxX(bubbleFrame) -
25, CGRectGetMaxY(bubbleFrame) - 14.5) controlPoint1: CGPointMake
(CGRectGetMaxX(bubbleFrame) - 12, CGRectGetMaxY(bubbleFrame) - 20.32)
controlPoint2: CGPointMake(CGRectGetMaxX(bubbleFrame) - 17.82, CGRectGetMaxY
(bubbleFrame) - 14.5)];                               //以3个点画一段曲面
    [bubblePath addLineToPoint: CGPointMake(CGRectGetMinX(arrowFrame) +
40.5, CGRectGetMaxY(arrowFrame) - 13.5)];
    [bubblePath addLineToPoint: CGPointMake(CGRectGetMinX(arrowFrame) +
29.5, CGRectGetMaxY(arrowFrame) - 0.5)];
    [bubblePath addLineToPoint: CGPointMake(CGRectGetMinX(arrowFrame) +
18.5, CGRectGetMaxY(arrowFrame) - 13.5)];
    [bubblePath addLineToPoint: CGPointMake(CGRectGetMinX(bubbleFrame) +
26.5, CGRectGetMaxY(bubbleFrame) - 14.5)];
    [bubblePath addCurveToPoint: CGPointMake(CGRectGetMinX(bubbleFrame) +
13.5, CGRectGetMaxY(bubbleFrame) - 27.5) controlPoint1: CGPointMake
(CGRectGetMinX(bubbleFrame) + 19.32, CGRectGetMaxY(bubbleFrame) - 14.5)
controlPoint2: CGPointMake(CGRectGetMinX(bubbleFrame) + 13.5, CGRect
GetMaxY(bubbleFrame) - 20.32)];                       //以3个点画一段曲面
    [bubblePath addLineToPoint: CGPointMake(CGRectGetMinX(bubbleFrame) +
13.5, CGRectGetMinY(bubbleFrame) + 28.5)];
    [bubblePath addCurveToPoint: CGPointMake(CGRectGetMinX(bubbleFrame) +
26.5, CGRectGetMinY(bubbleFrame) + 15.5) controlPoint1: CGPointMake
(CGRectGetMinX(bubbleFrame) + 13.5, CGRectGetMinY(bubbleFrame) + 21.32)
controlPoint2: CGPointMake(CGRectGetMinX(bubbleFrame) + 19.32, CGRectGetMinY
(bubbleFrame) + 15.5)];
    [bubblePath addLineToPoint: CGPointMake(CGRectGetMaxX(bubbleFrame) -
25, CGRectGetMinY(bubbleFrame) + 15.5)];
    [bubblePath addCurveToPoint: CGPointMake(CGRectGetMaxX(bubbleFrame) -
12, CGRectGetMinY(bubbleFrame) + 28.5) controlPoint1: CGPointMake
(CGRectGetMaxX(bubbleFrame) - 17.82, CGRectGetMinY(bubbleFrame) + 15.5)
controlPoint2: CGPointMake(CGRectGetMaxX(bubbleFrame) - 12, CGRectGetMinY
(bubbleFrame) + 21.32)];                              //以3个点画一段曲面
    [bubblePath closePath];
    CGContextSaveGState(context);
    CGContextSetShadowWithColor(context, outerShadowOffset, outerShadow
BlurRadius, outerShadow.CGColor);                          //设置阴影
    CGContextBeginTransparencyLayer(context, NULL);
    [bubblePath addClip];
    CGRect bubbleBounds = CGPathGetPathBoundingBox(bubblePath.CGPath);
    CGContextDrawLinearGradient(context,
```

```
bubbleGradient,CGPointMake(CGRectGetMidX(bubbleBounds),
CGRectGetMinY(bubbleBounds)),CGPointMake(CGRectGetMidX(bubbleBounds),
CGRectGetMaxY(bubbleBounds)),0);                      //绘制线的渐变
CGContextEndTransparencyLayer(context);
}
```

(6) 创建一个基于 UIView 类的 SliderLeft 类。

(7) 打开 SliderLeft.m 文件,编写代码。使用 drawRect:方法实现绘制左边的用于滑动的滑块。程序代码如下:

```
- (void)drawRect:(CGRect)rect
{
    CGColorSpaceRef colorSpace = CGColorSpaceCreateDeviceRGB();
    CGContextRef context = UIGraphicsGetCurrentContext();
    //颜色
    UIColor* color5 = [UIColor colorWithRed: 0.333 green: 0.902 blue: 0.004 alpha: 1];
    UIColor* gradientColor2 = [UIColor colorWithRed: 1 green: 1 blue: 1 alpha: 1];
    //渐变
    NSArray* gradient3Colors = [NSArray arrayWithObjects:(id)gradient
Color2.CGColor,(id)[UIColor colorWithRed: 0.333 green: 0.951 blue: 0.502
alpha: 1].CGColor,(id)color5.CGColor, nil];
    CGFloat gradient3Locations[] = {0, 0, 0.49};
    CGGradientRef gradient3 = CGGradientCreateWithColors(colorSpace,
(__bridge CFArrayRef)gradient3Colors, gradient3Locations);
    CGRect bubbleFrame = self.bounds;
    //绘制圆角矩形
    CGRect roundedRectangleRect = CGRectMake(CGRectGetMinX(bubbleFrame),
CGRectGetMinY(bubbleFrame), CGRectGetWidth(bubbleFrame), CGRectGetHeight
(bubbleFrame));
    UIBezierPath* roundedRectanglePath = [UIBezierPath bezierPathWith
RoundedRect: roundedRectangleRect byRoundingCorners: UIRectCornerTopLeft
| UIRectCornerBottomLeft cornerRadii: CGSizeMake(5, 5)];
    [roundedRectanglePath closePath];
    CGContextSaveGState(context);
    [roundedRectanglePath addClip];
    //绘制线的渐变
    CGContextDrawLinearGradient(context, gradient3,
                        CGPointMake(CGRectGetMidX(roundedRectangleRect),
CGRectGetMinY(roundedRectangleRect)),
                        CGPointMake(CGRectGetMidX(roundedRectangleRect),
CGRectGetMaxY(roundedRectangleRect)),
                        0);
    CGContextRestoreGState(context);
    //设置颜色
    [[UIColor clearColor] setStroke];
    roundedRectanglePath.lineWidth = 0.5;
    [roundedRectanglePath stroke];
}
```

(8) 创建一个基于 UIView 类的 SliderRight 类。

(9) 打开 SliderRight.m 文件,编写代码。使用 drawRect:方法实现绘制右边的用于滑动的滑块。程序代码和 SliderLeft.m 文件中的代码一样。

(10) 创建一个基于 UIView 类的 Slider 类。

(11) 打开 Slider.h 文件,编写代码。实现头文件、宏定义、协议、属性和方法的声明。程序代码如下:

```
#import <UIKit/UIKit.h>
```

```
//头文件
#import <AVFoundation/AVFoundation.h>
#import "Bubble.h"
#import "SliderLeft.h"
#import "SliderRight.h"
//宏定义
#define SLIDER_BORDERS_SIZE 6.0f
#define BG_VIEW_BORDERS_SIZE 3.0f
@class Slider;
//协议
@protocol SliderDelegate <NSObject>
@optional
- (void)videoRange:(Slider *)videoRange didChangeLeftPosition:(CGFloat)
leftPosition rightPosition:
(CGFloat)rightPosition;
- (void)videoRange:(Slider *)videoRange didGestureStateEndedLeftPosition:
(CGFloat)leftPosition rightPosition:
(CGFloat)rightPosition;
@end
@interface Slider : UIView<SliderDelegate> //Slider 遵守的协议为 SliderDelegate
//属性
@property (nonatomic, weak) id <SliderDelegate> delegate;
@property (nonatomic) CGFloat leftPosition;
……
@property (nonatomic) Float64 durationSeconds;
@property (nonatomic, strong) Bubble *popoverBubble;
//方法
- (id)initWithFrame:(CGRect)frame videoUrl:(NSURL *)videoUrl;
- (void)setPopoverBubbleSize: (CGFloat) width height:(CGFloat)height;
@end
```

（12）打开 Slider.m 文件，编写代码。在此文件中，代码实现了两个功能：绘制一个视频修剪控件，然后实现拖动和计算时间。使用的方法如表 2-14 所示。

表 2-14　Slider.m文件中方法总结

方　　法	功　　能
initWithFrame:videoUrl:	视频修剪控件的初始化
setPopoverBubbleSize:height:	设置气泡的尺寸
setMaxGap:	设置两个滑块的最大间隙
setMinGap:	设置两个滑块的最小间隙
delegateNotification	消息通知
handleLeftPan:	拖动左边的滑块
handleRightPan:	拖动右边的滑块
layoutSubviews	布局
getMovieFrame	获取视频的框架
leftPosition	左边的位置
rightPosition	右边的位置
hideBubble:	气泡的隐藏
setTimeLabel	设置时间标签
trimDurationStr	设置时间字符串
trimIntervalStr	设置时间间隔字符串
timeToStr:	时间转换字符串
isRetina	获取 Retina 显示技术

这里需要讲解几个重要的方法（其他方法请读者参考源代码）。要实现视频修剪控件的绘制，需要使用 initWithFrame:videoUrl:、setPopoverBubbleSize:height:、layoutSubviews、getMovieFrame 和 isRetina 方法实现。其中，initWithFrame:videoUrl:方法实现视频修剪控件的初始化。程序代码如下：

```objc
- (id)initWithFrame:(CGRect)frame videoUrl:(NSURL *)videoUrl{
    self = [super initWithFrame:frame];
    if (self) {
        _frame_width = frame.size.width;
        int thumbWidth = ceil(frame.size.width*0.05);
        //创建并设置背景
        _bgView = [[UIControl alloc] initWithFrame:
CGRectMake(thumbWidth-BG_VIEW_BORDERS_SIZE, 0,
frame.size.width-(thumbWidth*2)+BG_VIEW_BORDERS_SIZE*2, frame.size.height)];
        _bgView.layer.borderColor = [UIColor grayColor].CGColor;
                                                                //设置边框颜色
        _bgView.layer.borderWidth = BG_VIEW_BORDERS_SIZE;       //设置线宽
        [self addSubview:_bgView];
        _videoUrl = videoUrl;
        //创建并设置上边界
        _topBorder = [[UIView alloc] initWithFrame:CGRectMake(0, 0,
        frame.size.width, SLIDER_BORDERS_SIZE)];
        _topBorder.backgroundColor = [UIColor colorWithRed: 0.996 green:
0.951 blue: 0.502 alpha: 1];
        [self addSubview:_topBorder];
        //创建并设置下边界
        _bottomBorder = [[UIView alloc] initWithFrame:CGRectMake(0, frame.
size.height-SLIDER_BORDERS_SIZE, frame.size.width, SLIDER_BORDERS_SIZE)];
        _bottomBorder.backgroundColor = [UIColor colorWithRed: 0.992 green:
0.902 blue: 0.004 alpha: 1];
        [self addSubview:_bottomBorder];                        //添加视图
        //创建并设置左边的滑块
        _leftThumb = [[SliderLeft alloc] initWithFrame:CGRectMake(0, 0,
        thumbWidth, frame.size.height)];
        _leftThumb.contentMode = UIViewContentModeLeft;
        _leftThumb.userInteractionEnabled = YES;
        _leftThumb.clipsToBounds = YES;
        _leftThumb.backgroundColor = [UIColor clearColor];      //设置背景颜色
        _leftThumb.layer.borderWidth = 0;
        [self addSubview:_leftThumb];
        UIPanGestureRecognizer *leftPan = [[UIPanGestureRecognizer alloc]
initWithTarget:self action:@selector(handleLeftPan:)];
        [_leftThumb addGestureRecognizer:leftPan];              //添加手势
        //创建并设置右边的滑块
        _rightThumb = [[SliderRight alloc] initWithFrame:CGRectMake(0, 0,
        thumbWidth, frame.size.height)];
        _rightThumb.contentMode = UIViewContentModeRight;
        _rightThumb.userInteractionEnabled = YES;
        _rightThumb.clipsToBounds = YES;
        _rightThumb.backgroundColor = [UIColor clearColor]; //设置背景颜色
        [self addSubview:_rightThumb];
        UIPanGestureRecognizer *rightPan = [[UIPanGestureRecognizer alloc]
initWithTarget:self action:@selector(handleRightPan:)];
        [_rightThumb addGestureRecognizer:rightPan];            //添加手势
        _rightPosition = frame.size.width;
        _leftPosition = 0;
        //创建并设置中间的视图
```

```
        _centerView = [[UIView alloc] initWithFrame:CGRectMake(0, 0,
        frame.size.width, frame.size.height)];
        _centerView.backgroundColor = [UIColor clearColor];
        [self addSubview:_centerView];
        //创建并设置气泡
        _popoverBubble = [[Bubble alloc] initWithFrame:CGRectMake(0, -50, 100, 50)];
        _popoverBubble.alpha = 0;                              //设置透明度
        _popoverBubble.backgroundColor = [UIColor clearColor];
        [self addSubview:_popoverBubble];
        //创建并设置气泡中的标签
        _bubleText = [[UILabel alloc] initWithFrame:_popoverBubble.frame];
        _bubleText.font = [UIFont boldSystemFontOfSize:20];
        _bubleText.backgroundColor = [UIColor clearColor];   //设置背景颜色
        _bubleText.textColor = [UIColor blackColor];
        _bubleText.textAlignment = NSTextAlignmentCenter;     //对齐方式
        [_popoverBubble addSubview:_bubleText];
        [self getMovieFrame];
    }
    return self;
}
```

setPopoverBubbleSize:height:方法,对气泡的尺寸进行设置。程序代码如下:

```
-(void)setPopoverBubbleSize: (CGFloat) width height:(CGFloat)height{
    CGRect currentFrame = _popoverBubble.frame;         //获取气泡的框架
    currentFrame.size.width = width;                    //设置气泡的宽度
    currentFrame.size.height = height;                  //设置气泡的高度
    currentFrame.origin.y = -height;
    _popoverBubble.frame = currentFrame;                //设置气泡的框架
    currentFrame.origin.x = 0;
    currentFrame.origin.y = 0;
    _bubleText.frame = currentFrame;
}
```

layoutSubviews方法,对视图中的子视图进行重新布局。程序代码如下:

```
- (void)layoutSubviews
{
    CGFloat inset = _leftThumb.frame.size.width / 2;
    //左边、右边滑块的中心位置
    _leftThumb.center = CGPointMake(_leftPosition+inset, _leftThumb.frame.
size.height/2);
    _rightThumb.center = CGPointMake(_rightPosition-inset, _rightThumb.frame.
size. height/2);
    //上、下边界的框架
    _topBorder.frame = CGRectMake(_leftThumb.frame.origin.x + _leftThumb.
frame.size.width, 0, _rightThumb.frame.origin.x - _leftThumb.frame.
origin.x - _leftThumb.frame.size.width/2, SLIDER_BORDERS_SIZE);
    _bottomBorder.frame   =   CGRectMake(_leftThumb.frame.origin.x   +
_leftThumb.frame.size.width, _bgView.frame.size.height-SLIDER_BORDERS_SIZE,
_rightThumb.frame.origin.x - _leftThumb.frame.origin.x - _leftThumb.frame.
size.width/2, SLIDER_BORDERS_SIZE);
    //中间视图的框架
    _centerView.frame = CGRectMake(_leftThumb.frame.origin.x + _leftThumb.
frame.size.width, _centerView.frame.origin.y, _rightThumb.frame.origin.x
- _leftThumb.frame.origin.x - _leftThumb.frame.size.width, _centerView.
frame.size.height);
    CGRect frame = _popoverBubble.frame;
    frame.origin.x = _centerView.frame.origin.x+_centerView.frame.size.
```

```
        width/2-frame.size.width/2;
    _popoverBubble.frame = frame;
}
```

getMovieFrame 方法，获取视频的框架。程序代码如下：

```
-(void)getMovieFrame{
    //获取资源
    AVAsset *myAsset = [[AVURLAsset alloc] initWithURL:_videoUrl options:nil];
    //生成图片控制器
    self.imageGenerator = [AVAssetImageGenerator assetImageGenerator
WithAsset:myAsset];
    //判断，并设置最大尺寸
    if ([self isRetina]){
        self.imageGenerator.maximumSize = CGSizeMake(_bgView.frame.size.
width*2, _bgView.frame.size.height*2);                    //设置最大尺寸
    } else {
        self.imageGenerator.maximumSize = CGSizeMake(_bgView.frame.size.width,
_bgView.frame.size.height);
    }
    int picWidth = 49;
    NSError *error;
    CMTime actualTime;
    CGImageRef halfWayImage = [self.imageGenerator copyCGImageAtTime:
kCMTimeZero actualTime:&actualTime error:&error];        //获取指定时间的图片
    if (halfWayImage != NULL) {
        UIImage *videoScreen;
        if ([self isRetina]){
            videoScreen = [[UIImage alloc] initWithCGImage:halfWayImage
scale:2.0 orientation:UIImageOrientationUp];            //实例化对象
        } else {
            videoScreen = [[UIImage alloc] initWithCGImage:halfWayImage];
        }
        UIImageView *tmp = [[UIImageView alloc] initWithImage:videoScreen];
        [_bgView addSubview:tmp];                        //添加视图
        picWidth = tmp.frame.size.width;                 //获取宽度
        CGImageRelease(halfWayImage);
    }
    _durationSeconds = CMTimeGetSeconds([myAsset duration]);
    int picsCnt = ceil(_bgView.frame.size.width / picWidth);
    NSMutableArray *allTimes = [[NSMutableArray alloc] init];
    int time4Pic = 0;
    //遍历
    for (int i=1; i<picsCnt; i++){
        time4Pic = i*picWidth;
        CMTime timeFrame = CMTimeMakeWithSeconds(_durationSeconds*time4Pic/_bgView.
frame.size.width, 600);                                  //获取视频时间
        [allTimes addObject:[NSValue valueWithCMTime:timeFrame]];//添加对象
    }
    NSArray *times = allTimes;
    __block int i = 1;
    //创建视频中指定时间的图片
    [self.imageGenerator       generateCGImagesAsynchronouslyForTimes:times
completionHandler:^(CMTime  requestedTime,  CGImageRef  image,  CMTime
actualTime,AVAssetImageGeneratorResult result, NSError *error) {
        //判断 result 是否为 AVAssetImageGeneratorSucceeded
        if (result == AVAssetImageGeneratorSucceeded) {
            UIImage *videoScreen;
```

```
        if ([self isRetina]){
            videoScreen = [[UIImage alloc] initWithCGImage:image scale:
2.0 orientation:UIImageOrientationUp];              //实例化对象
        } else {
            videoScreen = [[UIImage alloc] initWithCGImage:image];
        }
        UIImageView *tmp = [[UIImageView alloc] initWithImage:videoScreen];
        int all = (i+1)*tmp.frame.size.width;
        CGRect currentFrame = tmp.frame;                //获取宽度
        currentFrame.origin.x = i*currentFrame.size.width;
        if (all > _bgView.frame.size.width){
            int delta = all - _bgView.frame.size.width;
            currentFrame.size.width -= delta;           //设置宽度
        }
        tmp.frame = currentFrame;
        i++;
        dispatch_async(dispatch_get_main_queue(), ^{
            [_bgView addSubview:tmp];                   //添加对象
        });
    }
    if (result == AVAssetImageGeneratorFailed) {
        NSLog(@"Failed with error: %@", [error localizedDescription]);
                                                        //输出
    }
    if (result == AVAssetImageGeneratorCancelled) {
        NSLog(@"Canceled");                             //输出
    }
}];
```

实现拖动的功能需要使用 setMaxGap:、setMinGap:、leftPosition、rightPosition、handleLeftPan:和 handleRightPan:方法。其中，setMaxGap:方法对两个滑块之间的最大间隙进行设置。程序代码如下：

```
-(void)setMaxGap:(NSInteger)maxGap{
    _leftPosition = 0;
    _rightPosition = _frame_width*maxGap/_durationSeconds;
    _maxGap = maxGap;
}
```

handleLeftPan:方法，通过拖动手势实现左边滑块的滑动。程序代码如下：

```
- (void)handleLeftPan:(UIPanGestureRecognizer *)gesture
{
    //判断手势的状态
    if (gesture.state == UIGestureRecognizerStateBegan || gesture.state ==
UIGestureRecognizerStateChanged) {
        CGPoint translation = [gesture translationInView:self];   //获取点
        _leftPosition += translation.x;
        //判断左边滑块的位置是否小于0
        if (_leftPosition < 0) {
            _leftPosition = 0;
        }
        //判断两个滑块的位置
        if ((_rightPosition-_leftPosition <= _leftThumb.frame.size.width+
_rightThumb.frame.size.width) ||((self.maxGap > 0) && (self.right
Position-self.leftPosition > self.maxGap)) ||((self.minGap > 0) && (self.
```

```
rightPosition-self.leftPosition < self.minGap))){
        _leftPosition -= translation.x;
    }
    [gesture setTranslation:CGPointZero inView:self];
    [self setNeedsLayout];                                    //重新绘制
    [self delegateNotification];                              //调用消息通知
}
_popoverBubble.alpha = 1;
[self setTimeLabel];
if (gesture.state == UIGestureRecognizerStateEnded){
    [self hideBubble:_popoverBubble];
}
}
```

handleRightPan:方法通过拖动手势实现右边滑块的滑动。程序代码如下:

```
- (void)handleRightPan:(UIPanGestureRecognizer *)gesture
{
    //判断手势的状态
    if (gesture.state == UIGestureRecognizerStateBegan || gesture.state ==
UIGestureRecognizerStateChanged) {
        CGPoint translation = [gesture translationInView:self];
        _rightPosition += translation.x;
        //判断右边滑块的位置是否小于0
        if (_rightPosition < 0) {
            _rightPosition = 0;
        }
        //判断右边滑块的位置是否超出了框架的宽度
        if (_rightPosition > _frame_width){
            _rightPosition = _frame_width;
        }
        //判断两个滑块是否重合,或间隙是否小于等于0
        if (_rightPosition-_leftPosition <= 0){
            _rightPosition -= translation.x;
        }
        //判断两个滑块的位置
        if ((_rightPosition-_leftPosition <= _leftThumb.frame.size.width+_
rightThumb.frame.size.width) ||((self.maxGap > 0) && (self.rightPosition
-self.leftPosition > self.maxGap)) ||((self.minGap > 0) && (self.right
Position-self.leftPosition < self.minGap))){
            _rightPosition -= translation.x;
        }
        [gesture setTranslation:CGPointZero inView:self];
        [self setNeedsLayout];                                //重写绘制
        [self delegateNotification];
    }
    _popoverBubble.alpha = 1;                                 //设置透明度
    [self setTimeLabel];
    //判断手势识别器对象的状态
    if (gesture.state == UIGestureRecognizerStateEnded){
        [self hideBubble:_popoverBubble];
    }
}
```

计算两个滑块滑动的时间，并显示在气泡中，需要使用 setTimeLabel、trimDurationStr、trimIntervalStr 和 timeToStr:方法实现。其中，trimIntervalStr 方法对字符串的时间间隔进行设置。程序代码如下：

```
-(NSString *)trimIntervalStr{
    NSString *from = [self timeToStr:self.leftPosition];    //实例化对象
    NSString *to = [self timeToStr:self.rightPosition];
    return [NSString stringWithFormat:@"%@ - %@", from, to];    //返回
}
```

hideBubble:方法实现气泡的隐藏动画。程序代码如下：

```
- (void)hideBubble:(UIView *)popover
{
    //动画
    [UIView animateWithDuration:0.4
                     delay:0
                   options:UIViewAnimationCurveEaseIn | UIViewAnimationOption
                AllowUserInteraction
                animations:^(void) {
                    _popoverBubble.alpha = 0;    //设置透明度
                }
                completion:nil];
    //判断
    if ([_delegate respondsToSelector:@selector(videoRange:didGestureStateEndedLeftPosition:rightPosition:)]){
        [_delegate videoRange:self didGestureStateEndedLeftPosition:self.leftPosition rightPosition:self.rightPosition];
    }
}
```

（13）打开 ViewController.h 文件，编写代码。实现头文件、对象、属性和动作的声明。程序代码如下：

```
#import <UIKit/UIKit.h>
//头文件
#import <MediaPlayer/MediaPlayer.h>
#import "Slider.h"
@interface ViewController : UIViewController<SliderDelegate>{
    Slider *mySlider;
}
//属性
@property (strong, nonatomic) NSString *originalVideoPath;
@property (strong, nonatomic) NSString *tmpVideoPath;
@property (nonatomic) CGFloat startTime;
@property (nonatomic) CGFloat stopTime;
@property (strong, nonatomic) AVAssetExportSession *exportSession;
//方法
- (IBAction)play:(id)sender;
- (IBAction)selectPlay:(id)sender;
@end
```

（14）打开 Main.storyboard 文件，对 View Controller 视图控制器的设计界面进行设计，效果如图 2.20 所示。

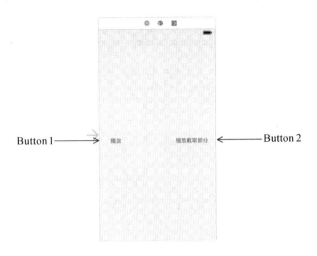

图 2.20　实例 16 View Controller 视图控制器的设计界面

需要添加的视图、控件以及对它们的设置如表 2-15 所示。

表 2-15　实例 16 视图、控件设置

视图、控件	属 性 设 置	其 他
设计界面	Background：浅紫色	
Button1	Title：播放	与动作 play 关联
Button2	Title：播放截取部分	与动作 selectplay 关联

（15）打开 ViewController.m 文件，编写代码。使用的方法如表 2-16 所示。

表 2-16　ViewController.m 文件中方法总结

方　　法	功　　能
viewDidLoad	将创建的视频修剪控件添加到当前视图中
videoRange: didChangeLeftPosition: rightPosition:	设置视图的范围
play:	实现视频的播放
playMovie:	创建一个视频播放器，用来播放视频
selectPlay:	修剪视频，并播放
deleteTmpFile	删除文件

代码分为：将创建的视频修剪控件添加到当前的视图中；播放视频；播放修剪的视频。其中，将创建的视频修剪控件添加到当前的视图中，并为视频修剪控件初始化视频，是通过 viewDidLoad 方法实现的。程序代码如下：

```
- (void)viewDidLoad
{
    NSString *tempDir = NSTemporaryDirectory();      //获取临时文件的目录路径
    self.tmpVideoPath = [tempDir stringByAppendingPathComponent:@"tmpMov.m4v"];
    NSBundle *mainBundle = [NSBundle mainBundle];
    self.originalVideoPath = [mainBundle pathForResource: @"1" ofType: @"m4v"];
    NSURL *videoFileUrl = [NSURL fileURLWithPath:self.originalVideoPath];
```

```objc
    //对视频修剪控件进行创建，并设置
    mySlider = [[Slider alloc] initWithFrame:CGRectMake(10, 100, self.
view.frame.size.width-20, 50) videoUrl:videoFileUrl ];
    mySlider.bubleText.font = [UIFont systemFontOfSize:12];//设置文本的字体
    [mySlider setPopoverBubbleSize:120 height:60];
    mySlider.topBorder.backgroundColor = [UIColor colorWithRed:0.333
green: 0.951 blue: 0.502 alpha: 1];                            //设置背景颜色
    mySlider.bottomBorder.backgroundColor = [UIColor colorWithRed:0.333
green: 0.902 blue: 0.004 alpha: 1];
    mySlider.delegate = self;
    [self.view addSubview:mySlider];                           //添加视图对象
    [super viewDidLoad];
    // Do any additional setup after loading the view, typically from a nib.
}
```

播放视频需要使用 playMovie:和 play:方法共同实现。其中，在 playMovie:方法中，实现创建一个用来播放视频的视频播放器。程序代码如下：

```objc
-(void)playMovie: (NSString *) path{
    NSURL *url = [NSURL fileURLWithPath:path];
    MPMoviePlayerViewController *theMovie = [[MPMoviePlayerViewController
alloc] initWithContentURL:url];                                //创建播放器
    [self presentMoviePlayerViewControllerAnimated:theMovie];
    theMovie.moviePlayer.movieSourceType = MPMovieSourceTypeFile;
    [theMovie.moviePlayer play];
}
```

实现播放修剪部分的视频，需要使用 videoRange:didChangeLeftPosition: rightPosition:、selectPlay:和 deleteTmpFile 方法实现。其中，deleteTmpFile 方法实现文件的删除，程序代码如下：

```objc
-(void)deleteTmpFile{
    NSURL *url = [NSURL fileURLWithPath:self.tmpVideoPath];
    NSFileManager *fm = [NSFileManager defaultManager];
    BOOL exist = [fm fileExistsAtPath:url.path];   //获取指定文件是否存在
    NSError *err;
    if (exist) {
        [fm removeItemAtURL:url error:&err];       //删除指定文件
        if (err) {
            NSLog(@"file remove error, %@", err.localizedDescription );//输出
        }
    } else {
        NSLog(@"no file by that name");                              //输出
    }
}
```

selectPlay:方法实现播放用户修剪（截取）后的视频。程序代码如下：

```objc
- (IBAction)selectPlay:(id)sender {
    [self deleteTmpFile];
    NSURL *videoFileUrl = [NSURL fileURLWithPath:self.originalVideoPath];
                                                               //获取视频
    AVAsset *anAsset = [[AVURLAsset alloc] initWithURL:videoFileUrl options:nil];
    NSArray *compatiblePresets = [AVAssetExportSession exportPresets
CompatibleWithAsset:anAsset];
    if ([compatiblePresets containsObject:AVAssetExportPresetMediumQuality]) {
        self.exportSession = [[AVAssetExportSession alloc]initWithAsset:
    anAsset presetName:
```

```objc
AVAssetExportPresetPassthrough];                          //创建导出会话
    NSURL *furl = [NSURL fileURLWithPath:self.tmpVideoPath];
    self.exportSession.outputURL = furl;                  //新文件路径
    self.exportSession.outputFileType = AVFileTypeQuickTimeMovie;
                                                          //新文件类型
    CMTime start = CMTimeMakeWithSeconds(self.startTime, anAsset.duration.timescale);
    CMTime duration = CMTimeMakeWithSeconds(self.stopTime-self.startTime, anAsset.duration.timescale);
    CMTimeRange range = CMTimeRangeMake(start, duration); //截取的时间
    self.exportSession.timeRange = range;
    //开始真正工作
    [self.exportSession exportAsynchronouslyWithCompletionHandler:^{
        switch ([self.exportSession status]) {
            //失败
            case AVAssetExportSessionStatusFailed:
                NSLog(@"Export failed: %@", [[self.exportSession error] localizedDescription]);
                break;
            //取消
            case AVAssetExportSessionStatusCancelled:
                NSLog(@"Export canceled");
                break;
            //成功
            default:
                dispatch_async(dispatch_get_main_queue(), ^{
                    [self playMovie:self.tmpVideoPath];
                });
                break;
        }
    }];
}
```

【代码解析】

本实例的关键功能是视频的修剪功能。下面将详细讲解这个知识点。

视频的修剪功能在本实例中分为了 4 个步骤。

1. 获取视频路径

```objc
NSURL *videoFileUrl = [NSURL fileURLWithPath:self.originalVideoPath];
                                                          //获取视频的地址
AVAsset *anAsset = [[AVURLAsset alloc] initWithURL:videoFileUrl options:nil];
```

2. 创建导出会话

```objc
self.exportSession = [[AVAssetExportSession alloc]initWithAsset:anAsset presetName:
AVAssetExportPresetPassthrough];
```

3. 设置新视频的配置数据、文件路径等

```objc
NSURL *furl = [NSURL fileURLWithPath:self.tmpVideoPath];
self.exportSession.outputURL = furl;                      //新文件路径
self.exportSession.outputFileType = AVFileTypeQuickTimeMovie;
                                                          //新文件类型
CMTime start = CMTimeMakeWithSeconds(self.startTime, anAsset.duration.timescale);
```

```
CMTime duration = CMTimeMakeWithSeconds(self.stopTime-self.startTime,
anAsset.duration.timescale);
CMTimeRange range = CMTimeRangeMake(start, duration);    //截取的时间
self.exportSession.timeRange = range;
```

4. 实现修剪工作

```
[self.exportSession exportAsynchronouslyWithCompletionHandler:^{
    switch ([self.exportSession status]) {
        case AVAssetExportSessionStatusFailed:
            NSLog(@"Export failed: %@", [[self.exportSession error]
            localizedDescription]);
            break;
        case AVAssetExportSessionStatusCancelled:
            NSLog(@"Export canceled");
            break;
        default:
            dispatch_async(dispatch_get_main_queue(), ^{
            [self playMovie:self.tmpVideoPath];
            });
            break;
    }
}];
```

第 3 章 开 关 类

对于 iPhone 或者 iPad 而言,用户除了单击操作外,还有拖动和单击并存的操作。对于这种操作,常用于开关控件中。本章将主要讲解开关控件的有关实例。

实例 17 自定义开关的外观

【实例描述】

本实例实现的功能是显示一个外观发生改变的开关。用户可以通过拖动或者单击的方式实现开关状态的切换,并显示不同效果。其中,使用标签显示开关当前状态信息。运行效果如图 3.1 所示。

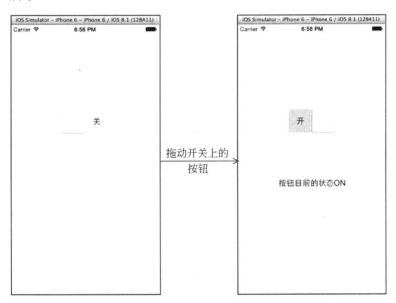

图 3.1 实例 17 运行效果

【实现过程】

(1)创建一个项目,命名为"自定义开关控制器的外观"。
(2)创建一个基于 UIControl 类的 Switch 类。
(3)打开 Switch 文件,编写代码。实现对象、实例变量、属性以及方法的声明。程序代码如下:

```
#import <UIKit/UIKit.h>
@interface Switch : UIControl{
```

```objc
    //对象
    UILabel *label1;
    UILabel *label2;
    UIView *background;
    UIView *knob;
    //实例变量
    double startTime;
    BOOL isAnimating;
}
//属性
@property (nonatomic, assign) BOOL on;
@property (nonatomic, strong) UIColor *inactiveColor;
@property (nonatomic, strong) UIColor *activeColor;
……
@property (nonatomic,strong) UILabel *label1;
@property (nonatomic,strong) UILabel *label2;
//方法
- (void)setOn:(BOOL)on animated:(BOOL)animated;
- (BOOL)isOn;
- (void)showOn:(BOOL)animated;
- (void)showOff:(BOOL)animated;
- (void)setup;                                    //创建开关
@end
```

（4）打开 Switch.m 文件，编写代码。使用的方法如表 3-1 所示。

表 3-1 Switch.m文件中方法总结

方　　法	功　　能
init	初始化
initWithCoder:	初始化实例化对象
initWithFrame:	自定义开关的初始化
setup	创建开关
beginTrackingWithTouch: withEvent:	开始触摸
continueTrackingWithTouch: withEvent:	继续触摸
endTrackingWithTouch: withEvent:	结束触摸
cancelTrackingWithEvent:	取消触摸
setInactiveColor:	设置不活动的颜色
setOnColor:	设置开关打开的颜色
setBorderColor:	设置边框的颜色
setKnobColor:	设置开关上按钮的颜色
setShadowColor:	设置阴影的颜色
setOn:	设置开关是开还是关
setOn:animated:	设置开关的状态，并设置过渡动画
isOn	获取开关是否为开或关的状态
showOn:	显示开关的开位置，并带有动画
showOff:	显示开关的关闭位置，并带有动画

在本文件中代码分为创建开关控件和实现开关控件中的开、关功能两个部分。这里需要讲解几个重要的方法（其他方法请读者参考源代码）。其中创建开关控件由 init、initWithCoder:、initWithFrame:、setup、setInactiveColor:、setOnColor:、setBorderColor:、setKnobColor:、setShadowColor:、setOn:、setOn:animated:、showOn:和 showOff:方法共同

实现。其中 initWithFrame:方法对自定义开关进行初始化。程序代码如下：

```
- (id)initWithFrame:(CGRect)frame
{
   CGRect initialFrame;
   if (CGRectIsEmpty(frame)) {
       initialFrame = CGRectMake(0, 0, 50, 30);         //设置框架
   }
   else {
       initialFrame = frame;                             //设置框架
   }
   self = [super initWithFrame:initialFrame];
   if (self) {
      [self setup];
   }
   return self;
}
```

setup 方法实现自定义开关控件的创建。程序代码如下：

```
- (void)setup {
   self.on = NO;
   self.inactiveColor = [UIColor clearColor];
   self.activeColor = [UIColor colorWithRed:0.89f green:0.89f blue:0.89f
   alpha:1.00f];
   self.onColor = [UIColor colorWithRed:1.0f green:0.85f blue:0.39f alpha:
   1.00f];
   self.borderColor = [UIColor colorWithRed:0.89f green:0.89f blue:0.91f
   alpha:1.00f];                                        //设置边框颜色
   self.knobColor = [UIColor whiteColor];
   self.shadowColor = [UIColor grayColor];              //设置阴影的颜色
   //背景
   background = [[UIView alloc] initWithFrame:CGRectMake(0, 0, self.frame.
   size.width, self.frame.size.height)];
   background.backgroundColor = [UIColor clearColor];   //设置背景颜色
   background.layer.cornerRadius = self.frame.size.height * 0.5;
   background.layer.borderColor = self.borderColor.CGColor;//设置边框颜色
   background.layer.borderWidth = 1.0;                  //设置边框宽度
   background.userInteractionEnabled = NO;
   [self addSubview:background];
   //标签
   label1 = [[UILabel alloc] initWithFrame:CGRectMake(0, 0, self.frame.
   size.width - 
self.frame.size.height, self.frame.size.height)];
   label1.alpha = 1;                                    //设置透明度
   label1.textAlignment =NSTextAlignmentCenter;
   [self addSubview:label1];
   label2= [[UILabel alloc] initWithFrame:CGRectMake(self.frame.size.
height, 0, self.frame.size.width - self.frame.size.height, self.frame.
size.height)];
   label2.alpha = 1.0;
   label2.textAlignment =NSTextAlignmentCenter;         //设置对齐方式
   [self addSubview:label2];
   //开关上的按钮
   knob = [[UIView alloc] initWithFrame:CGRectMake(1, 1, self.frame.
   size.height - 2,
self.frame.size.height - 2)];
   knob.backgroundColor = self.knobColor;               //设置背景颜色
   knob.layer.cornerRadius = (self.frame.size.height * 0.5) - 1;
```

```
    knob.layer.shadowColor = self.shadowColor.CGColor;   //设置阴影的颜色
    knob.layer.shadowRadius = 2.0;
    knob.layer.shadowOpacity = 0.5;
    knob.layer.shadowOffset = CGSizeMake(0, 3);          //设置阴影的偏移量
    knob.layer.masksToBounds = NO;
    knob.userInteractionEnabled = NO;
    [self addSubview:knob];                              //添加视图对象
    background.layer.cornerRadius = 2;
    knob.layer.cornerRadius = 2;
    isAnimating = NO;
}
```

setOn:animated:方法对开关的动画进行设置。程序代码如下：

```
- (void)setOn:(BOOL)isOn animated:(BOOL)animated {
    on = isOn;
    //判断开关是否打开
    if (isOn) {
        [self showOn:animated];
    }
    else {
        [self showOff:animated];
    }
}
```

showOn:方法对开关的打开位置进行显示，并带有动画。程序代码如下：

```
- (void)showOn:(BOOL)animated {
    CGFloat normalKnobWidth = self.bounds.size.height - 2;
    CGFloat activeKnobWidth = normalKnobWidth + 5;
    //判断是否需要动画
    if (animated) {
        isAnimating = YES;                               //有动画效果
        //动画实现
        [UIView animateWithDuration:0.3 delay:0.0 options:UIViewAnimation
OptionCurveEaseOut|UIViewAnimationOptionBeginFromCurrentState animations:^{
            if (self.tracking)
                knob.frame = CGRectMake(self.bounds.size.width - (activeKnob
Width + 1), knob.frame.origin.y, activeKnobWidth, knob.frame.size.height);
                                                         //设置框架
            else
                knob.frame = CGRectMake(self.bounds.size.width - (normalKnob
Width + 1), knob.frame.origin.y, normalKnobWidth, knob.frame.size.height);
            background.backgroundColor = self.onColor;   //设置背景颜色
            background.layer.borderColor = self.onColor.CGColor;
        } completion:^(BOOL finished) {
            isAnimating = NO;                            //没有动画效果
        }];
    }
    else {
        if (self.tracking)
            knob.frame = CGRectMake(self.bounds.size.width - (activeKnob
Width + 1), knob.frame.origin.y, activeKnobWidth, knob.frame.size.height);
                                                         //设置框架
        else
            knob.frame = CGRectMake(self.bounds.size.width - (normalKnob
Width + 1), knob.frame.origin.y, normalKnobWidth, knob.frame.size.height);
        background.backgroundColor = self.onColor;       //设置背景颜色
        background.layer.borderColor = self.onColor.CGColor;
```

showOff:方法对开关的关闭位置进行显示,并带有动画。程序代码如下:

```
- (void)showOff:(BOOL)animated {
    CGFloat normalKnobWidth = self.bounds.size.height - 2;
    CGFloat activeKnobWidth = normalKnobWidth + 5;
    //判断是否需要动画
    if (animated) {
        isAnimating = YES;
        //动画的实现
        [UIView animateWithDuration:0.3 delay:0.0 options:
UIViewAnimationOptionCurveEaseOut|UIViewAnimationOptionBeginFromCurrent
State animations:^{
            if (self.tracking) {
                knob.frame = CGRectMake(1, knob.frame.origin.y, activeKnobWidth,
knob.frame.size.height);                                      //设置框架
                background.backgroundColor = self.activeColor;//设置背景颜色
            }
            else {
                knob.frame = CGRectMake(1, knob.frame.origin.y, normalKnobWidth,
knob.frame.size.height);
                background.backgroundColor = self.inactiveColor;
            }
            background.layer.borderColor = self.borderColor.CGColor;
                                                            //设置边框颜色
        } completion:^(BOOL finished) {
            isAnimating = NO;
        }];
    }
    else {
        if (self.tracking) {
            knob.frame = CGRectMake(1, knob.frame.origin.y, activeKnobWidth,
knob.frame.size.height);
            background.backgroundColor = self.activeColor;   //设置背景颜色
        }
        else {
            knob.frame = CGRectMake(1, knob.frame.origin.y, normalKnobWidth,
knob.frame.size.height);                                     //设置框架
            background.backgroundColor = self.inactiveColor;
        }
        background.layer.borderColor = self.borderColor.CGColor; //设置边框颜色
    }
}
```

这时一个开关控件就创建好了,但是此时的开关控件是无法实现开和关功能的,需要使用 beginTrackingWithTouch:withEvent: 、 continueTrackingWithTouch:withEvent: 、 endTrackingWithTouch:withEvent: 和 cancelTrackingWithEvent: 方法实现。其中,beginTrackingWithTouch:withEvent:方法实现触摸开始的功能。程序代码如下:

```
- (BOOL)beginTrackingWithTouch:(UITouch *)touch withEvent:(UIEvent *)event
{
    [super beginTrackingWithTouch:touch withEvent:event];
    startTime = [[NSDate date] timeIntervalSince1970];     //实例化日期对象
    CGFloat activeKnobWidth = self.bounds.size.height - 2 + 5;
    isAnimating = YES;                                      //具有动画效果
    //动画实现
```

```objc
    [UIView animateWithDuration:0.3 delay:0.0 options:UIViewAnimation
OptionCurveEaseOut|UIViewAnimationOptionBeginFromCurrentState animations:^{
    if (self.on) {
        knob.frame = CGRectMake(self.bounds.size.width - (activeKnobWidth +
1), knob.frame.origin.y, activeKnobWidth, knob.frame.size.height);
                                                                //设置框架
        background.backgroundColor = self.onColor;
    }
    else {
        knob.frame = CGRectMake(knob.frame.origin.x, knob.frame.origin.
y, activeKnobWidth, knob.frame.size.height);                    //设置框架
        background.backgroundColor = self.activeColor;
    }
    } completion:^(BOOL finished) {
        isAnimating = NO;
    }];
    return YES;
}
```

ontinueTrackingWithTouch:withEvent 方法实现手指在屏幕上移动的动作。程序代码如下：

```objc
- (BOOL)continueTrackingWithTouch:(UITouch *)touch withEvent:(UIEvent *)event {
    [super continueTrackingWithTouch:touch withEvent:event];
    //获取触摸的位置
    CGPoint lastPoint = [touch locationInView:self];
    if (lastPoint.x > self.bounds.size.width * 0.5)
        [self showOn:YES];
    else
        [self showOff:YES];
    return YES;
}
```

endTrackingWithTouch:withEvent:方法实现触摸结束。程序代码如下：

```objc
- (void)endTrackingWithTouch:(UITouch *)touch withEvent:(UIEvent *)event {
    [super endTrackingWithTouch:touch withEvent:event];
    double endTime = [[NSDate date] timeIntervalSince1970];
    double difference = endTime - startTime;
    BOOL previousValue = self.on;
    //判断用户是否轻拍开关或者保持轻拍很久
    if (difference <= 0.2) {
        CGFloat normalKnobWidth = self.bounds.size.height - 2;
        knob.frame = CGRectMake(knob.frame.origin.x, knob.frame.origin.y,
normalKnobWidth, knob.frame.size.height);
        [self setOn:!self.on animated:YES];
    }
    else {
        CGPoint lastPoint = [touch locationInView:self];      //获取触摸位置
        if (lastPoint.x > self.bounds.size.width * 0.5)
            [self setOn:YES animated:YES];
        else
            [self setOn:NO animated:YES];
    }
    if (previousValue != self.on)
        [self sendActionsForControlEvents:UIControlEventValueChanged];
}
```

cancelTrackingWithEvent:方法，实现取消触摸的功能。程序代码如下：

```
- (void)cancelTrackingWithEvent:(UIEvent *)event {
    [super cancelTrackingWithEvent:event];
    //判断开关是否打开
    if (self.on)
        [self showOn:YES];
    else
        [self showOff:YES];
}
```

(5)打开 ViewController.h 文件,编写代码。实现头文件、插座变量以及对象的声明。程序代码如下:

```
#import <UIKit/UIKit.h>
#import "Switch.h"                                       //头文件
@interface ViewController : UIViewController{
    IBOutlet UILabel *lab;                               //插座变量
    Switch *mySwitch;                                    //对象
}
@end
```

(6)打开 Main.storyboard 文件。在 View Controller 视图控制器的设计界面中放入一个标签控件。双击,将此标签的文本内容删除,将插座变量 lab 与此标签控件进行关联。

(7)打开 ViewController.m 文件,编写代码。实现创建自定义选择器对象并放入到当前的视图中,以及实现显示开关的状态功能。程序代码如下:

```
- (void)viewDidLoad
{
    mySwitch = [[Switch alloc] initWithFrame:CGRectMake(0, 0, 100, 50)];
    mySwitch.center = CGPointMake(self.view.bounds.size.width * 0.5,
    self.view.bounds.size.height * 0.5 - 80);            //设置中心点
    [mySwitch addTarget:self action:@selector(switchChanged:) forControlEvents:
UIControlEventValueChanged];                             //添加动作
    [self.view addSubview:mySwitch];
    mySwitch.label1.text=@"开";                          //设置文本内容
    mySwitch.label2.text=@"关";
    [super viewDidLoad];
    // Do any additional setup after loading the view, typically from a nib.
}
- (void)switchChanged:(Switch *)sender {
    NSString *a=[NSString stringWithFormat:@"按钮目前的状态%@",sender.on ?
@"ON" : @"OFF"];
    lab.text=a;                                          //设置文本内容
}
```

【代码解析】

由于本实例中的代码和方法非常多,为了方便读者的阅读,笔者绘制了一些执行流程图,如图 3.2 和图 3.3 所示。其中,单击运行按钮后一直到在模拟器上显示自定义开关,使用了 viewDidLoad、initWithFrame:、setup、setOn:、setOn:animated:、showOff:、setInactiveColor:、setOnColor:、setBorderColor:、setKnobColor:和 setShadowColor:方法共同实现。以下将实现,在界面上显示一个状态为关的按钮。它们的执行流程如图 3.2 所示。

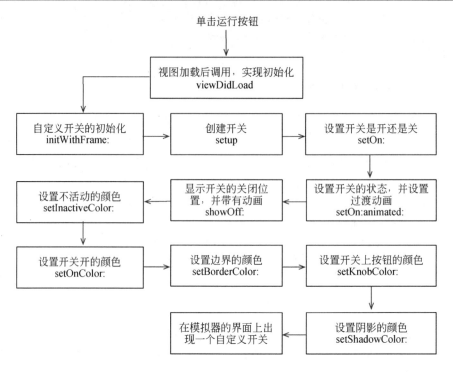

图 3.2 实例 17 程序执行流程 1

如果想要实现轻拍或者拖动开关上的按钮，并将当前的开关状态显示在界面上，需要使用 beginTrackingWithTouch:withEvent:、continueTrackingWithTouch:withEvent:、endTrackingWithTouch:withEvent:、showOn:、showOff:、setOn:animated:和 switchChanged:方法共同实现。以下将实现，通过拖动将按钮的当前状态关改变为开。它们的执行流程如图 3.3 所示。

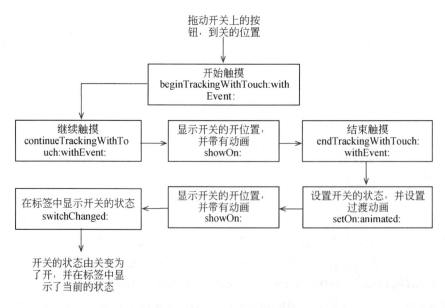

图 3.3 实例 17 程序执行流程 2

实例 18　实现滑块窗口滑动切换的效果

【实例描述】

本实例的功能是使用自定义的开关实现一个通过滑块实现切换效果的功能。运行效果如图 3.4 所示。

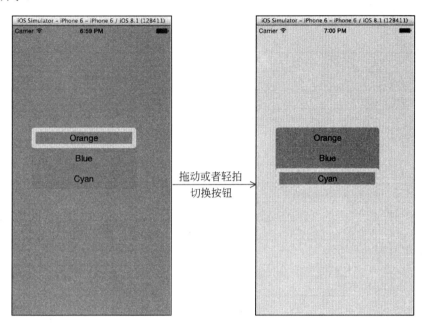

图 3.4　实例 18 运行效果

【实现过程】

在用户界面，首先显示一个自定义的开关，当用户滑动自定义开关上的滑块到指定的某一项时，实现背景颜色的切换。具体的实现步骤如下。

（1）创建一个项目，命名为"实现 UISwitch 窗口滑动切换效果"。

（2）创建一个基于 UIView 类的 Switch 类。

（3）打开 Switch.h 文件，编写代码。实现协议、属性和方法的声明。程序代码如下：

```
#import <UIKit/UIKit.h>
@class Switch;
//协议
@protocol SwitchDelegate <NSObject>
-(void)slideView:(Switch *)slideswitch switchChangedAtIndex:(NSUInteger)index;
@end
@interface Switch : UIView
//属性
@property(nonatomic,retain)UILabel *labelOne,*labelTwo,*labelThree;
@property(nonatomic,retain)UIButton *toggleButton;
@property(nonatomic,assign)id<SwitchDelegate> delegate;
//方法
- (void)setFrameVertical:(CGRect)frame withCornerRadius:(CGFloat)cornerRadius;
```

```
- (void)setText:(NSString *)text forTextIndex:(NSInteger *)number;
                                                        //设置字体
- (void)wasDraggedVertical:(UIButton *)button withEvent:(UIEvent *)event;
- (void)finishedDraggingVertical:(UIButton *)button withEvent:(UIEvent *)event;
- (void)setFrameBackgroundColor:(UIColor *)color;
- (void)setSwitchFrameColor:(UIColor *)color;
- (void)setTextColor:(UIColor *)color;                  //设置文本颜色
- (void)setSwitchBorderWidth:(CGFloat)width;
@end
```

（4）打开 Switch.m 文件，编写代码。使用的方法如表 3-2 所示。

表 3-2 Switch.m文件中方法总结

方 法	功 能
initWithFrame:	开关的初始化
hitTest:withEvent:	实现触碰测试
setFrameVertical: withCornerRadius:	设置垂直的框架
setSwitchBorderWidth:	设置框架的边框宽度
setFrameBackgroundColor:	设置框架的背景颜色
setSwitchFrameColor:	设置开关框架的颜色
didTapLabelOneWithGesture:	实现第一个标签的轻拍
didTapLabelTwoWithGesture:	实现第二个标签的轻拍
didTapLabelThreeWithGesture:	实现第三个标签的轻拍
finishedDraggingVertical:	按钮完成拖曳
wasDraggedVertical: withEvent:	按钮的拖曳
setText: forTextIndex:	设置文本

在本文件中代码分为了对开关的设置以及实现滑动切换的功能两个部分。这里需要讲解几个重要的方法（其他方法请读者参考源代码）。其中创建开关控件由 initWithFrame:、setFrameVertical: withCornerRadius:、setSwitchBorderWidth:、setFrameBackgroundColor:、setSwitchFrameColor: 和 setText: forTextIndex: 方法共同实现。其中 setFrameVertical:withCornerRadius: 方法，实现使用圆角半径对框架进行初始化。在此方法中代码分为三个部分：创建标签、创建手势识别器并添加到标签中、创建切换按钮。其中创建标签的程序代码如下：

```
    long double height;
    long double f=(long double)frame.size.height;
    height=f/3;
    //创建 3 个标签
    labelOne = [[UILabel alloc] initWithFrame:CGRectMake(frame.origin.x,
frame.origin.y, frame.size.width, height)];
    labelOne.textAlignment=NSTextAlignmentCenter;           //标签的对齐方式
    [labelOne.layer setBorderColor:[[UIColor darkGrayColor] CGColor]];
                                                            //设置边框颜色
    CAShapeLayer *maskLayerLeft = [CAShapeLayer layer];
    UIBezierPath *maskPathLeft=[UIBezierPath bezierPathWithRoundedRect:
labelOne.bounds byRoundingCorners:(UIRectCornerTopLeft | UIRectCornerTopRight)
cornerRadii:CGSizeMake(cornerRadius,cornerRadius)];
    maskLayerLeft.path = maskPathLeft.CGPath;               //设置路径
    labelOne.layer.mask = maskLayerLeft;
    [self addSubview:labelOne];                             //添加视图对象
```

```
    labelTwo = [[UILabel alloc] initWithFrame:CGRectMake(frame.origin.x,
frame.origin.y+height, frame.size.width, height)];
    labelTwo.textAlignment=NSTextAlignmentCenter;              //对齐方式
    [self addSubview:labelTwo];
    labelThree = [[UILabel alloc] initWithFrame:CGRectMake(labelTwo.frame.
origin.x, labelTwo.frame.origin.y+height, frame.size.width, height)];
    labelThree.textAlignment=NSTextAlignmentCenter;
    [labelThree.layer setBorderColor:[[UIColor darkGrayColor] CGColor]];
                                                       //设置边框的颜色
    CAShapeLayer *maskLayerRight = [CAShapeLayer layer];
    UIBezierPath *maskPathRight=[UIBezierPath bezierPathWithRoundedRect:
labelThree.bounds byRoundingCorners:(UIRectCornerBottomRight | UIRectCorner
BottomLeft) cornerRadii:CGSizeMake(cornerRadius,cornerRadius)];
    maskLayerRight.path = maskPathRight.CGPath;            //设置路径
    labelThree.layer.mask = maskLayerRight;
    [self addSubview:labelThree];                          //添加视图对象
    labelOne.userInteractionEnabled=YES;
    labelTwo.userInteractionEnabled=YES;
    labelThree.userInteractionEnabled=YES;
```

创建手势识别器,并将创建的手势识别器添加到创建的标签中。程序代码如下:

```
    UITapGestureRecognizer *tapGestureLabelone =[[UITapGestureRecognizer
alloc] initWithTarget:
    self action:@selector(didTapLabelOneWithGesture:)]; //创建轻拍手势识别器
    [labelOne addGestureRecognizer:tapGestureLabelone];
                             //将轻拍手势识别器添加到 labelOne 标签中
    UITapGestureRecognizer *tapGestureLabelTwo =[[UITapGestureRecognizer
alloc] initWithTarget:self action:@selector(didTapLabelTwoWithGesture:)];
    [labelTwo addGestureRecognizer:tapGestureLabelTwo]; //添加手势识别器对象
    UITapGestureRecognizer *tapGestureLabelThree =[[UITapGestureRecognizer
alloc] initWithTarget:self action:@selector(didTapLabelThreeWithGesture:)];
    [labelThree addGestureRecognizer:tapGestureLabelThree];
```

创建切换按钮的程序代码如下:

```
    toggleButton = [UIButton buttonWithType:UIButtonTypeCustom];
    [toggleButton setTitle:@"" forState:UIControlStateNormal]; //设置标题
    [toggleButton addTarget:self action:@selector(wasDraggedVertical:withEvent:)
        forControlEvents:UIControlEventTouchDragInside];      //添加动作
    [toggleButton addTarget:self action:@selector(finishedDraggingVertical:withEvent:)
        forControlEvents:UIControlEventTouchUpInside];        //添加动作
    toggleButton.frame = CGRectMake(frame.origin.x, frame.origin.y, frame.
        size.width, height);                                  //设置框架
    toggleButton.backgroundColor=[UIColor colorWithRed:0.0 green:0.0 blue:0.0
    alpha:0.0];                                               //设置背景颜色
    [toggleButton.layer setBorderWidth:4.0];
    [toggleButton.layer setBorderColor:[[UIColor lightGrayColor] CGColor]];
    toggleButton.layer.cornerRadius=cornerRadius;
    [self addSubview:toggleButton];                           //添加视图对象
```

setText:forTextIndex:方法,实现对文本的设置。程序代码如下:

```
- (void)setText:(NSString *)text forTextIndex:(NSInteger *)number
{
    int labelnumber=(int)number;
    //判断 labelnumber 是否为 1
    if(labelnumber==1)
    {
```

```
        labelOne.text=text;
        [self.delegate slideView:self switchChangedAtIndex:0];
    }
    //判断 labelnumber 是否为 2
    if(labelnumber==2)
    {
        labelTwo.text=text;
    }
    //判断 labelnumber 是否为 3
    if(labelnumber==3)
    {
        labelThree.text=text;
    }
}
```

要想实现滑动切换效果，需要使用 hitTest:withEvent:、didTapLabelOneWithGesture:、didTapLabelTwoWithGesture:、didTapLabelThreeWithGesture:和 finishedDraggingVertical:方法共同实现。其中，hitTest:withEvent:方法实现触碰测试，它可以找到触碰的第一响应者。程序代码如下：

```
-(UIView *)hitTest:(CGPoint)point withEvent:(UIEvent *)event
{
    NSEnumerator *reverseE = [self.subviews reverseObjectEnumerator];
    UIView *iSubView;
    while ((iSubView = [reverseE nextObject])) {
        UIView *viewWasHit = [iSubView hitTest:[self convertPoint:point
toView:iSubView] withEvent:event];                              //实例化对象
        if(viewWasHit) {
            return viewWasHit;                                  //返回
        }
    }
    return [super hitTest:point withEvent:event];
}
```

didTapLabelOneWithGesture:方法，对 LabelOne 的标签实现轻拍。程序代码如下：

```
-(void)didTapLabelOneWithGesture:(UITapGestureRecognizer *)tapGesture {
    [UIView animateWithDuration:0.6
                    animations:^{
                        [toggleButton setFrame:labelOne.frame];    //设置框架
                    }
                    completion:^(BOOL finished) {

                    }
    ];
    [self.delegate slideView:self switchChangedAtIndex:0];
}
```

finishedDraggingVertical:方法实现按钮完成拖动，并在所有选项和切换按钮之间找出最短距离的选项，将按钮以动画的形式移动到此选项上。程序代码如下：

```
- (void)finishedDraggingVertical:(UIButton *)button withEvent:(UIEvent *)event
{
    float diffone,difftwo,diffthree;
    //获取切换按钮和选项之间的距离
    diffone=fabsf(button.frame.origin.y-labelOne.frame.origin.y);
    difftwo=fabsf(labelTwo.frame.origin.y-button.frame.origin.y);
    diffthree=fabsf(labelThree.frame.origin.y-button.frame.origin.y);
```

```objectivec
//判断距离，并移动切换按钮到距离最近的选项上
if (diffone==difftwo) {
    [UIView animateWithDuration:0.6
                  animations:^{
                      [button setFrame:labelOne.frame];     //设置框架
                  }
                  completion:^(BOOL finished) {
                  }
     ];
}
if (difftwo==diffthree) {
    [UIView animateWithDuration:0.6
                  animations:^{
                      [button setFrame:labelTwo.frame];     //设置框架
                  }
                  completion:^(BOOL finished) {
                  }
     ];
}
if(diffone<difftwo && diffone<diffthree)
{
    [UIView animateWithDuration:0.6
                  animations:^{
                      [button setFrame:labelOne.frame];     //设置框架
                  }
                  completion:^(BOOL finished) {
                  }
     ];
}
if(difftwo<diffone && difftwo<diffthree)
{
    [UIView animateWithDuration:0.6
                  animations:^{
                      [button setFrame:labelTwo.frame];     //设置框架
                  }
                  completion:^(BOOL finished) {
                  }
     ];
}
if(diffthree<diffone && diffthree<difftwo)
{
    [UIView animateWithDuration:0.6
                  animations:^{
                      [button setFrame:labelThree.frame];   //设置框架
                  }
                  completion:^(BOOL finished) {
                  }
     ];
}
}
```

wasDraggedVertical:方法，实现对按钮的拖动。程序代码如下：

```objectivec
- (void)wasDraggedVertical:(UIButton *)button withEvent:(UIEvent *)event
{
    UITouch *touch = [[event touchesForView:button] anyObject];//实例化对象
    CGPoint previousLocation = [touch previousLocationInView:button];
                                                    //获取前一次触摸的位置
    CGPoint location = [touch locationInView:button];  //获取触摸的位置
    CGFloat delta_y = location.y - previousLocation.y;
```

```
        //设置中心位置
        button.center = CGPointMake(button.center.x ,
                            button.center.y +delta_y);
        //判断按钮的 y 值是否大于标签对象 labelThree 的 y 值
        if (button.frame.origin.y>labelThree.frame.origin.y) {
            [button setFrame:labelThree.frame];
        }
        //判断按钮的 y 值是否大于标签对象 labelOne 的 y 值
        if (button.frame.origin.y<labelOne.frame.origin.y) {
            [button setFrame:labelOne.frame];
        }
        //判断按钮的 y 值是否等于标签对象 labelOne 的 y 值
        if(button.frame.origin.y==labelOne.frame.origin.y)
        {
            [self.delegate slideView:self switchChangedAtIndex:0];
        }
        else
            //判断按钮的 y 值是否等于标签对象 labelTwo 的 y 值
            if (button.frame.origin.y==labelTwo.frame.origin.y)
            {
                [self.delegate slideView:self switchChangedAtIndex:1];
            }
            else
                //判断按钮的 y 值是否等于标签对象 labelThree 的 y 值
                if (button.frame.origin.y==labelThree.frame.origin.y)
                {
                    [self.delegate slideView:self switchChangedAtIndex:2];
                }
}
```

（5）打开 ViewController.h 文件，编写代码。实现头文件和属性等的声明。程序代码如下：

```
#import <UIKit/UIKit.h>
#import "Switch.h"
@interface ViewController : UIViewController<SwitchDelegate>
@property(nonatomic,retain) Switch *switchV;
@end
```

（6）打开 ViewController.m 文件，实现创建 Switch 类的对象并放入到当前的视图中，并实现当前视图的颜色的切换。程序代码如下：

```
- (void)viewDidLoad
{
    switchV=[[Switch alloc]init];
    switchV.delegate=self;
    [switchV setFrameVertical:(CGRectMake(40, 200, 210, 120)) withCornerRadius:5.0];
    [switchV setFrameBackgroundColor:[UIColor grayColor]];  //设置背景
    [switchV setSwitchFrameColor:[UIColor lightTextColor]];
    [switchV setSwitchBorderWidth:8.0];                     //设置边框的宽度
    //设置文本内容
    [switchV setText:@"Orange" forTextIndex:1];
    [switchV setText:@"Blue" forTextIndex:2];
    [switchV setText:@"Cyan" forTextIndex:3];
    [self.view addSubview:switchV];
    [super viewDidLoad];
    //Do any additional setup after loading the view, typically from a nib.
}
```

```
//实现切换
-(void)slideView:(Switch *)slideswitch switchChangedAtIndex:(NSUInteger)index
{
    //判断 index 是否等于 0
    if (index==0) {
        self.view.backgroundColor=[UIColor orangeColor];      //设置背景颜色
    }
    else
    //判断 index 是否等于 1
        if (index==1) {
            self.view.backgroundColor=[UIColor blueColor];
        }
        else
            //判断 index 是否等于 2
            if (index==2) {
                self.view.backgroundColor=[UIColor cyanColor];
            }
}
```

【代码解析】

本实例关键功能是颜色的相应切换。下面将详细讲解这个知识点。

当按钮到达不同的选项，就会出现不同的颜色切换。它的实现需要使用按钮的 y 值和每一个标签的 y 值进行判断，从而调用索引不同的 slideView: switchChangedAtIndex:方法。程序代码如下：

```
if(button.frame.origin.y==labelOne.frame.origin.y)
{
    [self.delegate slideView:self switchChangedAtIndex:0];
}
else
    if (button.frame.origin.y==labelTwo.frame.origin.y)
    {
        [self.delegate slideView:self switchChangedAtIndex:1];
    }
    else
        if (button.frame.origin.y==labelThree.frame.origin.y)
        {
            [self.delegate slideView:self switchChangedAtIndex:2];
        }
```

第 4 章 进度条类和指示器类效果

在 iPhone 或者 iPad 中，除了让用户实现操作的控件外，还有一些控件不需要用户进行操作。例如在打开网页时，网速比较慢，就会出现一个控件提示正在加载，让用户知道此网页正处于加载过程。在 iOS 中，这类控件一般有两种，一种是进度条，另一种是指示器。本章主要讲解这两类控件的相关实例。

实例 19 扁平带梯度效果的进度条

【实例描述】
本实例实现的功能是在界面显示两个扁平带梯度效果的进度条，其中一个具有加载动画。当用户拖动滑块控件上的滑块时，进度条中的进度就会加载到相应的位置。其中用于显示进度信息的是标签。运行效果如图 4.1 所示。

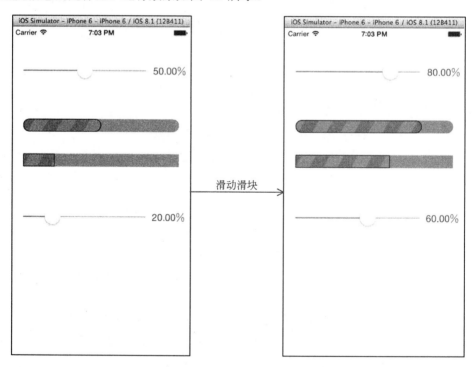

图 4.1 实例 19 运行效果

【实现过程】
（1）创建一个项目，命名为"扁平带梯度效果的进度条"。

（2）创建一个基于 UIView 类的 Progress 类。

（3）打开 Progress.h 文件，编写代码，实现属性的声明。程序代码如下：

```
#import <UIKit/UIKit.h>
@interface Progress:UIView
//属性的声明
@property (nonatomic) CGFloat progress;
@property (nonatomic, strong) UIColor *color;
……
@property (nonatomic) CGFloat progressToAnimateTo;
@end
```

（4）打开 Progress.m 文件，编写代码。实现绘制一个扁平带梯度效果的进度条，使用的方法如表 4-1 所示。

表 4-1 Progress.m文件中方法总结

方　　法	功　　能
initWithFrame:	扁平带梯度效果的进度条的初始化
initWithCoder:	初始化实例化对象
initialize	初始化
setAnimate:	进度条的进度加载动画
setProgress:	设置进度条的进度
incrementAnimatingProgress	增长动画
incrementOffset	增长偏量
drawRect:	扁平带梯度效果的进度条
drawProgressBackground:inRect:	绘制进度条的背景
drawProgress:withFrame:	绘制进度条的进度
drawStripes: inRect:	绘制条纹
animate	获取是否需要动画
setColor:	设置颜色
stripeWidth	获取条纹宽度
borderRadius	获取圆角

这里需要讲解几个重要的方法（其他方法请读者参考源代码）。添加的 initialize 方法实现对背景颜色的初始化。程序代码如下：

```
- (void)initialize {
    self.backgroundColor = [UIColor clearColor];
}
```

实现对扁平带梯度效果的进度条的绘制，需要绘制进度条的背景、绘制条纹（梯度效果）、绘制进度条的进度，以及绘制扁平带梯度效果的进度条。绘制条纹的背景，需要使用 borderRadius、setColor: 和 drawProgressBackground:inRect: 方法共同实现。其中，borderRadius 方法用来实现圆角效果，当绘制的进度条的背景需要圆角的效果时可使用此方法，如果绘制的进度条的背景不需要圆角效果，可以不使用此方法。程序代码如下：

```
- (NSNumber *)borderRadius {
  if (!_borderRadius) {
     return @(self.frame.size.height / 2.0);
```

```
        }
        return _borderRadius;                                    //返回边框半径
}
```

setColor:方法实现对颜色的设置。程序代码如下：

```
- (void)setColor:(UIColor *)color {
    _color = color;                                              //设置颜色
}
```

drawProgressBackground:inRect:方法对进度条的背景进行绘制。程序代码如下：

```
- (void)drawProgressBackground:(CGContextRef)context inRect:(CGRect)rect {
    CGContextSaveGState(context);
UIBezierPath *roundedRect = [UIBezierPath bezierPathWithRoundedRect:
    rect cornerRadius:
self.borderRadius.floatValue];                                   //根据矩形画带圆角的曲线
    CGContextSetFillColorWithColor(context, [UIColor colorWithRed:0.5f
    green:0.5f blue:0.8f alpha:1.00f].CGColor);                  //填充颜色
    [roundedRect fill];
}
```

绘制条纹（梯度效果）需要使用 drawStripes: inRect:和 stripeWidth 方法共同实现。
drawStripes: inRect:方法实现的功能是绘制条纹。程序代码如下：

```
-(void)drawStripes:(CGContextRef)context inRect:(CGRect)rect {
    CGContextSaveGState(context);                                //保存
    [[UIBezierPath bezierPathWithRoundedRect:rect cornerRadius:self.
    borderRadius.floatValue]
addClip];                                                        //根据矩形画带圆角的曲线
    CGContextSetFillColorWithColor(context, [[UIColor whiteColor]
colorWithAlphaComponent:0.2].CGColor);
    CGFloat xStart = self.offset, height = rect.size.height, width =
    self.stripeWidth;
    while (xStart < rect.size.width) {
        CGContextSaveGState(context);
        CGContextMoveToPoint(context, xStart, height);           //设置开始点
        CGContextAddLineToPoint(context, xStart + width * 0.25, 0);
                                                                 //设置结束点
        CGContextAddLineToPoint(context, xStart + width * 0.75, 0);
        CGContextAddLineToPoint(context, xStart + width * 0.50, height);
        CGContextClosePath(context);                             //关闭路径
        CGContextFillPath(context);
        CGContextRestoreGState(context);
        xStart += width;
    }
    CGContextRestoreGState(context);
}
```

stripeWidth 方法对条纹的宽度进行设置。程序代码如下：

```
- (CGFloat)stripeWidth {
    _stripeWidth = 50;
    return _stripeWidth;
}
```

绘制进度条的加载进度，需要 setProgress:和 drawProgress:withFrame:方法共同实现。

其中，setProgress:方法对进度条的进度进行设置。程序代码如下：

```
- (void)setProgress:(CGFloat)progress {
   self.progressToAnimateTo = progress;
   if (self.animationTimer) {
      [self.animationTimer invalidate];              //使时间定时器失效
   }
   self.animationTimer = [NSTimer scheduledTimerWithTimeInterval:0.0008
   target:self selector:
@selector(incrementAnimatingProgress) userInfo:nil repeats:YES];
                                                     //创建时间定时器
}
```

drawProgress:withFrame:方法对进度条的进度进行绘制。程序代码如下：

```
- (void)drawProgress:(CGContextRef)context withFrame:(CGRect)frame {
   CGRect rectToDrawIn = CGRectMake(0, 0, frame.size.width * self.progress,
   frame.size.height);
   CGRect insetRect = CGRectInset(rectToDrawIn, self.progress > 0.03 ? 0.5 :
   -0.5, 0.5);
   UIBezierPath *roundedRect = [UIBezierPath bezierPathWithRoundedRect:
   insetRect cornerRadius:
self.borderRadius.floatValue];                       //根据矩形画带圆角的曲线
   CGContextSetFillColorWithColor(context, self.color.CGColor);
   [roundedRect fill];
   if (self.progress != 1.0) {
      [self drawStripes:context inRect:insetRect];   //调用绘制的条纹
   }
   [roundedRect stroke];
}
```

进度条的背景、条纹和进度都绘制好后，就可以绘制扁平带梯度效果的进度条了。这时需要使用 drawRect:方法实现绘制扁平带梯度效果的进度条。程序代码如下：

```
- (void)drawRect:(CGRect)rect {
   CGContextRef context = UIGraphicsGetCurrentContext();
   [self drawProgressBackground:context inRect:rect];   //调用绘制的背景
   if (self.progress > 0) {
      [self drawProgress:context withFrame:rect];       //调用绘制的进度
   }
}
```

为了使进度条的加载更为形象。本实例为进度条的加载进度添加了动画效果。此动画效果可以使用 setAnimate:、incrementAnimatingProgress、incrementOffset 和 animate 方法共同实现。setAnimate:方法对动画进行设置。程序代码如下：

```
- (void)setAnimate:(NSNumber *)animate {
   _animate = animate;
   if ([animate boolValue]) {
      self.timer = [NSTimer scheduledTimerWithTimeInterval:0.02 target:
      self selector:@selector(incrementOffset) userInfo:nil repeats:YES];
                                                     //创建时间定时器
   } else if (self.timer) {
      [self.timer invalidate];
```

    }
}
```

incrementAnimatingProgress 方法实现增长动画效果。程序代码如下：

```
- (void)incrementAnimatingProgress {
  if (_progress >= self.progressToAnimateTo-0.01 && _progress <=
  self.progressToAnimateTo+0.01) {
      _progress = self.progressToAnimateTo;
      [self.animationTimer invalidate];              //让定时器失效
      [self setNeedsDisplay];
  } else {
      _progress = (_progress < self.progressToAnimateTo) ? _progress + 0.01 :
      _progress - 0.01;
      [self setNeedsDisplay];
  }
}
```

incrementOffset 方法对进度条的增长偏量进行设置。程序代码如下：

```
- (void)incrementOffset {
  if (self.offset >= 0) {
      self.offset = -self.stripeWidth;              //设置偏移量
  } else {
      self.offset += 1;
  }
  [self setNeedsDisplay];
}
```

animate 方法对动画的使用进行获取。程序代码如下：

```
- (NSNumber *)animate {
  if (_animate == nil) {
      return @YES;
  }
  return _animate;
}
```

（5）打开 ViewController.h 文件，编写代码。实现头文件、插座变量、对象以及动作的声明。程序代码如下：

```
#import <UIKit/UIKit.h>
#import "Progress.h"                                //头文件
@interface ViewController : UIViewController{
    //滑块的插座变量
    IBOutlet UISlider *slider1;
    IBOutlet UISlider *slider2;
    //标签的插座变量
    IBOutlet UILabel *label1;
    IBOutlet UILabel *label2;
    //对象
    Progress *progressview1;
    Progress *progressview2;
}
//动作
```
```

```
- (IBAction)change1:(id)sender;
- (IBAction)change2:(id)sender;
@end
```

（6）打开 Main.storyboard 文件，对 View Controller 视图控制器的设计界面进行设计，效果如图 4.2 所示。

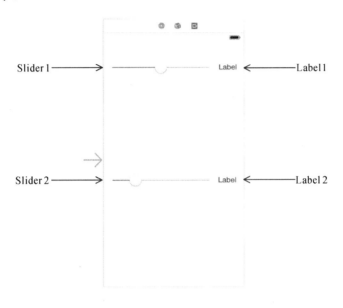

图 4.2　实例 19 View Controller 视图控制器的设计界面效果

需要添加的视图、控件以及对它们的设置如表 4-2 所示。

表 4-2　实例 19 视图、控件设置

视图、控件	属 性 设 置	其　　他
Slider1	Value：Current：0.5	与插座变量 slider1 关联 与动作 change1:关联
Slider2	Value：Current：0.2	与插座变量 slider2 关联 与动作 change2:关联
Label1	Color：紫色 Alignment：居中	与插座变量 label1 关联
Label2	Color：紫色 Alignment：居中	与插座变量 label2 关联

（7）打开 ViewController.m 文件，编写代码。此文件中，代码分为：创建扁平带梯度效果的进度条对象和滑动滑块使进度条的进度增长。在 viewDidLoad 方法中实现创建扁平带梯度效果的进度条对象。程序代码如下：

```
- (void)viewDidLoad
{
    label1.text=@"50.00%";
    label2.text=@"20.00%";
    //创建具有动画效果的进度条对象
    progressview1=[[Progress alloc] initWithFrame:CGRectMake(20, 160,
```

```
    self.view.frame.size.width-40, 22)];                              //实例化对象
    [self.view addSubview:progressview1];
    progressview1.color = [UIColor colorWithRed:0.0f green:0.5 blue:0.00f
    alpha:1.00f];                                                     //设置颜色
    progressview1.progress = 0.5;                                     //设置进度
    progressview1.animate = @YES;
    //创建没有动画的进度条对象
    progressview2=[[Progress alloc] initWithFrame:CGRectMake(20, 220,
    self.view.frame.size.width-40, 22)];
    [self.view addSubview:progressview2];                             //添加视图对象
    progressview2.color = [UIColor colorWithRed:0.5f green:0.5 blue:0.00f
    alpha:1.00f];
    progressview2.borderRadius=@0;
    progressview2.progress = 0.2;                                     //设置进度
    progressview2.animate = @NO;
    [super viewDidLoad];
    // Do any additional setup after loading the view, typically from a nib.
}
```

在 change1:和 change2:动作方法中实现滑动滑块使进度条的进度增长，并在标签中显示增长的内容。程序代码如下：

```
- (IBAction)change1:(id)sender {
    progressview1.progress=slider1.value;
    NSString *a=[NSString stringWithFormat:@"%0.2f%%",slider1.value*100];
    label1.text=a;                                                    //设置文本内容
}
- (IBAction)change2:(id)sender {
    progressview2.progress=slider2.value;
    NSString *a=[NSString stringWithFormat:@"%0.2f%%",slider2.value*100];
    label2.text=a;
}
```

【代码解析】

由于本实例中的代码和方法非常的多，为了方便读者的阅读，笔者绘制了一些执行流程图如图 4.3～4.5 所示。在本实例中创建了一个具有加载动画的进度条和一个没有加载动画的进度条，由于它们所执行的流程是一样的，为了方便，本书以第一个具有加载动画的进度条为例来让开发者看到它的执行流程。其中，扁平带梯度效果的进度条的显示，使用了 iewDidLoad、initWithFrame:、initialize、setColor:、setProgress:、setAnimate:、drawRect:、drawProgressBackground:inRect:和 borderRadius 方法共同实现。它们的执行流程如图 4.3 所示。

此时显示在模拟器上的进度条是没有梯度（条纹）效果的，梯度效果的显示是在进度条加载时出现的。如果实现进度条的加载，并且会伴有加载动画，需要使用 incrementAnimatingProgress、incrementOffset、stripeWidth、drawRect:、drawProgressBackground:inRect:、borderRadius、drawProgress:withFrame:和 drawStripes:inRect:方法共同实现。它们的执行流程如图 4.4 所示。

如果用户没有滑动界面的滑块，程序就会一直执行加载动画的方法，并且进度条的进度会到指定的位置。如果滑动了的界面上的滑块，那么此时它的执行流程和使用的方法也会发生改变。这时它需要的方法有：incrementAnimatingProgress、incrementOffset、drawRect:

drawProgressBackground:inRect:、borderRadius、drawProgress:withFrame:、drawStripes:inRect:、stripeWidth:、change1:和 setProgress:。它们的执行流程如图 4.5 所示。

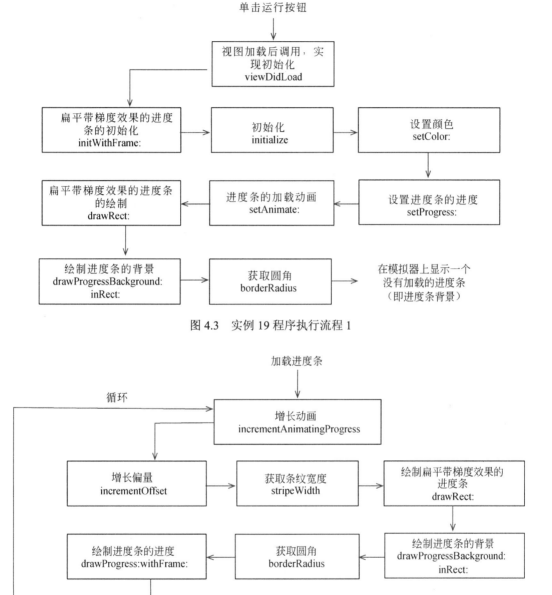

图 4.3 实例 19 程序执行流程 1

图 4.4 实例 19 程序执行流程 2

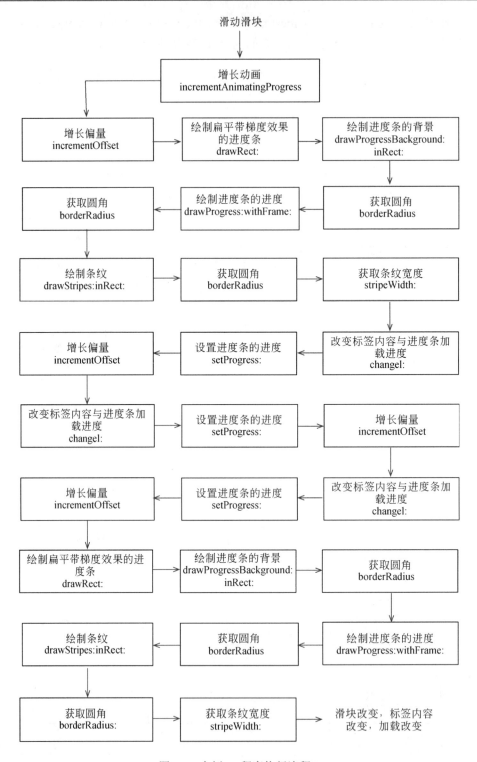

图 4.5　实例 19 程序执行流程 3

此时运行后的结果为，滑块控件中的滑块滑动到用户滑动的位置，标签中的内容都会改变。这时，进度条的加载进度只会实现固定长度的加载，而不会直接到达用户指定的位置。要想到达用户指定的位置，需要重复执行图 4.4 中的流程。

实例 20 扇形进度条

【实例描述】

本实例实现的功能是在界面出现一个加载的扇形进度条。运行效果如图 4.6 所示。

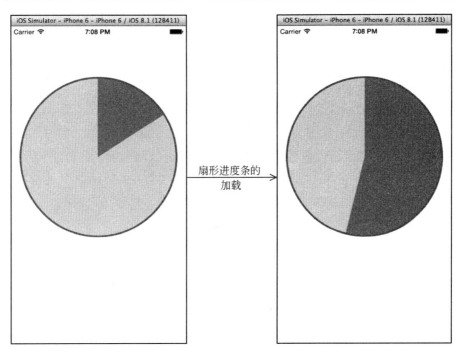

图 4.6 实例 20 运行效果

【实现过程】

（1）创建一个项目，命名为"扇形进度条"。
（2）创建一个基于 UIView 类的 RoundProgress 类。
（3）打开 RoundProgress.h 文件，编写代码。实现定义新的类型、宏定义、属性和方法的声明。程序代码如下：

```
#import <UIKit/UIKit.h>
//定义新的类型
typedef CGFloat (^ProgressBlock)();
//宏定义
#define UIColorMake(r, g, b, a) [UIColor colorWithRed:r / 255. green:g / 255. blue:b / 255. alpha:a]
@interface RoundProgress : UIView
//属性
@property(nonatomic) CGFloat frameWidth;
……
@property(nonatomic, copy) ProgressBlock block;
//方法
- (void)startWithBlock:(ProgressBlock)block;
@end
```

（4）打开 RoundProgress.m 文件，编写代码。实现绘制一个扇形进度条，使用的方法如表 4-3 所示。

表 4-3　RoundProgress.m文件中方法总结

方　　法	功　　能
initWithFrame:	扇形进度条的初始化
initWithCoder:	初始化实例化对象
setupParams	设置参数
startWithBlock:	进度条的进度加载
updateProgress	更新进度条的进度
drawRect:	绘制扇形进度条
drawFillPie: margin: color: percentage:	填充绘制扇形
drawFramePie:	绘制扇形的框架

在文件中，代码分为：初始化、绘制扇形进度条和实现进度加载。这里需要讲解几个重要的方法（其他方法请读者参考源代码）。扇形进度条的初始化需要 initWithFrame:、initWithCoder:和 setupParams 方法共同实现。其中，setupParams 方法对参数进行设置。程序代码如下：

```
- (void)setupParams {
    self.backgroundColor = [UIColor clearColor];
    self.frameWidth = 3;
    self.progressColor = UIColorMake(200, 100, 100, 1);     //进度条加载的颜色
    self.progressBackgroundColor = UIColorMake(0, 223, 244, 1);
                                                            //进度条背景颜色
    self.circleBackgroundColor = [UIColor redColor];        //圆环的背景颜色
}
```

在扇形的绘制中，分为绘制扇形的框架、填充绘制扇形以及绘制扇形进度条三个部分。其中，绘制扇形的框架需要使用 drawFramePie:方法实现。程序代码如下：

```
- (void)drawFramePie:(CGRect)rect {
    CGFloat radius = MIN(CGRectGetHeight(rect), CGRectGetWidth(rect)) * 0.5;
    CGFloat centerX = CGRectGetWidth(rect) * 0.5;
    CGFloat centerY = CGRectGetHeight(rect) * 0.5;
    [UIColorMake(155, 190, 225, 0.8) set];
    CGFloat fw = self.frameWidth + 1;
    //设置 frameRect 对象
    CGRect frameRect = CGRectMake(
                        centerX - radius + fw,
                        centerY - radius + fw,
                        (radius - fw) * 2,
                        (radius - fw) * 2);
    //根据矩形框的内切圆画曲线
    UIBezierPath *insideFrame = [UIBezierPath bezierPathWithOvalInRect:
    frameRect];
    insideFrame.lineWidth = 2;
    [insideFrame stroke];
}
```

drawFillPie:方法实现扇形的填充绘制。程序代码如下：

```
- (void)drawFillPie:(CGRect)rect margin:(CGFloat)margin color:(UIColor
*)color percentage:(CGFloat)percentage {
```

```
    CGFloat radius = MIN(CGRectGetHeight(rect), CGRectGetWidth(rect)) * 0.5 -
    margin;
    CGFloat centerX = CGRectGetWidth(rect) * 0.5;
    CGFloat centerY = CGRectGetHeight(rect) * 0.5;
    CGContextRef cgContext = UIGraphicsGetCurrentContext();    //获取当前图形
    CGContextSetFillColorWithColor(cgContext, [color CGColor]);
    CGContextMoveToPoint(cgContext, centerX, centerY);
    CGContextAddArc(cgContext, centerX, centerY, radius, (CGFloat) -M_PI_2,
    (CGFloat) (-M_PI_2 + M_PI * 2 * percentage), 0);            //绘制圆
    CGContextClosePath(cgContext);
    CGContextFillPath(cgContext);                               //填充路径
}
```

drawRect:方法实现绘制一个扇形进度条。程序代码如下：

```
- (void)drawRect:(CGRect)rect {
    [self drawFillPie:rect margin:0 color:self.circleBackgroundColor
    percentage:1];               //填充
    [self drawFramePie:rect];
    [self drawFillPie:rect margin:self.frameWidth color:self.
    progressBackgroundColor percentage:1];
    [self drawFillPie:rect margin:self.frameWidth color:self.progressColor
    percentage:self.progress];
}
```

此时的进度条是不会出现进度加载动画的，如果想要出现动画，需要使用 startWithBlock:方法实现。程序代码如下：

```
- (void)startWithBlock:(ProgressBlock)block {
    self.block = block;
    //创建时间定时器对象
    self.timer = [NSTimer scheduledTimerWithTimeInterval:1.0f / 30
                                        target:self
                                        selector:@selector(updateProgress)
                                        userInfo:nil
                                        repeats:YES];
}
```

updateProgress:方法实现进度的更新。程序代码如下：

```
- (void)updateProgress {
    self.progress = self.block();
    [self setNeedsDisplay];                                     //重新绘制
}
```

（5）打开 Main.storyboard 文件，在 View Controller 视图控制器的设计界面添加一个 View 空白视图，将 Class 设置为 RoundProgress。

（6）打开 ViewController.h 文件，编写代码。实现头文件和插座变量的声明。程序代码如下：

```
#import <UIKit/UIKit.h>
#import "RoundProgress.h"                                       //头文件
@interface ViewController : UIViewController{
    IBOutlet RoundProgress *timer1;             // RoundProgress 的插座变量
}
@end
```

（7）将设计界面的空白视图和插座变量 timer1 进行关联。

（8）打开 ViewController.m 文件，编写代码。实现扇形进度条的加载。程序代码如下：

```objc
- (void)viewDidLoad
{
    __block CGFloat i = 0;
    [timer1 startWithBlock:^CGFloat {
        return ((i++ >= 100) ? (i = 0) : i) / 100;
    }];
    [super viewDidLoad];
    // Do any additional setup after loading the view, typically from a nib.
}
```

【代码解析】

本实例关键功能是设置新的类型。下面就对这个知识点做一个详细的讲解。

在本实例中提到了定义一个新的类型，可以使用 typedef 关键字，其语法形式如下：

```
typedef 类型名称 类型标识符；
```

其中，类型名称表示已知数据类型名称；类型标识符表示新的类型名称。在代码中使用了 typedef 关键字定义了新的类型，代码如下：

```
typedef CGFloat (^ProgressBlock)();
```

其中，CGFloat 表示已知数据类型名称；(^ProgressBlock)() 表示新的类型名称。为了方便读者的阅读，笔者绘制了执行流程图，如图 4.7 所示。

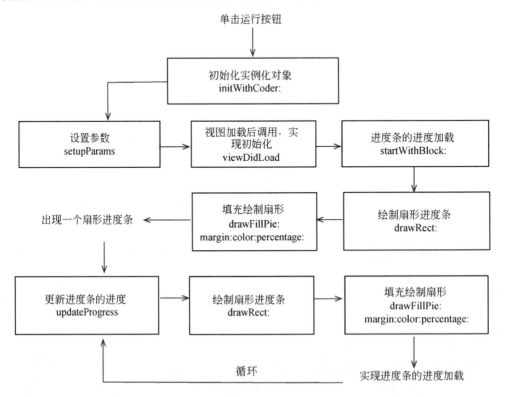

图 4.7　实例 20 程序执行流程

实例 21　环形进度条

【实例描述】

本实例实现的功能是在界面显示一个加载的环形进度条。运行效果如图 4.8 所示。

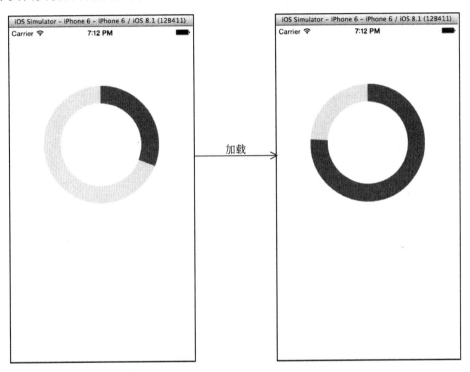

图 4.8　实例 21 运行效果

【实现过程】

（1）创建一个项目，命名为"环形进度条"。
（2）创建一个 CALayer 类的 Circularlayer 类。
（3）打开 Circularlayer.h 文件，编写代码，实现属性的声明。程序代码如下：

```
#import <QuartzCore/QuartzCore.h>
@interface Circularlayer : CALayer
@property(nonatomic, strong) UIColor *trackTintColor;
@property(nonatomic, strong) UIColor *progressTintColor;
@property(nonatomic) CGFloat thicknessRatio;
@property(nonatomic) CGFloat progress;
@end
```

（4）打开 Circularlayer.m 文件，编写代码。在此代码中会写入两个方法：needsDisplayForKey:和 drawInContext:方法。其中 needsDisplayForKey:方法实现显示关键字，将关键字和 progress 进行判断。程序代码如下：

```
+ (BOOL)needsDisplayForKey:(NSString *)key
{
```

```objc
//判断关键字是否为字符串progress
if ([key isEqualToString:@"progress"]) {
    return YES;
} else {
    return [super needsDisplayForKey:key];
}
}
```

drawInContext:方法对环形进度条进行绘制,在此代码中会创建两个圆和一个椭圆,由这3个图形就组成环形进度条。程序代码如下:

```objc
- (void)drawInContext:(CGContextRef)context
{
    CGRect rect = self.bounds;
    CGPoint centerPoint = CGPointMake(rect.size.width / 2.0f, rect.size.height / 2.0f);
    CGFloat radius = MIN(rect.size.height, rect.size.width) / 2.0f;
    CGFloat progress = MIN(self.progress, 1.0f - FLT_EPSILON);
    CGFloat radians = (progress * 2.0f * M_PI) - M_PI_2;
    CGContextSetFillColorWithColor(context, self.trackTintColor.CGColor);
                                                                //设置填充颜色
    CGMutablePathRef trackPath = CGPathCreateMutable();
    CGPathMoveToPoint(trackPath, NULL, centerPoint.x, centerPoint.y);
                                                                //设置开始点
    CGPathAddArc(trackPath, NULL, centerPoint.x, centerPoint.y, radius,
    2.0f * M_PI, 0.0f, TRUE);
    CGPathCloseSubpath(trackPath);
    CGContextAddPath(context, trackPath);                       //添加路径到上下文中
    CGContextFillPath(context);                                 //填充路径
    CGPathRelease(trackPath);
    CGContextSetFillColorWithColor(context, self.progressTintColor.CGColor);
    CGMutablePathRef progressPath = CGPathCreateMutable();
    CGPathMoveToPoint(progressPath, NULL, centerPoint.x, centerPoint.y);
    CGPathAddArc(progressPath, NULL, centerPoint.x, centerPoint.y, radius,
    3.0f * M_PI_2, radians, NO);                                //绘制圆
    CGPathCloseSubpath(progressPath);
    CGContextAddPath(context, progressPath);                    //添加路径到上下文中
    CGContextFillPath(context);
    CGContextSetBlendMode(context, kCGBlendModeClear);
    CGFloat innerRadius = radius * (1.0f - self.thicknessRatio);
    CGRect clearRect = (CGRect) {
        .origin.x = centerPoint.x - innerRadius,
        .origin.y = centerPoint.y - innerRadius,
        .size.width = innerRadius * 2.0f,
        .size.height = innerRadius * 2.0f
    };
    CGContextAddEllipseInRect(context, clearRect);              //绘制椭圆
    CGContextFillPath(context);
}
```

(5)创建一个基于UIView类的Circular类。

(6)打开Circular.h文件,编写代码。实现头文件、属性和方法的声明。程序代码如下:

```
#import <UIKit/UIKit.h>
#import "Circularlayer.h"                                    //头文件
@interface Circular : UIView
//属性
@property(nonatomic, strong) UIColor *trackTintColor;
@property(nonatomic, strong) UIColor *progressTintColor;
……
@property(nonatomic) CGFloat indeterminateDuration;
//方法
- (void)setProgress:(CGFloat)progress animated:(BOOL)animated;
@end
```

（7）打开 Circular.m 文件，编写代码。实现对进度条的一些设置，使用的方法如表 4-4 所示。

表 4-4　Circular.m文件中方法总结

方　　法	功　　能
initialize	设置初始值
layerClass	获取创建图层的类
circularProgressLayer	获取环形进度条的图层
init	环形进度条的初始化
progress	获取进度条的进度
setProgress:	设置进度条的进度
setProgress: animated:	设置进度条进度的动画
trackTintColor	获取轨道的颜色
setTrackTintColor:	设置轨道的颜色
progressTintColor	获取进度条中进度的颜色
setProgressTintColor:	设置进度条中进度的颜色
thicknessRatio	获取轨道的半径
setThicknessRatio:	设置轨道的半径

在文件中，代码分为：初始化设置和实现进度加载。这里需要讲解几个重要的方法（其他方法请读者参考源代码）。环形进度条的初始化需要 initialize、init、layerClass、circularProgressLayer、trackTintColor、setTrackTintColor:、progressTintColor、setProgressTintColor、thicknessRatio 和 setThicknessRatio 方法共同实现。其中，initialize 方法对进度条的初始值进行设置，程序代码如下：

```
+ (void) initialize
{
    if (self == [Circular class]) {
        id appearance = [Circular appearance];
        [appearance setTrackTintColor:[[UIColor redColor] colorWithAlphaComponent:
        0.3f]];
        [appearance setProgressTintColor:[UIColor blueColor]];   //进度的颜色
        [appearance setBackgroundColor:[UIColor clearColor]];
        [appearance setThicknessRatio:0.3f];                     //圆环的粗细
        [appearance setIndeterminateDuration:5.0f];
    }
}
```

init 方法对环形进度条进行初始化。程序代码如下：

```objc
- (id)init
{
    return [super initWithFrame:CGRectMake(0.0f, 0.0f, 40.0f, 40.0f)];
}
```

这时的进度条是不可以加载的，要实现进度条的加载需要使用 progress、setProgress:和 setProgress: animated:方法共同实现。其中 progress 方法实现了获取环形进度条的进度。程序代码如下：

```objc
- (CGFloat)progress
{
    return self.circularProgressLayer.progress;
}
```

setProgress:方法对环形进度条的进度进行设置。程序代码如下：

```objc
- (void)setProgress:(CGFloat)progress
{
    [self setProgress:progress animated:NO];
}
```

setProgress:animated:方法对进度条的进度动画进行设置。程序代码如下：

```objc
- (void)setProgress:(CGFloat)progress animated:(BOOL)animated
{
    CGFloat pinnedProgress = MIN(MAX(progress, 0.0f), 1.0f);
    if (animated) {
        CABasicAnimation *animation = [CABasicAnimation animationWithKeyPath:
        @"progress"];
        animation.duration = fabsf(self.progress - pinnedProgress);
                                                                //设置动画时间
        animation.timingFunction = [CAMediaTimingFunction functionWithName:
kCAMediaTimingFunctionEaseInEaseOut];                           //设置动画的速度
        animation.fromValue = [NSNumber numberWithFloat:self.progress];
                                                                //设置开始值
        animation.toValue = [NSNumber numberWithFloat:pinnedProgress];
                                                                //设置结束值
        [self.circularProgressLayer addAnimation:animation forKey:@"progress"];
    } else {
        [self.circularProgressLayer setNeedsDisplay];
    }
    self.circularProgressLayer.progress = pinnedProgress;
}
```

（8）打开 ViewController.h 文件，编写代码。实现头文件、对象以及属性的声明。程序代码如下：

```objc
#import <UIKit/UIKit.h>
#import "Circular.h"                                            //头文件
@interface ViewController : UIViewController{
    Circular *circular;                                         //对象
}
@property (strong, nonatomic) NSTimer *timer;                   //属性
@end
```

（9）打开 ViewController.m 文件，编写代码。实现创建环形进度条视图对象，以及实现进度条的进度改变的动画。在 viewDidLoad 方法中实现创建环形进度条对象。

```
- (void)viewDidLoad
{
    circular = [[Circular alloc] initWithFrame:CGRectMake(60.0f, 100.0f,
    200.0f, 200.0f)];
    [self.view addSubview:circular];
    [self startAnimation];                               //开始动画
    [super viewDidLoad];
    // Do any additional setup after loading the view, typically from a nib.
}
```

添加 startAnimation 和 progressChange 方法实现进度条的进度改变的动画。程序代码如下：

```
- (void)progressChange
{
    NSArray *progressViews = @[
                        circular,
                        ];
    //遍历
    for (Circular *progressView in progressViews) {
        CGFloat progress = ![self.timer isValid] ? 2 : progressView.progress
        + 0.01f;
        [progressView setProgress:progress animated:YES];      //设置进度
        if (progressView.progress >= 1.0f && [self.timer isValid]) {
            [progressView setProgress:0.f animated:YES];//调用设置进度的动画
        }
    }
}
- (void)startAnimation
{
    //创建时间定时器
    self.timer = [NSTimer scheduledTimerWithTimeInterval:0.03
                                        target:self
                                        selector:@selector(progressChange)
                                        userInfo:nil
                                        repeats:YES];
}
```

【代码解析】

由于本实例中的代码和方法非常的多，为了方便读者的阅读，笔者绘制了一些执行流程图如图 4.9～4.10 所示。其中，环形进度条的显示，使用了 iewDidLoad、initialize、layerClass、startAnimation:、setTrackTintColor:、circularProgressLayer、setProgressTintColor:、setThicknessRat:和 drawInContext:方法共同实现。它们的执行流程如图 4.9 所示。

执行完这个程序流程后，会在模拟器中显示一个加载完成的环形进度条。由于进度条的功能就是使用户看到加载的进度，所以想要看到环形进度条的加载需要使用 progress、circularProgressLayer、setProgress:animated:、drawInContext:和 progressChange 方法共同实现。它们的执行流程如图 4.10 所示。

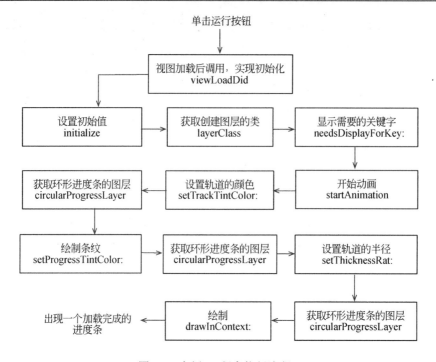

图 4.9 实例 21 程序执行流程 1

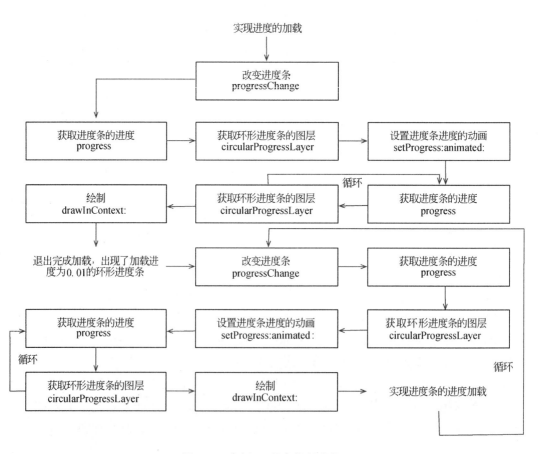

图 4.10 实例 21 程序执行流程 2

实例 22　具有范围的进度条

【实例描述】
　　本实例实现的功能是在界面显示一个进度条，当用户单击"加载进度"按钮后，进度条会加载。当加载结束，会出现两个指示器，用户可以拖动这两个指示器。运行效果如图4.11 所示。

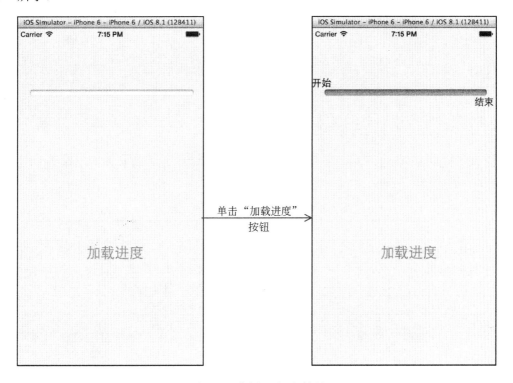

图 4.11　实例 22 运行效果

【实现过程】
（1）创建一个项目，命名为"具有范围的进度条"。
（2）添加图片 1.png 和 2.png 到创建项目的 Supporting Files 文件夹中。
（3）创建一个基于 UIView 类的 progress 类。
（4）打开 progress.h 文件，编写代码。实现定义数据结构、宏定义、对象、实例变量、属性和方法的声明。程序代码如下：

```
#import <UIKit/UIKit.h>
//定义数据结构
typedef enum {
    BJRSWPRMode=0 ,
    BJRSWPSMode,
} ProgressDisplayMode;
//宏定义
#define BJRANGESLIDER_THUMB_SIZE 32.0
@interface progress : UIView{
```

```
//对象
UIImageView *slider;
UIImageView *progressImage;
UIImageView *rangeImage;
UILabel *leftThumb;
UILabel *rightThumb;
//实例变量
CGFloat minValue;
CGFloat maxValue;
CGFloat currentProgressValue;
CGFloat leftValue;
CGFloat rightValue;
}
//属性
@property (nonatomic, assign) CGFloat minValue;
@property (nonatomic, assign) CGFloat maxValue;
……
@property (nonatomic, assign) BOOL showRange;
//方法
- (void)setDisplayMode:(ProgressDisplayMode)mode;
@end
```

（5）打开 progress.m 文件，编写代码。实现可以拖动，并具有范围的进度条的绘制，使用的方法如表 4-5 所示。

表 4-5　progress.m文件中方法总结

方　　法	功　　能
setLeftValue:	设置左边的值
setRightValue:	设置右边的值
setCurrentProgressValue:	设置当前进度的值
setMinValue:	设置最小值
setMaxValue:	设置最大值
minValue	获取最小值
maxValue	获取最大值
currentProgressValue	获取当前的进度值
leftValue	获取左边的值
rightValue	获取右边的值
setShowThumbs:	设置显示的指示器
showThumbs	获取显示的指示器
setShowProgress:	设置进度条进度的显示
showProgress	获取进度条进度的显示
setShowRange:	设置进度条范围的显示
showRange	获取进度条范围的显示
handleLeftPan:	拖动左边的指示器
handleRightPan:	拖动右边的指示器
setup	创建
initWithFrame:	具有范围的进度条的初始化
initWithCoder:	初始化实例化对象
layoutSubviews	布局
setDisplayMode:	设置显示的模式

这里需要讲解几个重要的方法（其他方法请读者参考源代码）。如果想要绘制一个具有范围并且可以滑动的进度条需要：绘制一个具有范围的进度条、对进度条的模式进行设置、实现拖动的动画这几部分。绘制一个具有范围的进度条，需要使用 setLeftValue:、setRightValue:、setMinValue:、setMaxValue:、minValue、maxValue、leftValue、rightValue、setup 和 layoutSubviews 方法共同实现。其中，setup 方法用来创建一个具有范围的进度条。程序代码如下：

```
- (void)setup {
    if (maxValue == 0.0) {
        maxValue = 100.0;
    }
    leftValue = minValue;
    rightValue = maxValue;
    //创建3个图像视图对象
    slider = [[UIImageView alloc] initWithImage:[[UIImage imageNamed:
    @"1.png"] stretchableImageWithLeftCapWidth:5 topCapHeight:4]];
    [self addSubview:slider];
    rangeImage = [[UIImageView alloc] initWithImage:[[UIImage imageNamed:
    @"2.png"] stretchableImageWithLeftCapWidth:5 topCapHeight:4]];
    [self addSubview:rangeImage];
    progressImage = [[UIImageView alloc] initWithImage:[[UIImage imageNamed:
    @"2.png"] stretchableImageWithLeftCapWidth:5 topCapHeight:4]];
    [self addSubview:progressImage];
    //创建并设置左边的范围指示器对象
    leftThumb = [[UILabel alloc] initWithFrame:CGRectMake(0, -BJRANGESLIDER_
    THUMB_SIZE, BJRANGESLIDER_THUMB_SIZE + 12, BJRANGESLIDER_THUMB_SIZE * 2)];
    leftThumb.text =@"开始";                              //设置文本内容
    leftThumb.userInteractionEnabled = YES;
    leftThumb.contentMode = UIViewContentModeCenter;
    [self addSubview:leftThumb];
    UIPanGestureRecognizer *leftPan = [[UIPanGestureRecognizer alloc]
    initWithTarget:self action:@selector(handleLeftPan:)];
                                                //实例化手势识别器对象
    [leftThumb addGestureRecognizer:leftPan];
    //创建并设置右边的指示器对象
      rightThumb = [[UILabel alloc] initWithFrame:CGRectMake(0, 0,
      BJRANGESLIDER_THUMB_SIZE + 12, BJRANGESLIDER_THUMB_SIZE * 2)];
    rightThumb.text=@"结束";
    rightThumb.userInteractionEnabled = YES;
    rightThumb.contentMode = UIViewContentModeCenter;
    [self addSubview:rightThumb];                       //添加视图对象
    UIPanGestureRecognizer *rightPan = [[UIPanGestureRecognizer alloc]
    initWithTarget:self action:@selector(handleRightPan:)];
    [rightThumb addGestureRecognizer:rightPan];
}
```

ayoutSubviews 方法对创建的进度条进行布局。程序代码如下：

```
- (void)layoutSubviews {
    CGFloat availableWidth = self.frame.size.width - BJRANGESLIDER_THUMB_
    SIZE;
    CGFloat inset = BJRANGESLIDER_THUMB_SIZE / 2;
```

```
CGFloat range = maxValue - minValue;                        //获取间距
CGFloat left = floorf((leftValue - minValue) / range * availableWidth);
CGFloat right = floorf((rightValue - minValue) / range * availableWidth);
//判断
if (isnan(left)) {
    left = 0;
}
if (isnan(right)) {
    right = 0;
}
//设置视图对象的框架
slider.frame = CGRectMake(inset, self.frame.size.height / 2 - 5,
availableWidth, 10);
CGFloat rangeWidth = right - left;
//判断是否显示范围
if ([self showRange]) {
    rangeImage.frame = CGRectMake(inset + left, self.frame.size.height /
    2 - 5, rangeWidth, 10);
}
//判断是否显示进度
if ([self showProgress]) {
    CGFloat progressWidth = floorf((currentProgressValue - leftValue) /
    (rightValue - leftValue) * rangeWidth);
    if (isnan(progressWidth)) {
        progressWidth = 0;
    }
    //设置图像视图对象的框架
    progressImage.frame = CGRectMake(inset + left, self.frame.size.
    height / 2 - 5, progressWidth, 10);
}
leftThumb.center = CGPointMake(inset + left, self.frame.size.height / 2 -
BJRANGESLIDER_THUMB_SIZE / 2);                          //设置左边指示器的中心点
rightThumb.center = CGPointMake(inset + right, self.frame.size.height / 2 +
BJRANGESLIDER_THUMB_SIZE / 2);                          //设置右边指示器的中心点
}
```

对进度条的模式进行设置，需要使用 setShowThumbs:、showThumbs、setShowProgress:、showProgress、setShowRange:、showRange 和 setDisplayMode:方法共同实现。其中，setDisplayMode:方法对进度条的显示模式进行设置，这样可以实现不同的进度条。程序代码如下：

```
- (void)setDisplayMode:(ProgressDisplayMode)mode {
    switch (mode) {
        case BJRSWPRMode:
            self.showThumbs = NO;
            self.showRange = NO;                        //不显示范围
            self.showProgress = YES;                    //显示进度
            progressImage.image = [[UIImage imageNamed:@"2.png"]
            stretchableImageWithLeftCapWidth:5 topCapHeight:4]; //设置图像
            break;
        default:
            self.showThumbs = YES;
            self.showRange = YES;
```

```
            self.showProgress = NO;                    //不显示进度
            break;
        }
    [self setNeedsLayout];
}
```

此时创建的进度条是不可以实现拖动的,如果需要实现拖动,需要使用 handleLeftPan: 和 handleRightPan:方法共同实现。其中 handleLeftPan:方法实现拖动左边的指示器。程序代码如下:

```
- (void)handleLeftPan:(UIPanGestureRecognizer *)gesture {
    //判断手势识别器的状态
    if (gesture.state == UIGestureRecognizerStateBegan || gesture.state == UIGestureRecognizerStateChanged) {
        CGPoint translation = [gesture translationInView:self];
        CGFloat range = maxValue - minValue;                       //设置间距
        CGFloat availableWidth = self.frame.size.width - BJRANGESLIDER_THUMB_SIZE;
        self.leftValue += translation.x / availableWidth * range;
        [gesture setTranslation:CGPointZero inView:self];
    }
}
```

handleRightPan:方法实现拖动右边的指示器。程序代码如下:

```
- (void)handleRightPan:(UIPanGestureRecognizer *)gesture {
    //判断手势识别器的状态
    if (gesture.state == UIGestureRecognizerStateBegan || gesture.state == UIGestureRecognizerStateChanged) {
        CGPoint translation = [gesture translationInView:self];
        CGFloat range = maxValue - minValue;                       //设置间距
        CGFloat availableWidth = self.frame.size.width - BJRANGESLIDER_THUMB_SIZE;
        self.rightValue += translation.x / availableWidth * range;
        [gesture setTranslation:CGPointZero inView:self];
        }
}
```

(6)打开 ViewController.h 文件,编写代码。实现头文件、插座变量、属性以及动作的声明。程序代码如下:

```
#import <UIKit/UIKit.h>
#import "progress.h"
@interface ViewController : UIViewController{
    IBOutlet progress *progress;                               //插座变量
}
@property (strong, nonatomic) IBOutlet progress *progress;
- (IBAction)load:(id)sender;
@end
```

(7)打开 Main.storyboard 文件,对 View Controller 视图控制器的设计界面进行设计,效果如图 4.12 所示。

需要添加的视图、控件以及对它们的设置如表 4-6 所示。

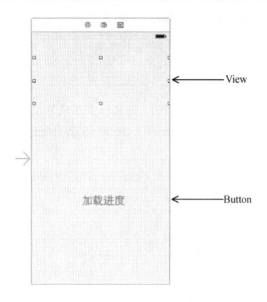

图 4.12　实例 22 View Controller 视图控制器的设计界面

表 4-6　实例 22 视图、控件设置

视图、控件	属 性 设 置	其　　他
设计界面	Background：浅粉色	调整大小
View	Class：progress Background：透明	与插座变量 progress 关联
Button	Title：关闭 Font：25.0	与动作 load:关联

（8）打开 ViewController.m 文件，编写代码，实现具有范围的进度条的加载功能。程序代码如下：

```
- (void)viewDidLoad
{
    [super viewDidLoad];
    [self.progress setDisplayMode:BJRSWPRMode];     //设置进度条的显示模式
    // Do any additional setup after loading the view, typically from a nib.
}
- (void)viewWillAppear:(BOOL)animated
{
    [super viewWillAppear:animated];
    [self.progress setDisplayMode:BJRSWPRMode];     //设置显示模式
}
//自动实现加载动画
- (void)recordDemoFire:(NSTimer*)timer {
    self.progress.currentProgressValue += 0.6;     //设置当前的进度值
    if (self.progress.currentProgressValue >= self.progress.maxValue) {
        [self.progress setDisplayMode:BJRSWPSMode];
        [timer invalidate];                         //让定时器失效
    }
}
//界面按钮被按下后调用
- (IBAction)load:(id)sender {
    [self.progress setDisplayMode:BJRSWPRMode];
```

```
    self.progress.currentProgressValue - 0;
    self.progress.leftValue =0;                              //设置左边的值
    self.progress.rightValue = self.progress.maxValue;       //设置右边的值
    [NSTimer scheduledTimerWithTimeInterval:0.01 target:self selector:
    @selector(recordDemoFire:) userInfo:nil repeats:YES];
}
```

【代码解析】

由于本实例中的代码和方法非常的多，为了方便读者的阅读，笔者绘制了一些执行流程图如图4.13～4.15所示。其中，进度条的显示，使用了initWithCoder:、setup、ViewDidLoad、setDisplayMode:、setShowThumbs:、setShowRange:、setShowProgress:、viewWillAppear:、layoutSubviews、showRange 和 showProgress 方法共同实现。它们的执行流程如图 4.13 所示。

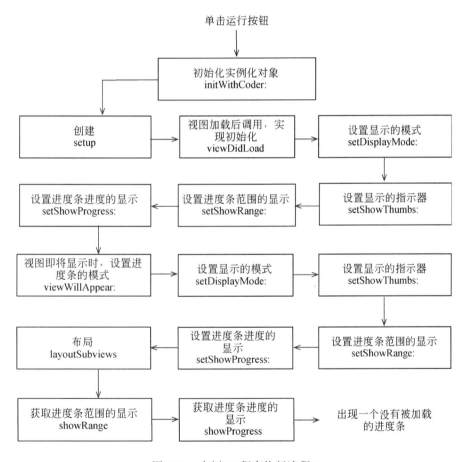

图 4.13　实例 22 程序执行流程 2

执行完这个程序流程后，会在模拟器中显示一个没有加载的进度条。如果想要看到进度条的加载过程，并且可以看到加载完成后范围的显示。需要使用 load:、setDisplayMode:、setShowThumbs:、setShowRange:、setShowProgress:、setCurrentProgressValue:、setLeftValue:、maxValue、setRightValue:、layoutSubviews、showRange、showProgress、recordDemoFire、currentProgressValue、setDisplayMode:、layoutSubviews、showRange 和 showProgress 方法共同实现。它们的执行流程如图 4.14 所示。

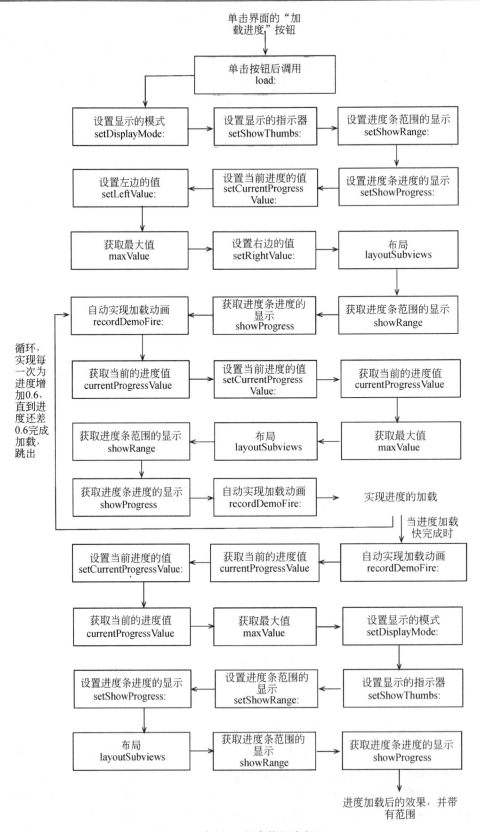

图 4.14 实例 22 程序执行流程 2

执行完此流程后，会在进度条的两端新增两个指示器。用户可以拖动这两个指示器到达进度条的任何位置，需要实现此功能，使用 handleLeftPan:、handleRightPan:、leftValue、rightValue、setLeftValue、setRightValue:、layoutSubviews、showRange 和 showProgress 方法共同实现。由于拖动左边的指示器和拖动右边的指示器的执行流程非常相似，所以，为了方便，本书以拖动左边的指示器到指定的位置为例，来让开发者看到它的执行流程，如图 4.15 所示。

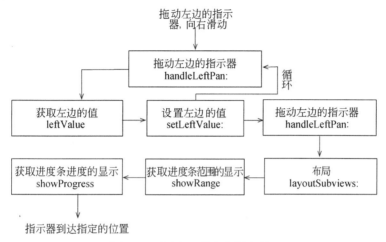

图 4.15　实例 22 程序执行流程 3

如果开发者想要看到拖动右边的指示器的执行流程，只需将 handleLeftPan:、leftValue、setLeftValue 改为 handleRightPan:、rightValue、setRightValue:即可，执行流程不变。

实例 23　文本上传进度条

【实例描述】

本实例实现的功能是，在界面显示一个文本上传的进度条。当用户单击"下载"按钮，会出现文本上传面板，还会实现进度条的加载。当单击"取消"按钮或者面板中的"取消"按钮，文本上传面板就会关闭。运行效果如图 4.16 所示。

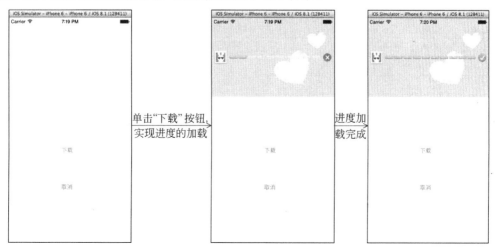

图 4.16　实例 23 运行效果

【实现过程】

（1）创建一个项目，命名为"文本上传进度条"。

（2）添加图片 1.jpg、2.jpg、3.jpg、4.jpg、5.jpg、6.jpg 和 7.jpg 到创建项目的 Supporting Files 文件夹中。

（3）创建一个基于 UIView 类的 progress 类。

（4）打开 progress.h 文件。编写代码，实现属性、方法以及协议的声明。程序代码如下：

```
#import <UIKit/UIKit.h>
@interface progress : UIView
//属性
@property (nonatomic, assign) CGFloat percentage;
@property (nonatomic, strong) UIImageView *thumbImageView;
……
@property (nonatomic, assign) id delegate;
//方法
- (void)showPanelInView:(UIView *)view;
+ (progress *)sharedInstance;
- (void)show;                              //显示
- (void)done;
- (void)hide;
@end
//协议
@protocol progressDelegate
- (void)progressPanelDidFinishLoad:(progress *)panel;
- (void)progressPanelDidCancelLoad:(progress *)panel;
@end
```

（5）打开 progress.m 文件，编写代码。实现文本上传进度条的功能。使用的方法如表 4-7 所示。

表 4-7　progress.m 文件中方法总结

方　　法	功　　能
init	文本上传进度条的初始化
isLoading	获取是否加载
show	以动画的形式显示文本上传面板
done	加载完成
sharedInstance	共享实例
setPercentage:	设置百分比
aa	停止指示器的加载
bb	出现一个按钮，表示上传成功
cancelUpload:	取消加载
showPanelInView:	对显示的文本上传面板进行设计
hide	以动画的形式隐藏文本上传面板

实现文本上传进度条的功能，需要设计一个进度条、实现进度条的进度加载和实现进度条的加载完成三个部分。在本实例，将进度条放入一个独特的面板中，采用了方块的形式显示进度条。那么，要实现对进度条以及它的存放面板的设计，需要使用 init、show、sharedInstance、showPanelInView:和 hide 方法共同实现。init 方法对面板中的内容进行设置，其中就包含了进度条。程序代码如下：

```objc
- (id)init {
    if (self = [super init]) {
        //创建按钮对象
        _cancelButton = [UIButton buttonWithType:UIButtonTypeCustom];
        _doneButton = [UIButton buttonWithType:UIButtonTypeCustom];
        _thumbImageView = [[UIImageView alloc] init];
        self.thumbImageView.backgroundColor = [UIColor grayColor];
                                                                    //设置背景颜色
        _progressImageViews = [[NSMutableArray alloc] initWithCapacity:10];
        //遍历
        for (int i = 0; i< 10; i++) {
            UIImageView *progressView = [[UIImageView alloc] initWithImage:
            [UIImage imageNamed:@"6.png"] highlightedImage:[UIImage imageNamed:
            @"7.png"]];                                             //实例化对象
            [progressView sizeToFit];
            [self.progressImageViews addObject:progressView];       //添加对象
        }
        self.backgroundColor = [UIColor clearColor];                //设置背景颜色
        //创建背景图像对象
        UIImageView *backgroundImageView = [[UIImageView alloc] initWithImage:
        [[UIImage imageNamed:@"2.jpg"] resizableImageWithCapInsets:
        UIEdgeInsetsMake(5, 5, 5, 5)] ];
        backgroundImageView.frame = CGRectMake(0, 0, panelWidth, panelHeight);
                                                                    //背景的设置
        [self addSubview:backgroundImageView];
        CGFloat left = 57;                                          //进度条的位置
        NSInteger count = 0;
        for (UIImageView *progressView in self.progressImageViews) {
            progressView.center = CGPointMake(left + 24 * count, panelHeight
            / 2);                                                   //设置中心位置
            [self addSubview:progressView];
            count++;
        }
        //创建一个活动指示器对象
        _waitingView = [[UIActivityIndicatorView alloc] initWithActivity-
        IndicatorStyle:
UIActivityIndicatorViewStyleGray];
        [_waitingView sizeToFit];
        [self addSubview: self.waitingView];                        //添加视图对象
        [self show];
    }
    return self;
}
```

show方法对文本上传面板进行显示，并会伴有动画效果。程序代码如下：

```objc
- (void)show {
    self.frame = CGRectMake(0, 0, panelWidth, panelHeight);
    CATransition *transition = [CATransition animation];            //创建
    transition.duration = 0.25;                                     //设置动画的时间
    transition.timingFunction = [CAMediaTimingFunction functionWithName:
kCAMediaTimingFunctionEaseInEaseOut];                               //设置动画的速度
    transition.type = kCATransitionPush;                            //设置动画的类型
    transition.subtype = kCATransitionFromBottom;
    [self.layer addAnimation:transition forKey:nil];
}
```

sharedInstance方法对实例进行共享。程序代码如下：

```objc
+ (progress *)sharedInstance {
  static dispatch_once_t pred;
  __strong static progress *sharedOverlay = nil;
  dispatch_once(&pred, ^{
      sharedOverlay = [[progress alloc] init];                //实例化对象
  });
  return sharedOverlay;
}
```

showPanelInView:方法对显示的文本上传面板进行设计。程序代码如下：

```objc
- (void)showPanelInView:(UIView *)view {
  [self.cancelButton addTarget:self action:@selector(cancelUpload:)
    forControlEvents:UIControlEventTouchUpInside];            //为"取消"按钮添加动作
  self.cancelButton.hidden = NO;
  [self.cancelButton setImage:[UIImage imageNamed:@"3"] forState:
    UIControlStateNormal];                                    //设置图像
  [self.cancelButton sizeToFit];
  self.cancelButton.center = CGPointMake(panelWidth - 18, panelHeight /
    2);                                                       //设置中心位置
  [self addSubview:self.cancelButton];
  [self.doneButton setImage:[UIImage imageNamed:@"4"] forState:
    UIControlStateNormal];                                    //设置图像
  self.doneButton.hidden = YES;                               //隐藏按钮
  [self.doneButton sizeToFit];
  self.doneButton.center = CGPointMake(panelWidth - 18, panelHeight / 2);
  [self addSubview:self.doneButton];                          //添加视图对象
  UIEdgeInsets recipeBackgroundImageViewInsets = UIEdgeInsetsMake(8, 8,
    8, 8);
  //创建图像视图
  UIImageView *iconShadowView = [[UIImageView alloc] initWithImage:
    [[UIImage imageNamed:@"5.png"] resizableImageWithCapInsets:
    recipeBackgroundImageViewInsets]];
  [iconShadowView sizeToFit];
  iconShadowView.center = CGPointMake(24, panelHeight / 2);//设置中心位置
  self.thumbImageView.frame = CGRectMake((iconShadowView.frame.size.
    width - 24) / 2, (iconShadowView.frame.size.width - 24) / 2, 24, 24);
                                                              //设置框架的大小
  [iconShadowView addSubview:self.thumbImageView];
  [self addSubview:iconShadowView];
  [view addSubview:self];
}
```

hide方法对文本上传面板进行隐藏，并会伴有动画效果。程序代码如下：

```objc
- (void)hide {
  [NSObject cancelPreviousPerformRequestsWithTarget:self];
  CATransition *transition = [CATransition animation];
  transition.duration = 0.25;                                 //设置时间
  transition.timingFunction = [CAMediaTimingFunction functionWithName:
    kCAMediaTimingFunctionEaseInEaseOut];
  transition.type = kCATransitionPush;                        //设置类型
  transition.subtype = kCATransitionFromTop;                  //设置方向
  [self.layer addAnimation:transition forKey:nil];
  self.frame = CGRectMake(0, -panelHeight * 2, panelWidth, panelHeight);
}
```

实现进度条的进度加载，需要使用 isLoading 和 setPercentage:方法共同实现。其中

isLoading 方法获取是否实现加载。程序代码如下：

```
- (BOOL)isLoading {
  return self.percentage > 0 && self.percentage < 1;
}
```

setPercentage:方法对进度条的百分比进行设置。程序代码如下：

```
- (void)setPercentage:(CGFloat)percentage {
  _percentage = percentage;
  NSInteger currentNumber = 10;
  for (UIImageView *progressView in self.progressImageViews) {
      if (percentage * 100 >= currentNumber) {
         [progressView setHighlighted:YES];            //经过的位置变亮
      } else {
         [progressView setHighlighted:NO];             //经过的位置不变亮
      }
      currentNumber += 10;
  }
  if (percentage * 100 >= 100) {
     self.waitingView.center = self.cancelButton.center;
     [self.waitingView startAnimating];                //开始动画
     self.cancelButton.hidden = YES;                   //隐藏按钮
     self.doneButton.hidden = YES;
     [self performSelector:@selector(aa) withObject:self afterDelay:5];
  } else {
     [self.waitingView stopAnimating];                 //结束动画
     self.cancelButton.hidden = NO;
     self.doneButton.hidden = YES;
  }
  if (self.isLoading && self.frame.origin.y < 0) {
     [self show];
  }
}
```

进度条的加载完成后，需要告诉用户，上传成功。这时需要使用 aa、bb 和 done 方法实现。其中 aa 方法实现活动指示器的停止以及对 bb 的调用。程序代码如下：

```
-(void)aa{
   [self.waitingView stopAnimating];                   //结束动画
   [self performSelector:@selector(bb) withObject:self afterDelay:0];
}
```

bb 方法实现出现"完成"按钮和对 done 方法的调用。程序代码如下：

```
-(void)bb{
   self.doneButton.hidden=NO;                          //不隐藏按钮
   [self done];
}
```

done 方法实现"取消"按钮的隐藏、活动指示器的动画停止，以及隐藏文本上传面板。程序代码如下：

```
- (void)done {
  self.cancelButton.hidden = YES;
  self.doneButton.hidden = NO;
  [self.waitingView stopAnimating];                    //结束动画
  [self performSelector:@selector(hide) withObject:nil afterDelay:1];
}
```

此时，在 progress.m 文件中的代码就可以算完成了。但是由于在文本上传面板中有一个"取消"按钮，所以，可以使用 cancelUpload:方法实现"取消"按钮的取消功能。程序代码如下：

```
- (void)cancelUpload:(id)sender {
    if (_delegate && [_delegate respondsToSelector:@selector
    (progressPanelDidCancelLoad:)]) {
        [_delegate progressPanelDidCancelLoad:self];
    }
}
```

（6）打开 ViewController.h 文件，编写代码。实现头文件、宏定义以及插座变量的声明。程序代码如下：

```
#import <UIKit/UIKit.h>
#import "progress.h"                                    //头文件
//宏定义
#define kNOTIFICATION_UPLOADMANAGER_DID_START_LOADING @"kNOTIFICATION_
UPLOADMANAGER_DID_START_LOADING"
#define kNOTIFICATION_UPLOADMANAGER_LOADING @"kNOTIFICATION_UPLOADMANAGER_
LOADING"
#define kNOTIFICATION_UPLOADMANAGER_DID_FINISH_LOADING @"kNOTIFICATION_
UPLOADMANAGER_DID_FINISH_LOADING"
#define kNOTIFICATION_UPLOADMANAGER_LOAD_WITH_ERROR @"kNOTIFICATION_
UPLOADMANAGER_LOAD_WITH_ERROR"
#define kNOTIFICATION_PHOTOS_UPLOAD_DONE @"kNOTIFICATION_PHOTOS_UPLOAD_DONE"
@interface ViewController : UIViewController{
    //插座变量
    IBOutlet UIButton *show;
    IBOutlet UIButton *hide;
}
@end
```

（7）打开 Main.storyboard 文件，对 View Controller 视图控制器的设计界面进行设计，效果如图 4.17 所示。

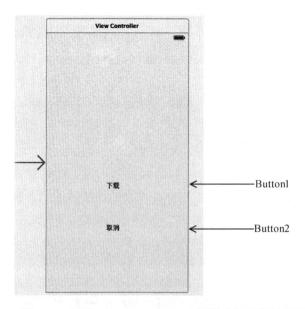

图 4.17　实例 23 View Controller 视图控制器的设计界面

需要添加的视图、控件以及对它们的设置如表 4-8 所示。

表 4-8 实例 23 视图、控件设置

视图、控件	属 性 设 置	其 他
设计界面	Background：米黄色	
Buttonn1	Title：下载	与插座变量 show 关联
Button2	Title：取消	与插座变量 hide 关联

（8）打开 ViewController.m 文件，编写代码。在此文件中，代码分为：视图加载、单击"下载"按钮实现文本上传面板的显示、单击"取消"按钮实现文本上传面板的隐藏。其中创建并设置文本上传进度条在 viewDidLoad 方法中实现。程序代码如下：

```
- (void)viewDidLoad
{
    __block progress *tempProgressPanel = [progress sharedInstance];
    //添加通知设置
    [[NSNotificationCenter defaultCenter] addObserverForName:
kNOTIFICATION_UPLOADMANAGER_DID_START_LOADING object:nil queue:
[NSOperationQueue mainQueue] usingBlock:^(NSNotification *note) {
        if ([note.object isKindOfClass:[UIImage class]]) {
            UIImage *uploadImage = (UIImage *)note.object;       //实例化对象
            tempProgressPanel.thumbImageView.image = uploadImage; //设置图像
            [tempProgressPanel.thumbImageView setNeedsLayout];
        }
    }];
    [[NSNotificationCenter defaultCenter] addObserverForName:
kNOTIFICATION_UPLOADMANAGER_LOADING object:nil queue:[NSOperationQueue
mainQueue] usingBlock:^(NSNotification *note) {
        if ([note.object isKindOfClass:[NSNumber class]]) {
            NSNumber *progress = (NSNumber *)note.object;        //实例化对象
            tempProgressPanel.percentage = progress.floatValue;
        }
    }];
    [[NSNotificationCenter defaultCenter] addObserverForName:
kNOTIFICATION_UPLOADMANAGER_LOAD_WITH_ERROR object:nil queue:
[NSOperationQueue mainQueue] usingBlock:^(NSNotification *note) {
        [tempProgressPanel hide];
    }];
    [[NSNotificationCenter defaultCenter] addObserverForName:
kNOTIFICATION_PHOTOS_UPLOAD_DONE object:nil queue:[NSOperationQueue
mainQueue] usingBlock:^(NSNotification *note) {
        [tempProgressPanel done];
    }];
    //添加面板视图
    [[progress sharedInstance] showPanelInView:self.view];
    [progress sharedInstance].delegate = self;
    //实现面板的隐藏
    [[progress sharedInstance] hide];
    //为按钮添加方法
    [show addTarget:self action:@selector(show:) forControlEvents:
```

```
    UIControlEventTouchUpInside];
    [hide addTarget:self action:@selector(hide:) forControlEvents:
UIControlEventTouchUpInside];
    [super viewDidLoad];
    // Do any additional setup after loading the view, typically from a nib.
}
```

show:方法实现单击"下载"按钮后,将文本上传面板显示。程序代码如下:

```
-(void)show:(id)sender {
    [progress sharedInstance].percentage=0.0;
    [[progress sharedInstance] show];
    [[NSNotificationCenter defaultCenter] postNotificationName:kNOTIFICATION_
UPLOADMANAGER_DID_START_LOADING object:[UIImage imageNamed:@"1.jpg"]];
    [self performSelector:@selector(a1) withObject:self afterDelay:0.5];
                                                        //0.5秒后执行动作 a1
}
```

a1~a8 方法实现了进度条进度的加载。程序代码如下:

```
-(void)a1{
    [progress sharedInstance].percentage = 0.1;
    [self performSelector:@selector(a2) withObject:self afterDelay:0.5];
}
……
-(void)a7{
    [progress sharedInstance].percentage = 0.8;
    [self performSelector:@selector(a8) withObject:self afterDelay:0.5];
}
-(void)a8{
    [progress sharedInstance].percentage = 1.0;
}
```

hide 方法实现单击"取消"按钮后,调用方法将文本上传面板进行隐藏。程序代码如下:

```
- (void)hide:(id)sender {
    [[progress sharedInstance] hide];
}
```

progressPanelDidCancelLoad:方法实现单击文本上传面板中的"取消"按钮,将隐藏文本上传面板。程序代码如下:

```
- (void)progressPanelDidCancelLoad:(progress *)panel {
    [panel hide];
}
```

【代码解析】

本实例关键功能是进度条的加载。下面就对这个知识点做一个详细的讲解。

加载进度条,主要是通过 performSelector:withObject:afterDelay:方法实现的。在本实例中它会在 0.5 秒调用 performSelector:后面的参数,也就是方法。加载进度条的执行流程如图 4.18 所示。

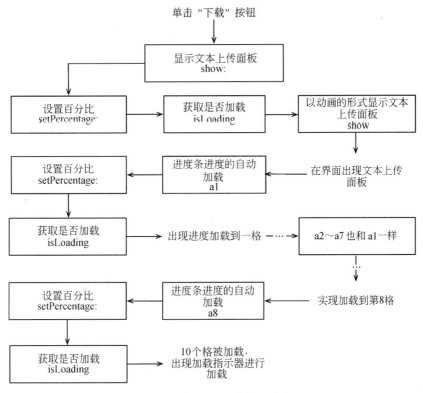

图 4.18　实例 23 程序执行流程

实例 24　倒计时进度条

【实例描述】
　　本实例的功能是使用进度条实现像电影中的倒计时动画的效果。当用户单击"播放"按钮，就会出现倒计时动画。运行效果如图 4.19 所示。

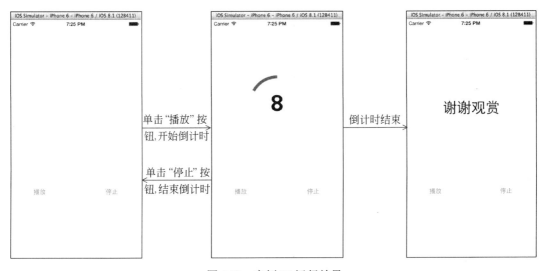

图 4.19　实例 24 运行效果

【实现过程】

（1）创建一个项目，命名为"倒计时进度条"。

（2）创建一个基于 UIView 类的 countview 类。

（3）打开 countview.h 文件，编写代码。实现宏定义、协议、对象、实例变量、属性和方法的声明。程序代码如下：

```objc
#import <UIKit/UIKit.h>
//宏定义
#define kCircleSegs 30
@class countview;
//协议
@protocol countviewDelegate <NSObject>
- (void)counterDownFinished:(countview *)circleView;
@end
@interface countview : UIView{
    //对象
    NSString *mTimeFormatString;
    UIColor *mNumberColor;
    UIFont *mNumberFont;
    UIColor *mCircleColor;
    //实例变量
    CGFloat mCircleBorderWidth;
    float mTimeInSeconds;
    float mTimeInterval;
    float mCircleTimeInterval;
    BOOL mIsRunning;
    int mCircleSegs;
    BOOL mCircleIncre;
}
//属性
@property (nonatomic, assign) id<countviewDelegate> delegate;
@property (nonatomic, assign) BOOL circleIncre;
……
@property (nonatomic, assign) float circleTimeInterval;
//方法
- (void)setup;
- (void)update:(id)sender;
- (void)updateTime:(id)sender;
- (void)startWithSeconds:(float)seconds;
- (void)stop;
@end
```

（4）打开 countview.m 文件，编写代码。此代码中的方法如表 4-9 所示。

表 4-9　countview.m 文件中方法总结

方　　法	功　　能
initWithFrame:	初始化
drawRect:	绘制
setup	设置
startWithSeconds:	设置秒数
stop	停止
update:	更新
updateTime:	更新时间

在文件中，代码分为：初始化、绘制和设置以及更新。这里需要讲解几个重要的方法（其他方法请读者参考源代码）。绘制和设置一个圆环并在圆环内显示数字，需要使用 drawRect:和 setup 方法共同实现。其中，drawRect:方法对圆环和数字进行绘制。程序代码如下：

```
- (void)drawRect:(CGRect)rect {
    CGContextRef context = UIGraphicsGetCurrentContext();
    float radius = CGRectGetWidth(rect)/2.0f - self.circleBorderWidth/2.0f;
    float angleOffset = M_PI_2;
    CGContextSetLineWidth(context, self.circleBorderWidth);
    CGContextBeginPath(context);
    //绘制圆
    CGContextAddArc(context,
                CGRectGetMidX(rect), CGRectGetMidY(rect),
                radius,
                (mCircleSegs + 1)/(float)kCircleSegs*M_PI*2 - angleOffset,
                2*M_PI - angleOffset,
                0);
    CGContextSetStrokeColorWithColor(context, self.circleColor.CGColor);
                                                                //设置颜色
    CGContextStrokePath(context);                               //绘制路径
    CGContextSetLineWidth(context, 1.0f);                       //设置线宽
    NSString *numberText=[NSString stringWithFormat:@"%.0f", mTimeInSeconds];
    NSDictionary *dis=[NSDictionary dictionaryWithObjectsAndKeys:self.
    numberFont, NSFontAttributeName, nil];
    CGSize sz = [numberText sizeWithAttributes:dis];
    [numberText drawInRect:CGRectInset(rect, (CGRectGetWidth(rect) -
    sz.width)/2.0f, (CGRectGetHeight(rect) - sz.height)/2.0f)
    withAttributes:dis];
}
```

setup 方法对绘制的圆环和数字进行设置。程序代码如下：

```
- (void)setup {
    mIsRunning = NO;
    self.numberColor = [UIColor blackColor];
    self.numberFont = [UIFont fontWithName:@"Courier-Bold" size:60.0f];
    self.circleColor = [UIColor blueColor];              //圆环的颜色
    self.circleBorderWidth = 7;                          //圆环的宽度
    self.circleTimeInterval = 1.0f;
    mTimeInSeconds = 20;
    mTimeInterval = 2;
    mCircleSegs = 0;
    self.backgroundColor = [UIColor whiteColor];         //设置背景颜色
}
```

设置好圆环以及圆环内的时间后，需要对它们进行定期更新，以实现倒计时的功能，需要使用 startWithSeconds:、updateTime:和 update:方法。其中，startWithSeconds:方法对秒数进行设置。程序代码如下：

```
- (void)startWithSeconds:(float)seconds {
    //判断 seconds 是否大于 0
    if (seconds > 0)
```

```objc
    {
        mTimeInSeconds = seconds;
        mIsRunning = YES;
        mCircleSegs = 0;
        [self update:nil];                                          //更新
        [self updateTime:nil];                                      //更新时间
    }
}
```

updateTime:方法用来更新显示的时间。程序代码如下：

```objc
- (void)updateTime:(id)sender {
    //判断 mIsRunning 是否为 true
    if (mIsRunning)
    {
        mTimeInSeconds -= mTimeInterval;
        if (fabs(mTimeInSeconds) < 1e-4)
        {
            mCircleSegs = (kCircleSegs - 1);
            mTimeInSeconds = 0;
            [self.delegate counterDownFinished:self];
            //调用 counterDownFinished:方法
        }
        else
        {
            [NSTimer scheduledTimerWithTimeInterval:mTimeInterval
                                    target:self
                                    selector:@selector(updateTime:)
                                    userInfo:nil
                                    repeats:NO];        //创建定时器
        }
    }
}
```

update:方法对圆环以及时间总体进行更新。程序代码如下：

```objc
- (void)update:(id)sender {
    if (mIsRunning)
    {
        mCircleSegs = (mCircleSegs + 1) % kCircleSegs;
        //判断倒计时是否结束
        if (fabs(mTimeInSeconds) < 1e-4)
        {
            mCircleSegs = (kCircleSegs - 1);
            mTimeInSeconds = 0;
            [self.delegate counterDownFinished:self];
            //调用 counterDownFinished:方法
        }
        else
        {
            [NSTimer scheduledTimerWithTimeInterval:self.circleTimeInterval/kCircleSegs
                                    target:self
```

```
                            selector:@selector(update:)
                            userInfo:nil
                            repeats:NO];              //创建时间定时器
    }
    [self setNeedsDisplay];
    }
}
```

（5）创建一个基于 NSObject 类的 count 类。

（6）打开 count.h 文件，编写代码。实现加载头文件、定义数据结构、宏定义、方法的声明以及协议的使用。程序代码如下：

```
#import <Foundation/Foundation.h>
#import "countview.h"                                 //头文件
//定义数据结构
typedef enum {
    CircleDownCounterTypeIntegerDecre = 0,
} CircleDownCounterType;
#define kDefaultCounterSize CGSizeMake(138,138)       //宏定义
@class countview;
@interface count: NSObject<countviewDelegate>
//方法
+ (count *)showCircleDownWithSeconds:(float)seconds onView:(UIView *)view
withSize:(CGSize)size andType:(CircleDownCounterType)type;
+ (CGRect)frameOfCircleViewOfSize:(CGSize)size inView:(UIView *)view;
+ (void)removeCircleViewFromView:(UIView *)view;
@end
```

（7）打开 count.m 文件编写代码。showCircleDownWithSeconds:onView:withSize:andType: 方法，对创建的 countview 视图的显示秒数、所在的视图、尺寸和类型进行设置。程序代码如下：

```
+ (countview *)showCircleDownWithSeconds:(float)seconds onView:(UIView
*)view withSize:(CGSize)size andType:(CircleDownCounterType)type {
    [self removeCircleViewFromView:view];             //移除视图
    countview *circleView = [[countview alloc] initWithFrame:[self
frameOfCircleViewOfSize:size inView:view]];           //实例化对象
    [view addSubview:circleView];                     //添加视图对象
    circleView.delegate = [self circleDownCounter];
    [circleView startWithSeconds:seconds];
    return circleView;
}
```

frameOfCircleViewOfSize:方法对 countview 视图进行框架大小的获取。程序代码如下：

```
+ (CGRect)frameOfCircleViewOfSize:(CGSize)size inView:(UIView *)view {
    return CGRectInset(view.bounds,
                (CGRectGetWidth(view.bounds) - size.width)/2.0f,
                (CGRectGetHeight(view.bounds) - size.height)/2.0f);
}
```

circleViewInView:方法获取 View 视图中的 countview 视图。程序代码如下：

```
+ (countview *)circleViewInView:(UIView *)view {
   for (UIView *subView in [view subviews])
   {
      if ([subView isKindOfClass:[countview class]])
      {
         return (countview *)subView;                    //返回子视图
      }
   }
   return nil;
}
```

circleDownCounter 方法获取 count 类的对象。程序代码如下：

```
+ (count *)circleDownCounter {
   if (!sharedDownCounter)
   {
      sharedDownCounter = [[count alloc] init];         //实例化对象
   }
   return sharedDownCounter;
}
```

removeCircleViewFromView:方法实现在视图中，将 countview 视图移除。程序代码如下：

```
+ (void)removeCircleViewFromView:(UIView *)view {
   countview *circleView = [self circleViewInView:view];
   if (circleView)
   {
      [circleView removeFromSuperview];                 //移除视图
   }
}
```

counterDownFinished:方法实现将 countview 视图删除。程序代码如下：

```
- (void)counterDownFinished:(countview *)circleView {
   [circleView removeFromSuperview];
}
```

（8）打开 ViewController.h 文件，编写代码。实现插座变量和动作的声明。程序代码如下：

```
#import <UIKit/UIKit.h>
@interface ViewController : UIViewController{
   IBOutlet UIView *containerView;                      //空白视图的插座变量
   IBOutlet UILabel *label;                             //标签的插座变量
}
//动作
- (IBAction)aa:(id)sender;
- (IBAction)bb:(id)sender;
@end
```

（9）打开 Main.storyboard 文件，对 View Controller 视图控制器的设计界面进行设计，效果如图 4.20 所示。

需要添加的视图、控件以及对它们的设置如表 4-10 所示。

第 4 章 进度条类和指示器类效果

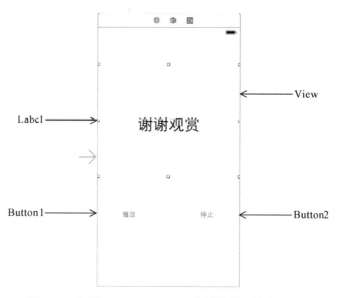

图 4.20　实例 24View Controller 视图控制器的设计界面

表 4-10　实例 24 视图、控件设置

视图、控件	属 性 设 置	其　　他
View		与插座变量 containerView 关联
Label	Text：谢谢观赏	与插座变量 label 关联
Button1	Title：播放	与动作 aa 关联
Button2	Title：停止	与动作 bb 关联

（10）打开 ViewController.m 文件，编写代码。实现在单击按钮后，出现一个倒计时的动画效果。程序代码如下：

```
- (void)viewDidLoad
{
    label.hidden=YES;                                    //隐藏标签
    [super viewDidLoad];
    // Do any additional setup after loading the view, typically from a nib.
}
- (CGSize)sizeFromSegControl {
    return kDefaultCounterSize;
}
//单击"播放"按钮后，调用的动作
- (IBAction)aa:(id)sender {
    [containerView setHidden:NO];
    //显示倒计时
    [count showCircleDownWithSeconds:10.0f
                              onView:containerView
                            withSize:[self sizeFromSegControl]
                             andType:CircleDownCounterTypeIntegerDecre];
    label.hidden=NO;
}
//单击"停止"按钮后，调用的动作
- (IBAction)bb:(id)sender {
    [containerView setHidden:YES];                       //隐藏 containerView
}
```

【代码解析】

由于本实例中的代码和方法非常的多,为了方便读者的阅读,笔者绘制了执行流程图如图 4.21 所示。

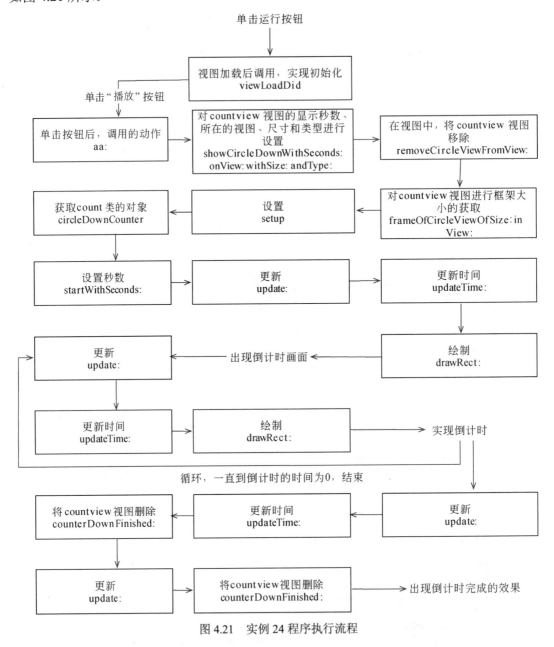

图 4.21 实例 24 程序执行流程

实例 25 带进度条的工具栏

【实例描述】

本实例实现的功能是在工具栏中显示一个进度条。当用户单击"显示"按钮,会在界面的下方显示一个工具栏,并且工具栏中的进度条实现加载的功能。运行效果如图 4.22 所示。

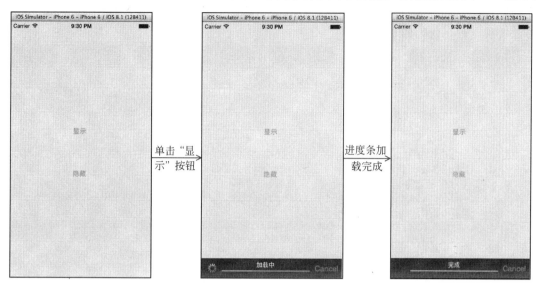

图 4.22　实例 25 运行效果

【实现过程】

（1）创建一个项目,命名为"带进度条的工具栏"。

（2）创建一个基于 UIToolbar 类的 Toolbar 类。

（3）打开 Toolbar.h 文件,编写代码。实现定义新的类型、协议、对象、属性和方法的声明。程序代码如下：

```
#import <UIKit/UIKit.h>
typedef void(^ToolbarCompletionHandler)(BOOL success);      //定义新的类型
@class Toolbar;
//协议
@protocol ToolbarDelegate <NSObject>
@optional
- (void)didCancelButtonPressed:(Toolbar*)toolbar;
@end
@interface Toolbar : UIToolbar{
//对象
@private
    id <ToolbarDelegate> __weak _actionDelegate;
    UIBarButtonItem* _stopButtonItem;
    UIActivityIndicatorView* _activityIndicator;
    UILabel* _statusLabel;                                  //声明标签对象
    UIProgressView* _progressBar;
}
//属性
@property (nonatomic, strong) UIBarButtonItem *stopButtonItem;
@property (nonatomic, strong) UIActivityIndicatorView *activityIndicator;
@property (nonatomic, strong) UILabel *statusLabel;
……
@property(nonatomic, copy) ToolbarCompletionHandler hideHandler;
//方法
- (void)show:(BOOL)animated completion:(void (^)(BOOL finished))completion;
- (void)hide:(BOOL)animated completion:(void (^)(BOOL finished))completion;
@end
```

（4）打开 Toolbar.m 文件，编写代码。在此文件中，代码分为 3 个部分：对工具栏中的内容进行设置、隐藏和显示工具栏。其中，对工具栏内容的设置，需要使用 initWithFrame:方法实现。程序代码如下：

```
- (id)initWithFrame:(CGRect)frame
{
    self = [super initWithFrame:frame];
    if (self) {
        //创建活动指示器对象
        self.activityIndicator = [[UIActivityIndicatorView alloc]
initWithActivityIndicatorStyle:UIActivityIndicatorViewStyleWhite];
        UIBarButtonItem *activityIndicatorItem = [[UIBarButtonItem alloc]
initWithCustomView:self.activityIndicator];
        //标签对象的创建与设置
        self.statusLabel = [[UILabel alloc] initWithFrame:CGRectMake(0, 2,
200, 20)];
        self.statusLabel.font = [UIFont boldSystemFontOfSize:14.0];
        self.statusLabel.backgroundColor = [UIColor clearColor];
        self.statusLabel.textColor = [UIColor whiteColor];
        self.statusLabel.shadowColor = [UIColor blackColor];//设置阴影的颜色
        self.statusLabel.shadowOffset = CGSizeMake(0, -1);
        self.statusLabel.textAlignment = NSTextAlignmentCenter;
        //进度条对象的创建与设置
        self.progressBar = [[UIProgressView alloc] initWithFrame:CGRectMake
(0, 25, 200, 10)];
        self.progressBar.progress=1.0;
        //视图对象的创建与设置
        UIView *statusView = [[UIView alloc] initWithFrame:CGRectMake(0, 0,
200, 40)];
        [statusView addSubview:self.statusLabel];              //添加视图对象
        [statusView addSubview:self.progressBar];
        //按钮条目对象的创建
        UIBarButtonItem *statusItem = [[UIBarButtonItem alloc] initWithCustomView:
statusView];
        UIBarButtonItem *flexSpace = [[UIBarButtonItem alloc] initWithBarButton-
SystemItem:UIBarButtonSystemItemFlexibleSpace target:nil action:nil];
        self.stopButtonItem = [[UIBarButtonItem alloc] initWithBarButtonSystemItem:
UIBarButtonSystemItemCancel target:self action:@selector
(didCancelButtonPressed:)];                                    //实例化对象
        [self setBackgroundImage:nil
            forToolbarPosition:UIToolbarPositionAny
                 barMetrics:UIBarMetricsDefault];              //设置图像
        self.autoresizingMask = UIViewAutoresizingFlexibleTopMargin |
UIViewAutoresizingFlexibleWidth;
        self.translucent = YES;
        self.barStyle = UIBarStyleBlackTranslucent;            //设置工具栏的风格
        //向工具栏中添加内容
        self.items = @[activityIndicatorItem, flexSpace, statusItem,
flexSpace, self.stopButtonItem];
    }
    return self;
}
```

要实现工具栏的显示和隐藏，需要使用 show:completion:和 hide:completion:方法共同实现。其中 show: completion:方法实现以动画的形式显示工具栏。程序代码如下：

```
- (void)show:(BOOL)animated completion:(void (^)(BOOL finished))completion {
    self.showHandler = completion;
```

```
    self.stopButtonItem.enabled = YES;
    [self.activityIndicator startAnimating];              //开始动画
    //动画
    [UIView animateWithDuration:0.4 animations:^{
        self.frame = CGRectMake(0, self.superview.bounds.size.height - 44,
self.superview.bounds.size.width, 44);                    //设置框架
    } completion:^(BOOL finished) {
        if (self.showHandler!=nil) {
            self.showHandler(YES);
        }
        self.showHandler = nil;
    }];
}
```

hide: completion:方法实现以动画的形式隐藏工具栏。程序代码如下:

```
- (void)hide:(BOOL)animated completion:(void (^)(BOOL finished))completion {
    self.hideHandler = completion;
    self.stopButtonItem.enabled = NO;
    [self.activityIndicator stopAnimating];               //结束动作
    //动画效果
    [UIView animateWithDuration:0.4 delay:1.0 options:UIViewAnimation-
OptionLayoutSubviews animations:^{
        self.frame = CGRectMake(0, self.superview.bounds.size.height,
        self.superview.bounds.size.width, 44);            //设置框架
    } completion:^(BOOL finished) {
        if (self.hideHandler!=nil) {
            self.hideHandler(YES);
        }
        self.hideHandler = nil;
    }];
}
```

由于本实例在工具栏中添加了一个取消按钮,所以 didCancelButtonPressed:方法实现了单击取消按钮后实现的动作。程序代码如下:

```
- (void)didCancelButtonPressed:(id*)sender {
    if ([_actionDelegate respondsToSelector:@selector(didCancelButton-
Pressed:)]) {
        [_actionDelegate performSelector:@selector(didCancelButtonPressed:)
        withObject:self];
    }
}
```

(5) 打开 ViewController.h 文件,编写代码。实现头文件、对象、实例变量、属性以及动作的声明。程序代码如下:

```
#import <UIKit/UIKit.h>
#import "Toolbar.h"                                       //头文件
@interface ViewController : UIViewController<ToolbarDelegate>{
    NSTimer *timer;                                       //对象
    float i;                                              //实例变量
}
@property (strong, nonatomic) Toolbar *statusToolbar;     //属性
//动作
- (IBAction)start;
- (IBAction)stop;
@end
```

（6）打开 Main.storyboard 文件，对 View Controller 视图控制器的设计界面进行设计，效果如图 4.23 所示。

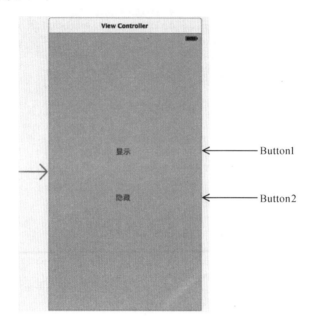

图 4.23 实例 25 View Controller 视图控制器的设计界面

需要添加的视图、控件以及对它们的设置如表 4-11 所示。

表 4-11 实例 25 视图、控件设置

视图、控件	属 性 设 置	其 他
View		与插座变量 containerView 关联
Label	Text：谢谢观赏	与插座变量 label 关联
Button	Title：按钮	与动作 aa 关联

（7）打开 ViewController.m 文件，编写代码。在此文件中，代码分为两个部分：创建工具栏；实现单击按钮后的动作。创建工具栏对象，需要在 viewDidLoad 中实现。程序代码如下：

```
- (void)viewDidLoad
{
    [super viewDidLoad];
    // Do any additional setup after loading the view, typically from a nib.
    CGRect statusToolbarFrame = CGRectMake(0, self.view.bounds.size.height,
    self.view.bounds.size.width, 44);
    self.statusToolbar = [[Toolbar alloc] initWithFrame:statusToolbarFrame];
                                                                    //实例化对象
    self.statusToolbar.actionDelegate = self;
    [self.view addSubview:self.statusToolbar];           //添加视图对象
}
```

实现单击按钮后的动作，这本实例中一个有 3 个按钮，分别为"显示"、"隐藏"和 Cancel 按钮。单击"显示"按钮后，会出现工具栏，并且工具栏中的进度条实现加载。要实现此功能，需要使用 start 和 aa 方法共同实现。其中 star 方法，实现单击按钮后，出现

工具栏。程序代码如下：

```
- (IBAction)start{
    self.statusToolbar.statusLabel.text = @"加载中";
    self.statusToolbar.progressBar.progress=0.0;
    [self.statusToolbar show:YES completion:^(BOOL finished) {
    }];
    timer=[NSTimer scheduledTimerWithTimeInterval:0.1 target:self selector:
    @selector(aa) userInfo:nil repeats:YES];                //创建时间定时器
}
```

aa方法实现进度条的加载功能。程序代码如下：

```
-(void)aa{
    self.statusToolbar.progressBar.progress+=0.01;
    if(self.statusToolbar.progressBar.progress<=0.99) {
        [self.statusToolbar.progressBar setProgress:self.statusToolbar.
        progressBar.progress animated:YES];                 //设置进度
        self.statusToolbar.progressBar.tintColor=[UIColor yellowColor];
    }else{
        [timer invalidate];                                 //让定时器失效
        self.statusToolbar.statusLabel.text = @"完成";
        self.statusToolbar.activityIndicator.hidden=YES;
        [self performSelector:@selector(stop) withObject:self afterDelay:1];
    }
}
```

单击"隐藏"按钮后，会将显示的工具栏隐藏。程序代码如下：

```
- (IBAction)stop{
    [self.statusToolbar hide:YES completion:^(BOOL finished) {
    }];
}
```

单击Cancel按钮后，实现隐藏工具栏。程序代码如下：

```
- (void)didCancelButtonPressed:(Toolbar *)toolbar {
    [self stop];
}
```

【代码解析】

本实例关键功能是进度条的进度颜色的改变，以及进度条的加载。下面依次讲解这两个知识点。

1. 进度条的进度颜色的改变

改变进度条进度的颜色，可以通过UIView的tintColor属性实现。其语法形式如下：

```
@property(nonatomic, retain) UIColor *tintColor;
```

在代码中，使用了tintColor属性对进度条的颜色进行改变，代码如下：

```
self.statusToolbar.progressBar.tintColor=[UIColor yellowColor];
```

2. 进度条进度的加载

实现进度条进度的加载，主要是通过定时器动画完成，代码如下：

```
timer=[NSTimer scheduledTimerWithTimeInterval:0.1 target:self selector:
```

```
@selector(aa) userInfo:nil repeats:YES];
```

此代码每隔 0.1 秒调用一次 aa 方法。aa 方法中有一个重要的知识,就是对进度条中进度值的设置,实现此功能,需要使用 UIProgressView 类中的 setProgress:animated:方法,其语法形式如下:

```
- (void)setProgress:(float)progress animated:(BOOL)animated;
```

其中,(float)progress 表示进度的值;(BOOL)animated 表示动画的变化,如果为 YES,表示改变时有动画,如果是 NO,表示改变时没有动画。在代码中,使用了 setProgress:animated:方法对进度条进度值进行了改变,代码如下:

```
[self.statusToolbar.progressBar setProgress:self.statusToolbar.progressBar.progress animated:YES];
```

其中,self.statusToolbar.progressBar.progress 表示进度的值;YES 表示改变时有动画。

实例 26 变色的指示器

【实例描述】

本实例实现的功能是在界面显示一个具有多种颜色的指示器。当用户单击"开始"按钮后,就会出现此指示器,并实现加载动画。运行效果如图 4.24 所示。

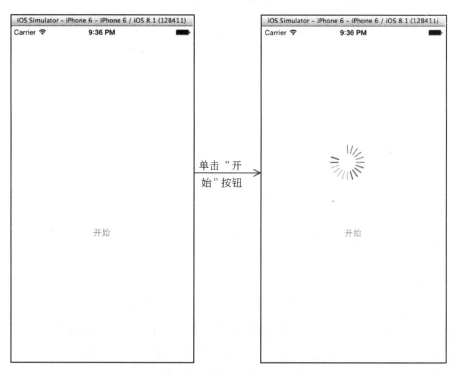

图 4.24 实例 26 运行结果

【实现过程】

单击"开始"按钮,显示指示器,并且在指示器加载的不同阶段,会出现不同的颜色。

具体的实现步骤如下。

（1）创建一个项目，命名为"变色的 Activity Indicator"。
（2）创建一个基于 UIView 类的 Activityindicator 类。
（3）打开 Activityindicator.h 文件，编写代码。实现属性以及方法的声明。程序代码如下：

```
#import <UIKit/UIKit.h>
@interface Activityindicator : UIView
@property (nonatomic, strong) NSTimer *timer;              //属性
-(void) startAnimating;                                    //方法
@end
```

（4）打开 Activityindicator.m 文件，编写代码。在此文件中，代码分为两个部分，一个是绘制指示器，另一个是实现指示器的加载动画。其中绘制指示器，需要添加 pointOnInnerCirecleWithAngel: 、pointOnOuterCirecleWithAngel:方法，以及文件自带的 initWithFrame:和 drawRect:方法共同实现。getColorForStage:方法实现获取指示器阶段的颜色。程序代码如下：

```
-(UIColor *) getColorForStage:(int) currentStage WithAlpha:(double) alpha
{
    int max = 20;
    int cycle = currentStage % max;
    if (cycle < max/4) {
        return [UIColor colorWithRed:1.0 green:0.0 blue:0.0 alpha:alpha];
                                                           //返回红色
    } else if (cycle < max/4*2) {
        return [UIColor colorWithRed:0.0 green:1.0 blue:0.0 alpha:alpha];
                                                           //返回绿色
    } else if (cycle < max/4*3) {
        return [UIColor colorWithRed:0.0 green:0.0 blue:1.0 alpha:alpha];
                                                           //返回蓝色
    } else {
        return [UIColor grayColor];                        //返回灰色
    }
}
```

pointOnInnerCirecleWithAngel:方法实现获取圆内部的点。程序代码如下：

```
-(CGPoint) pointOnInnerCirecleWithAngel:(int) angel
{
    double r = self.frame.size.height/2/2;
    double cx = self.frame.size.width/2;
    double cy = self.frame.size.height/2;
    double x = cx + r*cos(M_PI/10*angel);
    double y = cy + r*sin(M_PI/10*angel);
    return CGPointMake(x, y);                              //返回点
}
```

pointOnOuterCirecleWithAngel:方法实现的功能是获取圆外部的点。程序代码如下：

```
-(CGPoint) pointOnOuterCirecleWithAngel:(int) angel
{
    double r = self.frame.size.height/2;
    double cx = self.frame.size.width/2;
    double cy = self.frame.size.height/2;
    double x = cx + r*cos(M_PI/10*angel);
    double y = cy + r*sin(M_PI/10*angel);
```

```
    return CGPointMake(x, y);                                    //返回点
}
```

drawRect:方法实现绘制一个指示器,并在不同的阶段填充不同的颜色。程序代码如下:

```
-(void) drawRect:(CGRect)rect
{
    CGPoint point;
    CGContextRef ctx = UIGraphicsGetCurrentContext();
    CGContextSetLineWidth(ctx, 2.0);                             //设置线宽
    for (int i = 1 ; i<=18; ++i) {
        CGContextSetStrokeColorWithColor(ctx, [[self getColorForStage:
        stage+i WithAlpha:0.1 *i] CGColor]);                     //设置填充颜色
        point = [self pointOnOuterCirecleWithAngel:stage+i];
        CGContextMoveToPoint(ctx, point.x, point.y);             //设置开始点
        point = [self pointOnInnerCirecleWithAngel:stage+i];
        CGContextAddLineToPoint( ctx, point.x, point.y);         //设置结束点
        CGContextStrokePath(ctx);
    }
    stage++;
}
```

实现指示器的加载动画,需要使用 startAnimating 方法实现,它的程序代码如下:

```
-(void) startAnimating
{
    self.timer=[NSTimer scheduledTimerWithTimeInterval:0.1 target:self selector:
@selector(setNeedsDisplay) userInfo:nil repeats:YES];            //创建时间定时器
    self.hidden = NO;
    stage++;
}
```

(5)打开 ViewController.h 文件,编写代码。实现头文件、对象和动作的声明。程序代码如下:

```
#import <UIKit/UIKit.h>
#import "Activityindicator.h"                                    //头文件
@interface ViewController : UIViewController{
    Activityindicator *ac;                                       //对象
}
- (IBAction)load:(id)sender;                                     //动作
@end
```

(6)打开 Main.storyboard 文件,在 View Controller 视图控制器的设计界面中,添加 Button 按钮控件,将 Title 设置为"开始"。将此按钮和动作 load:进行关联。

(7)打开 ViewController.m 文件,编写代码,显示一个正在加载的变色的指示器。程序代码如下:

```
- (IBAction)load:(id)sender {
    ac = [[Activityindicator alloc] initWithFrame:CGRectMake(120, 200, 60,
    60)];
    [self.view addSubview:ac];
    [ac startAnimating];                                         //调用加载动画
}
```

【代码解析】

本实例关键功能是指示器的颜色变化和指示器的加载。下面依次讲解这两个知识点。

1. 指示器的颜色变化

在本实例中指示器会在不同的阶段显示不同的颜色,它使用的方法是 getColorForStage: 方法,此方法将指示器分为了 4 个阶段,对于每一个阶段,采用了不同的颜色,代码如下:

```
if (cycle < max/4) {                                        //第一阶段
    return [UIColor colorWithRed:1.0 green:0.0 blue:0.0 alpha:alpha];
} else if (cycle < max/4*2) {                               //第二阶段
    return [UIColor colorWithRed:0.0 green:1.0 blue:0.0 alpha:alpha];
} else if (cycle < max/4*3) {                               //第三阶段
    return [UIColor colorWithRed:0.0 green:0.0 blue:1.0 alpha:alpha];
} else {                                                    //第四阶段
    return [UIColor grayColor];
}
```

2. 指示器的加载

对于指示器的加载,本实例采用了定时器实现。在定时器中每隔 0.1 秒就会执行一次 setNeedsDisplay 方法,代码如下:

```
self.timer=[NSTimer scheduledTimerWithTimeInterval:0.1 target:self
selector:@selector(setNeedsDisplay) userInfo:nil repeats:YES];
```

其中,setNeedsDisplay 方法实现的功能是对视图中的内容进行重新的绘制,它会自动去调用 drawRect 方法。

实例 27 仿 Facebook web 上正在加载中的效果

【实例描述】

本实例实现的功能是模仿 Facebook web 上指示器加载的效果。当单击停止按钮,指示器会隐藏,并出现一个运行按钮,当单击运行按钮,会出现一个加载的指示器。运行效果如图 4.25 所示。

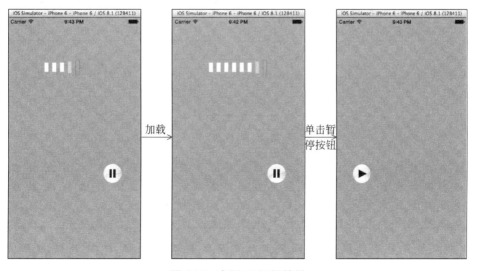

图 4.25 实例 27 运行效果

【实现过程】

在 Facebook web 中，指示器是由多个矩形构成的。所以要绘制矩形作为指示器，之后再实现加载动画。具体的实现步骤如下。

（1）创建一个项目，命名为"仿 Facebook web 上正在加载中的效果"。

（2）添加图片 1.png 和 2.png 到创建项目的 Supporting Files 文件夹中。

（3）创建一个基于 UIView 类的 Activityindicator 类。

（4）打开 Activityindicator.h 文件，编写代码。实现对象、实例变量、属性和方法的声明。程序代码如下：

```
#import <UIKit/UIKit.h>
@interface Activityindicator : UIView{
    //对象
    NSTimer*   _timer;
    //实例变量
    BOOL       _isAnimating;
    BOOL       _hidesWhenStopped;
    NSInteger  _currentStep;
    NSInteger  _dotCount;
    CGFloat    _duration;
}
//属性
@property (nonatomic, assign)BOOL hidesWhenStopped;
@property (nonatomic, assign)NSInteger dotCount;
@property (nonatomic, assign)CGFloat duration;
//方法
- (void)startAnimating;
- (void)stopAnimating;
- (BOOL)isAnimating;
- (id)init;
@end
```

（5）打开 Activityindicator.m 文件，编写代码，实现仿 Facebook web 上加载中的效果。使用的方法如表 4-12 所示。

表 4-12　Activityindicator.m文件中方法总结

方法	功能
setDefaultProperty	设置默认值
initWithFrame:	指示器的初始化
init	设置框架并获取
startAnimating	加载动画
stopAnimating	结束动画
isAnimating	获取是否需要动画
currentBorderColor:	设置当前的边框颜色
currentInnerColor:	设置当前的内部颜色
currentRect:	设置当前的矩形区域
repeatAnimation	重复
drawRect:	绘制

在文件中，代码分为：绘制指示器和时间指示器的加载动画。这里需要讲解几个重要的地方（其他方法请读者参考源代码）。绘制指示器需要使用 setDefaultProperty、 initWithFrame:、init、currentBorderColor:、currentInnerColor:、currentRect:和 drawRect:方法共同实现。其中 setDefaultProperty 方法对一些默认的属性进行设置。程序代码如下：

```objc
- (void)setDefaultProperty
{
    _currentStep = 0;                              //指示器加载的当前阶段
    _dotCount = 8;
    _isAnimating = NO;                             //是否需要动画
    _duration = 1.0f;                              //设置时间
    _hidesWhenStopped = YES;
}
```

currentBorderColor:方法对当前的边框颜色进行设置。程序代码如下：

```objc
- (UIColor*)currentBorderColor:(NSInteger)index
{
    //判断加载的当前阶段和当前索引的关系
    if (_currentStep == index) {
        return [UIColor colorWithRed:82.0f/255.0f
                               green:111.0f/255.0f
                                blue:167.0f/255.0f
                               alpha:1];           //返回颜色
    } else if (_currentStep < index) {
        return [UIColor clearColor];
    } else {
        if (_currentStep - index == 1) {
            return [UIColor colorWithRed:158.0f/255.0f
                                   green:172.0f/255.0f
                                    blue:203.0f/255.0f
                                   alpha:1];       //返回颜色
        } else {
            return [UIColor colorWithRed:239.0f/255.0f
                                   green:242.0f/255.0f
                                    blue:246.0f/255.0f
                                   alpha:1];       //返回颜色
        }
    }
}
```

currentInnerColor:方法对当前的内部颜色进行设置。程序代码如下：

```objc
- (UIColor*)currentInnerColor:(NSInteger)index
{
    if (_currentStep == index) {
        return [UIColor colorWithRed:140.0f/255.0f
                               green:158.0f/255.0f
                                blue:195.0f/255.0f
                               alpha:1];           //返回颜色
    } else if (_currentStep < index) {
        return [UIColor clearColor];
    } else {
        if (_currentStep - index == 1) {
            return [UIColor colorWithRed:189.0f/255.0f
                                   green:198.0f/255.0f
                                    blue:219.0f/255.0f
                                   alpha:1];       //返回颜色
        } else {
            return [UIColor colorWithRed:244.0f/255.0f
                                   green:246.0f/255.0f
```

```
                         blue:249.0f/255.0f
                         alpha:1];                                          //返回颜色
   }
 }
}
```

currentRect:方法对当前的矩形区域进行设置。程序代码如下：

```
- (CGRect)currentRect:(NSInteger)index
{
    if (_currentStep == index) {
        return CGRectMake(self.frame.size.width/(_dotCount*2+1),
                          0,
                          self.frame.size.width/(_dotCount*2+1),
                          self.frame.size.height);                          //返回框架
    } else if (_currentStep < index) {
        return CGRectMake(self.frame.size.width/(_dotCount*2+1),
                          self.frame.size.height/5.0,
                          self.frame.size.width/(_dotCount*2+1),
                          self.frame.size.height*3.0/5.0);                  //返回框架
    } else {
        if (_currentStep - index == 1) {
            return CGRectMake(self.frame.size.width/(_dotCount*2+1),
                              self.frame.size.height/10.0,
                              self.frame.size.width/(_dotCount*2+1),
                              self.frame.size.height*4.0/5.0);              //返回框架
        } else {
            return CGRectMake(self.frame.size.width/(_dotCount*2+1),
                              self.frame.size.height/5.0,
                              self.frame.size.width/(_dotCount*2+1),
                              self.frame.size.height*3.0/5.0);              //返回框架
        }
    }
}
```

drawRect:方法实现对指示器的绘制。程序代码如下：

```
- (void)drawRect:(CGRect)rect
{
    CGContextRef context = UIGraphicsGetCurrentContext();
    for (int i = 0; i < _dotCount; i++) {
        [[self currentInnerColor:i] setFill];                               //设置填充的颜色
        [[self currentBorderColor:i] setStroke];
        CGMutablePathRef path = CGPathCreateMutable();
        CGRect rect1 = [self currentRect:i];
        CGPathAddRect(path, NULL, rect1);                                   //绘制矩形
        CGContextBeginPath(context);
        CGContextAddPath(context, path);
        CGContextSetLineWidth(context, 1);                                  //设置线宽
        CGContextClosePath(context);
        CGContextDrawPath(context, kCGPathFillStroke);
        CGContextTranslateCTM(context, self.frame.size.width/_dotCount, 0);
        CGPathRelease(path);
    }
}
```

绘制好指示器后，就可以对指示器加载动画效果了。想要实现此功能需要使用startAnimating、stopAnimating、isAnimating、repeatAnimation 和 drawRect:方法共同实现。

其中，startAnimating 方法实现加载动画效果。程序代码如下：

```objectivec
- (void)startAnimating
{
    if (_isAnimating) {
        return;
    }
    _timer = [NSTimer scheduledTimerWithTimeInterval:_duration/(_dotCount*2+1)
                                    target:self
                                    selector:@selector(repeatAnimation)
                                    userInfo:nil
                                    repeats:YES];            //创建时间定时器
    _isAnimating = YES;
    if (_hidesWhenStopped) {
        self.hidden = NO;
    }
}
```

stopAnimating 方法实现结束加载动画。程序代码如下：

```objectivec
- (void)stopAnimating
{
    if (_timer)
    {
        [_timer invalidate];                                 //让定时器失效
        _timer = nil;
    }
    _isAnimating = NO;
    if (_hidesWhenStopped) {
        self.hidden = YES;                                   //隐藏
    }
}
```

isAnimating 方法获取是否需要加载动画。程序代码如下：

```objectivec
- (BOOL)isAnimating
{
    return _isAnimating;
}
```

repeatAnimation 方法实现动画的重复。程序代码如下：

```objectivec
- (void)repeatAnimation
{
    _currentStep = ++_currentStep % (_dotCount*2+1);
    [self setNeedsDisplay];                                  //重新绘制
}
```

（6）打开 ViewController.h 文件，编写代码，实现头文件、对象、插座变量以及动作的声明。程序代码如下：

```objectivec
#import <UIKit/UIKit.h>
#import "Activityindicator.h"
@interface ViewController : UIViewController{
    Activityindicator *ac;
    //按钮的插座变量
    IBOutlet UIButton *button1;
    IBOutlet UIButton *button2;
}
```

```
//动作
- (IBAction)play:(id)sender;
- (IBAction)stop:(id)sender;
@end
```

(7)打开 Main.storyboard 文件,对 View Controller 视图控制器的设计界面进行设计,效果如图 4.26 所示。

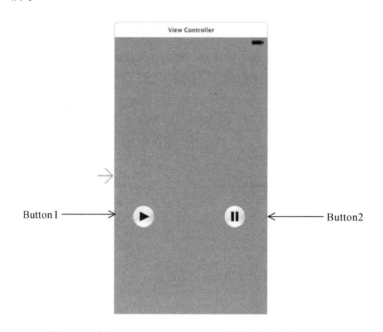

图 4.26　实例 27 View Controller 视图控制器的设计界面

需要添加的视图、控件以及对它们的设置如表 4-13 所示。

表 4-13　实例 27 视图、控件设置

视图、控件	属 性 设 置	其　　他
设计界面	Background:深灰色	
Button1	Title:（空） Background:2.png	与插座变量 button1 关联 与动作 play 关联
Button2	Title:（空） Background:1.png	与插座变量 button2 关联 与动作 stop 关联

(8)打开 ViewController.m 文件,编写代码。实现指示器对象的创建以及加载。程序代码如下:

```
- (void)viewDidLoad
{
    ac=[[Activityindicator alloc]initWithFrame:CGRectMake(80, 100, 150, 35)];
    [self.view addSubview:ac];
    [ac startAnimating];                                //开始动画
    [button1 setHidden:YES];
    [super viewDidLoad];
    // Do any additional setup after loading the view, typically from a nib.
}
//单击播放按钮,调用此方法
```

```
- (IBAction)play:(id)sender {
    [ac startAnimating];                              //开始动画
    [button1 setHidden:YES];
    [button2 setHidden:NO];
}
//单击暂停按钮，调用此方法
- (IBAction)stop:(id)sender {
    [ac stopAnimating];                               //结束动画
    [button1 setHidden:NO];
    [button2 setHidden:YES];
}
```

【代码解析】

本实例关键功能是指示器加载时的颜色、大小变化和指示器的加载。下面依次讲解这两个知识点。

1．指示器加载时的颜色变化

在本实例中指示器会将正在加载的阶段颜色、大小变得和没有加载以及加载过后的阶段不一样。它使用的方法是 currentBorderColor:、 currentInnerColor:和 currentRect:。这些方法将当前加载的阶段和当前的索引进行比较，从而将指示器分为了 4 个阶段。以 currentBorderColor:方法为例，程序代码如下：

```
if (_currentStep == index) {                         //当前加载的阶段
    return [UIColor colorWithRed:82.0f/255.0f
                    green:111.0f/255.0f
                    blue:167.0f/255.0f
                    alpha:1];
} else if (_currentStep < index) {                   //没有加载的阶段
    return [UIColor clearColor];
} else {
    if (_currentStep - index == 1) {                 //当前加载的前一个阶段
        return [UIColor colorWithRed:158.0f/255.0f
                        green:172.0f/255.0f
                        blue:203.0f/255.0f
                        alpha:1];
    } else {
        return [UIColor colorWithRed:239.0f/255.0f
                                                     //前一阶段加载以后再加载一个阶段
                        green:242.0f/255.0f
                        blue:246.0f/255.0f
                        alpha:1];
    }
}
```

2．指示器的加载

对于指示器的加载，本实例采用了定时器来实现。在定时器中每隔_duration/(_dotCount*2+1)秒就会执行一次 repeatAnimation 方法，代码如下：

```
_timer = [NSTimer scheduledTimerWithTimeInterval:_duration/(_dotCount*
2+1)
                            target:self
                            selector:@selector(repeatAnimation)
                            userInfo:nil
                            repeats:YES];
```

其中，repeatAnimation 方法实现的功能是将当前加载的阶段进行加载后，再对视图中的内容进行重新绘制，它会自动调用 drawRect 方法。

实例 28　Windows Phone 风格的指示器

【实例描述】
本实例实现的功能是模仿 Windows Phone 上的指示器加载效果。当用户单击"暂停"按钮，指示器就会隐藏，并显示一个"播放"按钮。当用户单击"播放"按钮，会出现一个加载的指示器。

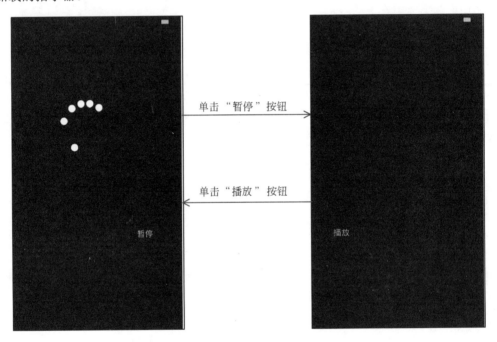

图 4.27　实例 28 运行效果

【实现过程】
在 Windows Phone 中，它的指示器是由多个圆组成的。所以要绘制圆，并实现它的加载。具体的实现步骤如下。
（1）创建一个项目，命名为"Windows Phone 风格的指示器"。
（2）创建一个基于 UIView 类的 Circle 类。
（3）打开 Circle.h 文件，编写代码，实现属性的声明。程序代码如下：

```
#import <UIKit/UIKit.h>
@interface Circle : UIView
@property (nonatomic,retain) UIColor *color;
@end
```

（4）打开 Circle.m 文件，编写代码。在这个文件中，代码分为两个部分：初始化和绘制圆。其中使用 drawRect:方法实现对圆的绘制。程序代码如下：

```
- (void)drawRect:(CGRect)rect
{
    CGContextRef context = UIGraphicsGetCurrentContext();
    [color set];
    CGContextFillEllipseInRect(context, CGRectMake(self.frame.origin.x,
    self.frame.origin.y, self.frame.size.width, self.frame.size.height));
                                                              //绘制椭圆
    CGContextAddArc(context, self.frame.size.width/2.0, self.frame.size.
    height/2.0, self.frame.size.height/2.0 -1.0, 0.0, 2.0 * M_PI, YES);
                                                              //绘制曲线
    CGContextDrawPath(context, kCGPathFill);
}
```

（5）创建一个基于 UIView 类的 WindowsActivityIndicatorView 类。

（6）打开 WindowsActivityIndicatorView.h 文件，编写代码。实现头文件、实例变量、对象、属性以及方法的声明。程序代码如下：

```
#import <UIKit/UIKit.h>
#import "Circle.h"                                            //头文件
@interface WindowsActivityIndicatorView : UIView{
    //实例变量
    BOOL isAnimating;
    int circleNumber;
    int maxCircleNumber;
    float circleSize;
    float radius;
    //对象
    UIColor *color;
    NSTimer *circleDelay;
}
//属性
@property (nonatomic,retain) UIColor *color;
//方法
-(id)initWithFrame:(CGRect)frame andColor:(UIColor*)theColor;
-(void)startAnimating;
-(void)stopAnimating;
-(BOOL)isAnimating;
@end
```

（7）打开 WindowsActivityIndicatorView.m 文件，编写代码。实现仿 Windows Phone 上加载中的效果。使用的方法如表 4-14 所示。

表 4-14 WindowsActivityIndicatorView.m 文件中方法总结

方　　法	功　　能
commonInit	初始化设置
initWithFrame:	使用框架对指示器初始化
initWithFrame: andColor:	使用框架和颜色对指示器初始化
initWithCoder:	初始化实例化对象
setFrame:	设置框架
startAnimating	动画的实现
nextCircle	创建圆
stopAnimating	停止动画
isAnimating	获取动画的有无

在文件中，代码分为初始化以及实现加载动画。这里需要讲解几个重要的方法（其他方法请读者参考源代码）。指示器的初始化需要使用 commonInit、initWithFrame:、initWithFrame: andColor:、initWithCoder:和 isAnimating 方法共同实现。其中 commonInit 方法对是否有动画以及圆的数目进行设置。程序代码如下：

```
- (void)commonInit
{
    isAnimating = NO;                       //无动画
    maxCircleNumber = 6;                    //圆的个数
}
```

initWithFrame:方法使用框架对指示器初始化，程序代码如下：

```
- (id)initWithFrame:(CGRect)frame
{
    return [self initWithFrame:frame andColor:[UIColor whiteColor]];
}
```

实现加载动画，需要使用 setFrame:、startAnimating 和 nextCircle 方法共同实现。其中，setFrame:方法对框架进行设置。程序代码如下：

```
- (void)setFrame:(CGRect)frame
{
    [super setFrame:frame];
    //判断动画的有无
    if (isAnimating)
    {
        [self stopAnimating];               //结束动画
        [self startAnimating];              //开始动画
    }
}
```

startAnimating 方法用于对加载动画的实现，代码如下：

```
-(void)startAnimating{
    //判断是否开始动画
    if (!isAnimating){
        isAnimating = YES;
        circleNumber = 0;
        radius = self.frame.size.width/2;
        //判断宽度是否大于高度
        if (self.frame.size.width > self.frame.size.height){
            radius = self.frame.size.height/2;
        }
        circleSize = 15*radius/55;
        circleDelay = [NSTimer scheduledTimerWithTimeInterval: 0.20 target:
        self selector: @selector(nextCircle) userInfo: nil repeats: YES];
                                                        //创建时间定时器
    }
}
```

nextCircle 方法实现对圆的创建。程序代码如下：

```
-(void)nextCircle{
    //判断圆的个数是否小于指定圆的个数
    if (circleNumber<maxCircleNumber){
        circleNumber ++;
        Circle *circle = [[Circle alloc] initWithFrame:CGRectMake((self.
```

```
    frame.size.width-circleSize)/2 - 1, self.frame.size.height-
    circleSize -1, circleSize +2, circleSize+2)];        //创建圆对象
    circle.color = color;
    circle.backgroundColor = [UIColor clearColor];        //圆的背景
    [self addSubview:circle];
    CGMutablePathRef circlePath = CGPathCreateMutable();
    CGPathMoveToPoint(circlePath, NULL, self.frame.size.width/2,
self.frame.size.height circleSize/2);                    //设置开始点
    CGPathAddArc(circlePath, NULL, self.frame.size.width/2, self.frame.
    size.height/2, radius-15/2,
M_PI_2, -M_PI_2*3, NO);                                  //绘制弧线
    CAKeyframeAnimation *circleAnimation = [CAKeyframeAnimation
    animationWithKeyPath:@"position"];
    circleAnimation.duration = 1.5;                       //设置圆的动画时间
    circleAnimation.timingFunction = [CAMediaTimingFunction
    functionWithControlPoints:0.15f :0.60f :0.85f :0.4f];
    [circleAnimation setCalculationMode:kCAAnimationPaced];
    circleAnimation.path = circlePath;                    //设置路径
    circleAnimation.repeatCount = HUGE_VALF;              //设置重复次数
    [circle.layer addAnimation:circleAnimation forKey:@"circleAnimation"];
                                                          //添加动画

    CGPathRelease(circlePath);
    } else {
    [circleDelay invalidate];
    }
}
```

如果想要停止加载动画,可以使用 stopAnimating 方法实现。程序代码如下:

```
-(void)stopAnimating{
    isAnimating = NO;
    for (UIView *v in self.subviews){
    [v removeFromSuperview];                              //移除视图
    }
}
```

(8) 打开 ViewController.h 文件,编写代码。实现头文件、对象、插座变量以及动作的声明。程序代码如下:

```
#import <UIKit/UIKit.h>
//头文件
#import "WindowsActivityIndicatorView.h"
@interface ViewController : UIViewController{
    WindowsActivityIndicatorView *windows;
    //按钮的插座变量
    IBOutlet UIButton *button1;
    IBOutlet UIButton *button2;
}
//动作
- (IBAction)play:(id)sender;
- (IBAction)pause:(id)sender;
@end
```

(9) 打开 Main.storyboard 文件,对 View Controller 控制器的设计界面进行设计,效果如图 4.28 所示。

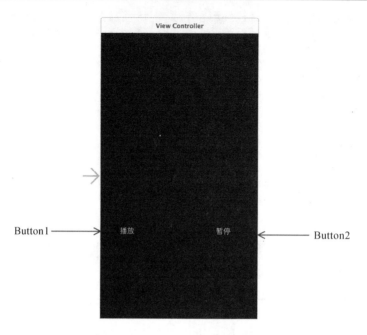

图 4.28　实例 28 View Controller 视图控制器的设计界面

需要添加的视图、控件以及对它们的设置如表 4-15 所示。

表 4-15　实例 28 视图、控件设置

视图、控件	属　性　设　置	其　　他
设计界面	Background：黑色	
Button1	Title：播放	与插座变量 button1 关联 与动作 play 关联
Button2	Title：暂停	与插座变量 button2 关联 与动作 pause 关联

（10）打开 ViewController.m 文件，编写代码。实现指示器对象的创建与加载。程序代码如下：

```
- (void)viewDidLoad
{
    [button1 setHidden:YES];
    windows= [[WindowsActivityIndicatorView alloc] initWithFrame:CGRectMake
    (90, 150, 100, 100)];
    [windows startAnimating];                        //实现动画的加载
    [self.view addSubview:windows];
    [super viewDidLoad];
    // Do any additional setup after loading the view, typically from a nib.
}
//实现加载动画
- (IBAction)play:(id)sender {
    [windows startAnimating];                        //开始动画
    [button1 setHidden:YES];
    [button2 setHidden:NO];
}
//实现动画的停止
- (IBAction)pause:(id)sender {
```

```
        [button1 setHidden:NO];
        [button2 setHidden:YES];
        [windows stopAnimating];                                    //结束动画
}
```

【代码解析】

本实例关键功能是指示器的加载动画。下面就对这个知识点做一个详细的讲解。

对于指示器的加载,本实例采用了定时器来实现。在定时器中每隔 0.20 秒就会执行一次 nextCircle 方法,代码如下:

```
circleDelay = [NSTimer scheduledTimerWithTimeInterval: 0.20 target:
selfselector: @selector(nextCircle) userInfo: nil repeats: YES];
```

其中,nextCircle 方法中实现了对圆的创建以及圆的动画效果。

实例 29 三个方块组成的指示器

【实例描述】

本实例实现的是由三个方块组成的指示器的加载效果。运行效果如图 4.29 所示。

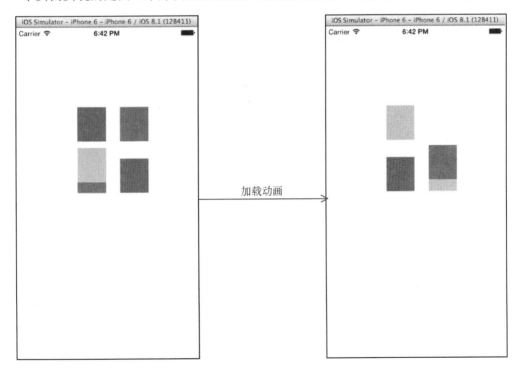

图 4.29 实例 29 运行效果

【实现过程】

(1) 创建一个项目,命名为"三个方块组成的指示器"。
(2) 创建一个基于 UIView 类的 ActivityIndicatorview 类。
(3) 打开 ActivityIndicatorview.h 文件,编写代码。实现宏定义、属性以及方法的声明。

程序代码如下：

```
#import <UIKit/UIKit.h>
//宏定义
#define AnimationTime       0.3
#define kDotSize            (CGSizeMake(0.4 * self.frame.size.width, 0.4
* self.frame.size.height))
#define LeftTopPosition     (CGPointMake(0, 0))
#define LeftTBottomPosition (CGPointMake(0, 0.6 * self.frame.size.height))
#define RightBottomPosition (CGPointMake(0.6 * self.frame.size.width,
0.6 * self.frame.size.height))
#define RightTopPosition    (CGPointMake(0.6 * self.frame.size.width, 0))
#define kDotColor           [UIColor colorWithRed:200/255.0 green:206/
255.0 blue:221/255.0 alpha:1.0]
@interface ActivityIndicatorview : UIView
//属性
@property (strong, nonatomic) UIView *dotView0,*dotView1,*dotView2;
@property (strong, nonatomic) NSArray *dotViews;
@property (assign, nonatomic) NSInteger dotIndex;
@property (strong, nonatomic) NSTimer *timer;
//方法
-(void)showView:(BOOL)show;
@end
```

（4）打开 ActivityIndicatorview.m 文件，编写代码。这里需要讲解几个重要的方法（其他方法请读者参考源代码）。添加 initView 方法实现对视图的初始化。程序代码如下：

```
- (void)initView
{
    self.backgroundColor = [UIColor clearColor];                    //设置背景
    self.clipsToBounds = YES;
    self.userInteractionEnabled = NO;
    _dotView0 = [[UIView alloc]initWithFrame:(CGRect){RightBottomPosition,
kDotSize}];
    _dotView0.backgroundColor = [UIColor greenColor];               //设置背景
    [self addSubview:_dotView0];                                    //添加视图对象_dotView0
    _dotView1 = [[UIView alloc]initWithFrame:(CGRect){LeftTBottomPosition,
kDotSize}];
    _dotView1.backgroundColor = [UIColor redColor];
    [self addSubview:_dotView1];                                    //添加视图对象_dotView1
    _dotView2 = [[UIView alloc]initWithFrame:(CGRect){LeftTopPosition,
kDotSize}];
    _dotView2.backgroundColor = [UIColor blueColor];
    [self addSubview:_dotView2];                                    //添加视图对象_dotView2
    _dotViews = @[_dotView0, _dotView1, _dotView2];
    _dotIndex = 0;
}
```

添加 showView:方法实现视图的显示。程序代码如下：

```
-(void)showView:(BOOL)show
{
    if (show) {
        self.hidden = NO;
        if (!_timer) {
            _timer=[NSTimer scheduledTimerWithTimeInterval:AnimationTime
            target:self selector:@selector(beginAnimation) userInfo:nil
            repeats:YES];                                           //创建时间定时器
        }
```

```
    }
    else{
        [_timer invalidate];                                    //让定时器失效
        _timer = nil;
        self.hidden = YES;
    }
}
```

beginAnimation 方法实现加载动画。程序代码如下：

```
-(void)beginAnimation
{
    [UIView animateWithDuration:AnimationTime delay:0 options:
    UIViewAnimationOptionCurveLinear animations:^{
        UIView *dotView = _dotViews[_dotIndex];                 //实例化对象
        [self moveDotViewToNextPosition:dotView];               //移动视图
        _dotIndex ++;
        _dotIndex = _dotIndex > 2 ? 0 : _dotIndex;
    } completion:nil];
}
```

moveDotViewToNextPosition:方法实现视图的位置移动。程序代码如下：

```
-(void)moveDotViewToNextPosition:(UIView*)dotView
{
    if (CGPointEqualToPoint(dotView.frame.origin, LeftTopPosition)) {
        dotView.frame = (CGRect){LeftTBottomPosition, dotView.frame.size};
                                                                //设置框架
    }
    else if (CGPointEqualToPoint(dotView.frame.origin, LeftTBottomPosition)) {
        dotView.frame = (CGRect){RightBottomPosition, dotView.frame.size};
                                                                //设置框架
    }
    else if (CGPointEqualToPoint(dotView.frame.origin, RightBottomPosition)) {
        dotView.frame = (CGRect){RightTopPosition, dotView.frame.size};
                                                                //设置框架
    }
    else if (CGPointEqualToPoint(dotView.frame.origin, RightTopPosition))
    {
        dotView.frame = (CGRect){LeftTopPosition, dotView.frame.size};
                                                                //设置框架
    }
}
```

（5）打开 ViewController.h 文件，编写代码，进行头文件和属性的声明。程序代码如下：

```
#import <UIKit/UIKit.h>
#import "ActivityIndicatorview.h"
@interface ViewController : UIViewController
@property (strong, nonatomic) IBOutlet ActivityIndicatorview *ac;
@end
```

（6）打开 Main.storyboard 文件，在设计界面中放入一个 View 空白视图，将 Class 设置为 ActivityIndicatorview。将此视图与属性声明的插座变量 ac 进行关联。

（7）打开 ViewController.m 文件，编写代码，实现指示器的加载。程序代码如下：

```
- (void)viewDidLoad
{
    [super viewDidLoad];
```

```
    // Do any additional setup after loading the view, typically from a nib.
    [ac showView:YES];
}
```

【代码解析】

本实例关键功能是指示器加载时方块的移动。下面就对这个知识点做一个详细的讲解。

指示器的加载动画的移动，本实例采用了订制边界的方法。当三个方块中的其中一个方块到达边界位置，它就是做出相应的移动。程序代码如下：

```
if (CGPointEqualToPoint(dotView.frame.origin, LeftTopPosition)) {
                                                //到达指定左边向上的指定位置
    dotView.frame = (CGRect){LeftTBottomPosition, dotView.frame.size};
}
else if (CGPointEqualToPoint(dotView.frame.origin, LeftTBottomPosition)) {
                                                //到达指定左边向下的指定位置
    dotView.frame = (CGRect){RightBottomPosition, dotView.frame.size};
}
else if (CGPointEqualToPoint(dotView.frame.origin, RightBottomPosition)) {
                                                //到达指定右边向下的指定位置
    dotView.frame = (CGRect){RightTopPosition, dotView.frame.size};
}
else if (CGPointEqualToPoint(dotView.frame.origin, RightTopPosition)) {
                                                //到达指定右边向上的指定位置
    dotView.frame = (CGRect){LeftTopPosition, dotView.frame.size};
}
```

实例 30　三个点的指示器

【实例描述】

本实例的功能是由三个点实现指示器的加载效果，运行效果如图 4.30 所示。

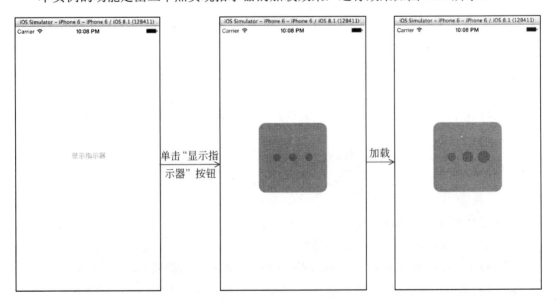

图 4.30　实例 30 运行效果

【实现过程】

(1) 创建一个项目,命名为"三个点指示器"。

(2) 创建一个基于 UIView 类的 circle 类。

(3) 打开 circle.m 文件编写代码,实现初始化以及点的绘制。要实现点的绘制,需要使用 drawRect:方法。程序代码如下:

```
- (void)drawRect:(CGRect)rect {
   CGContextRef context = UIGraphicsGetCurrentContext();
   CGContextSetFillColorWithColor(context, [UIColor redColor].CGColor);
   CGContextFillEllipseInRect(context, rect);       //填充指定矩形中的椭圆
}
```

(4) 创建一个基于 UIView 类的 HUD 类。

(5) 打开 HUD.h 文件,编写代码。实现头文件、对象、属性以及方法的声明。程序代码如下:

```
#import <UIKit/UIKit.h>
#import "circle.h"                                   //头文件
@interface HUD : UIView{
   //对象
   circle *leftCircle;
   circle *middleCircle;
   circle *rightCircle;
}
//属性
@property BOOL animating;
@property float interval;
@property int pointDiameter;
//方法
- (id)initWithFrame:(CGRect)frame withPointDiameter:(int)diameter withInterval: (float)interval;
@end
```

(6) 打开 HUD.m 文件,编写代码,实现三个点的指示器。添加 initWithFrame:withPointDiameter:withInterval:方法实现指示器的初始化。程序代码如下:

```
- (id)initWithFrame:(CGRect)frame withPointDiameter:(int)diameter withInterval:(float)interval {
   if ((self = [super initWithFrame:frame])) {
      self.interval = interval;
      self.pointDiameter = diameter;
      self.backgroundColor = [UIColor colorWithRed:0 green:0 blue:0 alpha:0.5];                                //设置背景颜色
      self.layer.cornerRadius = 15;                  //圆角半径
      leftCircle = [self addCircleViewWithXOffsetFromCenter:-35];
      middleCircle = [self addCircleViewWithXOffsetFromCenter:0];
      rightCircle = [self addCircleViewWithXOffsetFromCenter:35];
      self.animating = YES;
      NSArray *circles = @[leftCircle, middleCircle, rightCircle];
      //为视图添加动画
      [self animateWithViews:circles index:0 delay:self.interval*2 offset:self.pointDiameter];
      [self animateWithViews:circles index:1 delay:self.interval*3 offset:self.pointDiameter];
      [self animateWithViews:circles index:2 delay:self.interval*4 offset:self.pointDiameter];
```

```
    }
    return self;
}
```

addCircleViewWithXOffsetFromCenter:方法实现使用偏移量来添加圆。程序代码如下：

```
- (circle *)addCircleViewWithXOffsetFromCenter:(float)offset {
    CGRect rect = CGRectMake(0, 0, self.pointDiameter, self.pointDiameter);
    circle *cir=[[circle alloc]initWithFrame:rect];
    cir.center = self.center;                               //设置中心位置
    cir.frame = CGRectOffset(cir.frame, offset, 0);         //设置框架
    [self addSubview:cir];
    return cir;
}
```

animateWithViews:index:offset:方法实现指示器的加载动画。程序代码如下：

```
- (void)animateWithViews:(NSArray*)circles index:(int)index delay:(float)
delay offset:(float)offset {
    UIView *view = ((circle*)[circles objectAtIndex:index]);
    [UIView animateWithDuration:0.2 delay:delay options:UIViewAnimation-
    OptionCurveEaseIn animations:^{
        view.frame = CGRectMake(view.frame.origin.x - offset/2,
                                view.frame.origin.y - offset/2,
                                view.frame.size.width + offset,
                                view.frame.size.height + offset);    //设置框架
    } completion:^(BOOL finished) {
        [UIView animateWithDuration:0.5 delay:0 options:UIViewAnimation-
        OptionCurveEaseOut animations:^{
            view.frame = CGRectMake(view.frame.origin.x + offset/2,
                                    view.frame.origin.y + offset/2,
                                    view.frame.size.width - offset,
                                    view.frame.size.height - offset);//设置框架
        } completion:^(BOOL finished) {
            if (self.animating) {
                if (index == 2) {
                    //为视图添加动画
                    [self animateWithViews:circles index:0 delay:self.
                    interval*2 offset:
self.pointDiameter];
                    [self animateWithViews:circles index:1 delay:self.
                    interval*3 offset:
self.pointDiameter];
                    [self animateWithViews:circles index:2 delay:self.
                    interval*4 offset:
self.pointDiameter];
                }
            }
        }];
    }];
}
```

（7）打开 ViewController.h 文件，编写代码。实现头文件、对象、插座变量以及动作的声明。程序代码如下：

```
#import <UIKit/UIKit.h>
#import "HUD.h"
@interface ViewController : UIViewController{
    HUD *hub;
```

```
    IBOutlet UIButton *button;                    //插座变量
}
- (IBAction)show:(id)sender;
@end
```

（8）打开 Main.storyboard 文件，在设计界面中放入一个 Button 按钮，将 Title 设置为"显示指示器"，将此按钮和插座变量 button 进行关联，并将此按钮和动作 show:进行关联。

（9）打开 ViewController.m 文件，编写代码。实现单击按钮后，出现三个点的指示器。程序代码如下：

```
- (IBAction)show:(id)sender {
  button.hidden=YES;
  hub=[[HUD alloc]initWithFrame:CGRectMake(0, 0, 150, 150) withPointDiameter:
  16 withInterval:0.25];
  hub.center=self.view.center;                    //设置中心位置
  [self.view addSubview:hub];
}
```

【代码解析】

本实例关键功能是三个点加载时的放大缩小动画。下面就对这个知识点做一个详细的讲解。

三个点放大缩小的动画，可以使用 UIView 的 animateWithDuration: delay: options: animations: completion:方法，其语法形式如下：

```
+ (void)animateWithDuration:(NSTimeInterval)duration delay:(NSTimeInterval)
delay options:
(UIViewAnimationOptions)options animations:(void (^)(void))animations
completion:
(void (^)(BOOL finished))completion;
```

其中，它们的参数说明如下。

- ❑ (NSTimeInterval)duration 表示动画的持续时间。
- ❑ (NSTimeInterval)delay 表示动画开始之前的延续时间。
- ❑ (UIViewAnimationOptions)options 表示一种位掩码，用来描述额外的选项，默认情况下，它是 UIViewAnimationOptionCurveEaseInOut 选项。
- ❑ (void (^)(void))animations 表示包含视图属性动态改变的块。
- ❑ (void (^)(BOOL finished))completion 表示当动画终止后所执行的块。它有一个 BOOL 参数，用来指定动画是否运行完结。

在代码中，使用了 animateWithDuration: delay: options: animations: completion:方法对三个点实现放大缩小动画，代码如下：

```
[UIView animateWithDuration:0.2 delay:delay options:UIViewAnimation-
OptionCurveEaseIn animations:^{
  view.frame = CGRectMake(view.frame.origin.x - offset/2,
                  view.frame.origin.y - offset/2,
                  view.frame.size.width + offset,
                  view.frame.size.height + offset);
} completion:^(BOOL finished) {
  [UIView animateWithDuration:0.5 delay:0 options:UIViewAnimation-
  OptionCurveEaseOut animations:^{
    view.frame = CGRectMake(view.frame.origin.x + offset/2,
```

```
                                view.frame.origin.y + offset/2,
                                view.frame.size.width - offset,
                                view.frame.size.height - offset);
        } completion:^(BOOL finished) {
            if (self.animating) {
                if (index == 2) {
                    [self animateWithViews:circles index:0 delay:self.
                    interval*2 offset:
self.pointDiameter];
                    [self animateWithViews:circles index:1 delay:self.
                    interval*3 offset:
self.pointDiameter];
                    [self animateWithViews:circles index:2 delay:self.
                    interval*4 offset:
self.pointDiameter];
                }
            }
        }];
}];
```

其中，它们的参数表示如表4-16所示。

表4-16 参数说明

参　　数	表　　示
0.2	动画的持续时间
delay	动画开始之前的延续时间
UIViewAnimationOptionCurveEaseIn	一种位掩码，用来描述额外的选项
view.frame = CGRectMake(view.frame.origin.x - offset/2, view.frame.origin.y - offset/2,view.frame.size.width + offset, view.frame.size.height + offset)	设置视图属性的框架

[UIView animateWithDuration:0.5 delay:0 options:UIViewAnimationOption-CurveEaseOut animations:^{
　　　view.frame = CGRectMake(view.frame.origin.x + offset/2, view.frame.origin.y + offset/2,view.frame.size.width - offset,view.frame.size.height - offset);
}
	视图属性动态改变的块

completion:^(BOOL finished) {
　　if (self.animating) {
　　　　if (index == 2) {
　　　　　　[self animateWithViewS:circles index:0 delay:self.interval*2 offset:self.pointDiameter];
　　　　　　[self animateWithViewS:circles index:1 delay:self.interval*3 offset:self.pointDiameter];
　　　　　　[self animateWithViewS:circles index:2 delay:self.interval*4 offset:self.pointDiameter];
　　　　}
　　}
}];
	当动画终止后所执行的块

实例 31　多个方块组成的指示器

【实现描述】

本实例实现的功能是一个由多个方块组成的指示器的加载效果，运行效果如图 4.31 所示。

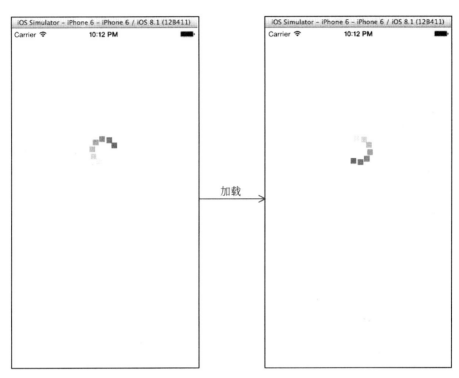

图 4.31　实例 31 运行效果

【实现过程】

（1）创建一个项目，命名为"由多个方块组成的指示器"。

（2）创建一个基于 UIView 类的 Indicator 类。

（3）打开 Indicator.h 文件，编写代码。实现对象、属性以及方法的声明。程序代码如下：

```
#import <UIKit/UIKit.h>
@interface Indicator : UIView{
    //对象
    NSTimer *animateTimer;
    NSMutableArray *sviews;
}
//属性
@property (nonatomic,assign) NSUInteger numOfObjects;
@property (nonatomic,assign) CGSize objectSize;
@property (nonatomic,retain) UIColor *color;
//方法
-(void)startAnimating;
@end
```

（4）打开 Indicator.m 文件，编写代码。实现指示器的绘制以及实现的加载。使用的方法如表 4-17 所示。

表 4-17　Indicator.m 文件中方法总结

方　　法	功　　能
initWithFrame:	指示器的初始化
setNumOfObjects:	设置方块对象的个数
setObjectSize:	设置方块对象的尺寸
pointWithDistance:	设置点之间的距离
startAnimating	实现动画
next	实现指示器的加载

要实现指示器的绘制需要使用 initWithFrame:、setNumOfObjects:、setObjectSize: 和 pointWithDistance: 方法共同实现。其中，initWithFrame: 方法对指示器实现初始化。程序代码如下：

```
- (id)initWithFrame:(CGRect)frame
{
    self = [super initWithFrame:frame];
    if (self) {
        sviews = [NSMutableArray array];
        _numOfObjects = 10;
        //遍历
        for (int i=0; i<_numOfObjects; ++i) {
            UIView *sview = [[UIView alloc] init];
            sview.tag = i;
            [sviews addObject:sview];                  //添加对象
            [self addSubview:sview];                   //添加视图对象
        }
        self.objectSize = CGSizeMake(10, 10);          //尺寸
        self.backgroundColor = [UIColor clearColor];   //设置背景颜色
        animateTimer = nil;
        _color = [UIColor blueColor];
    }
    return self;
}
```

使用 setNumOfObjects: 方法设置方块对象的个数。程序代码如下：

```
-(void)setNumOfObjects:(NSUInteger)numOfObjects1
{
    _numOfObjects = numOfObjects1;
    [sviews makeObjectsPerformSelector:@selector(removeFromSuperview)];
    [sviews removeAllObjects];                        //移除所有的sviews中的对象
    //遍历
    for (int i=0; i<_numOfObjects; ++i) {
        UIView *sview = [[UIView alloc] init];
        sview.tag = i;                                 //设置tag值
        [sviews addObject:sview];
        [self addSubview:sview];
    }
    self.objectSize = _objectSize;
}
```

setObjectSize:方法设置方块的大小。程序代码如下:

```
-(void)setObjectSize:(CGSize)objectSize1
{
    _objectSize = objectSize1;
    float r = MIN(self.frame.size.width, self.frame.size.height)/2;
    float wh = MAX(_objectSize.width, _objectSize.height);
    //遍历
    for (int i=0; i<sviews.count; ++i) {
        UIView *subview = sviews[i];
        subview.frame = CGRectMake(0, 0, _objectSize.width, _objectSize.height);
                                                                    //设置框架
        subview.center = [self pointWithDistance:(r-wh/2) Angel:M_PI_4+
        M_PI*2/_numOfObjects * i];
    }
}
```

pointWithDistance:方法设置点与点之间的距离。程序代码如下:

```
-(CGPoint)pointWithDistance:(float)r  Angel:(float)angel
{
    CGPoint c = CGPointMake(self.frame.size.width/2, self.frame.size.
    height/2);                                                      //设置点
    float dy = r*sin(angel);
    float dx = r*cos(angel);
    return CGPointMake(c.x+dx, c.y+dy);
}
```

这时一个指示器就绘制好了,但是此指示器就不会实现加载,要想实现加载,需要使用 startAnimating 和 next 方法共同实现。其中,会在 startAnimating 方法中创建一个定时器。程序代码如下:

```
-(void)startAnimating
{
    animateTimer = [NSTimer scheduledTimerWithTimeInterval:0.5/
    (_numOfObjects/2.) target:self selector:@selector(next)
    userInfo:nil repeats:YES];                              //创建时间定时器
}
```

next 方法实现指示器的加载。实现代码如下:

```
-(void)next
{
    for (int i=0; i<sviews.count; ++i) {
        UIView *subview = sviews[i];
        subview.tag = subview.tag-1;                        //设置 tag 值
        if (subview.tag<0) {
            subview.tag = _numOfObjects-1;
        }
        float alpha = (subview.tag-_numOfObjects/4.+1)/((float)_numOfObjects);
        if (alpha<0) {
            alpha = 0;                                      //设置透明度
        }
        subview.alpha = alpha;
        subview.backgroundColor = _color;                   //设置背景颜色
```

    }
}

（5）打开 ViewController.m 文件，编写代码。实现指示器对象的创建以及加载。程序代码如下：

```
- (void)viewDidLoad
{
    Indicator*indicator = [[Indicator alloc]initWithFrame:CGRectMake(135,
    180, 50, 50)];
    [indicator startAnimating];                    //实现加载
    [self.view addSubview:indicator];
    [super viewDidLoad];
    // Do any additional setup after loading the view, typically from a nib.
}
```

【代码解析】

本实例关键功能是指示器的自动动画。下面就对这个知识点做一个详细的讲解。

如果想要实现具有很多方块的指示器的加载，其实是通过 next 方法中对颜色透明度的变化实现的。程序代码如下：

```
for (int i=0; i<sviews.count; ++i) {
    UIView *subview = sviews[i];
    subview.tag = subview.tag-1;
    if (subview.tag<0) {
        subview.tag = _numOfObjects-1;
    }
    float alpha = (subview.tag-_numOfObjects/4.+1)/((float)_numOfObjects);
    if (alpha<0) {
        alpha = 0;
    }
    subview.alpha = alpha;
    subview.backgroundColor = _color;
}
```

定时器会每隔 0.5/(_numOfObjects/2.)秒，调用一次 next 方法，从而实现加载动画。程序代码如下：

```
animateTimer = [NSTimer scheduledTimerWithTimeInterval:0.5/(_numOfObjects/
2.) target:self selector:@selector(next) userInfo:nil repeats:YES];
```

实例 32 仿 Yahoo 天气应用的加载效果

【实例描述】

本实例实现的功能是模拟 Yahoo 天气应用的加载效果，当加载结束后，会出现一个具有动画效果的图片。运行效果如图 4.32 所示。

【实例实现】

（1）创建一个项目，命名为"仿 Yahoo 天气应用的加载效果"。

（2）添加图片 1.png 和 2.png 到创建项目的 Supporting Files 文件夹。

（3）创建一个基于 UIView 类的 load 类。

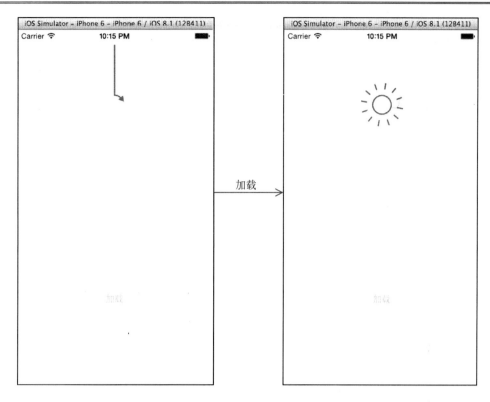

图 4.32　实例 32 运行效果

（4）打开 load 类，编写代码。实现协议、实例变量、属性和方法的声明。程序代码如下：

```
#import <UIKit/UIKit.h>
//协议
@protocol loadDelegate <NSObject>
@optional
- (void)startLoading;
- (void)stopLoading;
@end
@interface load : UIView{
    //实例变量
    float radius;
    CGPoint center;
    ……
    float _spring;
    int _springDirection;
    int _counter;
}
@property (assign, nonatomic) id<loadDelegate> delegate;           //属性
//方法
- (void)didScroll:(float)offset;
- (void)stopLoading;
@end
```

（5）打开 load.m 文件，编写代码，实现仿 Yahoo 天气应用的加载效果。使用的方法如表 4-18 所示。

表 4-18 load.m文件中方法总结

方　　法	功　　能
Radian2Degree	获取弧度
Degree2Radian	获取度
initWithFrame:	指示器的初始化
didScroll:	滚动
stopLoading	停止加载
drawRect:	绘制

在文件中，代码分为：绘制指示器和实现加载。这里需要讲解几个重要的方法（其他方法请读者参考源代码）。绘制指示器需要使用 Radian2Degree、Degree2Radian、initWithFrame:和 drawRect:方法共同实现。其中 initWithFrame:方法实现指示器的初始化。程序代码如下：

```
- (id)initWithFrame:(CGRect)frame {
  self = [super initWithFrame:frame];
  if (self) {
      self.backgroundColor = [UIColor clearColor];
      center = self.center;
      radius = 15;
      arrowEdgeLength = 10;                    //箭头
      radianceDegree = 30;
      radianceOffset = 5;
      radianceMinLength = 5;                   //圆上旋转的线
      radianceMaxLength = 15;
      sprintMax = 1;
      rotationSpeed = 1.0f;
      _startAngle = -90;                       //设置开始角度
      _endAngle = -90;                         //设置结束角度
      _startY = -(2.0f * M_PI * radius);
      _spring = 0;
      _springDirection = 0;
  }
  return self;
}
```

drawRect:方法实现指示器的绘制。在此方法中，代码分为两个部分：绘制线和圆弧、判断开始线 y 的位置，根据开始线的 Y 轴的不同，实现不同的效果。绘制直线和圆弧，程序代码如下：

```
CGContextRef ctx = UIGraphicsGetCurrentContext();
CGContextSaveGState(ctx);
CGContextSetShouldAntialias(ctx, YES);                    //设置锯齿的开关
CGContextSetAllowsAntialiasing(ctx, YES);
CGContextBeginPath(ctx);
CGContextSetStrokeColorWithColor(ctx, [UIColor blueColor].CGColor);
                                                          //线的颜色
CGContextSetLineWidth(ctx, 2);
CGContextMoveToPoint(ctx, center.x, center.y - radius + _startY);
CGContextAddLineToPoint(ctx, center.x, center.y - radius);
CGContextAddArc(ctx, center.x, center.y, radius, Degree2Radian
(_startAngle), Degree2Radian(_endAngle), NO);             //绘制弧线
CGPoint point = CGContextGetPathCurrentPoint(ctx);
CGContextDrawPath(ctx, kCGPathStroke);
```

判断开始线 y 的位置，从而实现不同的效果。当开始线 y 的位置小于 0 时，程序代码如下：

```
float degree = 0;
if ((point.x < center.x && point.y > center.y) || (point.x > center.x
&& point.y > center.y)) {
    degree = -90;
} else {
    degree = 90;
}
float k = -(point.x - center.x) / (point.y - center.y);
CGContextTranslateCTM(ctx, point.x, point.y);           //平移坐标轴
CGContextRotateCTM(ctx, Degree2Radian(degree) + atan(k));
CGContextMoveToPoint(ctx, 0, -1.0f / sqrtf(3) * arrowEdgeLength);
                                                         //开始点
CGContextAddLineToPoint(ctx, -arrowEdgeLength / 2.0f, sqrtf(3) /
6.0f * arrowEdgeLength);
CGContextAddLineToPoint(ctx, arrowEdgeLength / 2.0f, sqrtf(3) / 6.0f
* arrowEdgeLength);
CGContextClosePath(ctx);                                 //关闭路径
CGContextSetFillColorWithColor(ctx, [UIColor blueColor].CGColor);
                                                         //填充颜色
CGContextFillPath(ctx);
```

开始线 y 的位置大于 0 时，程序代码如下：

```
CGContextTranslateCTM(ctx, center.x, center.y);
//绘制圆周的线
for (int i = 0; i < (Radian2Degree(M_PI * 2) / radianceDegree); i++) {
    if (i > 0) {
        CGContextRotateCTM(ctx, Degree2Radian(radianceDegree));
                                                         //旋转坐标轴
    } else {
        CGContextRotateCTM(ctx, Degree2Radian(_counter * rotationSpeed));
    }
    //判断_springDirection 是否为 0
    if (_springDirection == 0) {
        _spring += 0.005f;
        //判断_spring 是否大于等于 sprintMax
        if (_spring >= sprintMax) {
            _springDirection = 1;
        }
    }
    //判断_springDirection 是否为 1
    if (_springDirection == 1) {
        _spring -= 0.005f;
        if (_spring <= 0) {
            _springDirection = 0;
        }
    }
    if (i % 2 == 1) {
        CGContextMoveToPoint(ctx, 0, -radius - radianceOffset -
        radianceMinLength +
_spring);                                    //开始点
        CGContextAddLineToPoint(ctx, 0, -radius - radianceOffset -
        radianceMaxLength -
_spring);                                    //结束点
    } else {
```

```
            CGContextMoveToPoint(ctx, 0, -radius - radianceOffset -
            radianceMinLength -
_spring);
            CGContextAddLineToPoint(ctx, 0, -radius - radianceOffset -
            radianceMaxLength +
_spring);
        }
    }
    CGContextDrawPath(ctx, kCGPathStroke);                              //绘制
}
    CGContextRestoreGState(ctx);
```

这时的指示器还是不会出现加载动画,要实现加载动画需要使用 didScroll:方法。程序代码如下:

```
- (void)didScroll:(float)offset {
    //判断开始线 y 的位置
    if (_startY == 0) {
        if (_delegate && [_delegate respondsToSelector:@selector(startLoading)]) {
            [_delegate startLoading];                                   //开始加载
        }
        _startY = 1;
    } else if (_startY == 1) {
        if (_counter == (NSUIntegerMax - 1)) {
            _counter = 10;
        }
        _counter++;
        [self setNeedsDisplay];                                         //重新绘制
    } else {
        _startY += offset;
        float deltaAngle = Radian2Degree(offset / radius);
        _endAngle += deltaAngle;
        if (roundf(_startY) >= 0) {
            _endAngle = 270;
            _startY = 0;
        }
        [self setNeedsDisplay];                                         //重新绘制
    }
}
```

如果想要指示器停止加载动画,需要使用 stopLoading 方法实现。程序代码如下:

```
- (void)stopLoading {
    _endAngle = -90;
    _startY = -(2.0f * M_PI * radius);
    [self setNeedsDisplay];
    if (_delegate && [_delegate respondsToSelector:@selector(stopLoading)]) {
        [_delegate stopLoading];                                        //调用停止加载的方法
    }
}
```

(6)打开 ViewController.h 文件,编写代码。实现头文件、插座变量、对象以及动作的声明。程序代码如下:

```
#import <UIKit/UIKit.h>
#import "load.h"                                                        //头文件
@interface ViewController : UIViewController<loadDelegate>{
    //对象
    load *indicator;
```

```
    NSTimer *ticker;
    NSTimer *stopTimer;
    //插座变量
    IBOutlet UIButton *button;              //按钮的插座变量
    IBOutlet UIView *vv;                    //空白视图的插座变量
    IBOutlet UIImageView *imv;              //图像视图的插座变量
}
动作
- (IBAction)loading:(id)sender;
@end
```

（7）打开 Main.storyboard 文件，对 View Controller 视图控制器的设计界面进行设计，效果如图 4.33 所示。

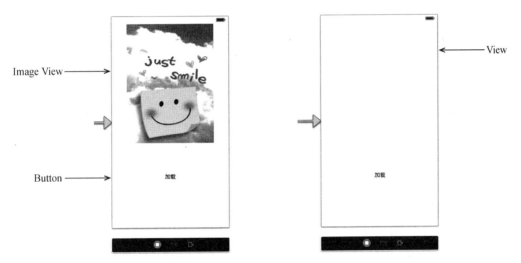

图 4.33　实例 32 View Controller 视图控制器的设计界面

需要添加的视图、控件以及对它们的设置如表 4-19 所示。

表 4-19　实例 32 视图、控件设置

视图、控件	属 性 设 置	其　　他
Image View	Image：1.png	与插座变量 imv 关联
Button	Title：加载	与插座变量 button 关联 与动作 loading:方法关联
View		与插座变量关联

（8）打开 ViewController.m 文件，编写代码。在文件中代码分为三个部分：创建指示器、单击按钮实现加载动画，以及加载结束显示图片。其中在 viewDidLoad 方法中实现指示器对象的创建，并显示。程序代码如下：

```
- (void)viewDidLoad
{
    [super viewDidLoad];
    // Do any additional setup after loading the view, typically from a nib.
    indicator=[[load alloc]initWithFrame:CGRectMake(0, 0, vv.bounds.size.
    width,
    vv.bounds.size.height-100)];                                    //实例化
```

```objc
    indicator.delegate = self;
    [vv addSubview:indicator];                              //添加视图对象
    button.enabled=NO;
    ticker = [NSTimer scheduledTimerWithTimeInterval:1.0f / 33.0f
    target:self selector:@selector(tick) userInfo:nil repeats:YES];
                                                            //创建时间定时器
}
```

单击按钮后,实现加载动画,需要 loading:方法外,还需要添加 tick 方法共同实现。程序代码如下:

```objc
- (IBAction)loading:(id)sender {
    ticker = [NSTimer scheduledTimerWithTimeInterval:1.0f / 33.0f
    target:self selector:@selector(tick) userInfo:nil repeats:YES];
                                                            //创建时间定时器
    button.enabled = NO;
    [vv setHidden:NO];
}
- (void)tick {
    [indicator didScroll:0.8f];
}
```

在加载一定的时间后,会停止加载,并显示一个具有动画效果的图片。要实现此功能,需要添加 stop、startLoading、stopLoading、animate1 和 animate2 方法共同实现。程序代码如下:

```objc
- (void)stop {
    [indicator stopLoading];                                //调用停止加载方法
}
- (void)startLoading {
    stopTimer = [NSTimer scheduledTimerWithTimeInterval:5 target:self
    selector:@selector(stop) userInfo:nil repeats:NO] ;     //创建定时器
}
- (void)stopLoading {
    [ticker invalidate];                                    //让定时器失效
    ticker = nil;
    [stopTimer invalidate];
    stopTimer = nil;
    button.enabled = YES;
    [vv setHidden:YES];                                     //隐藏
    [self performSelector:@selector(animate1) withObject:self afterDelay:1];
}
//实现显示动态的图片
-(void)animate1{
    imv.image=[UIImage imageNamed:@"2.png"];                //设置图像
    [self performSelector:@selector(animate2) withObject:self afterDelay:1];
}
-(void)animate2{
    imv.image=[UIImage imageNamed:@"1.png"];                //设置图像
    [self performSelector:@selector(animate1) withObject:self afterDelay:1];
}
```

【代码解析】

本实例关键功能是指示器加载时,图像的变化。下面就对这个知识点做一个详细的讲解。

在本实例中指示器使用的方法是 didScroll:，此方法将指示器的加载分为了三个阶段，这些阶段是根据开始线 y 的位置进行判断的。代码如下：

```
if (_startY == 0) {
    //第二阶段实现调用协议中的 startLoading 方法
        if (_delegate && [_delegate respondsToSelector:@selector
        (startLoading)]) {
            [_delegate startLoading];
        }
        _startY = 1;
} else if (_startY == 1) {
    //第三阶段实现旋转
        if (_counter == (NSUIntegerMax - 1)) {
            _counter = 10;
        }
        _counter++;
        [self setNeedsDisplay];
} else {
    //第一阶段实现将线段绘制为圆
        _startY += offset;
        float deltaAngle = Radian2Degree(offset / radius);
        _endAngle += deltaAngle;
        if (roundf(_startY) >= 0) {
            _endAngle = 270;
            _startY = 0;
        }
        [self setNeedsDisplay];
    }
}
```

当开始线 y 的位置小于 0 时，将线段绘制为圆。当开始线 y 的位置等于 0 时，判断 startLoading 方法是否存在，如果存在就调用 startLoading 方法。并将开始线 y 的位置设置为 1，实现旋转。

实例 33 消息提示指示器

【实例描述】

本实例实现的功能是在界面显示一个消息指示器，如果用户没有在文本框中输入内容，直接单击"显示信息"按钮，会显示"请输入内容"的指示器；如果在文本框中输入了内容，就会出现一个和文本框具有相同内容的指示器。其中，用于文本输入的是一个文本框，运行效果如图 4.34 所示。

【实现过程】

当用户单击"显示信息"按钮后，会判断文本框中的内容是否为空，如果为空，将显示一个"请输入内容"的指示器，如果不为空，会显示一个和文本框具有相同内容的指示器。具体的实现步骤如下：

（1）创建一个项目，命名为"消息提示指示器"。

（2）创建一个基于 UIView 类的 indicator 类。

图 4.34　实例 33 运行效果

（3）打开 indicator.h 文件，编写代码。实现定义数据结构、宏定义、属性和方法的声明。程序代码如下：

```
#import <UIKit/UIKit.h>
//定义数据结构
typedef enum {
    kWTShort = 1,
} WToastLength;
//宏定义
#define TABBAR_OFFSET 44.0f
#define RGB(a, b, c) [UIColor colorWithRed:(a / 255.0f) green:(b / 255.0f) blue:(c / 255.0f) alpha:1.0f]
#define RGBA(a, b, c, d) [UIColor colorWithRed:(a / 255.0f) green:(b / 255.0f) blue:(c / 255.0f) alpha:d]
@interface indicator : UIView
//属性
@property (nonatomic) NSInteger length;
//方法
+ (void)showWithText:(NSString *)text;
+ (void)showWithText:(NSString *)text length:(WToastLength)length;
@end
```

（4）打开 indicator.m 文件，编写代码，实现创建消息提示指示器。使用的方法如表 4-20 所示。

表 4-20　indicator.m 文件中方法总结

方　　法	功　　能
initWithFrame:	指示器的初始化
__show	显示指示器的动画
__hide	隐藏指示器的动画
__createWithText:	创建带有文本的指示器
__location	指示器显示的位置
showWithText:	使用文本实现指示器
showWithText: length:	使用文本和长度显示指示器

这里需要讲解几个重要的方法（其他方法请读者参考源代码）。__show 方法实现指示器显示的动画。程序代码如下：

```
- (void)__show {
    [UIView animateWithDuration:0.2f
                     animations:^{
                         self.alpha = 1.0f;                      //设置透明度
                     }
                     completion:^(BOOL finished) {
                         [self performSelector:@selector(__hide) withObject:
                          nil afterDelay:
_length];                                                        //执行 __hide 方法
                     }];
}
```

__hide 方法实现指示器隐藏的动画。程序代码如下：

```
- (void)__hide {
    [UIView animateWithDuration:0.8f
                     animations:^{
                         self.alpha = 0.0f;                      //设置透明度
                     }
                     completion:^(BOOL finished) {
                         [self removeFromSuperview];             //移除所有的视图
#if !__has_feature(objc_arc)
                         [self release];
#endif
                     }];
}
```

__createWithText:方法实现带有文本指示器的创建。程序代码如下：

```
+ (indicator *)__createWithText:(NSString *)text {
    float screenWidth, screenHeight;
    CGSize screenSize = [UIScreen mainScreen].bounds.size;
                                                                 //获取整个屏幕的尺寸
    screenWidth = MIN(screenSize.width, screenSize.height);
    screenHeight = MAX(screenSize.width, screenSize.height);
    float x = 10.0f;
    float width = screenWidth - x * 2.0f;
    //创建并设置标签对象
    UILabel *textLabel = [[UILabel alloc] init];
    textLabel.backgroundColor = [UIColor clearColor];            //设置背景颜色
    textLabel.textAlignment = NSTextAlignmentCenter;             //设置对齐方式
    textLabel.font = [UIFont systemFontOfSize:14];               //设置字体
    textLabel.textColor = RGB(255, 255, 255);
    textLabel.numberOfLines = 0;                                 //设置行数
    textLabel.lineBreakMode=NSLineBreakByWordWrapping;
    NSDictionary *dis=[NSDictionary dictionaryWithObjectsAndKeys:
    textLabel.font,
NSFontAttributeName, nil];                                       //实例化字典对象
    CGSize sz=[text boundingRectWithSize:CGSizeMake(width - 20.0f,
    9999.0f)options:
NSStringDrawingUsesFontLeading attributes:dis context:nil].size;
    CGRect tmpRect = CGRectZero;
    tmpRect.size.width = width;                                  //设置宽度
    tmpRect.size.height = MAX(sz.height + 20.0f, 38.0f);         //设置高度
```

```
    //创建并设置指示器对象
    indicator *toast = [[indicator alloc] initWithFrame:tmpRect];
    toast.backgroundColor = RGBA(0, 0, 0, 0.8f);              //设置背景
    CALayer *layer = toast.layer;
    layer.masksToBounds = YES;
    layer.cornerRadius = 5.0f;
    textLabel.text = text;                                    //设置文本
    tmpRect.origin.x = floor((toast.frame.size.width - sz.width) / 2.0f);
    tmpRect.origin.y = floor((toast.frame.size.height - sz.height) / 2.0f);
    tmpRect.size = sz;
    textLabel.frame = tmpRect;                                //设置框架
    [toast addSubview:textLabel];              //将标签添加到指示器上
#if !__has_feature(objc_arc)
    [textLabel release];
#endif
    toast.alpha = 0.0f;                                       //设置透明度
    return toast;
}
```

__location 方法对指示器显示的位置进行设置。程序代码如下：

```
- (void)__location {
    CGSize screenSize = [UIScreen mainScreen].bounds.size;    //获取尺寸
    float x, y;
    float screenWidth, screenHeight;
    screenWidth = MIN(screenSize.width, screenSize.height);
    screenHeight = MAX(screenSize.width, screenSize.height);
    x = floor((screenWidth - self.bounds.size.width) / 2.0f);
    y = screenHeight - self.bounds.size.height - 15.0f - TABBAR_OFFSET;
    CGRect f = self.frame;                                    //获取框架
    f.origin = CGPointMake(x, y);                             //设置点
    self.frame = f;                                           //设置框架
}
```

showWithText: length:方法实现通过文本和长度显示指示器。程序代码如下：

```
+ (void)showWithText:(NSString *)text length:(WToastLength)length {
    indicator *toast = [indicator __createWithText:text];
    toast.length = 3;                                         //设置长度
    UIWindow *mainWindow = [[UIApplication sharedApplication] keyWindow];
    [mainWindow addSubview:toast];                //添加视图对象
    [toast __location];
    [toast __show];
}
```

（5）打开 ViewController.h 文件，编写代码，实现头文件、插座变量、对象以及动作的声明。程序代码如下：

```
#import <UIKit/UIKit.h>
#import "indicator.h"                                         //头文件
@interface ViewController : UIViewController{
    IBOutlet UITextField *tf;                                 //文本框的插座变量
    indicator *ind;                                           //对象
}
- (IBAction)show:(id)sender;                                  //动作
@end
```

（6）打开 Main.storyboard 文件，对 View Controller 视图控制器的设计界面进行设计，

效果如图 4.35 所示。

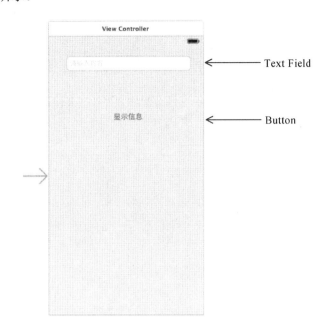

图 4.35　实例 33 View Controller 视图控制器的设计界面

需要添加的视图、控件以及对它们的设置如表 4-21 所示。

表 4-21　实例 33 视图、控件设置

视图、控件	属 性 设 置	其　　他
设计界面	Background：浅灰色	
Text Field	Placeholder：请输入内容	与插座变量 tf 关联
Button	Title：显示信息	与动作 show:关联

（7）打开 ViewController.m 文件，编写代码。实现用户单击按钮后，显示消息提示指示器。程序代码如下：

```
- (IBAction)show:(id)sender {
  [tf resignFirstResponder];
  NSString *text = tf.text;
  if (!text || ![text length]) {
      text = @"请输入内容!";                       //设置文本内容
  }
  [indicator showWithText:text];                  //显示文本
}
```

【代码解析】

本实例关键功能是指示器的显示、隐藏动画以及指示器中显示文本框内容。下面依次讲解这两个知识点。

1．显示和隐藏动画

指示器的显示和隐藏动画，可以使用 UIView 的 animateWithDuration: delay: options:

animations: completion:方法，程序代码如下：

```
    //显示动画
    [UIView animateWithDuration:0.2f
                     animations:^{
                         self.alpha = 1.0f;
                     }
                     completion:^(BOOL finished) {
                         [self performSelector:@selector(__hide) withObject:
                         nil afterDelay:
_length];
                     }];
    //隐藏动画
    [UIView animateWithDuration:0.8f
                     animations:^{
                         self.alpha = 0.0f;
                     }
                     completion:^(BOOL finished) {
                         [self removeFromSuperview];
#if !__has_feature(objc_arc)
                         [self release];
#endif
                     }];
}
```

2．指示器中显示文本框内容

指示器中文本的显示使用的方法是 showWithText，将文本框的内容保存在字符串中，然后通过调用 showWithText 方法实现指示器中显示文本内容的功能。程序代码如下：

```
    NSString *text = tf.text;
    ......
    [indicator showWithText:text];
```

第 5 章 选择器类效果

在 iPhone 或者 iPad 中,用户不停地做出一些选择。在 iOS 中,为用户提供了不同的选择器控件,如时间选择器、自定义选择器等。当前开发者也可以实现一些其他效果的选择器。本章将讲解一些关于选择器类的实例。

实例 34 时间设置器

【实例描述】

本实例实现的功能是使用日期选择器,实现对时间的设置。其中,用于显示当前时间的是标签,用于时间选择的是一个时间选择器。运行效果如图 5.1 所示。

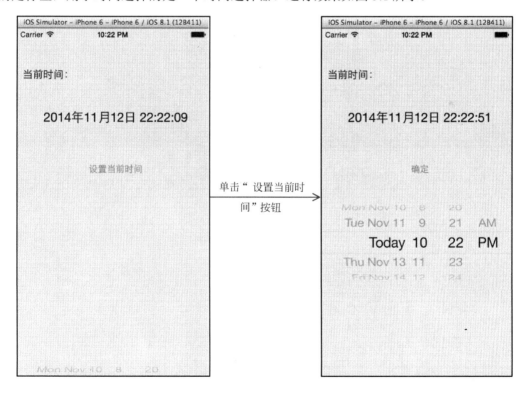

图 5.1 实例 34 运行效果

【实现过程】

当用户在日期选择器中进行选择时,被选择的日期就会出现在标签中。具体的实现步骤如下。

（1）创建一个项目，命名为"时间设置器"。

（2）打开 ViewController.h 文件，编写代码，实现插座变量和动作的声明。程序代码如下：

```
#import <UIKit/UIKit.h>
@interface ViewController : UIViewController{
    IBOutlet UILabel *label;              //标签的插座变量
    IBOutlet UIDatePicker *datepicker;    //日期选择器的插座变量
    //按钮的插座变量
    IBOutlet UIButton *button1;
    IBOutlet UIButton *button2;
}
//动作
- (IBAction)show:(id)sender;
- (IBAction)hide:(id)sender;
- (IBAction)select:(id)sender;
@end
```

（3）打开 Main.storyboard 文件，对 View Controller 视图控制器的设计界面进行设计，效果如图 5.2 所示。

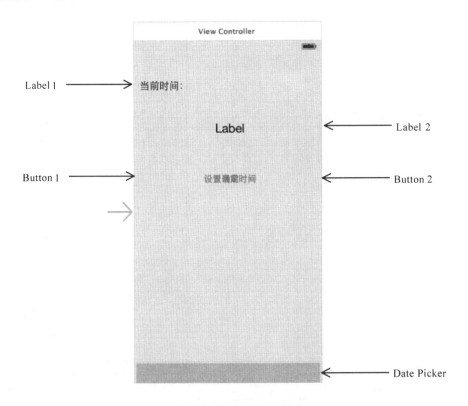

图 5.2　实例 34View Controller 视图控制器的设计界面

需要添加的视图、控件以及对它们的设置如表 5-1 所示。

表 5-1 实例 34 视图、控件设置

视图、控件	属 性 设 置	其 他
设计界面	Background：灰色	
Label1	Text：当前时间 Alignment：居中	
Label2	Font：System 21.0 Alignment：居中	与插座变量 label 关联
Button1	Title：设置当前时间	与插座变量 button1 关联 与动作 show:方法关联
Button2	Title：确定	与插座变量 button2 关联 与动作 hide:方法关联
Date Picker		与插座变量 datepicker 关联 与动作 select:方法关联

（4）打开 ViewController.m 文件，编写代码，实现一个时间设置器。在此文件中，代码分为 4 个部分：显示当前的时间；以动画的形式显示日期选择器；实现日期的选择；以动画的形式隐藏日期选择器。其中，viewDidLoad 方法实现在标签中显示当前的时间。程序代码如下：

```
- (void)viewDidLoad
{
    NSDateFormatter *formatt=[[NSDateFormatter alloc]init];
    [formatt setDateFormat:@"YYYY年MM月dd日 HH:mm:ss"];    //设置格式
    NSString *ten=[formatt stringFromDate:[NSDate date]];
    label.text=ten;            //设置标签中文本的内容
    [button2 setHidden:YES];
    [super viewDidLoad];
    // Do any additional setup after loading the view, typically from a nib.
}
```

show：方法实现以动画的形式显示日期选择器。程序代码如下：

```
- (IBAction)show:(id)sender {
    [button2 setHidden:NO];
    [button1 setHidden:YES];
    [UIView beginAnimations:@"" context:nil];
    [UIView setAnimationDuration:1];                    //设置动画所需时间
    datepicker.frame=CGRectMake(0, 269, 320, 164);      //设置日期选择器的框架
    [UIView commitAnimations];
}
```

hide：方法实现以动画的形式隐藏日期选择器。程序代码如下

```
- (IBAction)hide:(id)sender {
    [button1 setHidden:NO];
    [button2 setHidden:YES];
    [UIView beginAnimations:@"" context:nil];
    [UIView setAnimationDuration:1];                    //设置动画所需时间
    datepicker.frame=CGRectMake(0, 534, 320, 164);      //设置日期选择器的框架
    [UIView commitAnimations];
}
```

select：方法实现在日期选择器中选择日期的功能，并显示在标签中。程序代码如下：

```
- (IBAction)select:(id)sender {
    NSDateFormatter *formatt=[[NSDateFormatter alloc]init];
    [formatt setDateFormat:@"YYYY年MM月dd日 HH:mm:ss"];    //设置格式
    NSString *ten=[formatt stringFromDate:datepicker.date];
    label.text=ten;
}
```

【代码解析】

本实例关键功能是日期转换为字符串。下面就是这个知识点的详细讲解。日期转换为字符串，需要以下 3 个步骤。

1. 实例化一个对象

NSDateFormatter 类，实现了对格式化日期格式中的各种复杂处理。要想使用它，首先要对其进行初始化，一般使用的是 init 方法。在代码中，使用了 init 方法进行了对象的实例化。程序代码如下：

```
NSDateFormatter *formatt=[[NSDateFormatter alloc]init];
```

2. 设置格式

对日期格式进行的设置，使用了 NSDateFormatter 的 setDateFormat:方法。其语法形式如下：

```
- (void)setDateFormat:(NSString *)string;
```

其中，(NSString *)string 表示字符串，它是开发者希望日期转换为的字符串格式。在代码中，使用了 setDateFormat:方法对日期要转换为的字符串格式进行设置。程序代码如下：

```
[formatt setDateFormat:@"YYYY年MM月dd日 HH:mm:ss"];
```

3. 转换

日期转换为字符串，需要使用 NSDateFormatter 的 stringFromDate:方法实现。其语法形式如下：

```
- (NSString *)stringFromDate:(NSDate *)date;
```

其中，(NSDate *)date 表示日期，它是开发者想要转换为字符串的日期。在代码中，使用了 stringFromDate:方法将日期转换为字符串。程序代码如下：

```
NSString *ten=[formatt stringFromDate:[NSDate date]];
```

实例 35　闹　　铃

【实例描述】

本实例实现的是闹铃功能，当用户选择好时间后，到达此时间就会进行提醒。运行效果如图 5.3 所示。

第 5 章 选择器类效果

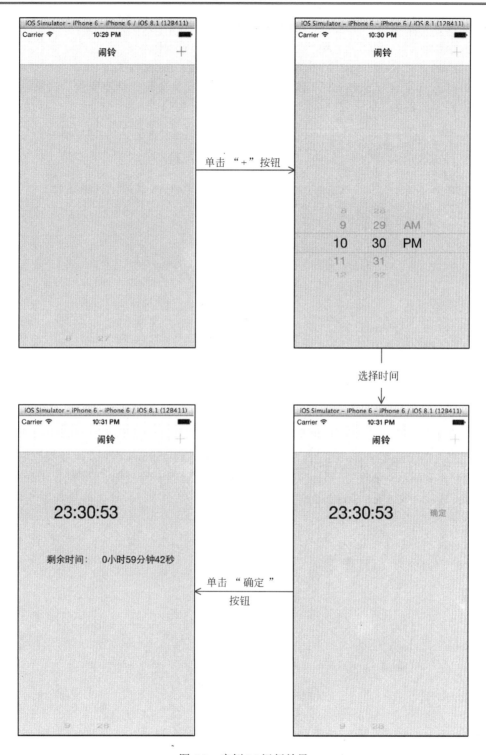

图 5.3 实例 35 运行效果

【实现过程】

当用户在日期选择器中选择好时间后，会出现一个"确定"按钮，用户单击此按钮后，就会按照选择的时间，实现倒计时。具体的实现步骤入如下。

(1)创建一个项目,命名为"闹铃"。
(2)添加 AVFoundation.framework 框架到创建的项目中。
(3)添加音频 music.mp3 到创建项目的 Supporting Files 文件夹中。
(4)打开 ViewController.h 文件,编写代码。实现头文件、插座变量、对象以及动作的声明。程序代码如下:

```
#import <UIKit/UIKit.h>
#import <AVFoundation/AVFoundation.h>
@interface ViewController : UIViewController{
    //插座变量
    IBOutlet UIDatePicker *datepicker;      //日期选择器的插座变量
    //标签的插座变量
    IBOutlet UILabel *label;
    IBOutlet UILabel *label1;
    IBOutlet UILabel *label2;
    IBOutlet UIButton *button;              //按钮的插座变量
    IBOutlet UIBarButtonItem *bar;          //UIToolbar 或 UINavigationBar 上
                                            //的按钮的插座变量
    //对象
    NSTimer *timer;
    AVAudioPlayer *audioplayer;
}
//动作
- (IBAction)push:(id)sender;
- (IBAction)select:(id)sender;
- (IBAction)exit:(id)sender;
@end
```

(5)打开 Main.storyboard 文件,从视图库中拖动 Navigation Controller 导航控制器到画布中。将 root view 设置为 View Controller 视图控制器中的视图,选择 Is Initial View Controller 选项。对 View Controller 视图控制器的设计界面进行设计。效果如图 5.4 所示。

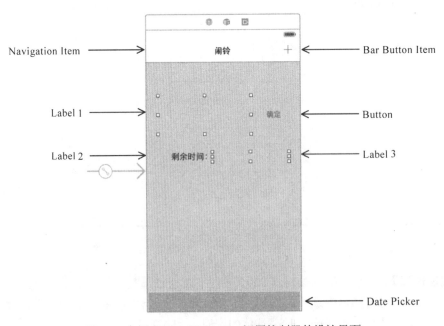

图 5.4 实例 35View Controller 视图控制器的设计界面

需要添加的视图、控件以及对它们的设置如表 5-2 所示。

表 5-2 实例 35 视图、控件设置

视图、控件	属 性 设 置	其 他
设计界面	Background：灰色	
Navigation Item	Title：闹铃	
Bar Button Item	Identifier：Add	与插座变量bar关联 与动作push:关联
Label1	Text：（空） Font：System 31.0 Alignment：居中	与插座变量label关联
Label2	Text：剩余时间	与插座变量label2关联
Label3	Text：（空）	与插座变量label1关联
Button	Title：确定	与插座变量button关联 与动作exit:关联
Date Picker	Mode：Time	与插座变量datepicker关联 与动作select:关联

（6）打开 ViewController.m 文件，编写代码，实现一个闹铃的功能。使用的方法如表 5-3 所示。

表 5-3 ViewController.m文件中方法总结

方 法	功 能
viewDidLoad	实现一些视图、控件的隐藏
push:	以动画的形式显示日期选择器
select:	实现时间的选择
exit:	以动画的形式隐藏日期选择器
timerFireMethod:	实现倒计时

这里需要讲解几个重要的方法（其他方法请读者参考源代码）。push:方法实现以动画的形式显示日期选择器。程序代码如下：

```
- (IBAction)push:(id)sender {
  [UIView beginAnimations:@"" context:nil];
  [UIView setAnimationDuration:1];          //设置动画所需时间
  datepicker.frame=CGRectMake(0, 300, 320, 164); //设置日期选择器的框架
  [UIView commitAnimations];
  [bar setEnabled:NO];
}
```

select：方法实现在日期选择器中日期的选择。程序代码如下：

```
- (IBAction)select:(id)sender {
  NSDateFormatter *formatt=[[NSDateFormatter alloc]init];
  [formatt setDateFormat:@"HH:mm:ss"];      //设置格式
  NSString *ten=[formatt stringFromDate:datepicker.date];
  label.text=ten;                            //设置标签中文本的内容
  [button setHidden:NO];
}
```

imerFireMethod:方法实现闹铃的倒计时功能。程序代码如下：

```objc
-(void)timerFireMethod:(NSTimer *)theTimer{
    //隐藏标签对象
    [label1 setHidden:NO];
    [label2 setHidden:NO];
    NSDate *date=[NSDate date];              //实例化日期对象
    NSDate *bb=datepicker.date;              //实例化日期对象
    NSCalendar*chineseClendar=[[NSCalendar alloc]initWithCalendarIdentifier:
    NSGregorianCalendar ];
    NSUInteger unitFlags = NSHourCalendarUnit | NSMinuteCalendarUnit |
    NSSecondCalendarUnit | NSDayCalendarUnit| NSMonthCalendarUnit |
    NSYearCalendarUnit;
    NSDateComponents*cps =[ chineseClendar components:unitFlags fromDate:
    date toDate:bb options:0];
    //设置文本内容
    label1.text=[NSString stringWithFormat:@"%d 小时%d 分钟%d 秒",[cps hour],
    [cps minute],[cps second]];
    //判断秒数是否为1
    if([label1.text isEqualToString:@"0 小时 0 分钟 0 秒"]){
        [timer invalidate];
        //音乐的播放
        NSString *path=[[NSBundle mainBundle]pathForResource:@"music" ofType:
        @"mp3"];
        NSURL *url=[NSURL fileURLWithPath:path];
        audioplayer=[[AVAudioPlayer alloc]initWithContentsOfURL:url error:nil];
        audioplayer.numberOfLoops=-1;        //设置循环次数
        [audioplayer prepareToPlay];
        //判断音乐是否正在播放
        if (![audioplayer isPlaying]) {
            [audioplayer play];
        }
    }
    if ([cps second]<0) {
        [timer invalidate];                  //让时间定时器失效
        UIAlertView *alert=[[UIAlertView alloc]initWithTitle:@"你设置的时间
        已经过时"message:nil delegate:nil cancelButtonTitle:@"OK"
        otherButtonTitles: nil];             //创建警告视图
        [alert show];
        [bar setEnabled:YES];
    }
}
```

【代码解析】

本实例关键功能是倒计时的实现。下面就是这个知识点的详细讲解。

倒计时的实现需要使用 NSCalendar 的 components:fromDate:toDate:options:方法。其语法形式如下：

```objc
- (NSDateComponents *)components:(NSUInteger)unitFlags fromDate:(NSDate
*)startingDate toDate:(NSDate *)resultDate options:(NSUInteger)opts;
```

其中，(NSUInteger)unitFlags 表示指定组件返回的 NSDateComponents 常量；(NSDate *)startingDate 表示开始的时间；(NSDate *)resultDate 表示结束的时间；options:(NSUInteger)opts 表示一个关于计算的选项，一般设置为 0。NSDateComponents 的常量如表 5-4 所示。

第 5 章 选择器类效果

表 5-4 NSDateComponents 的常量

常 量	说 明
NSHourCalendarUnit	指定小时
NSEraCalendarUnit	指定纪元
NSMinuteCalendarUnit	指定分钟
NSSecondCalendarUnit	指定秒
NSDayCalendarUnit	指定天
NSMonthCalendarUnit	指定月
NSYearCalendarUnit	指定年
NSWeekCalendarUnit	指定周
NSWeekdayCalendarUnit	指定工作日

在代码中，使用了 NSCalendar 的 components:fromDate:toDate:options:方法实现了倒计时功能。程序代码如下：

```
NSDateComponents *cps =[ chineseClendar components:unitFlags fromDate:date
toDate:bb options:0];
```

其中，unitFlags 表示指定组件返回的 NSDateComponents 常量；date（当前时间）表示开始的时间；bb（用户指定的时间）表示结束的时间；0 表示关于计算的选项。

实例 36　城市经纬度查询

【实现描述】

本实例实现的功能是在自定义选择器中，实现城市经纬度查询的功能。其中，用于经纬度显示的是标签。运行效果如图 5.5 所示。

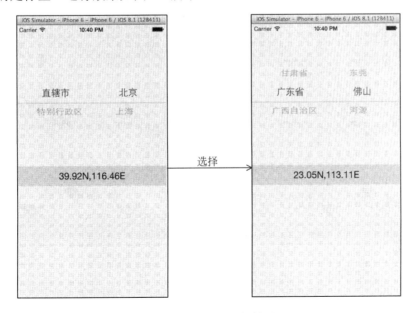

图 5.5　实例 36 运行效果

【实现描述】

（1）创建一个项目，命名为"城市经纬度查询"。

（2）添加"数据.plist"文件到创建项目的 Supporting Files 文件夹中。在此文件中存放了每一个省的经纬度数据。

（3）打开 ViewController.m 文件，编写代码。实现插座变量、对象以及实例变量的声明。程序代码如下：

```
#import <UIKit/UIKit.h>
@interface ViewController : UIViewController{
    IBOutlet UILabel *label;            //标签的插座变量
    IBOutlet UIPickerView *pv;          //自定义选择器的插座变量
    NSArray *array;
    NSInteger row0Component;
    NSInteger row1Component;
}
@end
```

（4）打开 Main.storyboard 文件，对 View Controller 视图控制器的设计界面进行设计，效果如图 5.6 所示。

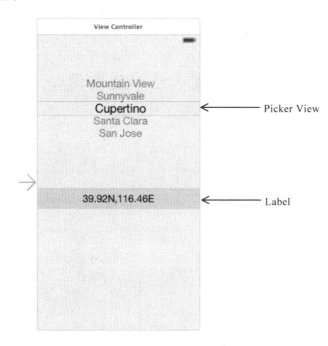

图 5.6　实例 36View Controller 视图控制器的设计界面

需要添加的视图、控件以及对它们的设置如表 5-5 所示。

表 5-5　实例 36 视图、控件设置

视图、控件	属性设置	其他
设计界面	Background：浅紫色	
Picker View		与插座变量pv关联 dataSource与View Controller关联 delegate与View Controller关联

第 5 章　选择器类效果

续表

视图、控件	属性设置	其他
Label	Text：39.92N,116.46E Font：System 20.0 Alignment：居中 Background：浅蓝色	与插座变量label关联

（5）打开 ViewController.m 文件，编写代码，实现选择城市及显示经纬度。使用的方法如表 5-6 所示。

表 5-6　ViewController.m文件中方法总结

方　　法	功　　能
viewDidLoad	视图加载后调用，实现初始化
numberOfComponentsInPickerView:	返回显示的列数
pickerView: numberOfRowsInComponent:	返回当前显示的行数
pickerView:widthForComponent:	返回列宽度
pickerView:rowHeightForComponent:	返回列中每行的高度
pickerView:viewForRow: forComponent:reusingView:	定义在自定义选择器中显示的视图
pickerView:didSelectRow:inComponent:	实现行的选择

在本实例中代码可以分为两部分：填充自定义选择器以及行选择的实现。其中，要实现自定义选择器的内容填充，需要使用 viewDidLoad、numberOfComponentsInPickerView:、pickerView: numberOfRowsInComponent:、pickerView:widthForComponent:、pickerView:rowHeightForComponent:和 pickerView:viewForRow: forComponent:reusingView:方法。程序代码如下：

```
- (void)viewDidLoad
{
    //获取数据源
    NSString* path=[[NSBundle mainBundle]pathForResource:@"数据" ofType:
    @"plist"];
    array=[[NSArray alloc] initWithContentsOfFile:path];
    [super viewDidLoad];
    // Do any additional setup after loading the view, typically from a nib.
}
//返回显示的列数
-(NSInteger)numberOfComponentsInPickerView:(UIPickerView *)pickerView
{
    return 2;
}
//返回当前显示的行数
-(NSInteger)pickerView:(UIPickerView *)pickerView numberOfRowsInComponent:
(NSInteger)component
{
    //判断 component 是否为 0
    if(0==component)
        return array.count;             //返回数组的个数
    else
    {
        row0Component=[pickerView selectedRowInComponent:0];
                                        //指定列当前选择的行数
```

```objc
        return [[[array objectAtIndex:row0Component] objectForKey:@"Cities"]
        count];
    }
}
//返回列宽度
-(CGFloat)pickerView:(UIPickerView*)pickerView widthForComponent:(NSInteger)
component
{
    return 150;
}
//返回列中每行的高度
-(CGFloat)pickerView:(UIPickerView*)pickerView rowHeightForComponent:(NSInteger)
component
{
    return 40;
}
//定义在自定义选择器中显示的视图
-(UIView*)pickerView:(UIPickerView *)pickerView viewForRow:(NSInteger)row
forComponent:
(NSInteger)component reusingView:(UIView *)view
{
    CGFloat width=[self pickerView:pickerView widthForComponent:component];
                              //获取宽度
    CGFloat height=[self pickerView:pickerView rowHeightForComponent:
    component];                   //获取高度
    UILabel* la;
    if(nil==view)
    {
        la=[[UILabel alloc] initWithFrame:CGRectMake(0, 0, width, height)];
                              //实例化标签对象
    }
    else
        la=(UILabel*)view;
    if(0==component)
    {
        la.text=[[array objectAtIndex:row]objectForKey:@"State"];
                                                //设置标签显示的文本
        la.font=[UIFont fontWithName:@"Arial" size:20]; //设置字体
        la.textAlignment=NSTextAlignmentCenter;
    }
    else
    {
        row0Component=[pickerView selectedRowInComponent:0];
                                                //指定列当前选择的行数
        la.text=[[[[array objectAtIndex:row0Component]objectForKey:@"Cities"]
        objectAtIndex:row]objectForKey:@"city"];     //设置标签显示的文本
        la.font=[UIFont fontWithName:@"Arial" size:20];
        la.textAlignment=NSTextAlignmentCenter;
        la.textColor=[UIColor blueColor];
    }
    return la;
}
```

pickerView:didSelectRow:inComponent:方法实现在选择某一行后，出现对象的经纬度。程序代码如下：

```objc
-(void)pickerView:(UIPickerView *)pickerView didSelectRow:(NSInteger)row
inComponent:
```

```
(NSInteger)component
{
    //实现在第一栏中选择省后，显示对应的市
    if(0==component)
    {
        [pickerView reloadComponent:1];
        [pickerView selectRow:0 inComponent:1 animated:NO];        //选择值
    }
    //实现选择市后，出现对应的经纬度
    row0Component=[pickerView selectedRowInComponent:0];//指定列当前选择的行数
    row1Component=[pickerView selectedRowInComponent:1];
    NSNumber* numlon=[[[array objectAtIndex:row0Component] objectForKey:@
            "Cities"]objectAtIndex:row1Component] objectForKey:@"lon"];
    NSNumber* numlat=[[[array objectAtIndex:row0Component] objectForKey:@
            "Cities"] objectAtIndex:row1Component] objectForKey:@"lat"];
    label.text=[NSString stringWithFormat:@"%@N,%@E",numlat,numlon];
                                                    //设置标签文本的内容
}
```

【代码解析】

本实例关键功能是自定义选择器中行选择的实现。下面就是这个知识点的详细讲解。

自定义选择器中行选择的实现需要使用 UIPickerViewDelegate 的 pickerView:didSelectRow:inComponent:方法，此方法是自定义选择器实现操作的委托方法。其语法形式如下：

```
- (void)pickerView:(UIPickerView *)pickerView didSelectRow:(NSInteger)row
inComponent:(NSInteger)component;
```

其中，(UIPickerView *)pickerView 表示选择器请求的数据对象；(NSInteger)row 表示选择的行；(NSInteger)component 表示选择的列。在代码中，使用了 UIPickerViewDelegat 的 c pickerView:didSelectRow:inComponent:方法实现行的选择。程序代码如下：

```
-(void)pickerView:(UIPickerView *)pickerView didSelectRow:(NSInteger)row
inComponent:
(NSInteger)component
{
    //实现在第一栏中选择省后，显示对应的市
    if(0==component)
    {
        [pickerView reloadComponent:1];
        [pickerView selectRow:0 inComponent:1 animated:NO];
    }
    //实现选择市后，出现对应的经纬度
    row0Component=[pickerView selectedRowInComponent:0];
    row1Component=[pickerView selectedRowInComponent:1];
    NSNumber*numlon=[[[array objectAtIndex:row0Component] objectForKey:@"Cities"]
                    objectAtIndex:row1Component]objectForKey:@"lon"];
    NSNumber*numlat=[[[array objectAtIndex:row0Component] objectForKey:@"Cities"]
                    objectAtIndex:row1Component]objectForKey:@"lat"];
    label.text=[NSString stringWithFormat:@"%@N,%@E",numlat,numlon];
}
```

实例 37　定制多选功能选择器

【实例描述】

本实例的功能是实现一个多选功能的选择器，并且可以将选择的内容显示在警告视图中。运行效果如图 5.7 所示。

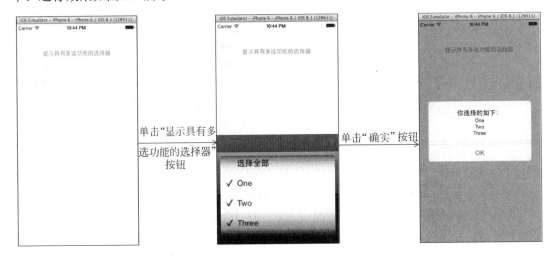

图 5.7　实例 37 运行效果

【实现过程】

（1）创建一个项目，命名为"定制多选功能选择器"。

（2）添加图片 1.png、2.png、3.png、4.png、5.png、6.png 和 7.png 到创建项目的 Supporting Files 文件夹中。

（3）创建一个基于 UITableViewCell 类的 PickerViewCell 类。

（4）打开 PickerViewCell.h 文件，编写代码，实现实例变量、属性以及方法的声明。程序代码如下：

```objc
#import <UIKit/UIKit.h>
@interface PickerViewCell : UITableViewCell{
    @private BOOL selectionState_;                              //私有的实例变量
}
@property (nonatomic, assign) BOOL selectionState;              //属性
- (id)initWithReuseIdentifier:(NSString *)reuseIdentifier;      //方法
@end
```

（5）打开 PickerViewCell.m 文件，编写代码，实现对选择器中单元格的设置及布局。使用的方法如表 5-7 所示。

表 5-7　PickerViewCell.m 文件方法总结

方　　法	功　　能
initWithReuseIdentifier:	选择器单元格的初始化
prepareForReuse	准备一个要即将出列的单元格

方　　法	功　　能
selectionState	判断选择的状态
setSelectionState:	设置选择状态
ayoutSubviews	重新布局

这里需要讲解几个重要的方法（其他方法请读者参考源代码）。其中，initWithReuse Identifier:方法对选择器单元格进行初始化。程序代码如下：

```
- (id)initWithReuseIdentifier:(NSString *)reuseIdentifier {
  if ((self = [super initWithStyle:UITableViewCellStyleDefault reuse
   Identifier:reuseIdentifier])) {
      selectionState_ = NO;
      self.textLabel.font = [UIFont boldSystemFontOfSize:21];//设置字体
  }
  return self;
}
```

setSelectionState:方法对选择的状态进行设置。程序代码如下：

```
- (void)setSelectionState:(BOOL)selectionState {
  selectionState_ = selectionState;
  //判断选择的状态是否为 YES
  if (selectionState_ != NO) {
      //当单元格处于被选择的状态
      self.imageView.image = [UIImage imageNamed:@"1.png"];
      self.imageView.highlightedImage = [UIImage imageNamed:@"2.png"];
      self.textLabel.textColor = [UIColor redColor];     //设置文本的颜色
  }
  else {
      //当单元格处于没有选择的状态
      self.imageView.image = [UIImage imageNamed:@"2.png"];
      self.imageView.highlightedImage = [UIImage imageNamed:@"1.png"];
                                                    //设置视图的高亮图像
      self.textLabel.textColor = [UIColor blackColor];   //设置文本的颜色
  }
  [self.imageView setNeedsDisplay];
  [self.textLabel setNeedsDisplay];
  [self setNeedsLayout];
}
```

layoutSubviews 方法对视图中的子视图进行重新布局。程序代码如下：

```
- (void)layoutSubviews {
  [super layoutSubviews];
  self.imageView.frame = CGRectMake(15, 12, 18, 18);
  self.textLabel.frame = CGRectMake(44, 9, self.frame.size.width - 54, 24);
}
```

（6）创建一个基于 UIView 类的 PickerView 类。

（7）打开 PickerView.h 文件，编写代码。实现头文件、协议、对象和属性的声明。程序代码如下：

```
#import <UIKit/UIKit.h>
//头文件
#import "PickerViewCell.h"
```

```
@class PickerView;
//协议
@protocol PickerViewDelegate <NSObject>
-(NSInteger)numberOfRowsForPickerView:(PickerView *)pickerView;
-(NSString*)pickerView:(PickerView*)pickerView textForRow:(NSInteger)row;
-(BOOL)pickerView:(PickerView*)pickerView selectionStateForRow:(NSInteger)row;
@optional
- (void)pickerView:(PickerView *)pickerView didCheckRow:(NSInteger)row;
- (void)pickerView:(PickerView *)pickerView didUncheckRow:(NSInteger)row;
@end
@interface PickerView : UIView<UITableViewDataSource,UITableViewDelegate>{
    @private UITableView *internalTableView_;          //私有的对象
}
//属性
@property (nonatomic, assign) id<PickerViewDelegate> delegate;
@property (nonatomic, copy) NSString *allOptionTitle;
@end
```

（8）打开 PickerView.m 文件，编写代码，实现对自定义选择器的的创建。使用的方法如表 5-8 所示。

表 5-8　PickerView.m文件中方法总结

方　　法	功　　能
initWithFrame:	选择器的初始化
tableView:numberOfRowsInSection:	返回行数
tableView:cellForRowAtIndexPath:	设置每一行单元格显示的内容和格式
tableView:didSelectRowAtIndexPath:	选择行
scrollViewDidEndDecelerating:	用户拖动，之后继续滚动调用
scrollViewDidEndDragging: willDecelerate:	用户拖动并停止调用

这里方法主要可以分为：初始化、实现内容填充和实现选择这几部分。其中初始化的实现使用的是 initWithFrame:方法。程序代码如下：

```
- (id)initWithFrame:(CGRect)frame {
  if ((self = [super initWithFrame:CGRectMake(frame.origin.x, frame.origin.y,
     320, 216)])) {
      self.backgroundColor = [UIColor blackColor];
      self.clipsToBounds = YES;
      self.allOptionTitle = NSLocalizedString(@"选择全部", @"All option
       title");
      //创建并设置表视图对象
      internalTableView_ = [[UITableView alloc] initWithFrame:CGRectMake
      (10, -2, 300, 218) style:UITableViewStylePlain];       //实例化对象
      internalTableView_.delegate = self;                    //设置委托
      internalTableView_.dataSource = self;
      internalTableView_.separatorStyle = UITableViewCellSeparatorStyleNone;
                                                             //设置分隔线风格
      internalTableView_.showsVerticalScrollIndicator = NO;
      internalTableView_.scrollsToTop = NO;
      UIImage*backgroundImage=[[UIImage imageNamed:@"7.png"] stretchableImage
      WithLeftCapWidth:4 topCapHeight:0];                    //实例化对象
      internalTableView_.backgroundView = [[UIImageView alloc] initWithImage:
       backgroundImage];
      [self addSubview:internalTableView_];                  //添加视图对象
      //创建并设置阴影对象
```

```
    UIImageView *shadow = [[UIImageView alloc] initWithImage:[UIImage
    imageNamed:@"6.png"]];
    shadow.frame = internalTableView_.frame;              //设置框架
    [self addSubview:shadow];
    //实现边界的设置
    UIImageView *leftBorder = [[UIImageView alloc] initWithImage:[UIImage
    imageNamed:@"3.png"]];
    leftBorder.frame = CGRectMake(0, 0, 15, 216);
    [self addSubview:leftBorder];                         //添加视图对象
    UIImageView *rightBorder = [[UIImageView alloc] initWithImage:[UIImage
    imageNamed:@"5.png"]];
    rightBorder.frame = CGRectMake(self.frame.size.width - 15, 0, 15, 216);
    [self addSubview:rightBorder];
    //实例化对象
    UIImageView *middleBorder = [[UIImageView alloc] initWithImage:
                                 [[UIImage imageNamed:@"4.png"]
                                  stretchableImageWithLeftCapWidth:0
topCapHeight:10]];/middleBorder.frame = CGRectMake(15, 0, self.frame.
size.width - 30, 216);
    [self addSubview:middleBorder];
   }
    return self;
}
```

tableView:numberOfRowsInSection:方法实现对表视图显示的行数进行获取。程序代码如下：

```
-(NSInteger)tableView:(UITableView *)tableView  numberOfRowsInSection:
(NSInteger)section {
    //判断是否是操作所有的标题
    if (allOptionTitle)
        return [delegate numberOfRowsForPickerView:self] ? [delegate numberOfRows
        ForPickerView:self] + 4 : 0;
    else
        return [delegate numberOfRowsForPickerView:self] ? [delegate numberOfRows
        ForPickerView:self] + 3 : 0;
}
```

tableView:cellForRowAtIndexPath:方法实现每一行单元格显示的内容和格式的设置。程序代码如下：

```
- (UITableViewCell *)tableView:(UITableView *)aTableView cellForRowAtIndexPath:
(NSIndexPath *)indexPath {
    static NSString *CellIdentifier = @"ALPVCell";         //可重用单元格标识
    PickerViewCell *cell = [aTableView dequeueReusableCellWithIdentifier:
    CellIdentifier];
    //判断单元格是否为可重用的单元格
    if (cell == nil) {
        //如果没有重用
        cell = [[PickerViewCell alloc] initWithReuseIdentifier:CellIdentifier];
        //创建一个可重用的单元格对象
    }
    if (indexPath.row < 2 || indexPath.row >= ([delegate numberOfRowsFor
    PickerView:self] + (allOptionTitle ? 3 : 2))) {
        //空白单元格
        cell.textLabel.text = nil;                 //设置当前单元格显示的内容
        cell.selectionStyle = UITableViewCellSelectionStyleNone;
                                                   //设置选择单元格的风格
```

```
        }
        else {
            //设置内容和其他
            if (allOptionTitle && indexPath.row == 2) {
                cell.textLabel.text = allOptionTitle;              //设置标签的文本内容
                BOOL allSelected = YES;
                for (int i = 0; i < [self.delegate numberOfRowsForPickerView:self];
                    i++) {
                        if ([delegate pickerView:self selectionStateForRow:i] == NO) {
                            allSelected = NO;
                            break;
                        }
                    }
                cell.selectionState = allSelected;                 //设置单元格的状态
            }
            else {
                int actualRow = indexPath.row - (allOptionTitle ? 3 : 2);
                cell.textLabel.text = [delegate pickerView:self textForRow:
                    actualRow];//设置标签的文本内容
                cell.selectionState = [delegate pickerView:self selectionState
                    ForRow:actualRow];
            }
            cell.selectionStyle = UITableViewCellSelectionStyleBlue;
                                                                   //设置被选择行的风格
        }
    return cell;
}
```

tableView:didSelectRowAtIndexPath:方法,当选择某一行后调用,实现选择行的响应。程序代码如下:

```
-(void)tableView:(UITableView *)tableView didSelectRowAtIndexPath:
(NSIndexPath *)indexPath {
    if (indexPath.row > 1 && indexPath.row < ([delegate numberOfRowsForPic
    kerView:self] + (allOptionTitle ? 3 : 2))) {
        //设置选择的状态
        PickerViewCell *cell = (PickerViewCell *)[tableView cellForRowAtIndex
        Path:indexPath];
        cell.selectionState = !cell.selectionState;
        //通知委托
        int actualRow = indexPath.row - (allOptionTitle ? 3 : 2);
         //判断单元格的选择状态是否为 YES
        if (cell.selectionState != NO) {
          if ([self.delegate respondsToSelector:@selector(pickerView:
          didCheckRow:)])
              [delegate pickerView:self didCheckRow:actualRow];
                                              //调用 pickerView:didCheckRow:
        }
        else {
            //判断是否实现 pickerView:didUncheckRow:方法
            if ([self.delegate respondsToSelector:@selector(pickerView:
            didUncheckRow:)])
                [delegate pickerView:self didUncheckRow:actualRow];
        }
        //重复可见单元格,并更新它们
```

```
                for (PickerViewCell *aCell in tableView.visibleCells) {
                    int iterateRow = [tableView indexPathForCell:aCell].row - (allOption
                     Title ? 3 : 2);
                    if (allOptionTitle && iterateRow == -1) {
                        BOOL allSelected = YES;
                        //遍历
                        for (int i = 0; i < [self.delegate numberOfRowsForPickerView:
                        self]; i++) {
                            if ([delegate pickerView:self selectionStateForRow:i] == NO) {
                                allSelected = NO;           //所有的行没有被选择
                                break;
                            }
                        }
                        aCell.selectionState = allSelected;          //设置选择的状态
                    }
                    else if (iterateRow >= 0 && iterateRow < [delegate numberOfRows
                    ForPickerView:self]) {
                        if (iterateRow == actualRow)
                            continue;
                        //设置选择的状态
                        aCell.selectionState = [delegate pickerView:self selections
                        tateForRow:iterateRow];
                    }
                }
                [tableView setContentOffset:CGPointMake(0, tableView.rowHeight *
                (indexPath.row - 2)) animated:YES];          //通过动画设置内容视图的新位置
            }
            [tableView deselectRowAtIndexPath:indexPath animated:YES];
}
```

（9）创建一个基于 UIView 类的 MultiSelectPickerView 类。

（10）打开 MultiSelectPickerView.h 文件，编写代码，实现头文件、协议、属性和方法的声明。程序代码如下：

```
#import <UIKit/UIKit.h>
#import "PickerView.h"              //头文件
//协议
@protocol MultiSelectPickerViewDelegate <NSObject>
@required
-(void)returnChoosedPickerString:(NSMutableArray *)selectedEntriesArr;
@end
@interface MultiSelectPickerView : UIView<PickerViewDelegate>
//属性
@property (nonatomic, retain) NSArray *entriesArray;
@property (nonatomic, retain) NSArray *entriesSelectedArray;
……
@property (nonatomic, retain) UIToolbar *toolBar;
//方法
- (void)pickerShow;
@end
```

（11）打开 MultiSelectPickerView.m 文件，编写代码，实现一个具有多选效果的选择器。使用的方法如表 5-9 所示。

表 5-9 MultiSelectPickerView.m文件中方法总结

方 法	功 能
initWithFrame:	多选效果的选择器的初始化
pickerShow	以动画的形式显示选择器
pickerHide	以动画的形式隐藏选择器
confirmPickView	单击"确定"按钮实现的动作
numberOfRowsForPickerView:	返回行数
pickerView:textForRow:	返回显示的选择器的内容
pickerView:selectionStateForRow:	返回选择器的选择状态
pickerView:didCheckRow:	检测所有的行是否选中
pickerView:didUncheckRow:	检测所有的行是否没被选中

这里需要讲解几个重要的方法（其他方法请读者参考源代码）。其中，initWithFrame: 方法实现对多选效果的选择器进行初始化。程序代码如下：

```objc
- (id)initWithFrame:(CGRect)frame
{
    self = [super initWithFrame:frame];
    if (self) {
        //实例化对象
        self.selectionStatesDic = [[NSMutableDictionary alloc] initWithCapacity:16];
        self.selectedEntriesArr = [[NSMutableArray alloc] initWithCapacity:16];
        self.entriesArray = [[NSMutableArray alloc] initWithCapacity:16];
        self.entriesSelectedArray = [[NSMutableArray alloc] initWithCapacity:16];
    }
    return self;
}
```

pickerShow 方法实现以动画的形式显示选择器。程序代码如下：

```objc
- (void)pickerShow
{
    //遍历
    for (NSString *key in self.entriesArray){
        BOOL isSelected = NO;
        //遍历
        for (NSString *keyed in self.entriesSelectedArray) {
            if ([key isEqualToString:keyed]) {
                isSelected = YES;
            }
        }
        [self.selectionStatesDic setObject:[NSNumber numberWithBool:isSelected]
          forKey:key];                                      //设置对象
    }
    if (!self.pickerView) {
        self.pickerView = [[PickerView alloc] initWithFrame:CGRectMake
          (0,260, 320, 260)];                               //实例化对象
    }
    self.pickerView.delegate = self;
    [self addSubview:self.pickerView];                      //添加视图对象
```

```objc
//创建工具栏
NSMutableArray *items = [[NSMutableArray alloc] initWithCapacity:3];
UIBarButtonItem *confirmBtn = [[UIBarButtonItem alloc] initWithTitle:
@"确定" style:UIBarButtonItemStyleDone target:self action:@selector
(confirmPickView)];
UIBarButtonItem *flexibleSpaceItem = [[UIBarButtonItem alloc] initWith
BarButtonSystemItem:UIBarButtonSystemItemFlexibleSpace target:nil
action:nil];
UIBarButtonItem *cancelBtn = [[UIBarButtonItem alloc] initWithTitle:
@"取消" style:UIBarButtonItemStyleBordered target:self action:@selector
(pickerHide)];
//添加
[items addObject:cancelBtn];
[items addObject:flexibleSpaceItem];
[items addObject:confirmBtn];
//判断toolBar是否为空
if (self.toolBar==nil) {
    self.toolBar = [[UIToolbar alloc] initWithFrame:CGRectMake(0, self.
    pickerView.frame.origin.y - 44, 320, 44)];
}
self.toolBar.hidden = NO;
self.toolBar.barStyle = UIBarStyleBlackTranslucent;    //设置工具栏的风格
self.toolBar.items = items;
items = nil;
[self addSubview:self.toolBar];                        //添加视图对象
//实现动画
[UIView animateWithDuration:0.5 animations:^{
    self.pickerView.frame = CGRectMake(0, 44, 320, 260);
    self.toolBar.frame = CGRectMake(0, self.pickerView.frame.origin.
    y-44, 320, 44);
}];
}
```

pickerHide方法实现以动画的形式隐藏选择器。程序代码如下:

```objc
- (void)pickerHide
{
    [UIView animateWithDuration:0.5 animations:^{
        self.alpha = 0.0;
        self.pickerView.frame = CGRectMake(0, 260+44, 320, 260);//设置框架
        self.toolBar.frame = CGRectMake(0, self.pickerView.frame.origin.y-44,
         320, 44);                                              //设置框架
    }];
}
```

confirmPickView方法是单击工具栏上的"确定"按钮后实现的方法。程序代码如下:

```objc
-(void)confirmPickView
{
    //遍历
    for (NSString *row in [self.selectionStatesDic allKeys]) {
        if ([[self.selectionStatesDic objectForKey:row] boolValue]) {
            [self.selectedEntriesArr addObject:row];
        }
    }
```

```objc
//判断是否实现了returnChoosedPickerString:方法
if ([self.multiPickerDelegate respondsToSelector:@selector(returnChoosed
PickerString:)]) {
    [self.multiPickerDelegate returnChoosedPickerString:self.selected
    EntriesArr];
}
[self pickerHide];
}
```

pickerView:didCheckRow:和 pickerView:didUncheckRow:方法是用来检测选中的行的。程序代码如下：

```objc
- (void)pickerView:(PickerView *)pickerView didCheckRow:(NSInteger)row {
    //判断 row 是否等于-1
    if (row == -1)
        //遍历
        for (id key in [self.selectionStatesDic allKeys])
            [self.selectionStatesDic setObject:[NSNumber numberWithBool:YES]
            forKey:key];
    else
        [self.selectionStatesDic setObject:[NSNumber numberWithBool:YES]
        forKey:[self.entriesArray objectAtIndex:row]];    //设置对象
}
- (void)pickerView:(PickerView *)pickerView didUncheckRow:(NSInteger)row {
    if (row == -1)
        for (id key in [self.selectionStatesDic allKeys])
            [self.selectionStatesDic setObject:[NSNumber numberWithBool:NO]
            forKey:key];
    else
        [self.selectionStatesDic setObject:[NSNumber numberWithBool:NO]
        forKey:[self.entriesArray objectAtIndex:row]];    //设置对象
}
```

（12）打开 ViewController.h 文件，编写代码，实现头文件和对象的声明。程序代码如下：

```objc
#import <UIKit/UIKit.h>
#import "PickerView.h"
#import "MultiSelectPickerView.h"
@interface ViewController : UIViewController<MultiSelectPickerView
Delegate>{
    NSArray *entries;
    NSArray *entriesSelected;                //声明数组
    NSMutableDictionary *selectionStates;
    MultiSelectPickerView *multiPickerView;
}
@end
```

（13）打开 ViewController.m 文件，编写代码。在此文件中，代码分为：设置界面、获取数据以及显示选择的数据。其中，viewDidLoad 方法实现对界面的设置。程序代码如下：

```objc
- (void)viewDidLoad
{
    [super viewDidLoad];
```

```
   //获取数据源
   entries = [[NSArray alloc] initWithObjects:@"One", @"Two", @"Three",
   @"Four", @"Five",@"Six",nil];
   selectionStates = [[NSMutableDictionary alloc] init];
    // 配置是否选中状态
   for (NSString *key in entries){
      BOOL isSelected = NO;
      //遍历
      for (NSString *keyed in entriesSelected) {
          if ([key isEqualToString:keyed]) {
              isSelected = YES;
          }
      }
      [selectionStates  setObject:[NSNumber  numberWithBool:isSelected]
       forKey:key];                                       //设置对象
   }
   //创建并设置按钮对象
   UIButton *btn = [UIButton buttonWithType:UIButtonTypeRoundedRect];
   btn.frame = CGRectMake(20, 50, 280, 50);
   [btn setTitle:@"显示具有多选功能的选择器" forState:UIControlStateNormal];
                                                           //设置标题
   [btn  addTarget:self  action:@selector(getData)  forControlEvents:UI
   ControlEventTouchUpInside];
   [self.view addSubview:btn];
}
```

getData 方法实现获取数据。程序代码如下：

```
-(void) getData
{
   //单击后删除之前的 PickerView
   for (UIView *view in self.view.subviews) {
      if ([view isKindOfClass:[MultiSelectPickerView class]]) {
          [view removeFromSuperview];            //移除视图 view
      }
   }
   //创建并设置具有多选功能的选择器对象
   multiPickerView = [[MultiSelectPickerView alloc] initWithFrame:
   CGRectMake(0,[UIScreen mainScreen].bounds.size.height - 260-20, 320,
   260+44)];
   multiPickerView.entriesArray = entries;
   multiPickerView.entriesSelectedArray = entriesSelected;
   multiPickerView.multiPickerDelegate = self;
   [self.view addSubview:multiPickerView];        //添加视图对象
   [multiPickerView pickerShow];
}
```

returnChoosedPickerString:方法实现的功能是将选择的数据显示在警告视图中。程序代码如下：

```
-(void)returnChoosedPickerString:(NSMutableArray *)selectedEntriesArr
{
   NSString *dataStr = [selectedEntriesArr componentsJoinedByString:@"\n"];
   //创建并显示警告视图对象
   UIAlertView *alert=[[UIAlertView alloc]initWithTitle:@"你选择的如下："
```

```
             message:dataStr delegate:nil cancelButtonTitle:@"OK" otherButton
             Titles:nil];
             [alert show];
             // 再次初始化选中的数据
             entriesSelected = [NSArray arrayWithArray:selectedEntriesArr];
}
```

【代码解析】

由于本实例中的代码和方法非常的多，为了方便读者的阅读，笔者绘制了一些执行流程图如图 5.8~5.10 所示。其中，单击运行按钮后一直到在模拟器上显示选择器，需要使用 viewDidLoad、getData、initWithFrame:（MultiSelectPickerView.m）、pickerShow、initWithFrame:（PickerView.m）、tableView:numberOfRowsInSection:、numberOfRowsForPickerView:、tableView:cellForRowAtIndexPath:、initWithReuseIdentifier:、pickerView:selectionStateForRow:、setSelectionState:、pickerView:textForRow:和 layoutSubviews 方法共同实现。它们的执行流程如图 5.8 所示。

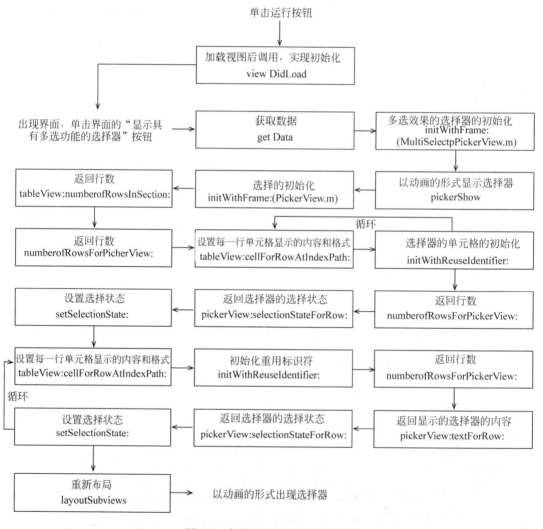

图 5.8 实例 37 程序执行流程 1

实现选择器内容的选择，可以有两个方法：选择选择器中的"选择全部"将选择器中的所有内容选中、选择选择器中的 One～Two。这里以选择选择器中的"选择全部"将所有内容选择实例，它的实现需要使用 layoutSubviews、tableView:didSelectRowAtIndexPath:、numberOfRowsForPickerView:、selectionState、pickerView:selectionStateForRow:和 pickerView:didCheckRow:方法共同实现。它们的执行流程如图 5.9 所示。

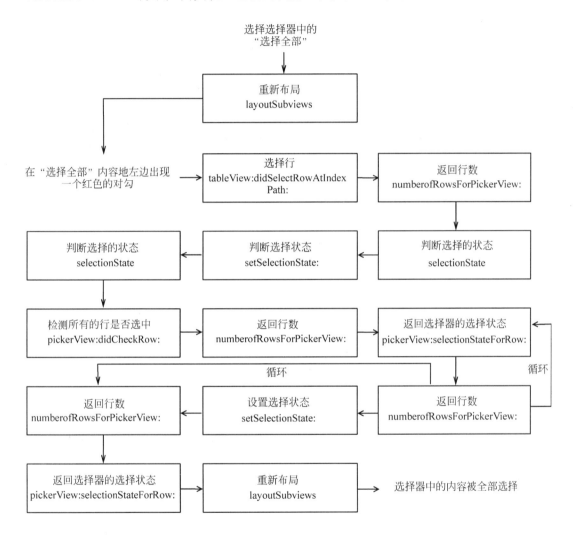

图 5.9　实例 37 程序执行流程 2

如果想要将选择的某一内容取消，需要使用 layoutSubviews、tableView:didSelectRowAtIndexPath:、numberOfRowsForPickerView:、selectionState、setSelectionState:、pickerView:didUncheckRow:和 pickerView:selectionStateForRow:、layoutSubviews、tableView:cellForRowAtIndexPath:、prepareForReuse:和 pickerView:textForRow:方法共同实现。这里以将 One 的内容取消为例，它的执行执行流程如图 5.10 所示。

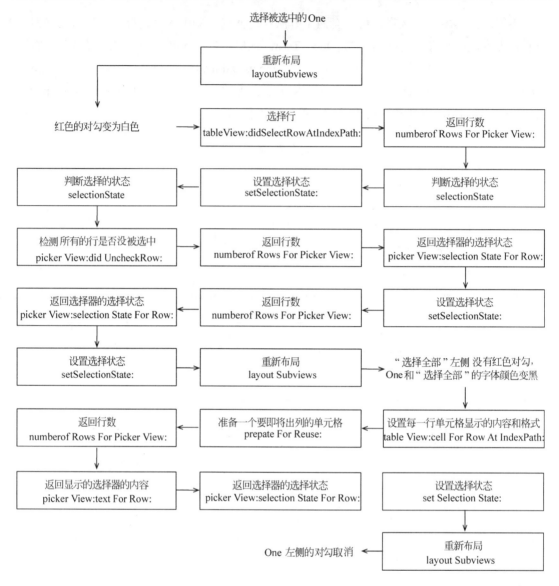

图 5.10　实例 37 程序执行流程 3

实例 38　转盘选择器

【实例描述】
本实例实现的功能是在界面显示一个转盘选择器，当用户选择转盘中的某一项时，会将选择的项显示在警告视图中。运行效果如图 5.11 所示。

【实现过程】
（1）创建一个项目，命名为"转盘选择器"。
（2）添加图片 1.png、2.png 和 3.png 到创建项目的 Supporting Files 文件夹中。
（3）创建一个基于 NSObject 类的 Clove 类。

第 5 章 选择器类效果

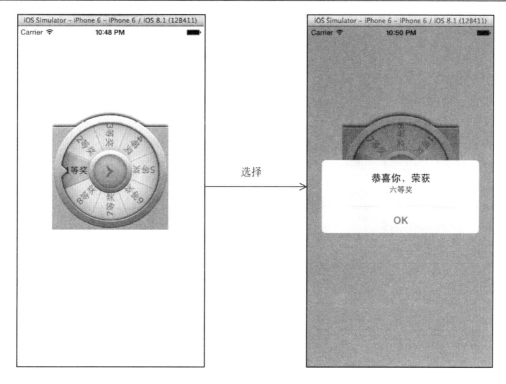

图 5.11 实例 38 运行效果

(4) 打开 Clove 类,编写代码。实现属性的声明。程序代码如下:

```
#import <Foundation/Foundation.h>
@interface Clove : NSObject
//属性
@property float minValue;
@property float maxValue;
@property float midValue;
@property int value;
@end
```

(5) 创建一个基于 UIControl 类的 WheelPickerView 类。

(6) 打开 WheelPickerView.h 类,编写代码。实现头文件、协议、属性以及方法的声明。程序代码如下:

```
#import <UIKit/UIKit.h>
#import "Clove.h"                        //头文件
//协议
@protocol Protocol <NSObject>
- (void) wheelDidChangeValue:(NSString *)newValue;
@end
@interface WheelPickerView : UIControl
//属性
@property (weak) id <Protocol> delegate;
@property (nonatomic, strong) UIView *container;
……
@property int currentValue;
//方法
- (id) initWithFrame:(CGRect)frame andDelegate:(id)del withSections:(int)sections
```

• 229 •

```
Number;
- (void)drawWheel;
- (float) calculateDistanceFromCenter:(CGPoint)point;
- (void) buildClovesEven;
- (void) buildClovesOdd;                    //创建转盘中的基数瓣
- (UIImageView *) getCloveByValue:(int)value;
- (NSString *) getCloveName:(int)position;
@end
```

（7）打开 WheelPickerView.m 类，编写代码。实现一个转盘选择器的效果。使用的方法如表 5-10 所示。

表 5-10　WheelPickerView.m 文件中方法总结

方　　法	功　　能
initWithFrame:andDelegate:withSections:	转盘选择器的初始化
drawWheel	绘制转盘
getCloveByValue:	通过值得到转盘中的当前瓣
buildClovesEven	创建转盘中的偶数瓣
buildClovesOdd	创建转盘中的奇数瓣
beginTrackingWithTouch:withEvent:	开始触摸
continueTrackingWithTouch:withEvent:	继续触摸
endTrackingWithTouch: withEvent:	结束触摸
getCloveName:	通过值得到当前瓣中的名称

要实现一个转盘选择器，需要分为 3 个部分才可以实现：绘制转盘、实现触摸进行选择以及获取当前选择的内容。这里需要讲解几个重要的方法（其他方法请读者参考源代码）。其中，绘制转盘需要使用 initWithFrame:andDelegate:withSections:、drawWheel、buildClovesEven 和 buildClovesOdd 方法。initWithFrame:andDelegate:withSections:方法实现对转盘选择器进行初始化。程序代码如下：

```
- (id) initWithFrame:(CGRect)frame andDelegate:(id)del withSections:
(int)sectionsNumber {
    if ((self = [super initWithFrame:frame])) {
        self.currentValue = 0;                      //设置当前的值
        self.numberOfSections = sectionsNumber;     //设置段数
        self.delegate = del;                        //设置协议
        [self drawWheel];
    }
    return self;
}
```

drawWheel 方法实现绘制一个转盘。程序代码如下：

```
- (void) drawWheel {
    //绘制转盘中的每一瓣
    container = [[UIView alloc] initWithFrame:self.frame]; //实例化对象
    CGFloat angleSize = 2*M_PI/numberOfSections;
    //遍历
    for (int i = 0; i < numberOfSections; i++) {
        UIImageView *im = [[UIImageView alloc] initWithImage:[UIImage image
        Named:@"2.png"]];
        im.layer.anchorPoint = CGPointMake(1.0f, 0.5f);       //设置锚点
        im.layer.position = CGPointMake(container.bounds.size.width/2.0-
        container.frame.origin.x,container.bounds.size.height/2.0-container.
```

```
        frame.origin.y);                                //设置位置
        im.transform = CGAffineTransformMakeRotation(angleSize*i);
        im.alpha = minAlphavalue;
        im.tag = i;
        if (i == 0) {
            im.alpha = maxAlphavalue;                   //设置透明度
        }
        //创建并设置转盘中每一瓣的标签内容
        UILabel *cloveImage = [[UILabel alloc] initWithFrame:CGRectMake(12,
        15, 70, 40)];
        cloveImage.font=[UIFont fontWithName:@"Verdana" size:15];
        cloveImage.text=[NSString stringWithFormat:@"%d 等奖",i+1];
                                                        //设置文本内容
        [im addSubview:cloveImage];
        [container addSubview:im];
    }
    container.userInteractionEnabled = NO;
    [self addSubview:container];
    cloves = [NSMutableArray arrayWithCapacity:numberOfSections];
    //创建并设置转盘的背景对象
    UIImageView *bg = [[UIImageView alloc] initWithFrame:self.frame];
                                                    //实例化对象
    bg.image = [UIImage imageNamed:@"1.png"];       //设置图像
    [self addSubview:bg];
    //创建并设置转盘的中心的视图对象
    UIImageView *mask = [[UIImageView alloc] initWithFrame:CGRectMake(0, 0,
    58, 58)];
    mask.image =[UIImage imageNamed:@"3.png"] ;
    mask.center = self.center;
    mask.center = CGPointMake(mask.center.x, mask.center.y+3);//设置中心位置
    [self addSubview:mask];
    //判断转盘的瓣数是否为偶数
    if (numberOfSections % 2 == 0) {
        [self buildClovesEven];
    } else {
        [self buildClovesOdd];
    }
    [self.delegate wheelDidChangeValue:[self getCloveName:currentValue]];
}
```

buildClovesEven 方法实现转盘的瓣数是偶数的创建。程序代码如下:

```
- (void) buildClovesEven {
    CGFloat fanWidth = M_PI*2/numberOfSections;
    CGFloat mid = 0;
    //遍历
    for (int i = 0; i < numberOfSections; i++) {
        Clove *clove = [[Clove alloc] init];
        clove.midValue = mid;           ///每一瓣的中间值
        clove.minValue = mid - (fanWidth/2);            //每一瓣的最小值
        clove.maxValue = mid + (fanWidth/2);            //每一瓣的最大值
        clove.value = i;
        if (clove.maxValue-fanWidth < - M_PI) {
            mid = M_PI;
            clove.midValue = mid;                       //每一瓣的中间值
            clove.minValue = fabsf(clove.maxValue);     //每一瓣的最小值
        }
        mid -= fanWidth;
```

```
        [cloves addObject:clove];              //添加对象
    }
}
```

要实现触摸并且选择，需要使用 getCloveByValue:、beginTrackingWithTouch:with Event:、continueTrackingWithTouch:withEvent:和 endTrackingWithTouch:withEvent:方法。getCloveByValue:方法实现通过值获取当前的瓣。程序代码如下：

```
- (UIImageView *) getCloveByValue:(int)value {
    UIImageView *res;
    NSArray *views = [container subviews];
    //遍历
    for (UIImageView *im in views) {
        if (im.tag == value)              //判断 im 的 tag 值是否等于 value
            res = im;
    }
    return res;
}
```

beginTrackingWithTouch:withEvent:方法实现开始触摸。程序代码如下：

```
- (BOOL)beginTrackingWithTouch:(UITouch *)touch withEvent:(UIEvent*)event {
    CGPoint touchPoint = [touch locationInView:self];        //获取位置
    float dx = touchPoint.x - container.center.x;            //获取 x 的分量
    float dy = touchPoint.y - container.center.y;            //获取 y 的分量
    deltaAngle = atan2(dy,dx);
    startTransform = container.transform;
    UIImageView *im = [self getCloveByValue:currentValue];
    im.alpha = minAlphavalue;
    return YES;
}
```

continueTrackingWithTouch:withEvent:方法实现继续触摸（手指在屏幕上移动）。程序代码如下：

```
-(BOOL)continueTrackingWithTouch:(UITouch*)touch      withEvent:(UIEvent*)
event
{
    CGPoint pt = [touch locationInView:self];             //获取触摸的位置
    float dx = pt.x - container.center.x;
    float dy = pt.y - container.center.y;
    float ang = atan2(dy,dx);
    float angleDifference = deltaAngle - ang;
    container.transform = CGAffineTransformRotate(startTransform, -angle
    Difference);                                          //旋转
    return YES;
}
```

endTrackingWithTouch: withEvent:方法实现触摸结束。程序代码如下：

```
- (void)endTrackingWithTouch:(UITouch*)touch withEvent:(UIEvent*)event
{
    CGFloat radians = atan2f(container.transform.b, container.transform.a);
    //获取反正切函数
    CGFloat newVal = 0.0;
    for (Clove *c in cloves) {
        //判断
        if (c.minValue > 0 && c.maxValue < 0) {
```

```
            if (c.maxValue > radians || c.minValue < radians) {
                if (radians > 0) {
                    newVal = radians - M_PI;            //设置新值
                } else {
                    newVal = M_PI + radians;
                }
                currentValue = c.value;                 //设置当前值
            }
        }
        else if (radians > c.minValue && radians < c.maxValue) {
            newVal = radians - c.midValue;              //设置新值
            currentValue = c.value;
        }
    }
}
//旋转动画
[UIView beginAnimations:nil context:NULL];
[UIView setAnimationDuration:0.2];
CGAffineTransform t = CGAffineTransformRotate(container.transform,
-newVal);
container.transform = t;
[UIView commitAnimations];
[self.delegate wheelDidChangeValue:[self getCloveName:currentValue]];
//设置被选择的瓣的效果
UIImageView *im = [self getCloveByValue:currentValue];
im.alpha = maxAlphavalue;                               //设置图像视图的透明度
}
```

getCloveName:方法实现通过值得到当前瓣中的名称。程序代码如下：

```
- (NSString *) getCloveName:(int)position {
    NSString *res = @"";
    switch (position) {
        case 0:
            res = @"一等奖";
            break;
        case 1:
            res = @"二等奖";
            break;
        case 2:
            res = @"三等奖";
            break;
        ……
            case 7:
            res = @"八等奖";
            break;
        default:
            break;
    }
    return res;
}
```

（8）打开 ViewController.m 文件，编写代码。实现将转盘添加到当前视图中，并且当值发生改变时，弹出警告视图。程序代码如下：

```
- (void)viewDidLoad
{
    [super viewDidLoad];
```

```
        // Do any additional setup after loading the view, typically from a nib.
        WheelPickerView *wheel = [[WheelPickerView alloc] initWithFrame:
        CGRectMake(0, 0, 200, 200)andDelegate:self withSections:8];//实例化对象
        wheel.center = CGPointMake(160, 240);          //设置中心位置
        [self.view addSubview:wheel];
}
//当值发生改变,弹出警告视图
- (void) wheelDidChangeValue:(NSString *)newValue {
    UIAlertView*alert=[[UIAlertView alloc]initWithTitle:@"恭喜你,荣获"
    message:newValue delegate:nil cancelButtonTitle:@"OK" otherButton
    Titles:nil];                                       //创建警告视图
    [alert show];
}
```

【代码解析】

本实例关键功能是转盘的转动与转盘中指示选择的内容。下面依次讲解这两个知识点。

1. 转盘的转动

转盘的转动,首先需要使用 CGAffineTransform 的 CGAffineTransformRotate 函数,对旋转的角度进行设置。

```
CGAffineTransform CGAffineTransformRotate (
    CGAffineTransform t,
    CGFloat angle
);
```

其中,CGAffineTransform t 表示要变换的对象;CGFloat angle 表示要旋转的弧度值。然后使用 UIView 的 transform 属性,实现 view 在其父视图中的呈现方式。其语法形式如下:

```
@property(nonatomic) CGAffineTransform transform;
```

在代码中,使用了 CGAffineTransformRotate 函数和 transform 属性实现了转盘的选择。程序代码如下:

```
CGAffineTransform t = CGAffineTransformRotate(container.transform, -newVal);
container.transform = t;
```

在代码中,使用 CGAffineTransformRotate 函数和 transform 属性时,还可以将它们合起来使用。程序代码如下:

```
container.transform = CGAffineTransformRotate(startTransform, -angleDifference);
```

2. 转盘中指示选择的内容

在转盘中由指针指向的瓣是被选中的,这时它的背景颜色会变深,告诉用户在此瓣中的数据是被选中的,它的实现需要调用 getCloveByValue:方法,然后使用 alpha 属性将透明度改变。程序代码如下:

```
UIImageView *im = [self getCloveByValue:currentValue];
im.alpha = maxAlphavalue;
```

实例 39 老 虎 机

【实例描述】
本实例的功能实现一个老虎机的效果。当用户单击 Start 按钮后，老虎机上的卷轴就开始滚动，在一定时间后停止。当老虎机上被选择的图形和指定的图形一致时，就赢得了比赛。运行效果如图 5.12 所示。

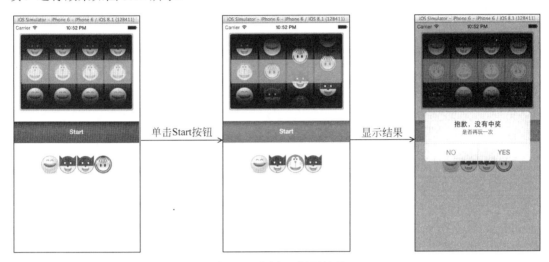

图 5.12 实例 39 运行效果

【实现过程】
（1）创建一个项目，命名为"老虎机"。
（2）添加图片 1.png、2.png、3.png、4.png、5.png、6.png 和 7.png 到创建项目的 Supporting Files 文件夹中。
（4）创建一个基于 UIView 类的 machine 类。
（5）打开 machine.h 文件，编写代码。实现协议、对象、属性和方法等的声明。程序代码如下：

```
#import <UIKit/UIKit.h>
@class machine;
//协议
@protocol machineDelegate <NSObject>
@optional
- (void)slotMachineWillStartSliding:(machine *)slotMachine;
- (void)slotMachineDidEndSliding:(machine *)slotMachine;
@end
@protocol machineDataSource <NSObject>
@required
- (NSUInteger)numberOfSlotsInSlotMachine:(machine *)slotMachine;
- (NSArray *)iconsForSlotsInSlotMachine:(machine *)slotMachine;
@optional
- (CGFloat)slotWidthInSlotMachine:(machine *)slotMachine;
- (CGFloat)slotSpacingInSlotMachine:(machine *)slotMachine;
@end
@interface machine : UIView{
```

```
@private
    //对象
    UIImageView *_backgroundImageView;
    ......
    NSArray *_currentSlotResults;
    __weak id<machineDataSource> _dataSource;
    UIEdgeInsets _contentInset;            //实例变量
}
//属性
@property (nonatomic) UIEdgeInsets contentInset;
......
@property (nonatomic, weak) id <machineDataSource> dataSource;
- (void)startSliding;                      //方法
@end
```

(6) 打开 machine.m 文件，编写代码，实现老虎机的效果。使用的方法如表 5-11 所示。

表 5-11   machine.m 文件中方法总结

方　　法	功　　能
initWithFrame:	老虎机的初始化
setBackgroundImage:	设置背景图片
setCoverImage:	设置覆盖的图片
contentInset	获取内容视图的边缘
setContentInset:	设置内容视图的边缘
slotResults	获取结果
setSlotResults:	设置结果
dataSource	获取数据来源
setDataSource:	设置数据来源
reloadData	刷新数据
startSliding	开始滚动

要实现老虎机的效果，需要 3 个部分实现：绘制老虎机、实现滚动以及获取数据。这里对几个重要方法进行讲解，其他方法请读者参考源代码。其中，绘制老虎机，需要使用 initWithFrame:、setBackgroundImage:、setCoverImage:、contentInset、setContentInset:、dataSource、setDataSource:和 reloadData 方法实现。initWithFrame:方法实现老虎机的初始化。程序代码如下：

```
- (id)initWithFrame:(CGRect)frame {
    self = [super initWithFrame:frame];
    if (self) {
        self.autoresizingMask = UIViewAutoresizingFlexibleLeftMargin | UIView
        AutoresizingFlexibleRightMargin;
        //创建并设置背景视图对象
        _backgroundImageView = [[UIImageView alloc] initWithFrame:frame];
        _backgroundImageView.contentMode = UIViewContentModeCenter;
        [self addSubview:_backgroundImageView];
        //创建并设置内容视图对象
        _contentView = [[UIView alloc] initWithFrame:frame];
#if SHOW_BORDER
        _contentView.layer.borderColor = [UIColor redColor].CGColor;
        _contentView.layer.borderWidth = 1;
#endif
        [self addSubview:_contentView];
```

```
        //创建并设置覆盖的视图对象（此视图用来实现选择）
        _coverImageView = [[UIImageView alloc] initWithFrame:frame];
        _coverImageView.contentMode = UIViewContentModeCenter;
        [self addSubview:_coverImageView];
        _slotScrollLayerArray = [NSMutableArray array];
        self.singleUnitDuration = 0.14f;
        _contentInset = UIEdgeInsetsMake(0, 0, 0, 0);    //指定内容视图的边缘
    }
    return self;
}
```

setContentInset:方法实现设置内容视图的边缘。程序代码如下：

```
- (void)setContentInset:(UIEdgeInsets)contentInset {
    _contentInset = contentInset;
    CGRect viewFrame = self.frame;
    //设置框架
    _contentView.frame = CGRectMake(_contentInset.left, _contentInset.top,
    viewFrame.size.width - _contentInset.left - _contentInset.right,
    viewFrame.size.height - _contentInset.top - _contentInset.bottom);
}
```

reloadData 方法实现对数据进行刷新。程序代码如下：

```
- (void)reloadData {
    if (self.dataSource) {
        for (CALayer *containerLayer in _contentView.layer.sublayers) {
            [containerLayer removeFromSuperlayer];    //将此图层从父视图中删除
        }
        _slotScrollLayerArray = [NSMutableArray array];
        //获取老虎机中的卷轴数量
        NSUInteger numberOfSlots = [self.dataSource numberOfSlotsInSlot
Machine:self];
        CGFloat slotSpacing = 0;
        //判断是否有slotSpacingInSlotMachine:方法
        if ([self.dataSource respondsToSelector:@selector(slotSpacing
InSlotMachine:)]) {
            slotSpacing = [self.dataSource slotSpacingInSlotMachine:self];
        }
        CGFloat slotWidth = _contentView.frame.size.width / numberOfSlots;
        //判断是否有slotWidthInSlotMachine:方法
        if ([self.dataSource respondsToSelector:@selector(slotWidthInSlot
Machine:)]) {
            slotWidth = [self.dataSource slotWidthInSlotMachine:self];
        }
        //添加并设置滚动的卷轴图层
        for (int i = 0; i < numberOfSlots; i++) {
            CALayer *slotContainerLayer = [[CALayer alloc] init];
            slotContainerLayer.frame = CGRectMake(i * (slotWidth + slotSpacing),
            0, slotWidth, _contentView.frame.size.height);    //设置框架
            slotContainerLayer.masksToBounds = YES;
            CALayer *slotScrollLayer = [[CALayer alloc] init];
            slotScrollLayer.frame = CGRectMake(0, 0, slotWidth, _contentView.
            frame.size.height);
#if SHOW_BORDER
            slotScrollLayer.borderColor = [UIColor greenColor].CGColor;
                                                          //设置边框颜色
            slotScrollLayer.borderWidth = 1;              //设置边框宽度
#endif
            [slotContainerLayer addSublayer:slotScrollLayer];
```

```objc
        [_contentView.layer addSublayer:slotContainerLayer];
        [_slotScrollLayerArray addObject:slotScrollLayer];    //添加对象
    }
    CGFloat singleUnitHeight = _contentView.frame.size.height / 3;
    NSArray *slotIcons = [self.dataSource iconsForSlotsInSlotMachine:
    self];
    NSUInteger iconCount = [slotIcons count];              //获取数组中的个数
    //为滚动的卷轴图层添加图片
    for (int i = 0; i < numberOfSlots; i++) {
        CALayer *slotScrollLayer = [_slotScrollLayerArray objectAtIndex:i];
                                                            //实例化对象
        NSInteger scrollLayerTopIndex = - (i + kMinTurn + 3) * iconCount;
                                                            //遍历
        for (int j = 0; j > scrollLayerTopIndex; j--) {
            UIImage *iconImage = [slotIcons objectAtIndex:abs(j) %
            numberOfSlots];                                 //实例化对象
            CALayer *iconImageLayer = [[CALayer alloc] init];
            NSInteger offsetYUnit = j + 1 + iconCount;
            iconImageLayer.frame = CGRectMake(0, offsetYUnit * single
            UnitHeight, slotScrollLayer.frame.size.width,singleUnitHeight);
                                                            //设置框架
            iconImageLayer.contents = (id)iconImage.CGImage;
            iconImageLayer.contentsScale = iconImage.scale;
            iconImageLayer.contentsGravity = kCAGravityCenter;
#if SHOW_BORDER
            iconImageLayer.borderColor = [UIColor redColor].CGColor;
                                                            //设置边框颜色
            iconImageLayer.borderWidth = 1;
#endif
            [slotScrollLayer addSublayer:iconImageLayer];   //添加图层
        }
    }
}
```

startSliding 方法实现老虎机中卷轴的滚动。程序代码如下：

```objc
- (void)startSliding {
    if (isSliding) {
        return;
    }
    else {
        isSliding = YES;
        //判断是否有 slotMachineWillStartSliding:方法
        if ([self.delegate respondsToSelector:@selector(slotMachine
        WillStartSliding:)]) {
            [self.delegate slotMachineWillStartSliding:self];
        }
        //获取每一个卷轴的图片
        NSArray *slotIcons = [self.dataSource iconsForSlotsInSlotMachine:
        self];
        NSUInteger slotIconsCount = [slotIcons count];//获取老虎图片的数量
        __block NSMutableArray *completePositionArray = [NSMutableArray
        array];
        //实现动画
        [CATransaction begin];
        [CATransaction setAnimationTimingFunction:[CAMediaTimingFunction
        functionWithName:kCAMediaTimingFunctionEaseInEaseOut]];    //定时
        [CATransaction setDisableActions:YES];           //关闭隐身动画
```

```objc
//当动画结束时被调用的块
[CATransaction setCompletionBlock:^{
    isSliding = NO;
    if ([self.delegate respondsToSelector:@selector(slotMachineDidEndSliding:)]) {
        [self.delegate slotMachineDidEndSliding:self];
    }
    //设置滚动的卷轴图层的位置
    for (int i = 0; i < [_slotScrollLayerArray count]; i++) {
        CALayer*slotScrollLayer=[_slotScrollLayerArray objectAtIndex:i];
        slotScrollLayer.position = CGPointMake(slotScrollLayer.position.x, ((NSNumber *)[completePositionArray objectAtIndex:i]).floatValue);                                    //设置位置
        NSMutableArray *toBeDeletedLayerArray = [NSMutableArray array];
        NSUInteger resultIndex = [[self.slotResults objectAtIndex:i] unsignedIntegerValue];
        NSUInteger currentIndex = [[_currentSlotResults objectAtIndex:i] unsignedIntegerValue];
        //将滚动的卷轴图层中的子图层放入 toBeDeletedLayerArray 数组中
        for (int j = 0; j < slotIconsCount * (kMinTurn + i) + resultIndex - currentIndex; j++) {
            CALayer *iconLayer = [slotScrollLayer.sublayers objectAtIndex:j];
            [toBeDeletedLayerArray addObject:iconLayer];
        }
        //设置 toBeDeletedLayerArray 数组中的对象
        for (CALayer *toBeDeletedLayer in toBeDeletedLayerArray) {
            CALayer *toBeAddedLayer = [CALayer layer];
            toBeAddedLayer.frame = toBeDeletedLayer.frame;
                                                                //设置框架
            toBeAddedLayer.contents = toBeDeletedLayer.contents;
            toBeAddedLayer.contentsScale = toBeDeletedLayer.contentsScale  ;                                //设置缩放
            toBeAddedLayer.contentsGravity = toBeDeletedLayer.contentsGravity;
            CGFloat shiftY = slotIconsCount * toBeAddedLayer.frame.size.height * (kMinTurn + i + 3);
            toBeAddedLayer.position = CGPointMake(toBeAddedLayer.position.x, toBeAddedLayer.position.y - shiftY);
                                                                //设置位置
            [toBeDeletedLayer removeFromSuperlayer];   //删除
            [slotScrollLayer addSublayer:toBeAddedLayer]; //添加图层
        }
        toBeDeletedLayerArray = [NSMutableArray array];
    }
    _currentSlotResults = self.slotResults;
    completePositionArray = [NSMutableArray array];
}];
static NSString * const keyPath = @"position.y";
//为每一个滚动卷轴图层数组实现滚动动画
for (int i = 0; i < [_slotScrollLayerArray count]; i++) {
    CALayer *slotScrollLayer = [_slotScrollLayerArray objectAtIndex:i];
                                                        //实例化对象
    NSUInteger resultIndex = [[self.slotResults objectAtIndex:i] unsignedIntegerValue];
    NSUInteger currentIndex = [[_currentSlotResults objectAtIndex:i] unsignedIntegerValue];
    NSUInteger howManyUnit = (i + kMinTurn) * slotIconsCount +
```

```
            resultIndex - currentIndex;
        CGFloat slideY=howManyUnit*(_contentView.frame.size.height/3);
        //实现滚动动画
        CABasicAnimation *slideAnimation = [CABasicAnimation animation
        WithKeyPath:keyPath];
        slideAnimation.fillMode = kCAFillModeForwards;
        slideAnimation.duration = howManyUnit * self.singleUnitDuration;
                                                            //设置时间
        slideAnimation.toValue = [NSNumber numberWithFloat:slotScroll
        Layer.position.y + slideY];
        slideAnimation.removedOnCompletion = NO;
        [slotScrollLayer addAnimation:slideAnimation forKey:@"slideAnimation"];
                                                            //添加动画
        [completePositionArray addObject:slideAnimation.toValue];
    }
    [CATransaction commit];
  }
}
```

要获取最后卷轴的结果,需要使用 slotResults 和 setSlotResults:方法实现。其中,setSlotResults:方法实现对卷轴结果的设置。程序代码如下:

```
- (void)setSlotResults:(NSArray *)slotResults {
  if (!isSliding) {
    _slotResults = slotResults;
        if (!_currentSlotResults) {
        NSMutableArray *currentSlotResults = [NSMutableArray array];
        //遍历
        for (int i = 0; i < [slotResults count]; i++) {
            [currentSlotResults addObject:[NSNumber numberWithUnsigned
            Integer:0]];    //添加对象
        }
        _currentSlotResults = [NSArray arrayWithArray:currentSlotResults];
    }
  }
}
```

(7)打开 ViewController.h 文件,编写代码,实现头文件、实例变量和对象等的声明。程序代码如下:

```
#import <UIKit/UIKit.h>
#import "machine.h"
@interface ViewController : UIViewController<machineDataSource,machine
Delegate>{
    NSUInteger slotOne;
    NSUInteger slotTwo;
    ……
    NSUInteger slotThreeIndex;          //声明实例变量
    NSUInteger slotFourIndex;
@private
    //声明对象
    machine *_slotMachine;
    UIButton *_startButton;
    ……
    UIImageView *_slotFourImageView;
    NSArray *_slotIcons;
}
@end
```

(8) 打开 ViewController.m 文件，编写代码。实现将老虎机添加到当前的视图中，并且可以实现游戏。使用的方法如表 5-12 所示。

表 5-12 ViewController.m文件中方法总结

方　　法	功　　能
viewDidLoad	视图加载后调用，实现初始化
start	单击按钮后，调用
slotMachineWillStartSliding:	老虎机开始滚动调用
slotMachineDidEndSliding:	老虎机结束滚动调用
compare	比较
alertView:clickedButtonAtIndex:	警告视图的响应
iconsForSlotsInSlotMachine:	获取老虎机卷轴中的图片
numberOfSlotsInSlotMachine:	获取卷轴的数量
slotWidthInSlotMachine:	获取卷轴的宽度
slotSpacingInSlotMachine:	获取卷轴之间的间隙

其中，viewDidLoad 方法实现创建一个老虎机对象，并将其添加到当前的视图中。程序代码如下：

```
- (void)viewDidLoad
{
    [super viewDidLoad];
    //获取卷轴的图片
    _slotIcons = [NSArray arrayWithObjects:[UIImage imageNamed:@"1.png"],
    [UIImage imageNamed:@"2.png"], [UIImage imageNamed:@"3.png"], [UIImage
    imageNamed:@"4.png"], nil];
    //创建并设置老虎机的对象
    _slotMachine = [[machine alloc] initWithFrame:CGRectMake(0, 0, 291, 193)];
    _slotMachine.center = CGPointMake(self.view.frame.size.width / 2, 120);
        //设置中心位置
    _slotMachine.autoresizingMask = UIViewAutoresizingFlexibleLeftMargin |
    UIViewAutoresizingFlexibleRightMargin;
    _slotMachine.contentInset = UIEdgeInsetsMake(5, 8, 5, 8);
    _slotMachine.backgroundImage = [UIImage imageNamed:@"5.png"];
    //设置背景图层
    _slotMachine.coverImage = [UIImage imageNamed:@"6.png"];
    _slotMachine.delegate = self;                    //设置委托
    _slotMachine.dataSource = self;
    [self.view addSubview:_slotMachine];
    //创建并设置按钮对象
    _startButton = [UIButton buttonWithType:UIButtonTypeCustom];
    UIImage *btnImageN = [UIImage imageNamed:@"7.png"];
    _startButton.frame = CGRectMake(0, 0, btnImageN.size.width, btnImageN.
    size.height);
    _startButton.center = CGPointMake(self.view.frame.size.width / 2, 270);
    //设置按钮的中心位置
    _startButton.autoresizingMask = UIViewAutoresizingFlexibleLeftMargin |
    UIViewAutoresizingFlexibleRightMargin | UIViewAutoresizingFlexibleBottom
    Margin;
    _startButton.titleLabel.font = [UIFont boldSystemFontOfSize:16.0f];
    //设置按钮中标题的字体
    [_startButton setBackgroundImage:btnImageN forState:UIControlState
    Normal];
```

```objc
[_startButton setTitle:@"Start" forState:UIControlStateNormal];
    //设置按钮标题
    [_startButton addTarget:self action:@selector(start) forControlEvents:UIControlEventTouchUpInside];
    [self.view addSubview:_startButton];
    //创建并设置放置内容的视图对象
    _slotContainerView = [[UIView alloc] initWithFrame:CGRectMake(0, 0, 180, 45)];
    _slotContainerView.autoresizingMask = UIViewAutoresizingFlexibleLeftMargin | UIViewAutoresizingFlexibleRightMargin;
    _slotContainerView.center = CGPointMake(self.view.frame.size.width / 2, 350);                                    //设置中心位置
    [self.view addSubview:_slotContainerView];
    //创建并设置卷轴视图对象
    _slotOneImageView = [[UIImageView alloc] initWithFrame:CGRectMake(0, 0, 45, 45)];
    _slotOneImageView.contentMode = UIViewContentModeCenter;
                                            //设置图片的显示方式
    _slotTwoImageView = [[UIImageView alloc] initWithFrame:CGRectMake(45, 0, 45, 45)];
    _slotTwoImageView.contentMode = UIViewContentModeCenter;
    _slotThreeImageView = [[UIImageView alloc] initWithFrame:CGRectMake(90, 0, 45, 45)];
    _slotThreeImageView.contentMode = UIViewContentModeCenter;
    _slotFourImageView = [[UIImageView alloc] initWithFrame:CGRectMake(135, 0, 45, 45)];
    _slotFourImageView.contentMode = UIViewContentModeCenter;
                                            //设置图片的显示方式
    //添加视图对象
    [_slotContainerView addSubview:_slotOneImageView];
    [_slotContainerView addSubview:_slotTwoImageView];
    [_slotContainerView addSubview:_slotThreeImageView];
    [_slotContainerView addSubview:_slotFourImageView];
    //设置指定的结果
    NSUInteger slotIconCount = [_slotIcons count];
    slotOne = abs(rand() % slotIconCount);
    slotTwo = abs(rand() % slotIconCount);
    slotThree = abs(rand() % slotIconCount);
    slotFour = abs(rand() % slotIconCount);
    //设置图像
    _slotOneImageView.image = [_slotIcons objectAtIndex:slotOne];
    _slotTwoImageView.image = [_slotIcons objectAtIndex:slotTwo];
    _slotThreeImageView.image = [_slotIcons objectAtIndex:slotThree];
    _slotFourImageView.image = [_slotIcons objectAtIndex:slotFour];
    // Do any additional setup after loading the view, typically from a nib.
}
```

start方法实现单击按钮后执行的动作。程序代码如下：

```objc
- (void)start {
    //设置选择的结果
    NSUInteger slotIconCount = [_slotIcons count];
    //产生随机数
    slotOneIndex = abs(rand() % slotIconCount);
    slotTwoIndex = abs(rand() % slotIconCount);
    slotThreeIndex = abs(rand() % slotIconCount);
    slotFourIndex = abs(rand() % slotIconCount);
    _slotMachine.slotResults = [NSArray arrayWithObjects:
                                [NSNumber numberWithInteger:slotOneIndex],
                                [NSNumber numberWithInteger:slotTwoIndex],
                                [NSNumber numberWithInteger:slotThreeIndex],
```

```
                        [NSNumber numberWithInteger:slotFourIndex],
                        nil];
    [_slotMachine startSliding];                    //开始滚动
}
```

compare 方法实现两个结果的比较，如果两个结果相同，就赢了；如果不相同，就输了。程序代码如下：

```
-(void)compare{
    if (slotOne==slotOneIndex&&slotTwo==slotTwoIndex&&slotThree==slotThree
    Index&& slotFour==slotFourIndex) {
        UIAlertView *alert=[[UIAlertView alloc]initWithTitle:@"恭喜你中了500
        万大奖" message:@"是否再玩一次" delegate:self cancelButtonTitle:@"NO"
        otherButtonTitles:@"YES", nil];                //创建警告视图
        [alert show];
    }else{
        UIAlertView *alert=[[UIAlertView alloc]initWithTitle:@"抱歉，没有中奖"
        message:@"是否再玩一次" delegate:self cancelButtonTitle:@"NO" other
        ButtonTitles:@"YES", nil];
        [alert show];
    }
}
```

alertView: clickedButtonAtIndex:方法实现对警告视图的响应。程序代码如下：

```
-(void)alertView:(UIAlertView *)alertView clickedButtonAtIndex:(NSInteger)
buttonIndex{
    NSString *b=[alertView buttonTitleAtIndex:buttonIndex];
    //判断选择的按钮的标题是否为 YES
    if([b isEqualToString:@"YES"]){
        _slotOneImageView.image = nil;
        _slotTwoImageView.image = nil;
        _slotThreeImageView.image =nil;
        _slotFourImageView.image =nil;
        [self viewDidLoad];
}
//判断选择的按钮的标题是否为 NO
    if([b isEqualToString:@"NO"]){
        exit(1);
    }
}
```

【代码解析】

本实例关键功能是结果的判断。下面就是这个知识点的详细讲解。

实现结果的判断，首先要设置指定的结果，在本实例中，采用了随机设置指定的结果。程序代码如下：

```
    NSUInteger slotIconCount = [_slotIcons count];
    slotOne = abs(rand() % slotIconCount);
    slotTwo = abs(rand() % slotIconCount);
    slotThree = abs(rand() % slotIconCount);
    slotFour = abs(rand() % slotIconCount);
    _slotOneImageView.image = [_slotIcons objectAtIndex:slotOne];
    _slotTwoImageView.image = [_slotIcons objectAtIndex:slotTwo];
    _slotThreeImageView.image = [_slotIcons objectAtIndex:slotThree];
    _slotFourImageView.image = [_slotIcons objectAtIndex:slotFour];
```

随后，在老虎机中生成随机结果，在本实例中，也是采用了随机的形式。程序代码

如下:

```
NSUInteger slotIconCount = [_slotIcons count];
slotOneIndex = abs(rand() % slotIconCount);
slotTwoIndex = abs(rand() % slotIconCount);
slotThreeIndex = abs(rand() % slotIconCount);
slotFourIndex = abs(rand() % slotIconCount);
_slotMachine.slotResults = [NSArray arrayWithObjects:
                    [NSNumber numberWithInteger:slotOneIndex],
                    [NSNumber numberWithInteger:slotTwoIndex],
                    [NSNumber numberWithInteger:slotThreeIndex],
                    [NSNumber numberWithInteger:slotFourIndex],
                    nil];
```

最后,将这两组用来存放随机数的值进行比较,看是否相等。程序代码如下:

```
if (slotOne==slotOneIndex&&slotTwo==slotTwoIndex&&slotThree==slotThreeIndex&& slotFour==slotFourIndex) {
    UIAlertView *alert=[[UIAlertView alloc]initWithTitle:@"恭喜你中了500万大奖" message:@"是否再玩一次" delegate:self cancelButtonTitle:@"NO" otherButtonTitles:@"YES", nil];
    [alert show];
}else{
    UIAlertView *alert=[[UIAlertView alloc]initWithTitle:@"抱歉,没有中奖" message:@"是否再玩一次" delegate:self cancelButtonTitle:@"NO" otherButtonTitles:@"YES", nil];
    [alert show];
}
```

第 6 章 视 图

在 iPhone 或者 iPad 中，用户所看到的都是视图。它是界面上的一个矩形区域。用户可以通过触摸视图上的对象与应用程序进行交互。本章将主要讲解视图的相关实例。

实例 40　更改空白视图的背景颜色

【实例描述】

backgroundColor 属性可以设置空白视图的背影颜色。本实例通过使用此属性，对空白视图的背影颜色进行设置。其中，用于颜色选择的是分段控件，用于显示颜色的是空白视图。运行效果如图 6.1 所示。

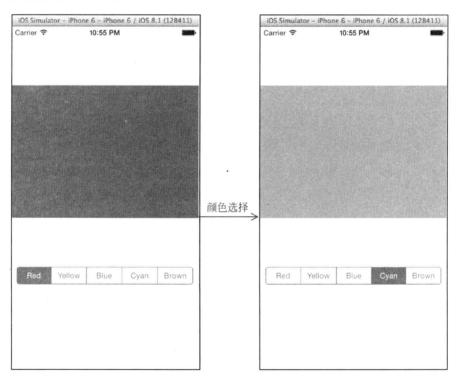

图 6.1　实例 40 运行效果

【实现过程】

当用户选择分段控件中的某一段，会在空白的视图出现对应的背景颜色。具体的实现步骤如下。

（1）创建一个项目，命名为"更改背景颜色"。

（2）打开 ViewController.h 文件，编写代码，实现插座变量和动作的声明。程序代码如下：

```
#import <UIKit/UIKit.h>
@interface ViewController : UIViewController{
    IBOutlet UIView *vv;                         //空白视图的插座变量
    IBOutlet UISegmentedControl *segmented;      //分段控件的插座变量
}
- (IBAction)changeColor:(id)sender;
@end
```

（3）打开 Main.storyboard 文件，对 View Controller 视图控制器的设计界面进行设计，效果如图 6.2 所示。

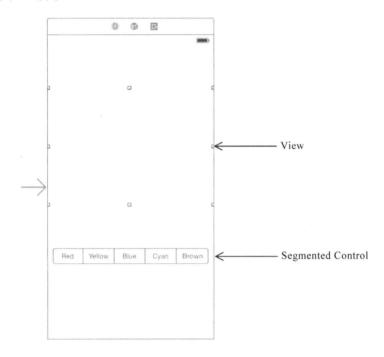

图 6.2　实例 40View Controller 视图控制器的设计界面

需要添加的视图、控件以及对它们的设置如表 6-1 所示。

表 6-1　实例 40 视图、控件设置

视图、控件	属 性 设 置	其　　他
View		调整大小 与插座变量vv关联
Segmented Control	Segment：5 Title：Red、Yellow、Blue、Cyan、Brown	与插座变量segmented关联 与动作changeColor:关联

（4）打开 ViewController.m 文件，编写代码，实现空白视图的颜色变化。程序代码如下：

```
- (IBAction)changeColor:(id)sender {
    NSInteger select=segmented.selectedSegmentIndex;    //获取选择的段的索引
    //判断选择的段的索引是否为 0
    if (select==0) {
        vv.backgroundColor=[UIColor redColor];
    }else if (select==1){                               //判断选择的段的索引是否为 1
        vv.backgroundColor=[UIColor yellowColor];
    }else if (select==2){                               //判断选择的段的索引是否为 2
        vv.backgroundColor=[UIColor blueColor];
    }else if (select==3){                               //判断选择的段的索引是否为 3
        vv.backgroundColor=[UIColor cyanColor];
    }else{
        vv.backgroundColor=[UIColor brownColor];
    }
}
```

【代码解析】

本实例关键功能是设置空白视图的背景颜色和分段控件所选段的索引号识别。下面依次讲解这两个知识点。

1．设置空白视图的背景颜色

要对空白视图的背景颜色进行设置，可以使用 UIView 的 backgroundColor 属性实现，其语法形式如下：

```
@property(nonatomic,copy)UIColor *backgroundColor;
```

在代码中，使用了 backgroundColor 属性对空白视图的背景颜色进行设置，代码如下：

```
vv.backgroundColor=[UIColor redColor];
```

2．分段控件所选段的索引号识别

分段控件所选段的索引号识别，是通过 UISegmentedControl 的 selectedSegmentIndex 属性实现的，其语法形式如下：

```
@property(nonatomic) NSInteger selectedSegmentIndex;
```

在代码中，使用了 selectedSegmentIndex 属性对分段控件所选段的索引号进行识别，代码如下：

```
NSInteger select=segmented.selectedSegmentIndex;
```

实例 41　关闭应用程序

【实例描述】

由于在 iOS 设备中有一个 Home 按钮，实现的功能是将正在运行的应用程序关闭，所以很多的开发者忽略了在应用程序中关闭程序。本实例就通过应用程序中的按钮将应用程序关闭，其中，用于显示数据的是表视图。运行效果如图 6.3 所示。

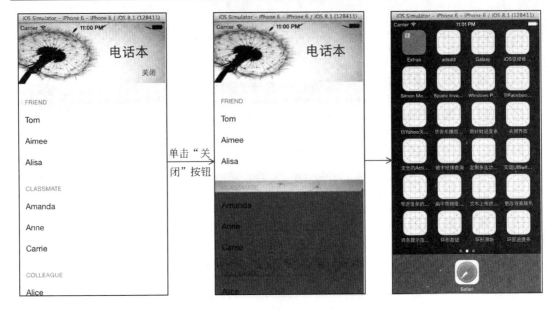

图 6.3　实例 41 运行效果

【实现过程】

当用户单击"关闭"按钮，会以动画的形式切换为黑色的空白视图，之后，将此应用程序退出。具体的实现步骤如下。

（1）创建一个项目，命名为"关闭应用程序"。

（2）添加图片 12.jpg 到创建项目的 Supporting Files 文件夹中。

（3）打开 Main.storyboard 文件，对 View Controller 控制器的设计界面进行设计，效果如图 6.4 所示。

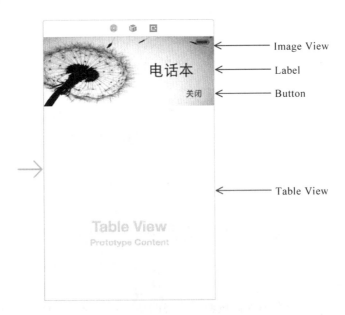

图 6.4　实例 41View Controller 视图控制器的设计界面

需要添加的视图、控件以及对它们的设置如表 6-2 所示。

表 6-2　实例 41 视图、控件设置

视图、控件	属性设置	其他
Image View	Image：1.jpg	调整大小
Label	Text：电话本 Font：System 28.0	调整大小
Button	Title：关闭	调整大小 将"关闭"按钮和ViewController.h文件进行动作aa:的声明和关联
Table View		调整大小 将dataSource连接到dock工作区的View Controller图标上

（4）创建一个基于 UIViewController 类的 FirstViewController 类。

（5）回到 Main.storyboard 文件，从视图库中拖动 View Controller 视图控制器到画布中。

（6）将 Class 设置为 FirstViewController。将 Storyboard ID 设置为 first，选择 Use Storyboard ID 复选框。

（7）单击 First View Controller 视图控制器的设计界面，将 Background 属性设置为黑色。

（8）打开 ViewController.h 文件，编写代码，实现头文件以及类对象的声明。程序代码如下：

```
#import <UIKit/UIKit.h>
#import "FirstViewController.h"          //头文件的声明
@interface ViewController : UIViewController{
    FirstViewController *first;
    NSArray *group;
    NSArray *friend;                     //实例化数组对象
    NSArray *classmate;
    NSArray *colleague;
}
- (IBAction)aa:(id)sender;
@end
```

（9）打开 ViewController.m 文件，编写代码。在此文件中，代码分为 3 个部分：数据的填充；视图的切换；应用程序的关闭。实现数据的填充的程序代码如下：

```
- (void)viewDidLoad
{
    //创建数组
    friend=[NSArray arrayWithObjects:@"Tom",@"Aimee",@"Alisa", nil];
    classmate=[NSArray arrayWithObjects:@"Amanda",@"Anne",@"Carrie",nil];
    colleague=[NSArray arrayWithObjects:@"Alice",@"Emily",@"Ailsa",nil];
    group=[NSArray arrayWithObjects:@"friend",@"classmate",@"colleague",nil];
    [super viewDidLoad];
    // Do any additional setup after loading the view, typically from a nib.
}
//设置行数
- (NSInteger)tableView:(UITableView *)tableView numberOfRowsInSection:
(NSInteger)section
{
    return 3;
}
- (UITableViewCell*)tableView:(UITableView *)tableView cellForRowAtIndexPath:
(NSIndexPath *)indexPath
{
```

```objc
    static NSString *CellIdentifier = @"Cell";
    UITableViewCell *cell = [tableView dequeueReusableCellWithIdentifier:
    CellIdentifier];
    //判断cell是否为空
    if (cell==nil)
    {
        cell = [[UITableViewCell alloc] initWithStyle:UITableViewCellSty
        leSubtitle reuseIdentifier:CellIdentifier];         //实例化cell
    }   //判断块的索引是否为0
    if (indexPath.section==0)
    {
        cell.textLabel.text = [friend objectAtIndex:indexPath.row];
    }
    //判断块的索引是否为1
    if (indexPath.section==1)
    {
        cell.textLabel.text = [classmate objectAtIndex:indexPath.row];
    }
    //判断块的索引是否为2
    if (indexPath.section==2)
    {
        cell.textLabel.text = [colleague objectAtIndex:indexPath.row];
    }
    return cell;
}
//设置节数
- (NSInteger )numberOfSectionsInTableView:(UITableView *)tableView
{
    return group.count;
}
//设置块的标题
- (NSString *)tableView:(UITableView *)tableView titleForHeaderInSection:
(NSInteger )section
{
    return [group objectAtIndex:section];
}
```

在aa:动作方法中实现视图的切换。程序代码如下：

```objc
- (IBAction)aa:(id)sender {
    first=[self.storyboard instantiateViewControllerWithIdentifier:@"first"];
    CATransition *t=[CATransition animation];
    t.duration=5;
    t.type=kCATransitionMoveIn;                //过渡类型
    t.subtype=kCATransitionFromTop;            //方向
    [self.view addSubview:first.view];
    [self.view.layer addAnimation:t forKey:nil];
    [self performSelector:@selector(bb) withObject:self afterDelay:5];
}
```

添加bb方法实现应用程序的关闭。程序代码如下：

```objc
//实现应用程序的关闭
-(void)bb{
    exit(1);
}
```

【代码解析】

本实例关键功能是视图CATransition动画切换和方法的延迟执行。下面依次讲解这两个知识点。

1．视图CATransition动画切换

视图的CATransition动画切换只要有两个内容：第一，视图的CATransition动画切换方式，可以使用CATransiton的type属性，其语法形式如下：

```
@property(copy) NSString *type;
```

其中，它的动画效果有两种，一种是公开的过渡动画，一种是非公开的过渡动画，如表6-3所示。

表6-3 过渡动画

公开的过渡动画	
公开动画效果	功能
kCATransitionFade	渐渐消失
kCATransitionMoveIn	覆盖进入
kCATransitionPush	推出
kCATransitionReveal	揭开
非公开的过渡动画	
非公开动画效果	功能
@"cube"	立方体
@"suckEffect"	吸收
@"oglFlip"	翻转
@"rippleEffect"	波纹
@"pageCurl"	卷页
@"cameraIrisHollowOpen"	镜头开
@"cameraIrisHollowClose"	镜头关

第二，对动画方向的设置使用的是CATransiton的subtype属性，其语法形式如下：

```
@property(copy) NSString *subtype;
```

其中，动画方向的设置有4种，如表6-4所示。

表6-4 动画方向

动 画 方 向	功　　能
kCATransitionFromRight	从过渡层右侧开始实现动画
kCATransitionFromLeft	从过渡层左侧开始实现动画
kCATransitionFromTop	从过渡层顶部开始实现动画
kCATransitionFromBottom	从过渡层底部开始实现动画

在代码中，使用了CATransition动画中的type属性进行动画效果的设置，代码如下：

```
t.type=kCATransitionMoveIn;
```

使用了subtype属性进行动画方向的设置，代码如下：

```
t.subtype=kCATransitionFromTop;
```

2. 方法的延迟执行

方法的延迟执行，主要是通过 performSelector:withObject:afterDelay:方法实现的。它是 NSObject 提供的方法。其语法形式如下：

```
-(void) performSelector:(SEL)aSelector withObject:(id)anArgument afterDelay:
(NSTimeInterval)delay;
```

其中，(SEL)aSelector 表示触发器所调用的方法；(id)anArgumen 在参数调用时使用；(NSTimeInterval)delay 表示消息发送的最短时间。在代码中，使用了 performSelector: withObject:afterDelay：方法表示方法的延迟执行，代码如下：

```
[self performSelector:@selector(bb) withObject:self afterDelay:5];
```

其中，@selector(bb)表示触发器所调用的方法；5 表示消息发送的最短时间。

实例 42 手 电 筒

【实例描述】

本实例实现的功能是一个手电筒的功能。调整亮度将会使整个手机屏幕跟着变亮或变暗，其中用于调整光亮暗的是滑块控件。运行效果如图 6.5 所示。

图 6.5 实例 42 运行效果

【实现过程】

当用户滑动滑块时，屏幕的亮度会发生变化。具体的实现步骤如下：

（1）创建一个项目，命名为"手电筒"。

（2）在 ViewController.h 文件中编写代码，实现插座变量和动作的声明。程序代码如下：

```
#import <UIKit/UIKit.h>
@interface ViewController : UIViewController{
    IBOutlet UISlider *slider;          //插座变量
}
```

```
- (IBAction)aa:(id)sender;
@end
```

(3)打开 Main.sstoryboard 文件,对设计界面进行设计,效果如图 6.6 所示。

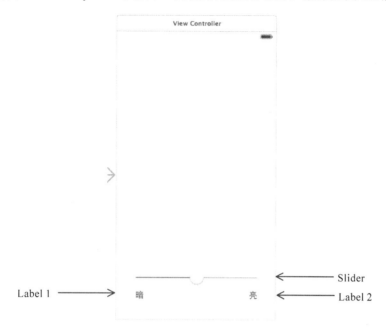

图 6.6 实例 42 设计界面效果

需要添加的视图、控件以及对它们的设置如表 6-5 所示。

表 6-5 实例 42 视图、控件设置

视图、控件	属 性 设 置	其　　他
Slider		调整大小 与插座变量slider关联 与动作aa:关联
Label1	Text:暗	
Label2	Text:亮	

(4)打开 ViewController.m 文件,编写代码,实现手电筒的功能。程序代码如下:

```
- (IBAction)aa:(id)sender {
    [[UIScreen mainScreen] setBrightness: slider.value];      //设置亮度
}
@end
```

【代码解析】

本实例关键功能是屏幕亮暗的设置。下面就对这个知识点做一个详细的讲解。屏幕的亮暗设置,可以使用 UIScreen 的 brightness 属性,其语法形式如下:

```
@property(nonatomic)CGFloat brightness;
```

在代码中,使用了 brightness 属性对屏幕的亮暗进行设置,代码如下:

```
[[UIScreen mainScreen] setBrightness: slider.value];
```

实例 43　旋转大挑战

【实例描述】

本实例实现的功能是在一定时间限制内旋转手机屏幕到指定的方向，如果成功会进入下一关；如果失败会结束游戏。其中，向用户显示时间和方向指令的都是标签。运行效果如图 6.7 所示。

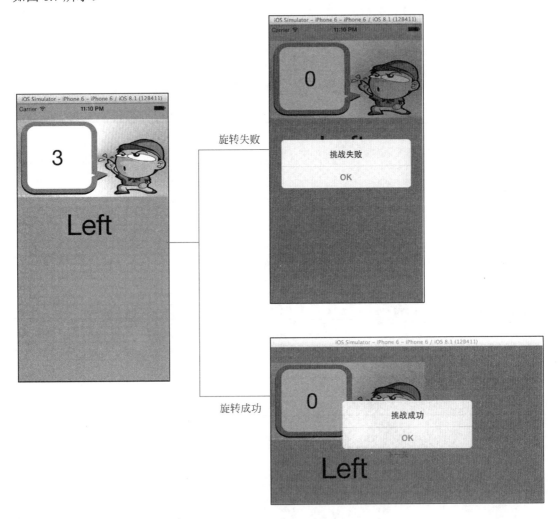

图 6.7　实例 43 运行效果

【实现过程】

用户根据标签中出现的方向指令旋转手机。当旋转的方向和标签中的方向一致时，就会进入下一关，否则，就会重玩本关。具体的实现步骤如下。

（1）创建一个项目，命名为"旋转大挑战"。

（2）添加图片 1.jpg 和 2.jpg 到创建项目的 Supporting Files 文件夹中。

(3) 打开 Main.storyboard 文件，对 View Controller 视图控制器的设计界面进行设计，效果如图 6.8 所示。

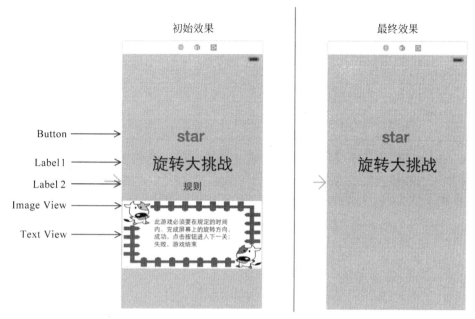

图 6.8 实例 43View Controller 视图控制器的设计界面

需要添加的视图、控件以及对它们的设置如表 6-6 所示。

表 6-6 实例 43 视图、控件设置

视图、控件	属性设置	其他
Button	Title：star Font：System Bold 36.0	
Label1	Text：旋转大挑战 Font：System 37.0	调整大小
Label2	Text：规则 Font：System 21.0 Aligment：居中	调整大小
Image View	Image：2.jpg	
Text View	Text：此游戏必须要在规定的时间内，完成屏幕上的旋转方向，成功，单击按钮进入下一关；失败，游戏结束	

(4) 打开 ViewController.h 文件，编写代码，实现插座变量的声明。程序代码如下：

```
#import <UIKit/UIKit.h>
@interface ViewController : UIViewController{
    //标签的插座变量
    IBOutlet UILabel *label1;
    IBOutlet UILabel *label2;
    IBOutlet UITextView *tv;          //文本视图的插座变量
    IBOutlet UIButton *button;        //按钮的插座变量
    IBOutlet UIImageView *image;      //图像视图的插座变量
}
@end
```

（5）将声明的插座变量和 View Controller 视图控制器设计界面上的视图以及控件相关联，如表 6-7 所示。

表 6-7 插座变量的关联

插座变量	关联的视图、控件
label	与 Label1 关联
label2	与 Label2 关联
tv	与 Text View 文本关联
button	与标题为 star 的按钮关联
image	与 Image View 图像视图进行关联

（6）将插座变量 label2、tv、image 视图的控件移动位置，一直到在设计界面上看不到，如图 6.8 所示的最终效果。

（7）打开 ViewController.m 文件，编写代码，实现游戏的开场动画。程序代码如下：

```
#import "ViewController.h"
@interface ViewController ()
@end
@implementation ViewController
- (void)viewDidLoad
{
    [button setHidden:YES];                     //实现隐藏
    [label2 setHidden:YES];
    [tv setHidden:YES];
    [image setHidden:YES];
    [self performSelector:@selector(aa) withObject:self afterDelay:5];
                                                //过5秒后执行aa动作
    [super viewDidLoad];
    // Do any additional setup after loading the view, typically from a nib.
}
//aa动作的实现
-(void)aa{
    [label1 setHidden:YES];
    [label2 setHidden:NO];
    [image setHidden:NO];
    [tv setHidden:NO];
    //动画
    [UIView beginAnimations:@"aaa" context:nil];
    [UIView setAnimationDuration:3.0];          //设置动画所需的时间
    label2.frame=CGRectMake(150, 205, 48,25);   //设置标签的框架
    tv.frame=CGRectMake(75, 265, 203, 86);
    image.frame=CGRectMake(0, 250, 325, 128);
    [UIView commitAnimations];
    [self performSelector:@selector(bb) withObject:self afterDelay:5];
}
//出现按钮
-(void)bb{
    [button setHidden:NO];
}
@end
```

（8）创建一个基于 UIViewController 类的 aaViewController 类。

（9）回到 Main.storyboard 文件，从视图库中拖动 View Controller 视图控制器到画布中。

（10）将 Class 设置为 aaViewController。

（11）单击 Aa View Controller 视图控制器的设计界面，对其设计界面进行设计，其效果如图 6.9 所示。

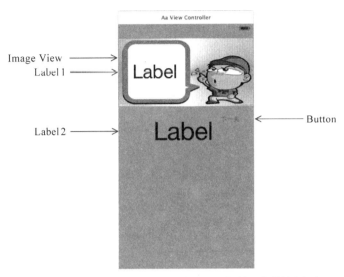

图 6.9　Aa View Controller 视图控制器的设计界面

需要添加的视图、控件以及对它们的设置如表 6-8 所示。

表 6-8　实例 43 视图、控件设置

视图、控件	属 性 设 置	其　　他
Image View	Image：1.jpg	
Label1	Font：System 44.0	调整大小
Label2	Font：System 59.0	
Button	Title：下一关	与aaViewController.h文件进行动作next:的声明和关联

（12）按住 Ctrl 键，将 View Controller 视图控制器设计界面上的 star 按钮拖动到 Aa View Controller 设计界面上。这时画布效果如图 6.10 所示。

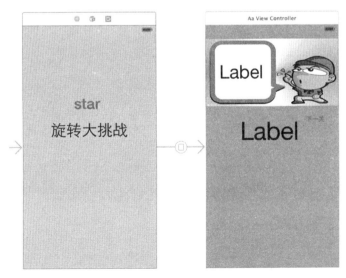

图 6.10　画布效果

(13) 打开 aaViewController.h 文件，编写代码，实现插座变量、时间定时器以及实例变量的声明。程序代码如下：

```
#import <UIKit/UIKit.h>
@interface aaViewController : UIViewController{
    IBOutlet UILabel *label3;              //标签的插座变量
    IBOutlet UILabel *label4;              //标签的插座变量
    NSTimer *t;                            //时间定时器
    int secondsCountDown;
    IBOutlet UIButton *butt;
    UIDeviceOrientation deviceOrientation;
}
- (IBAction)next:(id)sender;
@end
```

(14) 将声明的插座变量和 Aa View Controller 视图控制器设计界面上的视图以及控件相关联，如表 6-9 所示。

表 6-9　插座变量的关联

插 座 变 量	关联的视图、控件
label3	与label1标签关联
label4	与label2标签关联
butt	与标题为"下一关"的按钮进行关联

(15) 打开 aaViewController.m 文件，编写代码，在此文件中，代码分为 4 个部分：获取设备的方向、倒计时、设备在不同方向的向左旋转，以及设备在不同方向的向右旋转。在 viewDidLoad 方法中实现获取设备的方向，以及随机的方向向左、向右的文字。程序代码如下：

```
- (void)viewDidLoad
{
    deviceOrientation=[UIDevice currentDevice].orientation;
                                                            //获取当前设备的方向
    [butt setHidden:YES];
    secondsCountDown=5;
    t=[NSTimer scheduledTimerWithTimeInterval:1.0 target:self selector:
    @selector(time) userInfo:nil repeats:YES];             //创建时间定时器
    int i=arc4random()%2;                                   //产生随机值
    switch (i) {
        case 0:
            label4.text=@"Left";                            //设置标签的文本内容
            [self performSelector:@selector(aa) withObject:self afterDelay:5];
            break;
        default:
            label4.text=@"Right";
            [self performSelector:@selector(ac) withObject:self afterDelay:5];
                                                            //经过 5 秒后执行 ac
            break;
    }
    [super viewDidLoad];
    // Do any additional setup after loading the view.
}
```

添加 time 方法实现倒计时功能。程序代码如下：

```
-(void)time{
    secondsCountDown--;
    label3.text=[NSString stringWithFormat:@"%i",secondsCountDown];
    //判断 secondsCountDown 是否为 0
    if(secondsCountDown==0){
        [t invalidate];          //让定时器失效
    }
}
```

添加 aa 方法实现设备在不同方向的向左旋转。程序代码如下：

```
-(void)aa{
    UIDeviceOrientation d=[UIDevice currentDevice].orientation;
    //如果当前设备是横向的，并且 Home 按钮位于左侧
    if (deviceOrientation==UIInterfaceOrientationLandscapeLeft) {
        if (UIInterfaceOrientationMaskLandscapeLeft) {
            deviceOrientation=UIInterfaceOrientationPortrait;
            if(deviceOrientation==d){
                UIAlertView *a=[[UIAlertView alloc]initWithTitle:@"挑战成功"
                message:nil delegate:self cancelButtonTitle:@"OK" otherButton
                Titles:nil];          //创建警告视图
                [a show];
                [self performSelector:@selector(bb) withObject:self afterDelay:1];
            }else{
                UIAlertView *a=[[UIAlertView alloc]initWithTitle:@"挑战失败"
                message:nil delegate:nil cancelButtonTitle:@"OK" otherButton
                Titles:nil];          //创建警告视图
                [a show];
            }
        }
    //如果当前设备是纵向的
    }else if(deviceOrientation==UIInterfaceOrientationPortrait){
        if (UIInterfaceOrientationMaskLandscapeLeft) {
            deviceOrientation=UIInterfaceOrientationLandscapeRight;
            if(deviceOrientation==d){
                UIAlertView *a=[[UIAlertView alloc]initWithTitle:@"挑战成功"
                message:nil delegate:self cancelButtonTitle:@"OK" otherButton
                Titles:nil];          //创建警告视图
                [a show];
                [self performSelector:@selector(bb) withObject:self afterDelay:1];
            }else{
                UIAlertView *a=[[UIAlertView alloc]initWithTitle:@"挑战失败"
                message:nil delegate:nil cancelButtonTitle:@"OK" otherButton
                Titles:nil];          //创建警告视图
                [a show];
            }
        }
    //如果当前设备是横向的，并且 Home 按钮位于右侧
    }else if(deviceOrientation==UIInterfaceOrientationLandscapeRight){
        if (UIInterfaceOrientationMaskLandscapeLeft) {
            deviceOrientation=UIInterfaceOrientationPortraitUpsideDown;
            if(deviceOrientation==d){
                UIAlertView *a=[[UIAlertView alloc]initWithTitle:@"挑战成功"
                message:nil delegate:self cancelButtonTitle:@"OK" otherButton
                Titles:nil];          //创建警告视图
                [a show];
                [self performSelector:@selector(bb) withObject:self afterDelay:1];
            }else{
```

```
            UIAlertView *a=[[UIAlertView alloc]initWithTitle:@"挑战失败"
             message:nil delegate:nil cancelButtonTitle:@"OK" otherButton
             Titles:nil];                //创建警告视图
             [a show];
          }
      }
//如果当前设备是纵向的,并且 Home 按钮位于屏幕的上方
      }else{
          if (UIInterfaceOrientationMaskLandscapeLeft) {
             deviceOrientation=UIInterfaceOrientationLandscapeLeft;
             if(deviceOrientation==d){
                UIAlertView *a=[[UIAlertView alloc]initWithTitle:@"挑战成功"
                 message:nil delegate:self cancelButtonTitle:@"OK" otherButton
                 Titles:nil];                //创建警告视图
                 [a show];
                 [self performSelector:@selector(bb) withObject:self afterDelay:1];
             }else{
                 UIAlertView *a=[[UIAlertView alloc]initWithTitle:@"挑战失败"
                 message:nil delegate:nil cancelButtonTitle:@"OK" otherButton
                 Titles:nil];                //创建警告视图
                 [a show];
             }
          }
      }
}
```

添加 ac 方法实现设备在不同方向的向右旋转。程序代码如下:

```
-(void)ac{
    UIDeviceOrientation d=[UIDevice currentDevice].orientation;
    if (deviceOrientation==UIInterfaceOrientationLandscapeLeft) {
        if (UIInterfaceOrientationMaskLandscapeRight) {
            deviceOrientation=UIInterfaceOrientationPortraitUpsideDown;
            if(deviceOrientation==d){
                UIAlertView *a=[[UIAlertView alloc]initWithTitle:@"挑战成功"
                 message:nil delegate:self cancelButtonTitle:@"OK" otherButton
                 Titles:nil];                //创建警告视图
                 [a show];
                 [self performSelector:@selector(bb) withObject:self afterDelay:1];
            }else{
                UIAlertView *a=[[UIAlertView alloc]initWithTitle:@"挑战失败"
                 message:nil delegate:nil cancelButtonTitle:@"OK" otherButton
                 Titles:nil];                //创建警告视图
                 [a show];
            }
        }
    }else if(deviceOrientation==UIInterfaceOrientationPortrait){
        if (UIInterfaceOrientationMaskLandscapeRight) {
            deviceOrientation=UIInterfaceOrientationLandscapeLeft;
            if(deviceOrientation==d){
                UIAlertView *a=[[UIAlertView alloc]initWithTitle:@"挑战成功"
                 message:nil delegate:self cancelButtonTitle:@"OK" otherButton
                 Titles:nil];                //创建警告视图
                 [a show];
                 [self performSelector:@selector(bb) withObject:self afterDelay:1];
            }else{
                UIAlertView *a=[[UIAlertView alloc]initWithTitle:@"挑战失败"
                 message:nil delegate:nil cancelButtonTitle:@"OK" otherButton
                 Titles:nil];                //创建警告视图
```

```
                [a show];
            }
        }
    }else if(deviceOrientation==UIInterfaceOrientationLandscapeRight){
        if (UIInterfaceOrientationMaskLandscapeRight) {
            deviceOrientation=UIInterfaceOrientationPortrait;
            if(deviceOrientation==d){
                UIAlertView *a=[[UIAlertView alloc]initWithTitle:@"挑战成功"
                message:nil delegate:self cancelButtonTitle:@"OK" otherButton
                Titles:nil];           //创建警告视图
                [a show];
                [self performSelector:@selector(bb) withObject:self afterDelay:1];
            }else{
                UIAlertView *a=[[UIAlertView alloc]initWithTitle:@"挑战失败"
                message:nil delegate:nil cancelButtonTitle:@"OK" otherButton
                Titles:nil];           //创建警告视图
                [a show];
            }
        }
    }else{
        if (UIInterfaceOrientationMaskLandscapeRight) {
            deviceOrientation=UIInterfaceOrientationLandscapeRight;
            if(deviceOrientation==d){
                UIAlertView *a=[[UIAlertView alloc]initWithTitle:@"挑战成功"
                message:nil delegate:self cancelButtonTitle:@"OK" otherButton
                Titles:nil];           //创建警告视图
                [a show];
                [self performSelector:@selector(bb) withObject:self afterDelay:1];
            }else{
                UIAlertView *a=[[UIAlertView alloc]initWithTitle:@"挑战失败"
                message:nil delegate:nil cancelButtonTitle:@"OK" otherButton
                Titles:nil];           //创建警告视图
                [a show];
            }
        }
    }
}
```

【代码解析】

本实例关键功能是获取设备的当前方向和NSTimer动画。下面依次讲解这两个知识点。

1．获取设备的当前方向

要对设备的当前方向进行获取，可以使用UIDevice的orientation属性实现，其语法形式如下：

`@property(nonatomic, readonly) UIDeviceOrientation orientation;`

其中，设备的方向有多种，如表6-10所示。

表6-10　设备的方向

设备的方向	说　　明
UIDeviceOrientationUnknown	方向不能被确定
UIDeviceOrientationPortrait	纵向模式，Home键在屏幕的下方
UIDeviceOrientationPortraitUpsideDown	纵向模式，Home键在屏幕的上方
UIDeviceOrientationLandscapeLeft	横向模式，Home键在屏幕的右侧

续表

设备的方向	说　明
UIDeviceOrientationLandscapeRight	横向模式，Home 键在屏幕的左侧
UIDeviceOrientationFaceUp	方向朝上，屏幕的面向上，并与地面水平
UIDeviceOrientationFaceDown	方向朝上，屏幕的面向下，并与地面水平

由于设备的方向不同，所以可以根据不同的方向调整不同的屏幕方向。一般，有 4 种常量可以用来指定屏幕的方向，如表 6-11 所示。

表 6-11　屏幕的方向

屏　幕　方　向	功　　能
UIInterfaceOrientationLandscapeLeft	水平方向的旋转，但Home键在屏幕的左方
UIInterfaceOrientationLandscapeRight	水平方向的旋转，但Home键在屏幕的右方
UIInterfaceOrientationPortrait	垂直方向
UIInterfaceOrientationPortraitUpsideDown	垂直方向，但Home键在屏幕的上方

在代码中，使用了 rientation 属性获取设备的当前方向，代码如下：

```
deviceOrientation=[UIDevice currentDevice].orientation;
```

2. NSTimer 动画

NSTimer 是时间定时器，它可以每隔一段时间将图像进行更新一次，这样可以使图片有一种动态的感觉。一般使用 NSTimer 的 scheduledTimerWithTimeInterval: target: selector: repeats:方法对其进行创建，其语法形式如下：

```
NSTimer *对象名=[NSTimer scheduledTimerWithTimeInterval:(NSTimeInterval)
target:(id)selector:(SEL)userInfo:(id) repeats:(BOOL)];
```

其中，该方法的参数说明如下。

- ❑ scheduledTimerWithTimeInterval 用来指定两次触发所间隔的秒数。
- ❑ target 用来指定消息发送的对象。
- ❑ selector 用来指定触发器所调用的方法。
- ❑ userInfo 该参数可以设定为 nil，当定时器失效时，由用户指定的对象保留和释放该定时器。
- ❑ repeats 用来指定是否重复调用自身。

在代码中，使用了 scheduledTimerWithTimeInterval: target: selector: repeats:方法表示对 NSTimer 定时器的创建，代码如下：

```
t=[NSTimer   scheduledTimerWithTimeInterval:1.0   target:self   selector:
@selector(time) userInfo:nil repeats:YES];
```

其中，它的参数说明如下。

- ❑ 1.0 表示两次触发所间隔的秒数。
- ❑ self 表示消息发送的对象。
- ❑ @selector(time)表示触发器所调用的方法。
- ❑ YES 表示重复调用自身。

第 7 章 分 段 控 件

分段控件是一行可以单击的片段,每段非常像一个按钮。它为用户提供了一个在多个相关项之间选择的方法。本章将主要讲解分段控件的相关实例。

实例 44 滑块式分段控件

【实例描述】

本实例实现的功能是绘制一个具有滑块的分段控件,当滑块到达某一点后,背景颜色会发生改变。运行效果如图 7.1 所示。

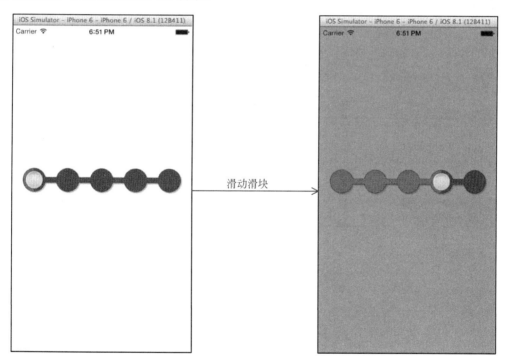

图 7.1 实例 44 运行效果

【实现过程】

(1)创建一个项目,命名为"滑块式分段控件"。
(2)添加图片 yuan.png 到创建项目的 Supporting Files 文件夹中。
(3)打开 segment.h 文件,编写代码,实现实例变量、对象、协议和方法的声明。程序代码如下:

```objc
#import <UIKit/UIKit.h>
@class segment;
//协议
@protocol segmentDelegate <NSObject>
@optional
- (void)timeSlider:(segment *)timeSlider didSelectPointAtIndex:(int)index;
@end
@interface segment : UIView{
//对象
    UIBezierPath *_drawPath;
    UIImageView * holderView;
NSMutableArray *_positionPoints;
//实例变量
CGContextRef _context;
    int _moveFinalIndex;
    bool firstTimeOnly;
}
//属性
@property (nonatomic, assign) float spaceBetweenPoints;
@property (nonatomic, assign) float numberOfPoints;
……
@property (nonatomic, assign) id<segmentDelegate> delegate;
//方法的声明
- (void)moveToIndex:(int)index;
- (CGPoint)positionForPointAtIndex:(int)index;
@end
```

（4）打开 segment.m 文件，编写代码。使用的方法如表 7-1 所示。

表 7-1　segment.m文件中的方法总结

方　　法	功　　能
initWithFrame:	滑块式分段控件的初始化
drawRect:	绘制
fillRect:withColor:onContext:	填充矩形
backgroundPath	路径
touchesEnded:withEvent:	触摸结束
handlePanGesture:	拖动手势
updatePositions	更新位置
getNearestLeftPossibleLocation	获取离滑块最近的左边的位置
getNearestRightPossibleLocation	获取离滑块最近的右边的位置
getMostLeftPossibleLocation	获取滑块式分段控件最左边的位置
getMostRightPossibleLocation	获取滑块式分段控件最右边的位置
moveToIndex:	移动索引
positionForPointAtIndex:	位置点的索引
setStrokeColor:	设置经过的颜色
setShadowColor:	设置阴影的颜色
setShadowSize:	设置阴影的尺寸
setShadowBlur:	设置阴影的模糊值
setStrokeSize:	设置经过的尺寸
setStrokeSizeForeground:	设置经过的尺寸的前景
setRadiusPoint:	设置半径点

续表

方 法	功 能
setNumberOfPoints:	设置点数
setHeightLine:	设置高度线
setRadiusCircle:	设置圆的半径
setSpaceBetweenPoints:	设置两个点之间的空间

这些方法可以分为3个部分：滑块式分段控件的绘制、滑块式分段控件设置以及滑块式分段控件的拖动和触摸。要实现滑块式分段控件的绘制需要使用 initWithFrame:、drawRect:、fillRect:和 backgroundPath:、setSpaceBetweenPoints:、setRadiusPoint:、setHeightLine:和 moveToIndex:方法共同实现。其中，initWithFrame:方法实现了滑块式分段控件的初始化。程序代码如下：

```
- (id)initWithFrame:(CGRect)frame {
    self = [super initWithFrame:frame];
    if (self) {
        [self setBackgroundColor:[UIColor clearColor]];      //设置背景色
        firstTimeOnly = TRUE;
        _spaceBetweenPoints = 40.0;
        _numberOfPoints = 5.0;                                //设置点数
        _heightLine = 10.0;
        _radiusPoint = 10.0;
        _shadowSize = CGSizeMake(2.0, 2.0);                   //设置警告的尺寸
        _shadowBlur = 2.0;
        _strokeSize = 1.0;
        _strokeColor = [UIColor blackColor];
        _shadowColor = [UIColor colorWithWhite:0.0 alpha:0.30];
                                                              //设置阴影的颜色
        _radiusCircle = 2.0;
        _moveFinalIndex = 1;
        _currentIndex = 0;                                    //设置当前的索引
        _touchEnabled = YES;
        _strokeColorForeground = [UIColor colorWithWhite:0.3 alpha:1.0];
        _strokeSizeForeground = 1.0;
        CGColorSpaceRef colorSpace = CGColorSpaceCreateDeviceRGB();
                                                              //实例化色彩空间
                                                              //实例化数组
        NSArray *gradientColors = [NSArray arrayWithObjects:
                        (id)[UIColor whiteColor].CGColor,
                        (id)[UIColor colorWithWhite : 0.793 alpha :
                        1.000].CGColor, nil];
        CGFloat gradientLocations[] = { 0, 1 };
        _gradientForeground = CGGradientCreateWithColors(colorSpace,
        (__bridge CFArrayRef)gradientColors, gradientLocations);
        _positionPoints = [NSMutableArray array];
        CGColorSpaceRelease(colorSpace);
        UIImage *imageToUse = [UIImage imageNamed:@"yuan.png"];
                                                              //实例化图像对象
        holderView = [[UIImageView alloc] initWithFrame:CGRectMake(0, 0,
        imageToUse.size.width, imageToUse.size.height)];//实例化图像视图对象
        [holderView setImage:imageToUse];                     //设置图像视图的图像
        [self addSubview:holderView];
        [self bringSubviewToFront:holderView];
        [holderView setUserInteractionEnabled:TRUE];
        [holderView setCenter:CGPointZero];                   //设置中心
```

```objc
    [self addGestureRecognizer:[[UIPanGestureRecognizer alloc]initWithTarget:
       self action:
@selector(handlePanGesture:)]];                    //添加手势
    }
    return self;
}
```

drawRect:方法实现对滑块式分段控件的绘制。程序代码如下:

```objc
- (void)drawRect:(CGRect)rect {
    _context = UIGraphicsGetCurrentContext();     //获取当前上下文
    _drawPath = [self backgroundPath];
    [_strokeColor setStroke];
    CGContextSaveGState(_context);
    //设置填充颜色
    CGContextSetShadowWithColor(_context,_shadowSize,_shadowBlur,
    _shadowColor.CGColor);
    [_drawPath setLineWidth:_strokeSize];         //设置线宽
    [_drawPath fill];                             //绘制
    [_drawPath stroke];
    [_drawPath addClip];                          //剪切上下文
    CGRect drawingRect = [self bounds];
    CGPoint center;
    if (CGPointEqualToPoint([holderView center], CGPointZero)) {
       center = [[_positionPoints objectAtIndex:_currentIndex]CGPoint
       Value];                                    //获取中心位置
    } else {
       center = [holderView center];              //获取中心位置
    }
    drawingRect = CGRectMake(drawingRect.origin.x, drawingRect.origin.y,
    center.x, drawingRect.size.height);
    //设置没有经过的颜色
    [self fillRect:drawingRect withColor:[UIColor redColor].CGColor
    onContext:_context];
    drawingRect = CGRectMake(drawingRect.origin.x + center.x, drawingRect.
    origin.y, self.frame.size.width, drawingRect.size.height);
    //设置经过的颜色
    [self fillRect:drawingRect withColor:[UIColor blueColor].CGColor
    onContext:_context];
    CGContextRestoreGState(_context);
    if (firstTimeOnly) [holderView setCenter:[[_positionPoints objectAtIndex:
    _currentIndex]CGPointValue]];
    firstTimeOnly = FALSE;
}
```

fillRect:withColor:onContext:方法实现对滑块式分段控件矩阵的填充。程序代码如下:

```objc
- (void)fillRect:(CGRect)rect withColor:(CGColorRef)color onContext:(CGContextRef)
currentGraphicsContext {
    CGContextAddRect(currentGraphicsContext, rect);          //绘制矩形
    CGContextSetFillColorWithColor(currentGraphicsContext, color);
                                                             //设置颜色
    CGContextFillRect(currentGraphicsContext, rect);
}
```

backgroundPath:方法实现对路径的设置。程序代码如下:

```objc
- (UIBezierPath *)backgroundPath {
    [_positionPoints removeAllObjects];                      //移出所有的对象
```

```objc
    UIBezierPath *path = [[UIBezierPath alloc] init];         //创建对象
    float angle = _heightLine / 2.0 / _radiusPoint;
    //遍历
    for (int i = 0; i < (_numberOfPoints - 2) * 2 + 2; i++) {
        int pointNbr = (i >= _numberOfPoints) ? (_numberOfPoints - 2) - (i
            - _numberOfPoints) : i;
        CGPoint centerPoint = CGPointMake(_radiusPoint + _spaceBetweenPoints
            * pointNbr + _radiusPoint * 2.0 * pointNbr + _strokeSize, _radiusPoint
            + _strokeSize);
        //实现判断
        if (i == 0) {
            [_positionPoints addObject:[NSValue valueWithCGPoint:
                centerPoint]];      //添加对象
            [path addArcWithCenter:centerPoint radius:_radiusPoint startAngle:
                angle endAngle:angle * -1.0 clockwise:YES];         //画弧线
            [path addLineToPoint:CGPointMake(centerPoint.x + _radiusPoint +
                _spaceBetweenPoints, centerPoint.y - _heightLine / 2.0)];
        } else if (i == _numberOfPoints - 1) {
            [_positionPoints addObject:[NSValue valueWithCGPoint:centerPoint]];
            [path addArcWithCenter:centerPoint radius:_radiusPoint startAngle:
                M_PI + angle endAngle:M_PI - angle clockwise:YES];   //画弧线
            [path addLineToPoint:CGPointMake(centerPoint.x - _radiusPoint -
                _spaceBetweenPoints - ((i == (_numberOfPoints - 2) * 2 + 1) ?
                (_radiusPoint * (1.0 - cosf(angle))) : 0), centerPoint.y +
                _heightLine / 2.0)];
        } else if (i < _numberOfPoints - 1) {
            [_positionPoints addObject:[NSValue valueWithCGPoint:centerPoint]];
            [path addArcWithCenter:centerPoint radius:_radiusPoint
                startAngle:M_PI + angle endAngle:angle * -1.0 clockwise:YES];
                    //画弧线
            [path addLineToPoint:CGPointMake(centerPoint.x + _radiusPoint +
                _spaceBetweenPoints, centerPoint.y - _heightLine / 2.0)];
        } else if (i >= _numberOfPoints) {
            [_positionPoints addObject:[NSValue valueWithCGPoint:centerPoint]];
            [path addArcWithCenter:centerPoint radius:_radiusPoint
                startAngle:angle endAngle:M_PI - angle clockwise:YES];//画弧线
            [path addLineToPoint:CGPointMake(centerPoint.x - _radiusPoint -
                _spaceBetweenPoints - ((i == (_numberOfPoints - 2) * 2 + 1) ?
                (_radiusPoint * (1.0 - cosf(angle))) : 0), centerPoint.y +
                _heightLine / 2.0)];
        }
    }
    return path;
}
```

setSpaceBetweenPoints:方法实现对两个位置之间的空间进行设置。程序代码如下:

```objc
- (void)setSpaceBetweenPoints:(float)spaceBetweenPoints {
    _spaceBetweenPoints = spaceBetweenPoints;           //设置两个位置之间的空间
    [self setNeedsDisplay];
}
```

setRadiusPoint:方法实现对半径的大小进行设置。程序代码如下:

```objc
- (void)setRadiusPoint:(float)radiusPoint {
    if (_radiusCircle > radiusPoint - 4) {
        radiusPoint = _radiusCircle + 4;
    }
    _radiusPoint = radiusPoint;
```

```
    [self setNeedsDisplay];                          //重写绘制
}
```

setHeightLine:方法实现对高度线进行设置。程序代码如下:

```
- (void)setHeightLine:(float)heightLine {
  //判断
  if (heightLine > _radiusPoint * 2) {
      heightLine = _radiusPoint * 2;                 //设置高度
  }
  _heightLine = heightLine;
  [self setNeedsDisplay];
}
```

moveToIndex:方法实现移动索引的功能。程序代码如下:

```
- (void)moveToIndex:(int)index {
  _moveFinalIndex = index;
  _currentIndex = index;                             //设置当前索引
  //判断
  if ([_positionPoints count] > index) [holderView setCenter:[[_position
   Points objectAtIndex:_currentIndex]CGPointValue]];
  [self setNeedsDisplay];
}
```

这时出现在界面的滑块式分段控件是不可以移动的,需要开发者为它实现移动。在本实例中,想要让滑块式分段控件实现移动有两种方法,一种是通过触摸;一种是通过拖动。touchesEnded: withEvent:方法可以实现触摸的手势。程序代码如下:

```
- (void)touchesEnded:(NSSet *)touches withEvent:(UIEvent *)event {
    [super touchesEnded:touches withEvent:event];
    //判断触摸是否结束
    if (_touchEnabled) {
       CGPoint touchPoint = [[touches anyObject] locationInView:self];
        //获取触摸的位置
       float x = touchPoint.x;
       x -= _strokeSize;
       //遍历
       for (int i = 0; i < [_positionPoints count]; i++) {
           CGPoint point = [[_positionPoints objectAtIndex:i] CGPointValue];
           double diggerence = fabs(point.x - x);
           //判断
           if (diggerence <= _radiusPoint) {
               //判断是否实现了timeSlider:didSelectPointAtIndex:方法
               if ([_delegate respondsToSelector:@selector(timeSlider:did
                SelectPointAtIndex:)]) {
                   [_delegate timeSlider:self didSelectPointAtIndex:i];
               }
               [self moveToIndex:i];
               return;
           }
       }
    }
}
```

要实现拖动,需要 handlePanGesture:、getNearestLeftPossibleLocation、getNearestRightPossibleLocation、getMostLeftPossibleLocation、getMostRightPossibleLocation 和 updatePositions 方法实现。handlePanGesture:方法实现的是拖动手势的动作,程序代码如下:

第7章 分段控件

```
- (void)handlePanGesture:(UIPanGestureRecognizer *)recogniser {
    //判断是否结束
    if (_touchEnabled) {
        CGPoint location = [recogniser locationInView:self];
        location.y = holderView.center.y;                    //获取中心位置的y值
        CGPoint leftMargin = [self getMostLeftPossibleLocation];
        CGPoint rightMargin = [self getMostRightPossibleLocation];
        if ((location.x >= leftMargin.x) && (location.x <= rightMargin.x)){
            [holderView setCenter:location];                 //设置中心位置
        }
        else if(location.x <= leftMargin.x){
            [holderView setCenter:leftMargin];               //设置中心位置
        }
        else if(location.x >= rightMargin.x){
            [holderView setCenter:rightMargin];              //设置中心位置
        }
        if ([recogniser state] == UIGestureRecognizerStateEnded) {
            [self updatePositions];
        }
    }
    [self setNeedsDisplay];                                  //重新绘制
}
```

getNearestLeftPossibleLocation:方法获取离滑块最近的左边的位置。程序代码如下：

```
- (CGPoint)getNearestLeftPossibleLocation {
    CGPoint location = CGPointZero;
    int difference = 10240;
    //遍历
    for (int i = 0; i < [_positionPoints count]; i++) {
        CGPoint point = [[_positionPoints objectAtIndex:i] CGPointValue];
        if (holderView.center.x >= point.x) {
            //判断location和CGPointZero是否为一个点
            if (CGPointEqualToPoint(location, CGPointZero)) {
                location = point;
            }
            if (difference > (holderView.center.x - point.x)) {
                difference = holderView.center.x - point.x;
                location = point;
            }
        }
    }
    return location;              //返回location
}
```

getNearestRightPossibleLocation方法获取离滑块最近的右边的位置。程序代码如下：

```
- (CGPoint)getNearestRightPossibleLocation {
    CGPoint location = CGPointZero;
    int difference = 10240;
    //遍历
    for (int i = 0; i < [_positionPoints count]; i++) {
        CGPoint point = [[_positionPoints objectAtIndex:i] CGPointValue];
        if (holderView.center.x <= point.x) {
            //判断location和CGPointZero是否为一个点
            if (CGPointEqualToPoint(location, CGPointZero)) {
                location = point;
            }
            if (difference > (point.x - holderView.center.x)) {
                difference = point.x - holderView.center.x;
```

```
            location = point;
        }
    }
}
    return location;                    //返回 location
}
```

getMostLeftPossibleLocation 方法获取滑块式分段控件最左边的位置。程序代码如下：

```
- (CGPoint)getMostLeftPossibleLocation {
    static CGPoint mostLeftFound;
    //判断 mostLeftFound 和 CGPointZero 是否为一个点
    if (CGPointEqualToPoint(mostLeftFound, CGPointZero)) {
        for (int i = 0; i < [_positionPoints count]; i++) {
            CGPoint point = [[_positionPoints objectAtIndex:i] CGPointValue];
            //判断 mostLeftFound 和 CGPointZero 是否为一个点
            if (CGPointEqualToPoint(mostLeftFound, CGPointZero)) {
                mostLeftFound = point;
            } else if (point.x < mostLeftFound.x) {
                mostLeftFound = point;
            }
        }
    }
    return mostLeftFound;               //返回 mostLeftFound
}
```

getMostRightPossibleLocation 方法获取滑块式分段控件最右边的位置。程序代码如下：

```
- (CGPoint)getMostRightPossibleLocation {
    static CGPoint mostRightFound;
    //判断 mostRightFound 和 CGPointZero 是否为一个点
    if (CGPointEqualToPoint(mostRightFound, CGPointZero)) {
        for (int i = 0; i < [_positionPoints count]; i++) {
            CGPoint point = [[_positionPoints objectAtIndex:i] CGPointValue];
            //判断 mostRightFound 和 CGPointZero 是否为一个点
            if (CGPointEqualToPoint(mostRightFound, CGPointZero)) {
                mostRightFound = point;
            } else if (point.x > mostRightFound.x) {
                mostRightFound = point;
            }
        }
    }
    return mostRightFound;              //返回 mostRightFound
}
```

updatePositions 方法实现当拖动手势结束后的滑块位置更新。程序代码如下：

```
- (void)updatePositions {
    CGPoint nearestLeft  = [self getNearestLeftPossibleLocation];
                                        //获取离左边最近的点
    CGPoint nearestRight = [self getNearestRightPossibleLocation];
                                        //获取离右边最近的点
    if ((holderView.center.x - nearestLeft.x) <(nearestRight.x - holderView.center.x))
        [holderView setCenter:nearestLeft];         //设置中心位置
    else
        [holderView setCenter:nearestRight];
    float x = holderView.center.x;      //获取 holderView 中心位置的 x 值
    x -= _strokeSize;
    //遍历
```

```
    for (int i = 0; i < [_positionPoints count]; i++) {
        CGPoint point = [[_positionPoints objectAtIndex:i] CGPointValue];
        if (fabs(point.x - x) <= _radiusPoint) {
            //判断是否实现了 timeSlider:didSelectPointAtIndex:方法
            if ([_delegate respondsToSelector:@selector(timeSlider:didSelect
            PointAtIndex:)]) {
                [_delegate timeSlider:self didSelectPointAtIndex:i];
            }
            [self moveToIndex:i];
            return;
        }
    }
}
```

(5) 打开 ViewController.h 文件，编写代码，实现头文件、宏定义以及对象的声明。程序代码如下：

```
#import <UIKit/UIKit.h>
#import "segment.h"
//宏定义
#define WIDTH   300
#define HEIGHT  54
#define X_POS   20
#define Y_POS   242
#define RADIUS_POINT  20
#define SPACE_BETWEEN_POINTS  20
#define SLIDER_LINE_WIDTH   9
#define IPHONE_4_SUPPORT    88
@interface ViewController : UIViewController<segmentDelegate>{
    segment *slidersegment;              //声明对象
}
@end
```

(6) 打开 ViewController.m 文件，编写代码。在此文件中，代码分为两个部分：滑块式分段控件对象的创建、滑块式分段控件的功能。在 viewDidLoad 方法中实现滑块式分段控件对象的创建。程序代码如下：

```
- (void)viewDidLoad
{
    CGRect slidersegmentFrame = CGRectNull;
    slidersegmentFrame = CGRectMake(X_POS,Y_POS+6.5,WIDTH,HEIGHT);
    slidersegment = [[segment alloc] initWithFrame:slidersegmentFrame];
                                                                //创建对象
    [slidersegment setDelegate:self];                           //设置委托
    [slidersegment moveToIndex:0];
    [slidersegment setSpaceBetweenPoints:SPACE_BETWEEN_POINTS];
    [slidersegment setRadiusPoint:RADIUS_POINT];
    [slidersegment setHeightLine:SLIDER_LINE_WIDTH];
    [self.view addSubview:slidersegment];                       //添加
    [super viewDidLoad];
    // Do any additional setup after loading the view, typically from a nib.
}
```

添加 timeSlider:didSelectPointAtIndex:方法实现滑块式分段控件的变色功能。程序代码如下：

```
- (void)timeSlider:(segment *)timeSlider didSelectPointAtIndex:(int)index{
    if (slidersegment.currentIndex==0) {              //判断当前的索引是否为 0
      [self.view setBackgroundColor:[UIColor redColor]];
    }else if(slidersegment.currentIndex==1){          //判断当前的索引是否为 1
      [self.view setBackgroundColor:[UIColor greenColor]];
    }else if(slidersegment.currentIndex==2){          //判断当前的索引是否为 2
      [self.view setBackgroundColor:[UIColor cyanColor]];
    }else if(slidersegment.currentIndex==3){          //判断当前的索引是否为 3
      [self.view setBackgroundColor:[UIColor orangeColor]];
    }else if(slidersegment.currentIndex==4){          //判断当前的索引是否为 4
      [self.view setBackgroundColor:[UIColor yellowColor]];
    }
}
```

【代码解析】

由于本实例中的代码和方法非常的多，为了方便读者的阅读，笔者绘制了一些执行流程图如图 7.2~7.4 所示。其中，单击运行按钮后一直到在模拟器上显示滑块式分段控件的界面，使用了 viewDidLoad、initWithFrame:、drawRect:、fillRect:withColor:onContext:和 backgroundPath:、setSpaceBetweenPoints:、setRadiusPoint:、setHeightLine:和 moveToIndex:方法。它们的执行流程如图 7.2 所示。

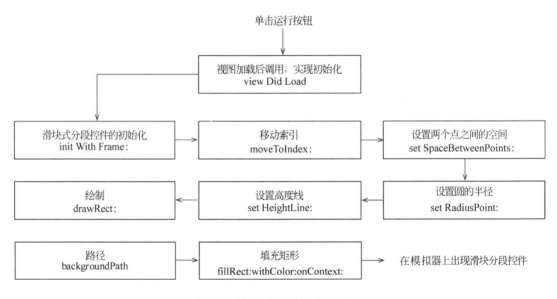

图 7.2 实例 44 程序执行流程 1

如果想要通过触摸的方法使分段控件上的滑块到达某一点，并且滑块走过的路径颜色会随之改变，要实现这些功能，需要使用 touchesEnded:withEvent:、timeSlider:didSelectPointAtIndex:、drawRect:、backgroundPath 和 fillRect:withColor:onContext:方法。它们的执行流程如图 7.3 所示。

如果想要通过拖动的方法使分段控件上的滑块到达某一点，并且滑块走过的路径颜色会随之改变，需要用到 handlePanGesture:、getMostLeftPossibleLocation、getMostRightPossibleLocation、updatePositions、getNearestLeftPossibleLocation、getNearestRightPossibleLocation、timeSlider:didSelectPointAtIndex:、moveToIndex:、drawRect:、backgroundPath 和 fillRect:withColor:onContext:方法。它们的执行流程如图 7.4 所示。

第 7 章 分段控件

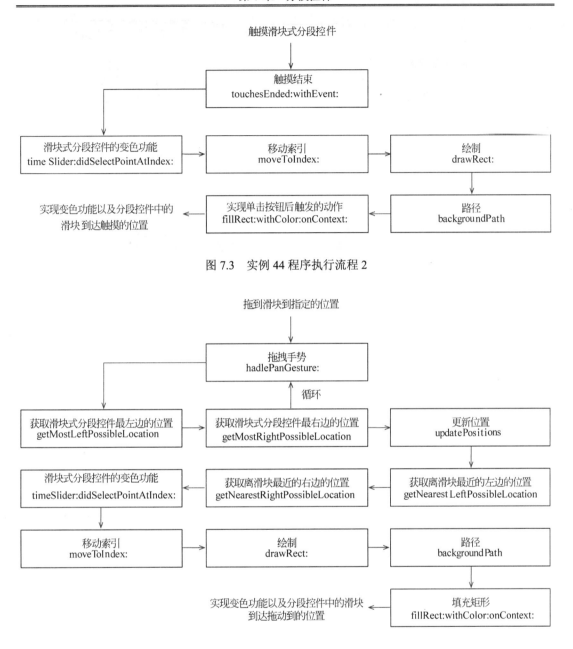

图 7.3 实例 44 程序执行流程 2

图 7.4 实例 44 程序执行流程 3

实例 45 开关式分段控件

【实例描述】

本实例实现的功能是绘制一个类似于 Switch 开关控件的分段控件。在此控件上，用户可以使用拖动和轻拍两种方式实现分段控件块的选择。其中，用来显示索引值发生改变的内容的是标签。运行效果如图 7.5 所示。

• 273 •

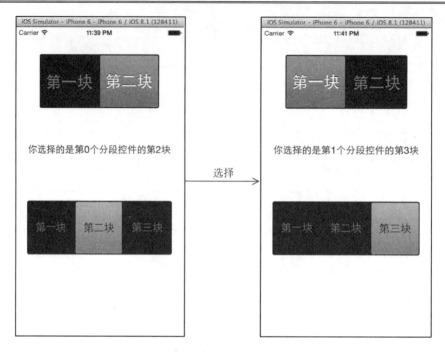

图 7.5 实例 45 运行效果

【实现过程】

（1）创建一个项目，命名为"开关式分段控件"。

（2）创建一个基于 UIView 类的 segmentedselect 类和一个基于 UIControl 类的 segment 类。

（3）打开 segmentedselect.h 文件，编写代码，实现头文件、属性和方法的声明。程序代码如下：

```
#import <UIKit/UIKit.h>
#import "segmented.h"
@class segmented;
@interface segmentedselect : UIView
//属性
@property (nonatomic, strong) UIImage *backgroundImage;
……
@property (strong, nonatomic, readonly) UILabel *secondLabel;
//方法
- (void)activate;
- (void)deactivate;
@end
```

（4）打开 segmentedselect.m 文件，编写代码，对开关式分段控件的选择部分进行设置以及绘制。使用的方法如表 7-2 所示。

表 7-2 segmentedselect.m 文件中的方法总结

方　　法	功　　能
initWithFrame:	初始化一些设置
label	设置第一个标签
secondLabel	设置第二个标签

第 7 章 分段控件

续表

方　　法	功　　能
font	获取字体
drawRect:	绘制选择的部分
setBackgroundImage:	设置背景图像
setTintColor:	设置颜色
setFont:	设置字体
setTextColor:	设置文本颜色
setTextShadowColor:	设置文本阴影颜色
setTextShadowOffset:	设置文本阴影的偏移
setShouldCastShadow:	获取图层的不透明度
setFrame:	设置框架
setSelected:	设置选择
activate	活动
deactivate	不活动
setShadowOffset:	设置阴影的偏移
setShadowColor:	设置阴影的颜色
setCastsShadow:	设置图层的不透明度

这里需要讲解几个重要的方法（其他方法请读者参考源代码）。添加的 initWithFrame: 方法用于对分段控件中的选择部分进行一些初始设置，代码如下：

```
- (id)initWithFrame:(CGRect)frame {
    self = [super initWithFrame:frame];
    if (self) {
        self.userInteractionEnabled = NO;
        self.backgroundColor = [UIColor clearColor];        //设置背景颜色
        self.textColor = [UIColor whiteColor];              //设置文本颜色
        self.textShadowColor = [UIColor blackColor];
        self.textShadowOffset = CGSizeMake(0, -1);
        self.tintColor = [UIColor grayColor];
        self.shouldCastShadow = YES;
    }
    return self;
}
```

label 和 secondLabel 方法是对被选择部分的字体和对齐方式等进行设置。程序代码如下：

```
- (UILabel*)label {
    if(label == nil) {
        label = [[UILabel alloc] initWithFrame:self.bounds];    //实例化对象
        label.textAlignment = NSTextAlignmentCenter;            //对齐方式
        label.font = self.font;
        label.backgroundColor = [UIColor clearColor];
        [self addSubview:label];
    }
    return label;
}
- (UILabel*)secondLabel {
    if(secondLabel == nil) {
        secondLabel = [[UILabel alloc] initWithFrame:self.bounds];
```

```objectivec
        secondLabel.textAlignment = NSTextAlignmentCenter;   //对齐方式
        secondLabel.font = self.font;                        //文本的字体
        secondLabel.backgroundColor = [UIColor clearColor];
        [self addSubview:secondLabel];
    }
    return secondLabel;
}
```

drawRect：方法是实现对被选择部分的绘制，程序代码如下：

```objectivec
- (void)drawRect:(CGRect)rect {
    //对背景图像以及是否选择进行判断
    if(self.backgroundImage && !self.selected)
        [self.backgroundImage drawInRect:rect];              //设置背景图像
    else if(self.highlightedBackgroundImage && self.selected)
        [self.highlightedBackgroundImage drawInRect:rect];   //设置高亮图像
    else {
        CGFloat cornerRadius = self.segmentedControl.cornerRadius;
        CGContextRef context = UIGraphicsGetCurrentContext();
        CGColorSpaceRef colorSpace = CGColorSpaceCreateDeviceGray();
                                                             //创建颜色空间
        //绘制渐变
        CGPathRef strokeRect = [UIBezierPath bezierPathWithRoundedRect:rect
         cornerRadius:cornerRadius-1.5].CGPath;
        CGContextAddPath(context, strokeRect);               //添加路径
        CGContextClip(context);
        CGContextSaveGState(context);                        //保存
        CGFloat strokeComponents[4] = {0.55, 1,   0.40, 1};
        if(self.selected) {
            strokeComponents[0]-=0.1;
            strokeComponents[2]-=0.1;
        }
        CGGradientRef strokeGradient = CGGradientCreateWithColorComponents
        (colorSpace, strokeComponents, NULL, 2);             //创建渐变对象
        CGContextDrawLinearGradient(context, strokeGradient,CGPointMake(0,0),
        CGPointMake(0,CGRectGetHeight(rect)), 0);            //填充渐变色
        CGGradientRelease(strokeGradient);                   //释放渐变对象
        //填充渐变
        CGPathRef fillRect = [UIBezierPath bezierPathWithRoundedRect:
        CGRectInset(rect, 1, 1) cornerRadius:cornerRadius-2.5].CGPath;
        CGContextAddPath(context, fillRect);                 //添加路径
        CGContextClip(context);
        CGFloat fillComponents[4] = {0.5, 1,   0.35, 1};
        //判断 selected
        if(self.selected) {
            fillComponents[0]-=0.1;
            fillComponents[2]-=0.1;
        }
        CGGradientRef fillGradient = CGGradientCreateWithColorComponents
        (colorSpace, fillComponents, NULL, 2);               //创建渐变对象
        CGContextDrawLinearGradient(context, fillGradient, CGPointMake(0,0),
        CGPointMake(0,CGRectGetHeight(rect)), 0);            //填充渐变色
        CGGradientRelease(fillGradient);
        CGColorSpaceRelease(colorSpace);                     //释放色彩空间
        CGContextRestoreGState(context);
        [self.tintColor set];
        UIRectFillUsingBlendMode(rect, kCGBlendModeOverlay);
    }
```

```
}
```

setFrame：方法是对被选择部分的框架进行设置。程序代码如下：

```
- (void)setFrame:(CGRect)newFrame {
    [super setFrame:newFrame];              //设置框架
    CGFloat posY = ceil((self.segmentedControl.height-self.font.point
    Size+self.font.descender)/2)+self.segmentedControl.titleEdgeInsets.
    top-self.segmentedControl.titleEdgeInsets.bottom-self.segmented
    Control.thumbEdgeInset.top+2;
    int pointSize = self.font.pointSize;
    if(pointSize%2 != 0)
        posY--;
    self.label.frame = self.secondLabel.frame = CGRectMake(0, posY, newFrame.
    size.width, self.font.pointSize);                //设置标签的框架
    self.layer.shadowOffset = CGSizeMake(0, 0);      //设置阴影的偏移量
    self.layer.shadowRadius = 1;
    self.layer.shadowColor = [UIColor blackColor].CGColor; //设置阴影的颜色
    self.layer.shadowPath = [UIBezierPath bezierPathWithRoundedRect:
    self.bounds  cornerRadius:self.segmentedControl.cornerRadius-1].CGPath;
                                                     //设置阴影的路径
    self.layer.shouldRasterize = YES;
}
```

setSelected:方法是对选择时和选择后进行的设置。程序代码如下：

```
- (void)setSelected:(BOOL)s {
    selected = s;
    if(selected && !self.segmentedControl.crossFadeLabelsOnDrag &&
!self.highlightedBackgroundImage)
        self.alpha = 0.8;          //设置透明度
    else
        self.alpha = 1;
    [self setNeedsDisplay];
}
```

activate 方法实现选择时的动作。程序代码如下：

```
- (void)activate {
    [self setSelected:NO];
    if(!self.segmentedControl.crossFadeLabelsOnDrag)
        self.label.alpha = 1;                        //设置透明度
}
```

deactivate 方法实现取消动作。程序代码如下：

```
- (void)deactivate {
    [self setSelected:YES];
    if(!self.segmentedControl.crossFadeLabelsOnDrag)
        self.label.alpha = 0;                        //设置透明度
}
```

（5）打开 segmented.h 文件，编写代码，实现头文件、属性和方法的声明。程序代码如下：

```
#import <UIKit/UIKit.h>
#import "segmentedselect.h"                          //头文件
@class segmentedselect;
@interface segmented : UIControl
```

```
//属性声明
@property (nonatomic, copy) void (^changeHandler)(NSUInteger newIndex);
……
@property (nonatomic, readwrite) CGFloat segmentWidth;
@property (nonatomic, readwrite) CGFloat thumbHeight;
//方法
- (segmented*)initWithSectionTitles:(NSArray*)titlesArray;
- (void)moveThumbToIndex:(NSUInteger)segmentIndex animate:(BOOL)animate;
@end
```

（6）打开 segmented.m 文件，编写代码。使用的方法如表 7-3 所示。

表 7-3　segmented.m文件中的方法总结

方　　法	功　　能
initWithSectionTitles:	初始化开关式分段控件
thumb	获取选择指示器
willMoveToSuperview:	设置将会移动的视图
drawRect:	绘制
beginTrackingWithTouch:withEvent:	开始触摸
continueTrackingWithTouch:withEvent:	继续触摸
endTrackingWithTouch:withEvent:	结束触摸
cancelTrackingWithEvent:	取消触摸
snap:	轻拍
updateTitles	更新标题
activate	动作
toggle	切换
moveThumbToIndex:animate:	移动选择指示器
setBackgroundImage:	设置背景
setSegmentPadding:	获取插入标题的边
setShadowOffset:	设置阴影的偏移
setShadowColor:	设置阴影的颜色

在这些方法中，可以分为 4 大部分：分段控件的绘制、绘制、轻拍和拖动。绘制分段控件需要 initWithSectionTitles:、thumb、willMoveToSuperview:、drawRect:、setBackgroundImage:、setSegmentPadding:、setShadowOffset:和 setShadowColor:方法实现。这里需要讲解几个重要的方法（其他方法请读者参考源代码）。initWithSectionTitles:方法实现对分段控件的初始化。程序代码如下：

```
- (id)initWithSectionTitles:(NSArray*)array {
    if (self = [super initWithFrame:CGRectZero]) {
        self.titlesArray = [NSMutableArray arrayWithArray:array];
                                                            //实例化对象
        self.thumbRects = [NSMutableArray arrayWithCapacity:[array count]];
        self.backgroundColor = [UIColor clearColor];        //设置背景颜色
        self.tintColor = [UIColor grayColor];
        self.clipsToBounds = YES;
        self.userInteractionEnabled = YES;     //是否接收用户的事件消息
        self.animateToInitialSelection = NO;
        self.clipsToBounds = NO;
        self.font = [UIFont boldSystemFontOfSize:15];       //设置字体
        self.textColor = [UIColor grayColor];
```

```objc
        self.textShadowColor = [UIColor blackColor];
        self.textShadowOffset = CGSizeMake(0, -1);      //设置文本阴影的偏移量
        self.titleEdgeInsets = UIEdgeInsetsMake(0, 10, 0, 10);
        self.thumbEdgeInset = UIEdgeInsetsMake(2, 2, 3, 2);
        self.height = 32.0;
        self.cornerRadius = 4.0;                        //圆角半径
        self.selectedIndex = 0;                         //被选择部分的索引
        self.thumb.segmentedControl = self;
    }
    return self;
}
```

thumb 方法实现获取选择指示器,有了此指示器用户可以看到选择的部分。程序代码如下:

```objc
- (segmentedselect *)thumb {
    if(thumb == nil)
        thumb = [[segmentedselect alloc] initWithFrame:CGRectZero];
                                                        //实例化对象
    return thumb;
}
```

willMoveToSuperview:方法对将会移动的视图进行设置。程序代码如下:

```objc
- (void)willMoveToSuperview:(UIView *)newSuperview {
    //判断 newSuperview 是否为空
    if(newSuperview == nil)
        return;
    //获取标题的个数
    int c = [self.titlesArray count];
    int i = 0;
    self.segmentWidth = 0;                              //设置分段的宽度
    //遍历
    for(NSString *titleString in self.titlesArray) {
        NSDictionary *dis=[NSDictionary dictionaryWithObjectsAndKeys:self.
        font,NSFontAttributeName, nil];
        CGFloat stringWidth = [titleString sizeWithAttributes:dis].width+
        (self.titleEdgeInsets.left+self.titleEdgeInsets.right+self.thumbEdge
        Inset.left+self.thumbEdgeInset.right);          //实例化字典对象 dis
        self.segmentWidth = MAX(stringWidth, self.segmentWidth);
    }
    self.segmentWidth = ceil(self.segmentWidth/2.0)*2;
    self.bounds = CGRectMake(0, 0, self.segmentWidth*c, self.height);
    //设置大小
    self.thumbHeight = self.thumb.backgroundImage ? self.thumb.background
    Image.size.height : self.height-(self.thumbEdgeInset.top+self.thumbEdgeInset.
    bottom);                                            //设置选择指示器的高度
    i = 0;
    //遍历
    for(NSString *titleString in self.titlesArray) {
        [self.thumbRects addObject:[NSValue valueWithCGRect:CGRectMake(self.
        segmentWidth*i+self.thumbEdgeInset.left, self.thumbEdgeInset.top,
        self.segmentWidth-(self.thumbEdgeInset.left*2), self.thumbHeight)]];
                                                        //添加对象
        i++;
    }
    self.thumb.frame = [[self.thumbRects objectAtIndex:0] CGRectValue];
                                                        //设置框架
    self.thumb.layer.shadowPath = [UIBezierPath bezierPathWithRounded
```

```
    Rect:self.thumb.bounds cornerRadius:2].CGPath;
    self.thumb.label.text = [self.titlesArray objectAtIndex:0];
    self.thumb.font = self.font;            //设置选择指示器的字体
    [self insertSubview:self.thumb atIndex:0];
    BOOL animateInitial = self.animateToInitialSelection;
    if(self.selectedIndex == 0)
        animateInitial = NO;
    [self moveThumbToIndex:selectedIndex animate:animateInitial];
                                            //调用移动指示器的方法
}
```

drawRect:方法实现分段控件的绘制。程序代码如下:

```
- (void)drawRect:(CGRect)rect {
    CGContextRef context = UIGraphicsGetCurrentContext();
    if(self.backgroundImage)
        [self.backgroundImage drawInRect:rect];
    else {
        CGColorSpaceRef colorSpace = CGColorSpaceCreateDeviceGray();
        //底部的光泽
        CGContextSetFillColorWithColor(context, [UIColor colorWithWhite:1
        alpha:0.1].CGColor);
        CGPathRef bottomGlossRect = [UIBezierPath bezierPathWithRounded
        Rect:CGRectMake(0, 0, rect.size.width, rect.size.height) corner
        Radius:self.cornerRadius].CGPath;
        CGContextAddPath(context, bottomGlossRect);      //添加路径
        CGContextFillPath(context);                      //填充
        CGPathRef roundedRect = [UIBezierPath bezierPathWithRounded
        Rect:CGRectMake(0, 0, rect.size.width, rect.size.height-1) corner
        Radius:self.cornerRadius].CGPath;
        CGContextAddPath(context, roundedRect);          //添加路径
        CGContextClip(context);
        //背景的颜色
        CGFloat components[4] = {0.10, 1, 0.12, 1};
        CGGradientRef gradient = CGGradientCreateWithColorComponents
        (colorSpace, components, NULL, 2);               //创建渐变对象
        CGContextDrawLinearGradient(context, gradient, CGPointMake(0,0),
        CGPointMake(0,CGRectGetHeight(rect)-1), 0);      //填充
        CGGradientRelease(gradient);
        [self.tintColor set];
        UIRectFillUsingBlendMode(rect, kCGBlendModeOverlay);
        //内部阴影
        CGContextAddPath(context, roundedRect);
        CGContextSetShadowWithColor(UIGraphicsGetCurrentContext(),
        CGSizeMake(0, 1), 1, [UIColor colorWithWhite:0 alpha:0.6].CGColor);
                                                //设置阴影的颜色
        CGContextSetStrokeColorWithColor(context, [UIColor colorWithWhite:0
        alpha:0.9].CGColor);
        CGContextStrokePath(context);                    //绘制
        CGColorSpaceRelease(colorSpace);
    }
    //绘制颜色
    CGContextSetShadowWithColor(context, self.textShadowOffset, 0, self.text
    ShadowColor.CGColor);
    [self.textColor set];
    CGFloat posY = ceil((CGRectGetHeight(rect)-self.font.pointSize+self.
    font.descender)/2)+self.titleEdgeInsets.top-self.titleEdgeInsets.bot
    tom;
    int pointSize = self.font.pointSize;            //获取字体的大小
```

```
    if(pointSize%2 != 0)
        posY--;
    int i = 0;
    //遍历
    for(NSString *titleString in self.titlesArray) {
       CGRect labelRect = CGRectMake((self.segmentWidth*i), posY, self.
       segmentWidth, self.font.pointSize);
       [titleString drawInRect:labelRect withFont:self.font lineBreakMode:
       UILineBreakModeClip alignment:NSTextAlignmentCenter];          //绘制
         i++;
    }
}
```

这时的分段控件是不可以进行选择的。在本实例中,实现分段控件的选择主要有两种方法,一种轻是拍;一种是通过触摸实现拖动。在本实例中会创建两个不同的分段控件,它实现的轻拍也是不同的。其中,第一个轻拍时不需要更改指示器上的标题,它的实现主要由此文件中的 beginTrackingWithTouch:withEvent:、thumb、endTrackingWithTouch: withEvent:、snap:、moveThumbToIndex:animate:和 activate 方法共同实现。beginTracking With Touch:withEvent:方法用于手指触摸到屏幕上实现开始触摸。程序代码如下:

```
-(BOOL)beginTrackingWithTouch:(UITouch*)touch withEvent:(UIEvent*)event {
    [super beginTrackingWithTouch:touch withEvent:event];
    CGPoint cPos = [touch locationInView:self.thumb];         //获取触摸位置
    self.activated = NO;
    self.snapToIndex = floor(self.thumb.center.x/self.segmentWidth);
    //判断点 cPos 是否包含在 self.thumb.bound 中
    if(CGRectContainsPoint(self.thumb.bounds, cPos)) {
        self.trackingThumb = YES;
        [self.thumb deactivate];
        self.dragOffset = (self.thumb.frame.size.width/2)-cPos.x;
    }
    return YES;
}
```

endTrackingWithTouch:withEvent:方法用于手指离开屏幕实现触摸结束。程序代码如下:

```
-(void)endTrackingWithTouch:(UITouch*)touch withEvent:(UIEvent *)event {
    [super endTrackingWithTouch:touch withEvent:event];
    CGPoint cPos = [touch locationInView:self];               //获取触摸位置
    CGFloat pMaxX = CGRectGetMaxX(self.bounds);
                                              //获取当前控件的 x 坐标值+宽度的数值
    CGFloat pMinX = CGRectGetMinX(self.bounds);//获取 y 坐标值+控件高度的值
    //判断
    if(!self.moved && self.trackingThumb && [self.titlesArray count] == 2)
        [self toggle];                        //调用 toggle 方法
    else if(!self.activated && cPos.x > pMinX && cPos.x < pMaxX) {
        self.snapToIndex = floor(cPos.x/self.segmentWidth);
        [self snap:YES];
    }
    else {
        CGFloat posX = cPos.x;                //获取当前触摸点的 x 值
        //判断
        if(posX < pMinX)
            posX = pMinX;
    //判断
```

```
    if(posX >= pMaxX)
        posX = pMaxX-1;
    self.snapToIndex = floor(posX/self.segmentWidth);
    [self snap:YES];
 }
}
```

snap:方法实现轻拍。程序代码如下:

```
- (void)snap:(BOOL)animated {
  [self.thumb deactivate];                              //取消动作
  if(self.crossFadeLabelsOnDrag)
      self.thumb.secondLabel.alpha = 0;                 //设置透明度
  int index;
  if(self.snapToIndex != -1)
      index = self.snapToIndex;
  else
      index = floor(self.thumb.center.x/self.segmentWidth);
  self.thumb.label.text = [self.titlesArray objectAtIndex:index];
      //设置文本内容
  if(self.changeHandler && self.snapToIndex != self.selectedIndex && !self.
  isTracking)
      self.changeHandler(self.snapToIndex);    //调用 changeHandler
  //判断 animated
  if(animated)
      [self moveThumbToIndex:index animate:YES];
  else
      self.thumb.frame = [[self.thumbRects objectAtIndex:index] CGRect
      Value];
}
```

moveThumbToIndex:animate:方法实现移动选择指示器。程序代码如下:

```
- (void)moveThumbToIndex:(NSUInteger)segmentIndex animate:(BOOL)animate {
  self.selectedIndex = segmentIndex;
  [self sendActionsForControlEvents:UIControlEventValueChanged];
  //判断 animate
  if(animate) {
     [self.thumb deactivate];
     //动画的实现
     [UIView animateWithDuration:0.2
                          delay:0
                        options:UIViewAnimationOptionCurveEaseOut
                     animations:^{
                        self.thumb.frame= [[self.thumbRects object
                        AtIndex: segmentIndex] CGRectValue];
                                                //设置框架
                        if(self.crossFadeLabelsOnDrag)
                           [self updateTitles];
                                        //调用更新标题的方法
                     }
                     completion:^(BOOL finished){
                        if (finished) {
                           [self activate];
                        }
                     }];
  }
  else {
     self.thumb.frame = [[self.thumbRects objectAtIndex:segmentIndex]
     CGRectValue];                            //设置框架
```

```
        [self activate];                            //调用实现动作的方法
    }
}
```

activate 方法实现指示器移动后的动作。程序代码如下：

```
- (void)activate {
    self.trackingThumb = self.moved = NO;
    self.thumb.label.text = [self.titlesArray objectAtIndex:self.
    selectedIndex];                                //设置文本内容
    void (^oldChangeHandler)(id sender) = [self valueForKey:@"selected
    SegmentChangedHandler"];
    if(oldChangeHandler)
        oldChangeHandler(self);
    //动画
    [UIView animateWithDuration:0.1
                          delay:0
                        options:UIViewAnimationOptionAllowUser
                        Interaction
                     animations:^{
                         self.activated = YES;
                         [self.thumb activate];   //调用指示器移动后的动作
                     }
                     completion:NULL];
}
```

第二个轻拍是需要更改指示器上的标题的。除了有第一种轻拍实现的方法外，还多了一个更新标题的方法 updateTitles。程序代码如下：

```
- (void)updateTitles {
    int hoverIndex = floor(self.thumb.center.x/self.segmentWidth);
    BOOL secondTitleOnLeft = ((self.thumb.center.x / self.segmentWidth) -
    hoverIndex) < 0.5;
    if (secondTitleOnLeft && hoverIndex > 0) {
        //设置透明度
        self.thumb.label.alpha = 0.5 + ((self.thumb.center.x / self.segment
        Width) - hoverIndex);
        //设置文本内容
        self.thumb.secondLabel.text = [self.titlesArray objectAtIndex:
        hoverIndex - 1];
        self.thumb.secondLabel.alpha = 0.5 - ((self.thumb.center.x / self.
        segmentWidth) - hoverIndex);
    }else if (hoverIndex + 1 < self.titlesArray.count) {
        self.thumb.label.alpha = 0.5 + (1 - ((self.thumb.center.x / self.
        segmentWidth) - hoverIndex));
        //设置文本内容
        self.thumb.secondLabel.text = [self.titlesArray objectAtIndex:hover
        Index + 1];
        self.thumb.secondLabel.alpha = ((self.thumb.center.x / self.segment
        Width) - hoverIndex) - 0.5;
    }
    self.thumb.label.text = [self.titlesArray objectAtIndex:hoverIndex];
    //设置指示器标签的文本
}
```

在本实例中，它实现的拖动的方法也是不同的。其中，第一种拖动是不更新标题的，它主要由此文件中的 beginTrackingWithTouch:withEvent:、thumb、continueTrackingWithTouch:withEvent、snap:、endTrackingWithTouch:withEvent 和 moveThumbToIndex:animate:

方法共同实现。其中 continueTrackingWithTouch:方法实现的功能是继续触摸，也就是手指在屏幕上移动。程序代码如下：

```
- (BOOL)continueTrackingWithTouch:(UITouch *)touch withEvent:(UIEvent *)event {
    [super continueTrackingWithTouch:touch withEvent:event];
    CGPoint cPos = [touch locationInView:self];           //获取触摸位置
    CGFloat newPos = cPos.x+self.dragOffset;              //获取新的位置
    CGFloat newMaxX = newPos+(CGRectGetWidth(self.thumb.frame)/2);
    CGFloat newMinX = newPos-(CGRectGetWidth(self.thumb.frame)/2);
    CGFloat buffer = 2.0; // to prevent the thumb from moving slightly too far
    CGFloat pMaxX = CGRectGetMaxX(self.bounds) - buffer;
    CGFloat pMinX = CGRectGetMinX(self.bounds) + buffer;
    //判断
    if((newMaxX > pMaxX || newMinX < pMinX) && self.trackingThumb) {
        self.snapToIndex = floor(self.thumb.center.x/self.segmentWidth);
        if(newMaxX-pMaxX > 10 || pMinX-newMinX > 10)
            self.moved = YES;
        [self snap:NO];
        if (self.crossFadeLabelsOnDrag)
            [self updateTitles];                          //调用更新标题的方法
    }
    else if(self.trackingThumb) {
        self.thumb.center = CGPointMake(cPos.x+self.dragOffset, self.thumb.center.y);
        self.moved = YES;
        if (self.crossFadeLabelsOnDrag)
            [self updateTitles];                          //调用更新标题的方法
    }
    return YES;
}
```

第二种拖动是更新标题，它的实现除了通过第一种拖动中使用的方法外，还使用了 updateTitles 方法。

（7）打开 ViewController.h 文件，编写代码，实现头文件、插座变量以及方法的声明。程序代码如下：

```
#import <UIKit/UIKit.h>
#import "segmented.h"                                     //头文件
@interface ViewController : UIViewController{
    IBOutlet UILabel *label1;                             //标签的插座变量
}
- (void)segmentedControlChangedValue:(segmented*)segmentedControl;
                                                          //方法
@end
```

（8）打开 Main.storyboard 文件，在 View Controller 视图控制器的设计界面中放入 Label 标签控件，将此控件的 Text 设置为"你选择的是第 0 个分段控件的第 1 块"，将 Font 设置为 System 18.0，将 Alignment 设置为居中对齐。将此标签控件和插座变量 label1 进行关联。

（9）打开 ViewController.m 文件，编写代码。在此文件中，代码分为两个部分：两个开关式分段控件对象的创建，以及选择分段控件对象中的某一块所产生的方法。在 viewDidLoad 方法中实现两个开关式分段控件对象的创建。程序代码如下：

```
- (void)viewDidLoad
{
[super viewDidLoad];
//第一个开关式分段控件的创建
    segmented *one = [[segmented alloc] initWithSectionTitles:[NSArray
    arrayWithObjects:@"第一块", @"第二块", nil]];      //初始化每一分段的标题
    one.center = CGPointMake(160, 100);
    [one addTarget:self action:@selector(segmentedControlChangedValue:)
    forControlEvents:UIControlEventValueChanged];    //实现方法的调用
    one.font = [UIFont boldSystemFontOfSize:30];     //设置字体
    one.selectedIndex=1;                             //设置被选择部分的索引
    one.height = 100;                                //设置高度
    one.thumb.tintColor = [UIColor orangeColor];     //设置指示器的颜色
    [self.view addSubview:one];
//第二个开关式分段控件的创建
    segmented *two = [[segmented alloc] initWithSectionTitles:[NSArray
    arrayWithObjects:@"第一块", @"第二块", @"第三块", nil]]; //实例化对象
    [two addTarget:self action:@selector(segmentedControlChangedValue:)
    forControlEvents:UIControlEventValueChanged];    //添加动作
    two.crossFadeLabelsOnDrag = YES;
    two.titleEdgeInsets=UIEdgeInsetsMake(0, 14, 0, 14);
    two.font = [UIFont fontWithName:@"Marker Felt" size:20];//设置字体
    two.height = 100;
    two.selectedIndex = 1;                           //设置选择块的索引
    two.thumb.tintColor = [UIColor cyanColor];
    two.thumb.textColor = [UIColor blackColor];
    two.thumb.textShadowColor = [UIColor colorWithWhite:1 alpha:0.5];
    two.thumb.textShadowOffset = CGSizeMake(0, 1);
    [self.view addSubview:two];                      //添加视图对象
    two.center = CGPointMake(160, 370);
    one.tag=0;
    two.tag=1;
}
```

segmentedControlChangedValue:方法实现当分段控件的值发生改变时，会在设计界面的 Label 标签中显示。程序代码如下：

```
- (void)segmentedControlChangedValue:(segmented*)segmentedControl {
    NSString *aa=[NSString stringWithFormat:@"你选择的是第%i 个分段控件的第%i
    块",segmentedControl.tag,segmentedControl.selectedIndex+1];
    label1.text=aa;
}
```

【代码解析】

由于本实例中的代码和方法非常的多，为了方便读者的阅读，笔者绘制了一些执行流程图如图 7.6～7.9 所示。其中，开关式分段控件的显示，使用了 ViewDidLoad、initWithSectionTitles:、thumb、initWithFrame:、setFrame:、willMoveToSuperview:、drawRect:

方法以及 segmentedselect.m 文件中对选择部分的设置方法。它们的执行流程如图 7.6 所示。

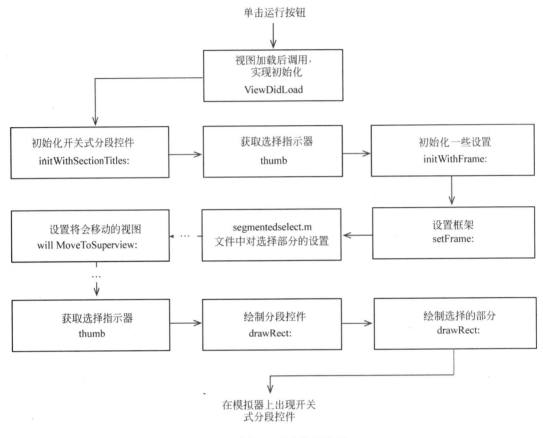

图 7.6 实例 45 程序执行流程 1

> 注意：图中省略号部分是对被选择部分的设置，例如标签、字体等。

通过轻拍的方法可以实现选择分段控件的某一块。在此实例中轻拍有两种方式。如果想要实现轻拍时不需要更改指示器上的标题，需要 beginTrackingWithTouch:、thumb、endTrackingWithTouch:、snap:、deactivate、setSelected:、label、secondLabel、moveThumbToIndex:、segmentedControlChangedValue:、setFrame: drawRect:和 activate 等方法实现。它们的执行流程如图 7.7 所示。

在执行完此程序后，分段控件上的指示器会移动到轻拍的块上。并且在 Label 标签中，会显示相应的内容。但是在指示器上不会出现内容，如果想要在指示器上出现被选块的标题，需要执行完以下程序，如图 7.8 所示。

通过触摸的方法可以实现选择分段控件的某一块。在此实例中触摸也是有两种方式。如果想要实现触摸时不需要更改指示器上的标题的，需要 beginTrackingWithTouch:、thumb、deactivate、setSelected:、label、secondLabel、continueTrackingWithTouch:、snap、setFrame:、font、endTrackingWithTouch:、moveThumbToIndex:和 segmentedControlChangedValue:等方法实现。它们的执行流程如图 7.9 所示。

第 7 章 分段控件

图 7.7 实例 45 程序执行流程 2

图 7.8 实例 45 程序执行流程 3

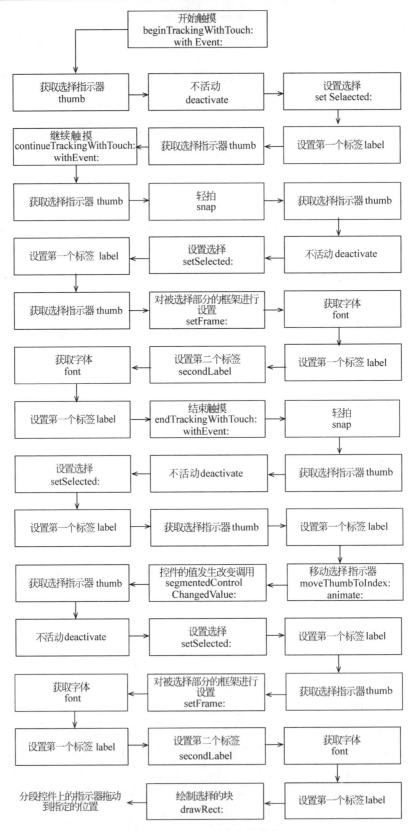

图 7.9 实例 45 程序执行流程 4

在执行完此程序后，分段控件上的指示器会移动到拖动到的块上。但是在指示器上不会出现内容。如果想要出现内容，在轻拍时需要执行在指示器上出现被选块的标题的程序流程图。

实例 46　自定义分段控件

【实例描述】

本实例实现的功能是一个自定义的分段控件。当用户选择其中的一段后，会出现对应的内容。运行效果如图 7.10 所示。

图 7.10　实例 46 运行效果

【实现过程】

（1）创建一个项目，命名为"自定义分段控件"。
（2）创建一个基于 UIControl 类的 segment 类。
（3）打开 segment.h 文件，编写代码，实现枚举、属性和方法的声明。程序代码如下：

```
#import <UIKit/UIKit.h>
//枚举
enum HMSelectionIndicatorMode {
    HMSelectionIndicatorResizesToStringWidth = 0,
};
@interface segmented : UIControl
//属性
@property (nonatomic, strong) NSArray *sectionTitles;
```

```
@property (nonatomic, copy) void (^indexChangeBlock)(NSUInteger index);
……
@property (nonatomic, readwrite) CGFloat segmentWidth;
//方法
- (id)initWithSectionTitles:(NSArray *)sectiontitles;
- (void)setSelectedIndex:(NSUInteger)index animated:(BOOL)animated;
@end
```

（4）打开 segment.m 文件，编写代码。在此文件中，代码分为两大部分：绘制分段控件和实现分段控件的选择操作。其中，分段控件的绘制需要 initWithFrame:、initWithSectionTitles:、setDefaults、frameForSelectionIndicator、updateSegmentsRects、setFrame: 和 drawRect:方法共同实现。initWithSectionTitles:方法，实现对自定义分段控件的初始化。程序代码如下：

```
- (id)initWithSectionTitles:(NSArray *)sectiontitles {
    self = [super initWithFrame:CGRectZero];
    if (self) {
        self.sectionTitles = sectiontitles;       //设置块的标题
        [self setDefaults];
    }
    return self;
}
```

setDefaults 方法对自定义分段控件的字体、字体颜色、背景颜色以及其他属性进行默认的设置。程序代码如下：

```
- (void)setDefaults {
    self.font = [UIFont fontWithName:@"STHeitiSC-Light" size:18.0f];
    self.textColor = [UIColor blackColor];              //设置文本的颜色
    self.backgroundColor = [UIColor whiteColor];
    self.selectionIndicatorColor = [UIColor colorWithRed:52.0f/255.0f
    green:181.0f/255.0f blue:229.0f/255.0f alpha:1.0f];//设置指示器颜色
    self.selectedIndex = 0;
    self.height = 60.0f;                                //设置高度
    self.selectionIndicatorHeight = 5.0f;
    self.selectionIndicatorMode = HMSelectionIndicatorResizesToStringWidth;
    self.selectedSegmentLayer = [CALayer layer];
}
```

frameForSelectionIndicator 方法对自定义分段控件中的指示器进行设置，此指示器是用来实现用户选择的。程序代码如下：

```
- (CGRect)frameForSelectionIndicator {
    CGFloat stringWidth = [[self.sectionTitles objectAtIndex:self.selectedIndex]
        sizeWithFont:self.font].width;                  //获取字符串的宽度
        CGFloat widthTillEndOfSelectedIndex = (self.segmentWidth * self.
        selectedIndex) + self.segmentWidth;
        CGFloat widthTillBeforeSelectedIndex = (self.segmentWidth * self.
        selectedIndex);
        CGFloat x = ((widthTillEndOfSelectedIndex - widthTillBeforeSelected
        Index) / 2) + (widthTillBeforeSelectedIndex - stringWidth / 2);
        return CGRectMake(x, 0.0, stringWidth, self.selectionIndicatorHeight);
                                                        //返回框架
}
```

updateSegmentsRects 方法对分段的框架进行更新。程序代码如下：

```objc
- (void)updateSegmentsRects {
    self.segmentWidth = self.frame.size.width / self.sectionTitles.count;
    self.height = self.frame.size.height;
}
```

setFrame:方法对自定义分段控件的框架进行设置。程序代码如下:

```objc
- (void)setFrame:(CGRect)frame {
    [super setFrame:frame];
    if (self.sectionTitles)
        [self updateSegmentsRects];             //调用 updateSegmentsRects 方法
    [self setNeedsDisplay];
}
```

drawRect:方法实现自定义分段控件的绘制。程序代码如下:

```objc
- (void)drawRect:(CGRect)rect
{
    [self.backgroundColor set];                 //背景颜色
    UIRectFill([self bounds]);                  //填充矩形
    [self.textColor set];                       //文本颜色
    [self.sectionTitles enumerateObjectsUsingBlock:^(id titleString, NSUInteger idx, BOOL *stop) {
        CGFloat stringHeight = [titleString sizeWithFont:self.font].height;
        CGFloat y = ((self.height - self.selectionIndicatorHeight) / 2) +
        (self.selectionIndicatorHeight - stringHeight / 2);
        CGRect rect = CGRectMake(self.segmentWidth * idx, y, self.segmentWidth, stringHeight);
#if __IPHONE_OS_VERSION_MIN_REQUIRED < 60000
        NSMutableParagraphStyle *style=[[NSMutableParagraphStyle defaultParagraphStyle] mutableCopy];     //实例化对象
        [style setLineBreakMode:NSLineBreakByClip];
        [style setAlignment:NSTextAlignmentCenter];    //对齐方式
        NSDictionary *dis=[[NSDictionary alloc]initWithObjectsAndKeys:style,NSParagraphStyleAttributeName,self.font,NSFontAttributeName,nil];
        [titleString drawInRect:rect withAttributes:dis];  //绘制
#else
        NSMutableParagraphStyle *style=[[NSMutableParagraphStyle defaultParagraphStyle] mutableCopy];
        [style setLineBreakMode:NSLineBreakByClipping];
        [style setAlignment:NSTextAlignmentCenter];       //对齐方法
        NSDictionary *dis=[[NSDictionary alloc] initWithObjectsAndKeys:style,NSParagraphStyleAttributeName,self.font,NSFontAttributeName, nil];
        [titleString drawInRect:rect withAttributes:dis];
#endif
        self.selectedSegmentLayer.frame = [self frameForSelectionIndicator];
                                                        //框架
        self.selectedSegmentLayer.backgroundColor = self.selectionIndicatorColor.CGColor;
        [self.layer addSublayer:self.selectedSegmentLayer]; //添加
    }];
}
```

这时一个自定义的分段控件就绘制好了。但是,这时它是不具备选择功能的。需要添加 touchesBegan:withEvent: 、setSelectedIndex: 和 setSelectedIndex:animated: 方法。其中

touchesBegan:withEvent:方法实现开启触摸的功能。程序代码如下：

```objc
- (void)touchesBegan:(NSSet *)touches withEvent:(UIEvent *)event {
    UITouch *touch = [[event allTouches] anyObject];
    CGPoint touchLocation = [touch locationInView:self];        //获取触摸位置
    //判断 touchLocation 是否包含在 self.bounds 中
    if (CGRectContainsPoint(self.bounds, touchLocation)) {
        NSInteger segment = touchLocation.x / self.segmentWidth;
        if (segment != self.selectedIndex) {
            [self setSelectedIndex:segment animated:YES];//设置被选块的索引
        }
    }
}
```

setSelectedIndex:方法对被选择的索引进行设置，也是就指示器所指向的块的设置。程序代码如下：

```objc
- (void)setSelectedIndex:(NSInteger)index {
    [self setSelectedIndex:index animated:NO];
}
```

setSelectedIndex:animated:方法对被选择索引的动画进行设置。程序代码如下：

```objc
- (void)setSelectedIndex:(NSUInteger)index animated:(BOOL)animated {
    _selectedIndex = index;
    //判断是否有动画效果
    if (animated) {
        //实现 CALayer 动画
        self.selectedSegmentLayer.actions = nil;
        [CATransaction begin];
        [CATransaction setAnimationDuration:0.15f];        //设置动画所需时间
        [CATransaction setCompletionBlock:^{
            if (self.superview)
                [self sendActionsForControlEvents:UIControlEventValueChanged];
            if (self.indexChangeBlock)
                self.indexChangeBlock(index);
        }];
        self.selectedSegmentLayer.frame = [self frameForSelectionIndicator];
        //设置框架
        [CATransaction commit];
    } else {
        //取消动画
        NSMutableDictionary *newActions = [[NSMutableDictionary alloc]
        initWithObjectsAndKeys:[NSNull null], @"position", [NSNull null],
        @"bounds", nil];
        self.selectedSegmentLayer.actions = newActions;
        self.selectedSegmentLayer.frame = [self frameForSelectionIndicator];
                                                            //设置框架
        if (self.superview)
            [self sendActionsForControlEvents:UIControlEventValueChanged];
        if (self.indexChangeBlock)
            self.indexChangeBlock(index);
    }
}
```

（5）打开 Main.storyboard 文件，在 View Controller 视图控制器的设计界面添加 Label 标签，将此标签的 Text 设置为"电视剧"，将 Font 设置为 System 50.0，将 Alignment 设

置为居中对齐。

（6）打开 ViewController.h 文件，编写代码，实现头文件和插座变量的声明。程序代码如下：

```
#import <UIKit/UIKit.h>
#import "segmented.h"
@interface ViewController : UIViewController{
    IBOutlet UILabel *label;                    //标签的插座变量
}
@end
```

（7）将插座变量 label 与设计界面的 Label 标签进行关联。

（8）打开 ViewController.m 文件，编写代码。在此文件中，代码分为两部分：自定义分段控件的创建以及分段控件被选择的块发生变化后产生的效果。在 viewDidLoad 方法中实现自定义分段控件对象的创建以及设置。程序代码如下：

```
- (void)viewDidLoad
{
    segmented *segmentedControl = [[segmented alloc] initWithSection
    Titles:@[@"电视剧", @"娱乐", @"新闻头条"]];                    //创建
    [segmentedControl setFrame:CGRectMake(10, 10, 300, 60)];    //设置框架
    [segmentedControl addTarget:self action:@selector(segmentedControl
    ChangedValue:) forControlEvents:UIControlEventValueChanged];//添加动作
    [segmentedControl setTextColor:[UIColor blueColor]];    //设置字体颜色
    [segmentedControl setSelectionIndicatorColor:[UIColor redColor]];
            //设置指示器颜色
    [self.view addSubview:segmentedControl];
    [super viewDidLoad];
    // Do any additional setup after loading the view, typically from a nib.
}
```

添加 segmentedControlChangedValue:方法实现分段控件被选择的块发生变化后，在标签中发生的改变。程序代码如下：

```
- (void)segmentedControlChangedValue:(segmented *)segmentedControl {
    if(segmentedControl.selectedIndex==0){
        label.text=@"电视剧";
    }else if (segmentedControl.selectedIndex==1){
        label.text=@"娱乐";
    }else{
        label.text=@"新闻头条";
    }
}
```

【代码解析】

由于本实例中的代码和方法非常的多，为了方便读者的阅读，笔者绘制了一些执行流程图如图 7.11~7.12 所示。其中，自定义分段控件的显示，使用了 ViewDidLoad、initWithSectionTitles:、setFrame:、setDefaults、setSelectedIndex:、setSelectedIndex:animated:、frameForSelectionIndicator、updateSegmentsRects 和 drawRect:方法共同实现。它们的执行流程如图 7.11 所示。

如果想要让自定义选择器实现选择，需要使用 touchesBegan:、setSelectedIndex:、frameForSelectionIndicator 和 segmentedControlChangedValue:方法共同实现。它们的执行流

程如图 7.12 所示。

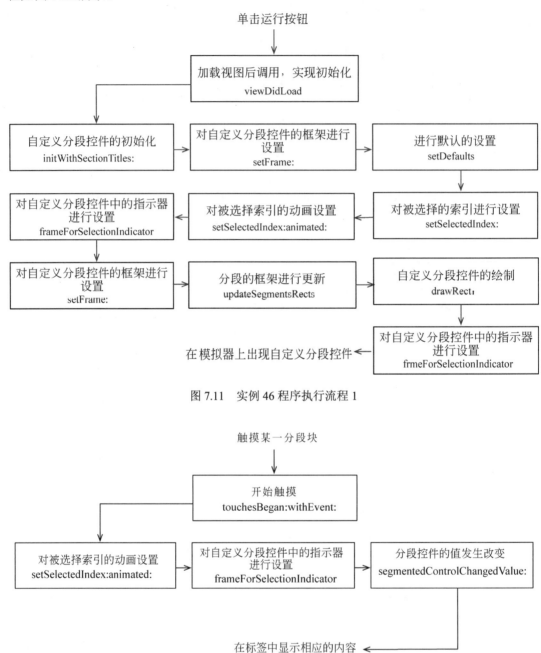

图 7.11　实例 46 程序执行流程 1

图 7.12　实例 46 程序执行流程 2

第 8 章 导 航 栏

在 iPhone 或者 iPad 中,导航栏可以实现视图切换的功能。它经常位于视图的顶端,一般是一个矩形视图。本章将主要讲解导航栏的相关实例。

实例 47 具有阴影的导航栏

【实例描述】

本实例的功能是在导航栏下方添加阴影效果,运行效果如图 8.1 所示。

图 8.1 实例 47 运行效果

【实现过程】

(1)创建一个项目,命名为"具有阴影的导航栏"。
(2)创建一个基于 UINavigationController 类的 Navigation 类。
(3)打开 Navigation.m 文件,在 viewDidLoad 方法中,编写代码,实现为导航栏添加阴影。程序代码如下:

```objc
- (void)viewDidLoad
{
    [super viewDidLoad];
    // Do any additional setup after loading the view.
    [[UINavigationBar appearance] setBackgroundColor:[UIColor blueColor]];
    self.navigationBar.layer.masksToBounds = NO;
    //设置阴影的高度
    self.navigationBar.layer.shadowOffset = CGSizeMake(0, 3);
    //设置透明度
    self.navigationBar.layer.shadowOpacity = 0.8;
    //设置阴影的位置
    self.navigationBar.layer.shadowPath =[UIBezierPath bezierPathWithRect:
    self.navigationBar.bounds].CGPath;
}
```

（4）打开 Main.storyboard 文件，将画布中的 View Controller 视图控制器删除，添加 Navigation Controller 导航控制器，将 Class 设置为 Navigation，将 Is Initial View Controller 复选框选中。

【代码解析】

本实例关键功能是为导航栏添加阴影。下面就是这个知识点的详细讲解。

1. 设置阴影的偏移量

要设置阴影的偏移量，需要使用 CALayer 的 shadowOffset 方法，其语法形式如下：

```objc
@property CGSize shadowOffset;
```

2. 设置阴影的透明度

透明度可以使阴影变浅或者变深，要设置阴影的透明度，需要使用 CALayer 的 shadowOpacity 方法，其语法形式如下：

```objc
@property float shadowOpacity;
```

3. 设置阴影的路径

本实例的阴影是在导航栏的下方，它的实现使用的是 CALayer 中的 shadowPath 属性，其语法形式如下：

```objc
@property CGPathRef shadowPath;
```

在代码中，使用了这 3 个属性进行了导航栏阴影的添加，代码如下：

```objc
self.navigationBar.layer.shadowOffset = CGSizeMake(0, 3);
self.navigationBar.layer.shadowOpacity = 0.8;
self.navigationBar.layer.shadowPath = [UIBezierPath bezierPathWithRect:
self.navigationBar.bounds].CGPath;
```

实例48 具有图片的导航栏

【实例描述】

本实例实现的功能是在界面显示一个具有图片的导航栏，轻拍导航栏图片和标题周围,标题会发生改变。运行效果如图 8.2 所示。

图 8.2 实例 48 运行效果

【实现过程】

在界面显示一个具有图片的导航栏后,用户轻拍导航栏的图标和标题周围,导航栏的标题会变为另一种颜色,轻拍结束,会还原颜色。具体的实现步骤如下。

(1)创建一个项目,命名为"具有图片的导航栏"。
(2)添加图片 1.png 到创建项目的 Supporting Files 文件夹中。
(3)创建一个基于 UIView 类的 Title 类。
(4)打开 Title.h 文件,编写代码,实现属性和方法的声明。程序代码如下:

```
#import <UIKit/UIKit.h>
@interface Title : UIView
//属性
@property (nonatomic, copy) NSString *title;
……
@property (nonatomic, retain) UILabel *titleLabel;
//方法
- (void)touchedDown:(id)sender;
- (void)touchedUpInside:(id)sender;
- (void)update;
@end
```

(5)打开 Title.m 文件,编写代码,实现在导航栏中添加图片的效果。使用的方法如表 8-1 所示。

表 8-1　Title.m文件中方法总结

方　　法	功　　能
initWithFrame:	标题视图的初始化
setTitle:	设置标题
title	获取标题
setImage:	设置图片
image	获取图片
setHighlighted:	设置Highlighted的状态
update	更新（设置显示的框架）
touchedDown:	按下
touchedUpInside:	弹起

可以将这些方法分为两个部分，其中，第一部分为实现在导航栏中添加图片；第二部分为触摸动作的实现。这里需要讲解几个重要的方法（其他方法请读者参考源代码）。实现在导航栏中添加图片，需要使用 initWithFrame:、setTitle:、title、setImage:、image、setHighlighted:和 update 方法共同实现。initWithFrame:方法实现标题视图的（导航栏中心）的初始化。程序代码如下：

```objc
- (id)initWithFrame:(CGRect)frame
{
    self = [super initWithFrame:frame];
    if (self) {
        self.margin = 6;
        self.imageViewSize = CGSizeMake(30, 30);
        //按钮
        UIButton *baseButton = [UIButton buttonWithType:UIButtonTypeCustom];
        baseButton.frame = self.bounds;
        [baseButton addTarget:self action:@selector(touchedDown:) forControlEvents:UIControlEventTouchDown];           //为按钮添加动作
        [baseButton addTarget:self action:@selector(touchedUpInside:) forControlEvents:UIControlEventTouchUpInside];   //为按钮添加动作
        [self addSubview:baseButton];
        //图片
        UIImageView *imageView = [[UIImageView alloc] initWithFrame:CGRectZero];
        imageView.clipsToBounds = YES;
        imageView.layer.cornerRadius = 4.0;
        imageView.layer.borderWidth = 1.0;              //设置边框宽度
        imageView.layer.masksToBounds = YES;
        imageView.layer.borderColor = [[UIColor colorWithWhite:0.35 alpha:1.0] CGColor];
        [baseButton addSubview:imageView];              //添加视图对象
        self.imageView = imageView;
        //标签
        UILabel *titleLabel = [[UILabel alloc] initWithFrame:CGRectZero];
        titleLabel.font = [UIFont boldSystemFontOfSize:20];
        titleLabel.textColor = [UIColor blueColor]; //设置文本颜色
        titleLabel.highlightedTextColor = [UIColor colorWithWhite:0.7 alpha:1.0];
        titleLabel.backgroundColor = [UIColor clearColor];   //设置背景颜色
        titleLabel.textAlignment = NSTextAlignmentCenter;
        titleLabel.shadowColor = [UIColor colorWithWhite:0.0 alpha:0.5];
        titleLabel.shadowOffset = CGSizeMake(0, -1);         //设置阴影偏移量
        titleLabel.lineBreakMode = NSLineBreakByTruncatingTail;
```

```
    [bascButton addSubview:titleLabel];
    self.titleLabel = titleLabel;                    //设置标题
    [self update];
    }
    return self;
}
```

update方法实现更新,即对显示的标签、图片的框架进行设置。程序代码如下:

```
- (void)update
{
    CGFloat margin = self.margin;
    //实例化对象
    CGSize imageViewSize = (self.image && self.highlightedImage) ? self.
    imageViewSize : CGSizeZero;
    CGSize actualTitleSize = [self.title sizeWithFont:[UIFont boldSystem
    FontOfSize:20] forWidth:self.bounds.size.width lineBreakMode:NSLine
    BreakByTruncatingTail];
    CGRect titleLabelFrame;
    CGRect imageViewFrame;
    //绘制框架
    titleLabelFrame = CGRectMake((self.bounds.size.width - actualTitleSize.
    width) / 2, 0, actualTitleSize.width, self.bounds.size.height);
    imageViewFrame = CGRectMake(titleLabelFrame.origin.x - (imageView/2,
    Size.width + margin), (self.bounds.size.height - imageViewSize.height)
    imageViewSize.width, imageViewSize.height);
    //设置框架
    self.titleLabel.frame = titleLabelFrame;
    self.imageView.frame = imageViewFrame;
}
```

要实现触摸动作,需要使用 touchedDown:和 touchedUpInside:方法实现。其中,touchedDown:方法实现按下的动作,touchedUpInside:方法实现弹起的动作。程序代码如下:

```
//按下
- (void)touchedDown:(id)sender
{
    self.highlighted = YES;
}
//弹起
- (void)touchedUpInside:(id)sender
{
    self.highlighted = NO;
}
```

(6)打开 Main.storyboard 文件,添加 Navigation Controller 导航控制器到画布中,将此控制器关联的根视图改为 View Controller 视图控制器。将 View Controller 视图控制器的设计界面设置为浅绿色。

(7)打开 ViewController.m 文件,编写代码。实现创建一个具有图片的导航栏对象,并将其添加到当前视图中。程序代码如下:

```
- (void)viewDidLoad
{
    [super viewDidLoad];
    // Do any additional setup after loading the view, typically from a nib.
    Title *titleView = [[Title alloc] initWithFrame:CGRectMake(0, 0, 170,
44)];
    titleView.image = [UIImage imageNamed:@"1.jpg"];         //设置图像
```

```
    titleView.title = @"我爱iOS";
    self.navigationItem.titleView = titleView;
    titleView = titleView;
}
```

【代码解析】

本实例关键功能是图片的添加。下面就是这个知识点的详细讲解。

本实例中，在添加图片之前，先创建了一个按钮对象，代码如下：

```
UIButton *baseButton = [UIButton buttonWithType:UIButtonTypeCustom];
……
[self addSubview:baseButton];
```

接着，创建了一个用于显示图片的图像视图，代码如下：

```
UIImageView *imageView = [[UIImageView alloc] initWithFrame:CGRectZero];
……
[baseButton addSubview:imageView];
self.imageView = imageView;
```

然后，在 update 中设置图像视图的框架，代码如下：

```
imageViewFrame = CGRectMake(titleLabelFrame.origin.x - (imageViewSize.width + margin), (self.bounds.size.height - imageViewSize.height) / 2, imageViewSize.width, imageViewSize.height);
```

最后，通过 setImage:和 image 方法，对在图像视图中要显示的图片进行设置。

实例 49 具有分段控件的导航栏

【实例描述】

本实例实现的功能是在导航栏上添加一个分段控件，并且在选择分段控件的某一段后，会有相应的背景颜色出现。运行效果如图 8.3 所示。

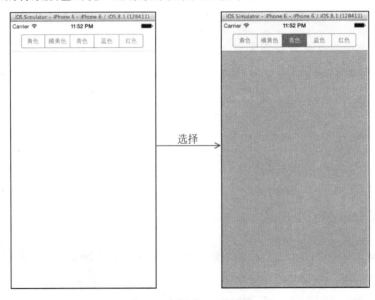

图 8.3 实例 49 运行效果

【实现过程】

（1）创建一个项目，命名为"具有分段控件的导航栏"。

（2）打开 Main.storyboard 文件，添加 Navigation Controller 导航控制器到画布中，将此控制器关联的根视图改为 View Controller 视图控制器。

（3）打开 ViewController.h 文件，编写代码，实现对象和动作的声明。程序代码如下：

```
#import <UIKit/UIKit.h>
@interface ViewController : UIViewController{
   UISegmentedControl *segmentedController;
}
- (IBAction)segmentAction:(id)sender;
@end
```

（4）打开 ViewController.m 文件，编写代码。在此文件中，代码分为了两部分：将分段控件添加到导航栏中和实现在分段控件中响应选择的段。其中，在 viewDidload 中进行初始化实现将分段控件添加到导航栏中。程序代码如下：

```
- (void)viewDidLoad
{
  NSArray *segmentButtons = [NSArray arrayWithObjects:@"黄色",@"橘黄色",
  @"青色",@"蓝色",@"红色", nil];
  segmentedController = [[UISegmentedControl alloc] initWithItems:
  segmentButtons];
  [segmentedController addTarget:self action:@selector(segmentAction:)
  forControlEvents:UIControlEventValueChanged];
  self.navigationItem.titleView = segmentedController;//设置导航栏的中心
  [super viewDidLoad];
  // Do any additional setup after loading the view, typically from a nib.
}
```

segmentAction:方法实现在分段控件中响应选择的段。程序代码如下：

```
- (IBAction)segmentAction:(id)sender {
NSInteger select=segmentedController.selectedSegmentIndex;
//实现响应
   if (select==0) {
      self.view.backgroundColor=[UIColor yellowColor];
   }else if (select==1){
      self.view.backgroundColor=[UIColor  orangeColor];
   }else if (select==2){
      self.view.backgroundColor=[UIColor cyanColor];
   }else if (select==3){
      self.view.backgroundColor=[UIColor blueColor];
   }else if (select==4){
      self.view.backgroundColor=[UIColor redColor];
   }
}
```

【代码解析】

本实例关键功能是为导航栏添加分段控件。下面就是这个知识点的详细讲解。

如果想要在导航栏中添加分段控件，需要使用 UINavigationItem 的 titleView 属性实现。其语法形式如下：

```
@property(nonatomic, retain) UIView *titleView;
```

在代码中,使用了 titleView 属性为导航栏添加分段控件,代码如下:

```
self.navigationItem.titleView = segmentedController;
```

实例 50　具有子标题的导航栏

【实例描述】

在导航栏中,如果一个标题无法表达完全整个意思,这时就需要使用子标题(副标题)进行解释说明。本实例实现的功能是为导航栏添加一个子标题,当用户想要添加子标题时,单击"+"按钮,在弹出的视图中,输入子标题单击"更新"按钮后,子标题就显示在了导航栏中。其中,用于用户进行输入的是文本框。运行效果如图 8.4 所示。

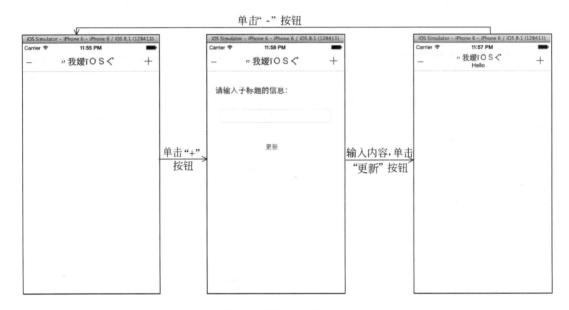

图 8.4　实例 50 运行效果

【实现过程】

(1) 创建一个项目,命名为"具有子标题的导航栏"。
(2) 创建一个基于 UIView 类的 ContentView 类。
(3) 打开 ContentView.h 文件,编写代码,实现协议和属性的声明。程序代码如下:

```
#import <UIKit/UIKit.h>
//协议
@protocol ContentViewDelegate <NSObject>
- (void)drawContent:(CGRect)rect;
@end
@interface ContentView : UIView
@property (nonatomic, weak) id <ContentViewDelegate> delegate;
    //属性
@end
```

(4) 打开 ContentView.m 文件,编写代码,在 drawRect:方法中,实现对协议 draw

Content:方法的调用。程序代码如下:

```
- (void)drawRect:(CGRect)rect
{
    if ([self.delegate respondsToSelector:@selector(drawContent:)]) {
        [self.delegate drawContent:rect];          //调用
    }
}
```

(5) 创建一个基于 UIView 类的 TitleView 类。

(6) 打开 TitleView.h 文件, 编写代码, 实现头文件和属性的声明。程序代码如下:

```
#import <UIKit/UIKit.h>
#import "ContentView.h"                           //头文件
@interface TitleView : UIView <ContentViewDelegate>
//属性
@property (nonatomic, copy) NSString *navigationBarTitle;
@property (nonatomic, copy) NSString *navigationBarSubtitle;
@property (nonatomic, copy) UIColor *navigationBarTitleFontColor;
@property (nonatomic, copy) UIColor *navigationBarSubtitleFontColor;
@property (nonatomic, strong) ContentView *contentView;
@end
```

(7) 打开 TitleView.m 文件, 编写代码, 实现导航栏标题和子标题的绘制。使用的方法如表 8-2 所示。

表 8-2 TitleView.m文件中方法总结

方　　法	功　　能
initWithCoder:	初始化实例化对象
initWithFrame:	初始化标题视图
setupContentView	创建内容视图
setNavigationBarTitle:	设置导航栏的标题
setNavigationBarSubtitle:	设置导航栏的子标题
drawContent:	绘制

这里需要讲解几个重要的方法(其他方法请读者参考源代码)。setupContentView 方法实现对放置标题与父标题的内容视图进行创建。程序代码如下:

```
- (void)setupContentView
{
    self.contentView = [[ContentView alloc] initWithFrame:self.frame];
    self.contentView.autoresizingMask = UIViewAutoresizingFlexibleHeight |
    UIViewAutoresizingFlexibleWidth;                //自动调整子视图与父视图中间的位置
    self.contentView.delegate = self;
    [self addSubview:self.contentView];
    self.backgroundColor = [UIColor clearColor];              //设置背景颜色
    self.contentView.backgroundColor = [UIColor clearColor];
    self.clipsToBounds = YES;
}
```

setNavigationBarSubtitle:方法实现对导航栏中的子标题进行设置。程序代码如下:

```
- (void)setNavigationBarSubtitle:(NSString *)navigationBarSubtitle
```

```objc
{
    if (![_navigationBarSubtitle isEqualToString:navigationBarSubtitle]) {
        if (navigationBarSubtitle.length && !_navigationBarSubtitle.length) {
            //向上的动画
            CATransition *transition = [CATransition animation];
            transition.duration = 0.4f;                          //设置动画所需时间
            transition.timingFunction = [CAMediaTimingFunction functionWith
            Name:kCAMediaTimingFunctionEaseInEaseOut];
            transition.type = kCATransitionPush;                 //设置动画类型
            transition.subtype = kCATransitionFromTop;           //设置动画方向
            [transition setValue:(id) kCFBooleanFalse forKey:kCATransitionFade];
            [self.contentView.layer addAnimation:transition forKey:nil];
        }
        else if (!navigationBarSubtitle.length && _navigationBarSubtitle.length) {
            //向下的动画
            CATransition *transition = [CATransition animation];
            transition.duration = 0.4f;                          //设置动画所需时间
            transition.timingFunction = [CAMediaTimingFunction functionWith
            Name:kCAMediaTimingFunctionEaseInEaseOut];
            transition.type = kCATransitionPush;                 //设置动画类型
            transition.subtype = kCATransitionFromBottom;        //设置动画方向
            [transition setValue:(id) kCFBooleanFalse forKey:kCATransition
            Fade];
            [self.contentView.layer addAnimation:transition forKey:nil];
        }
        _navigationBarSubtitle = navigationBarSubtitle;
        [self.contentView setNeedsDisplay];
    }
}
```

drawContent:方法实现对导航栏中的标题与子标题进行绘制。程序代码如下：

```objc
- (void)drawContent:(CGRect)rect
{
    if (self.navigationBarSubtitle.length) {
        //绘制标题与子标题
        CGRect titleRect = rect;
        titleRect.origin.y = 4;                  //设置 y 值
        titleRect.size.height = 20;              //设置高度
        [self.navigationBarTitleFontColor setFill];
        NSMutableParagraphStyle *style = [[NSParagraphStyle defaultParagraph
        Style] mutableCopy];
        [style setLineBreakMode:NSLineBreakByTruncatingTail];//设置换行模式
        [style setAlignment:NSTextAlignmentCenter];          //对齐方式
        NSDictionary *dic=[[NSDictionary alloc]initWithObjectsAndKeys:
        style,NSParagraphStyleAttributeName,[UIFont fontWithName:@"Helvetica"
        size:17],NSFontAttributeName ,nil];
        [self.navigationBarTitle drawInRect:titleRect withAttributes:dic];
                                                            //绘制
        //绘制子标题
        CGRect subtitleRect = rect;
```

```
            subtitleRect.origin.y = 24;
            subtitleRect.size.height = rect.size.height - 24;
            [self.navigationBarSubtitleFontColor setFill];
            NSMutableParagraphStyle *styl = [[NSParagraphStyle defaultParagraph
            Style] mutableCopy];
            [styl setLineBreakMode:NSLineBreakByTruncatingTail];//设置换行模式
            [styl setAlignment:NSTextAlignmentCenter];
            NSDictionary *di=[[NSDictionary alloc]initWithObjectsAndKeys:styl,
            NSParagraphStyleAttributeName,[UIFont fontWithName:@"Helvetica"
            size:13],NSFontAttributeName ,nil];
            [self.navigationBarSubtitle drawInRect:subtitleRect withAttributes:di];
                                                                //绘制
        }else {
            //设置标题
            CGRect titleRect = rect;
            titleRect.origin.y = (rect.size.height - 24.f) / 2.f;
            titleRect.size.height = 24.f;                       //设置高度
            [self.navigationBarTitleFontColor setFill];
            NSMutableParagraphStyle *style = [[NSParagraphStyle defaultParag
            raphStyle] mutableCopy];
            [style setLineBreakMode:NSLineBreakByTruncatingTail];//设置换行模式
            [style setAlignment:NSTextAlignmentCenter];
            NSDictionary *dic=[[NSDictionary alloc]initWithObjectsAndKeys:style,
            NSParagraphStyleAttributeName,[UIFont fontWithName:@"Helvetica"
            size:20],NSFontAttributeName ,nil];
            [self.navigationBarTitle drawInRect:titleRect withAttributes:dic];
            //绘制
        }
    }
```

（8）创建一个基于 UINavigationController 类的 Navigation 类。

（9）打开 Navigation.m 文件，编写代码，实现标题的添加，并显示。其中，标题的添加使用 addTitleViewToViewController:方法实现。程序代码如下：

```
- (void)addTitleViewToViewController:(UIViewController *)viewController
{
    if (!viewController.navigationItem.titleView) {
        CGFloat width = 0.95 * self.view.frame.size.width;
        TitleView *titleView = [[TitleView alloc] initWithFrame:CGRectMake(0,
        0, width, 44)];
        viewController.navigationItem.titleView = titleView;
        viewController.title = viewController.title.length ? viewController.
        title : viewController.navigationItem.title;           //设置标题
    }
}
```

viewDidLoad 方法，实现 addTitleViewToViewController:方法的调用。

（10）创建 UIViewController 类的分类 SubTitle。

（11）打开 UIViewController+SubTitle.h 文件，编写代码，实现属性的声明。程序代码如下：

```
#import <UIKit/UIKit.h>
@interface UIViewController (SubTitle)
@property (nonatomic, copy) NSString *subtitle;
@property (nonatomic, copy) NSString *title;
@end
```

(12)打开 UIViewController+SubTitle.m 文件,编写代码,实现对已有类 UIViewController 进行修改。使用的方法如表 8-3 所示。

表 8-3　UIViewController+SubTitle.m文件中方法总结

方　　法	功　　能
title	获取标题
setTitle:	设置标题
subtitle	获取子标题
setSubtitle:	设置子标题
updateTitleTo:	更新标题
updateSubtitleTo:	更新子标题

setTitle:方法实现对标题的设置。程序代码如下:

```
- (void)setTitle:(NSString *)title
{
    [self willChangeValueForKey:@"title"];
    objc_setAssociatedObject(self, &UIViewControllerTitleKey, title, OBJC_
    ASSOCIATION_COPY_NONATOMIC);                    //创建关联
    [self didChangeValueForKey:@"title"];
    [self updateTitleTo:title];                     //调用更新标题的方法
    self.navigationItem.title = title;              //设置导航栏的标题
}
```

setSubtitle:方法对标题进行设置。程序代码如下:

```
- (void)setSubtitle:(NSString *)subtitle
{
    [self willChangeValueForKey:@"subtitle"];
    objc_setAssociatedObject(self, &UIViewControllerSubtitleKey, subtitle,
    OBJC_ASSOCIATION_COPY_NONATOMIC);
    [self didChangeValueForKey:@"subtitle"];
    [self updateSubtitleTo:subtitle];               //更新子标题
}
```

updateTitleTo:和 updateSubtitleTo:方法实现对标题和子标题的更新。实现代码如下:

```
//更新标题
- (void)updateTitleTo:(NSString *)title
{
    //判断 self.navigationItem.titleView 是否为 TitleView 的成员
    if ([self.navigationItem.titleView isKindOfClass:[TitleView class]]) {
        [(TitleView *) self.navigationItem.titleView setNavigationBarTitle:
        title];
    }
}
//更新子标题
- (void)updateSubtitleTo:(NSString *)subtitle
{
```

```
//判断 self.navigationItem.titleView 是否为 TitleView 的成员
if ([self.navigationItem.titleView isKindOfClass:[TitleView class]]) {
    [(TitleView *) self.navigationItem.titleView setNavigationBar
     Subtitle:subtitle];
   }
}
```

（13）打开 ViewController.h 文件，编写代码，实现头文件、插座变量和动作的声明。程序代码如下：

```
#import <UIKit/UIKit.h>
#import "Navigation.h"                //头文件
#import "UIViewController+SubTitle.h"
@interface ViewController : UIViewController{
    //插座变量
    IBOutlet UIView *vv;
    IBOutlet UITextField *tf;
}
//动作
- (IBAction)add:(id)sender;
- (IBAction)update:(id)sender;
- (IBAction)delet:(id)sender;
@end
```

（14）打开 Main.storyboard 文件，添加 Navigation Controller 导航控制器到画布中。将导航控制器关联的根视图设置为 View Controller 视图控制器，并对 View Controller 视图控制器的设计界面进行设计，效果如图 8.5 所示。

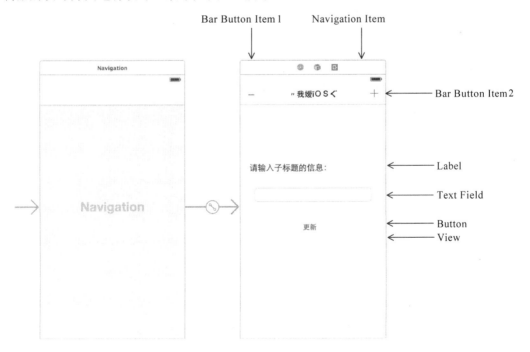

图 8.5　实例 50 View Controller 视图控制器设计界面

需要添加的视图、控件以及对它们的设置如表 8-4 所示。

表 8-4 实例 50 视图、控件设置

视图、控件	属 性 设 置	其 他
导航控制器		Class：Navigation
Navigation Item	Title："我爱ｉＯＳ"	
Bar Button Item1	Title：—	与动作delet:关联
Bar Button Item2	Identifier：Add	与动作add:关联
Label	Text：请输入子标题的信息：	与插座变量button2关联 与动作hide:关联
Text Field		与插座变量tf关联
View		与插座变量vv关联
Button	Title：更新	与动作update:关联

（15）将插座变量 vv 关联的 View 空白视图移动到设计界面看不到的位置。

（16）打开 ViewController.m 文件，编写代码。实现 delet:、add:和 update:方法的定义。其中，add:方法，实现单击"+"按钮，出现一个子视图。程序代码如下：

```
- (IBAction)add:(id)sender {
    [UIView beginAnimations:@"" context:nil];
    [UIView setAnimationDuration:0.3];
    vv.frame=CGRectMake(0, 80, 320, 222);          //设置框架
    [UIView commitAnimations];
}
```

update:方法，实现单击"更新"按钮，更新子标题，退出键盘等功能。程序代码如下：

```
- (IBAction)update:(id)sender {
self.subtitle=tf.text;          //将子标题的内容设置为文本框中的内容
//实现动画效果
    [UIView beginAnimations:@"" context:nil];
    [UIView setAnimationDuration:0.2];          //设置动画所需时间
    vv.frame=CGRectMake(0, 500, 320, 222);
    [UIView commitAnimations];
    //退出键盘
    [tf resignFirstResponder];
}
```

delet:方法，实现单击"—"按钮，将子标题去除的功能。程序代码如下：

```
- (IBAction)delet:(id)sender {
    tf.text = @"";
    self.subtitle = nil;
}
```

【代码解析】

由于本实例中的代码和方法非常的多，为了方便读者的阅读，笔者绘制了一些执行流程图，如图 8.6～8.7 所示。其中，单击运行按钮后一直到在模拟器上显示具有导航栏的界面，需要使用 viewDidLoad（Navigation.m）、addTitleViewToViewController:、initWithFrame:、setupContentView、title、setTitle:、updateTitleTo:、setNavigationBarTitle:、viewDidLoad（ViewController.m）、drawRect:和 drawContent:方法共同实现。它们的执行流程如图 8.6 所示。

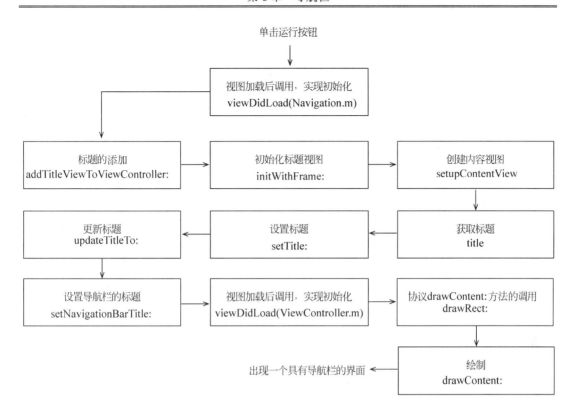

图 8.6 实例 50 程序执行流程 1

这时显示在界面的导航栏是没有标题的,需要添加,使用 add:、update:、setSubtitle:、updateSubtitleTo:、setNavigationBarSubtitle:和 drawContent:方法共同实现。它们的执行流程如图 8.7 所示。

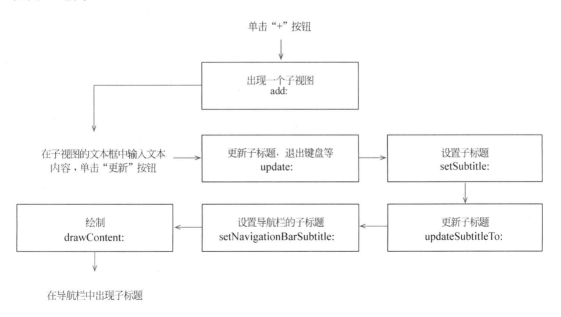

图 8.7 实例 50 程序执行流程 2

实例 51 上下滑动的导航栏

【实例描述】

本实例实现的功能是让导航栏可以实现上下滑动的效果。当鼠标向下滚动时,导航栏逐渐隐藏;当鼠标向上滑动时,导航栏逐渐显示,运行效果如图 8.8 所示。

图 8.8 实例 51 运行效果

【实现过程】

当鼠标向下滚动时,导航栏隐藏;当鼠标向上滚动时,导航栏显示。具体的实现步骤如下。

(1)创建一个项目,命名为"实现可以上下滑动的导航栏"。

(2)打开 ViewController.h 文件,编写代码,实现实例变量和属性的声明。程序代码如下:

```
#import <UIKit/UIKit.h>
@interface ViewController : UIViewController<UIScrollViewDelegate>{
    CGFloat lastOffsetY;
    BOOL isDecelerating;           //声明布尔类型的变量
}
@property (nonatomic, strong) IBOutlet UIScrollView *scrollForHideNavigation;
@end
```

(3)打开 ViewController.m 文件,编写代码,实现导航栏的上下滚动。使用的方法如表 8-5 所示。

表 8-5　ViewController.m文件中方法总结

方　　法	功　　能
scrollViewWillBeginDecelerating:	滑动将要减速调用
scrollViewDidEndDecelerating:	用户拖动，之后继续滚动调用
scrollViewDidScroll:	实现滚动

这里需要讲解几个重要的方法（其他方法请读者参考源代码）。scrollViewDidScroll:方法实现滚动。程序代码如下：

```
- (void)scrollViewDidScroll:(UIScrollView *)scrollView {
  if(self.scrollForHideNavigation != scrollView)
     return;
  if(scrollView.frame.size.height >= scrollView.contentSize.height)
     return;
  if(scrollView.contentOffset.y > -self.navigationController.navigation
  Bar.frame.size.height && scrollView.contentOffset.y < 0)
     //设置滚动视图中内容视图所在的坐标点
     scrollView.contentInset = UIEdgeInsetsMake(-scrollView.content
     Offset.y, 0, 0, 0);
  else if(scrollView.contentOffset.y >= 0)
     //设置滚动视图中内容视图所在的坐标点
     scrollView.contentInset = UIEdgeInsetsMake(0, 0, 0, 0);
if(lastOffsetY < scrollView.contentOffset.y && scrollView.contentOffset.y
>= -self.navigationController.navigationBar.frame.size.height){//moving up
   //判断导航栏是否隐藏
      if(self.navigationController.navigationBar.frame.size.height + self.
      navigationController.navigationBar.frame.origin.y > 0){
         //不隐藏
         float newY = self.navigationController.navigationBar.frame.origin.y -
         (scrollView.contentOffset.y - lastOffsetY);
         if(newY < -self.navigationController.navigationBar.frame.size.height)
            newY = -self.navigationController.navigationBar.frame.size.
            height;                       //获取导航栏的高度
         self.navigationController.navigationBar.frame = CGRectMake
         (self.navigationController.navigationBar.frame.origin.x,newY,
         self.navigationController.navigationBar.frame.size.width,self.
         navigationController.navigationBar.frame.size.height);//设置框架
      }
   }else if(self.navigationController.navigationBar.frame.origin.y
   <[UIApplication sharedApplication].statusBarFrame.size.height && (self.
   scrollForHideNavigation.contentSize.height > self.scrollForHideNavigation.
   contentOffset.y + self.scrollForHideNavigation.frame.size.height)){
         //不显示
         float newY = self.navigationController.navigationBar.frame.
         origin.y + (lastOffsetY - scrollView.contentOffset.y);
         if(newY > [UIApplication sharedApplication].statusBarFrame.size.
         height)
            //获取状态栏的高度
            newY = [UIApplication sharedApplication].statusBarFrame.size.
            height;
         self.navigationController.navigationBar.frame = CGRectMake
         (self.navigationController.navigationBar.frame.origin.x,newY,
```

```
                self.navigationController.navigationBar.frame.size.width,self.
                navigationController. navigationBar.frame.size.height);
                                                                            //设置框架
    }
    lastOffsetY = scrollView.contentOffset.y;
}
```

（4）创建一个基于 ViewController 类的 SecondViewController 类。

（5）打开 SecondViewController.h 文件，编写代码，实现属性的声明。程序代码如下：

```
#import "ViewController.h"
@interface SecondViewController : ViewController<UITableViewDataSource,
UITableViewDelegate>
@property (nonatomic, weak) IBOutlet UITableView *table;
@end
```

（6）打开 Main.storyboard 文件，添加 Navigation Controller 导航控制器到画布中。将此控制器关联的根视图设置为 View Controller 视图控制器。对视图控制器的设计界面进行设计。效果如图 8.9 所示。

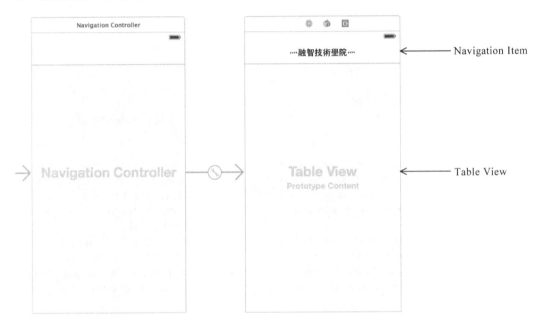

图 8.9　实例 51 视图控制器的设计界面

需要添加的视图、控件以及对它们的设置如表 8-6 所示。

表 8-6　实例 51 视图、控件设置

视图、控件	属性设置	其他
Navigation Item	Title：▱▱融智技術堡院▱▱	
Table View		dataSource 与 SecondViewController 关联 delegate 与 SecondViewController 关联 与插座变量 scrollForHideNavigation 关联 与插座变量 table 关联

（7）打开 SecondViewController.m 文件，编写代码，实现表视图内容的填充功能。程序代码如下：

```
- (void)viewDidLoad
{
    [super viewDidLoad];
    // Do any additional setup after loading the view.
    self.navigationController.navigationBar.barStyle = UIBarStyleBlack
    Translucent;   //设置导航栏的风格
}
//获取行数
- (NSInteger)tableView:(UITableView *)tableView numberOfRowsInSection:
(NSInteger)section{
    return 100;
}
//填充内容
- (UITableViewCell *)tableView:(UITableView *)tableView cellForRowAtIndex
Path:(NSIndexPath *)indexPath{
    UITableViewCell *cell = [tableView dequeueReusableCellWithIdentifier:
    @"cell"];
    if(!cell) cell = [[UITableViewCell alloc] initWithStyle:UITableView
    CellStyleDefault reuseIdentifier:@"cell"];
    cell.textLabel.text = [NSString stringWithFormat:@"第%d行", indexPath.
    row];                                                        //设置文本内容
    return cell;
}
```

【代码解析】

本实例关键功能是导航栏的上下滚动。下面就是这个知识点的详细讲解。

要实现导航栏的上下滚动，只需要使用 UIView 的 frame 属性进行设置，其语法形式如下：

```
@property(nonatomic) CGRect frame;
```

在代码中，使用了 frame 属性为导航栏的框架进行了设置，从而实现了上下滚动的效果，代码如下：

```
self.navigationController.navigationBar.frame = CGRectMake(self.
navigationController.navigationBar.
frame.origin.x,newY,self.navigationController.navigationBar.frame.size.
width,self.navigationController.navigationBar.frame.size.height);
```

实例 52 具有下拉菜单的导航栏

【实例描述】

本实例实现的功能是为导航栏添加一个下拉菜单。当单击导航栏上的标题或按钮周围，会显示一个下拉菜单；当再次单击导航栏上的标题或按钮周围，显示的下拉菜单就会隐藏。运行效果如图 8.10 所示。

图 8.10　实例 52 运行效果

【实现过程】

（1）创建一个项目，命名为"具有下拉菜单的导航栏"。

（2）创建一个 UIColor 的分类 Color 类。

（3）打开 UIColor+Color.h 文件，编写代码，实现方法的声明。程序代码如下：

```
#import <UIKit/UIKit.h>
@interface UIColor (Color)
+ (UIColor *)color:(UIColor *)color_ withAlpha:(float)alpha_;
@end
```

（4）打开 UIColor+Color.m 文件，编写代码，对相应的方法进行实现。程序代码如下：

```
+ (UIColor *)color:(UIColor *)color_ withAlpha:(float)alpha_
{
    UIColor *uicolor = color_;
    CGColorRef colorRef = [uicolor CGColor];
    int numComponents = CGColorGetNumberOfComponents(colorRef);
                                                    //获取颜色分量的个数
    CGFloat red = 0.0;
    CGFloat green = 0.0;
    CGFloat blue = 0.0;
    CGFloat alpha = 0.0;
    //判断颜色分量的个数是否为 4
    if (numComponents == 4)
    {
        const CGFloat *components = CGColorGetComponents(colorRef);
                                                    //获取颜色分量的个数
        red = components[0];
        green = components[1];
        blue = components[2];
        alpha = components[3];
    }
    return [UIColor colorWithRed:red green:green blue:blue alpha:alpha_];
        //返回颜色
}
```

（5）创建一个基于 NSObject 类的 MenuConfiguration 类。

(6) 打开 MenuConfiguration.h 文件，编写代码，实现方法的声明。程序代码如下：

```
#import <Foundation/Foundation.h>
@interface MenuConfiguration : NSObject
+ (float)menuWidth;
+ (float)itemCellHeight;
……
+ (UIColor *)selectionColor;
@end
```

(7) 打开 MenuConfiguration.m 文件，编写代码，对声明的方法进行实现。这些方法以及实现的功能如表 8-7 所示。

表8-7 MenuConfiguration.m文件中方法总结

方 法	功 能
menuWidth	获取下拉菜单的宽度
itemCellHeight	获取菜单中菜单项的高度
animationDuration	获取动画持续的时间
backgroundAlpha	获取背景的透明度
menuAlpha	获取菜单的透明度
bounceOffset	反弹的值
arrowImage	获取图片
arrowPadding	标题和三角的距离
itemsColor	下拉菜单中菜单项的背景颜色
mainColor	获取界面的颜色
selectionSpeed	选择的速度
itemTextColor	菜单项的文本颜色
selectionColor	选择的颜色

(8) 创建一个基于 UIView 类的 CellSelection 类。

(9) 打开 CellSelection.h 文件，编写代码，实现属性和方法的声明。程序代码如下：

```
#import <UIKit/UIKit.h>
@interface CellSelection : UIView
@property (nonatomic, strong) UIColor *baseColor;
- (id)initWithFrame:(CGRect)frame andColor:(UIColor *)baseColor_;
@end
```

(10) 打开 CellSelection.m 文件，编写代码，实现菜单项选择状态的初始化以及绘制菜单项的选择状态。其中，选择状态的绘制，使用 drawRect:方法实现。程序代码如下：

```
- (void)drawRect:(CGRect)rect
{
    [super drawRect:rect];
    CGFloat hue;
    CGFloat saturation;
    CGFloat brightness;
    CGFloat alpha;
    if([self.baseColor getHue:&hue saturation:&saturation brightness:&brightness alpha:&alpha]){
        brightness -= 0.35;
    }
    //创建颜色对象
```

```
UIColor * highColor =     [UIColor redColor];
UIColor * lowColor = [UIColor colorWithHue:hue saturation:saturation
brightness:brightness alpha:alpha];
CAGradientLayer * gradient = [CAGradientLayer layer];
[gradient setFrame:[self bounds]];                       //设置框架
[gradient setColors:[NSArray arrayWithObjects:(id)[highColor CGColor],
(id)[lowColor CGColor], nil]];
[[self layer] addSublayer:gradient];                     //为图层添加渐变
[self setNeedsDisplay];
}
```

（11）创建一个基于 UITableViewCell 类的 MenuCell 类。

（12）打开 MenuCell.h 文件，编写代码，实现头文件、属性和方法的声明。程序代码如下：

```
#import <UIKit/UIKit.h>
//头文件
#import "CellSelection.h"
#import "MenuConfiguration.h"
#import "UIColor+Color.h"
@interface MenuCell : UITableViewCell
@property (nonatomic, strong) CellSelection *cellSelection;          //属性
- (void)setSelected:(BOOL)selected withCompletionBlock:(void (^)())completion;
                                                                     //方法
@end
```

（13）打开 MenuCell.m 文件，编写代码，实现绘制菜单中的菜单项，使用的代码可以分为两部分：初始化菜单项和设置选择。其中，实现菜单项的初始化，需要使用 initWithStyle:reuseIdentifier:方法实现。程序代码如下：

```
- (id)initWithStyle:(UITableViewCellStyle)style reuseIdentifier:(NSString
*)reuseIdentifier
{
    self = [super initWithStyle:style reuseIdentifier:reuseIdentifier];
    if (self) {
        self.contentView.backgroundColor = [UIColor color:[MenuConfiguration
        itemsColor] withAlpha:[MenuConfiguration menuAlpha]];
        //设置标签
        self.textLabel.textColor = [MenuConfiguration itemTextColor];
                                                            //设置文本颜色
        self.textLabel.textAlignment = NSTextAlignmentCenter;
        self.textLabel.shadowColor = [UIColor darkGrayColor];
        self.textLabel.shadowOffset = CGSizeMake(0.0, -1.0);
        self.selectionStyle = UITableViewCellEditingStyleNone;
                                                            //设置选择后的风格
        //创建并设置菜单项的选择状态对象
        self.cellSelection = [[CellSelection alloc] initWithFrame:self.
        bounds andColor:[MenuConfiguration selectionColor]];
        [self.cellSelection.layer setCornerRadius:6.0]; //设置圆角半径
        [self.cellSelection.layer setMasksToBounds:YES];
        self.cellSelection.alpha = 0.0;
        [self.contentView insertSubview:self.cellSelection belowSubview:self.
        textLabel];
    }
    return self;
}
```

使用 setSelected:withCompletionBlock:方法实现对选择的设置。程序代码如下：

```
- (void)setSelected:(BOOL)selected withCompletionBlock:(void (^)())completion
{
    float alpha = 0.0;
    //判断是否选择
    if (selected) {
    //选择
        alpha = 1.0;
    } else {
    //没有选择
        alpha = 0.0;
    }
    [UIView animateWithDuration:[MenuConfiguration selectionSpeed] animations:^{
        self.cellSelection.alpha = alpha;                 //设置透明度
    } completion:^(BOOL finished) {
        completion();
    }];
}
```

(14)创建一个基于 UIView 类的 MenuTable 类。

(15)打开 MenuTable.h 文件，编写代码，实现头文件、协议、实例变量、对象、属性和方法的声明。程序代码如下：

```
#import <UIKit/UIKit.h>
//头文件
#import "MenuCell.h"
#import "MenuConfiguration.h"
#import "UIColor+Color.h"
#import "CellSelection.h"
//协议
@protocol MenuDelegate <NSObject>
- (void)didBackgroundTap;
- (void)didSelectItemAtIndex:(NSUInteger)index;
@end
@interface MenuTable : UIView <UITableViewDataSource,UITableViewDelegate>{
    //实例变量
    CGRect endFrame;
    CGRect startFrame;
    NSIndexPath *currentIndexPath;            //对象
}
//属性
@property (nonatomic, weak) id <MenuDelegate> menuDelegate;
@property (nonatomic, strong) UITableView *table;
@property (nonatomic, strong) NSArray *items;
//方法
- (id)initWithFrame:(CGRect)frame items:(NSArray *)items;
- (void)show;                                  //显示
- (void)hide;
@end
```

(16)打开 MenuTable.m 文件，编写代码，实现菜单效果。实现的方法如表 8-8 所示。

表 8-8　MenuTable.m 文件中方法总结

方　　法	功　　能
initWithFrame:items:	菜单列表的初始化
Show	显示菜单
Hide	隐藏菜单

续表

方　　法	功　　能
bounceAnimationDuration	获取反弹动画持续时间
addFooter	添加页脚
removeFooter	移除页脚
onBackgroundTap:	轻拍背景
numberOfSectionsInTableView:	获取表视图的块
tableView:heightForRowAtIndexPath:	获取表中每行的高度
tableView: numberOfRowsInSection:	获取行数
tableView: cellForRowAtIndexPath:	获取某一行的数据
tableView: didSelectRowAtIndexPath:	实现选择行

要实现菜单效果，首先要对菜单内容填充、对菜单初始化、实现菜单的显示隐藏以及实现菜单中菜单项的选择。这里需要讲解几个重要的方法（其他方法请读者参考源代码）。其中，要实现菜单内容的填充，需要使用 numberOfSectionsInTableView:、tableView: numberOfRowsInSection:和 tableView: cellForRowAtIndexPath:方法实现。程序代码如下：

```objc
//获取表视图的块
- (NSInteger)numberOfSectionsInTableView:(UITableView *)tableView
{
    return 1;
}
//获取行数
- (NSInteger)tableView:(UITableView *)tableView numberOfRowsInSection:
(NSInteger)section
{
    return self.items.count;
}
//获取某一行的数据
- (UITableViewCell *)tableView:(UITableView *)tableView cellForRowAtIndexPath:
(NSIndexPath *)indexPath
{
    static NSString *CellIdentifier = @"Cell";
    MenuCell *cell = (MenuCell *)[tableView dequeueReusableCellWith
    Identifier:@"Cell"];
    //判断cell是否为空
    if (cell == nil) {
        cell = [[MenuCell alloc] initWithStyle:UITableViewCellStyleDefault
        reuseIdentifier:CellIdentifier];
    }
    cell.textLabel.text = [self.items objectAtIndex:indexPath.row];
                                                          //设置文本内容
    return cell;
}
```

菜单列表的初始化，使用的是 initWithFrame:方法。程序代码如下：

```objc
- (id)initWithFrame:(CGRect)frame items:(NSArray *)items
{
    self = [super initWithFrame:frame];
    if (self) {
        self.items = [NSArray arrayWithArray:items];
        //设置背景颜色
        self.layer.backgroundColor = [UIColor color:[MenuConfiguration
```

```
        mainColor] withAlpha:0.0].CGColor;
        self.clipsToBounds = YES;
        endFrame = self.bounds;
        startFrame = endFrame;
        startFrame.origin.y -= self.items.count*[MenuConfiguration item
CellHeight];
        //创建并设置表视图对象
        self.table = [[UITableView alloc] initWithFrame:startFrame style:UITable
ViewStylePlain];
        self.table.delegate = self;
        self.table.dataSource = self;
        self.table.backgroundColor = [UIColor clearColor];
        self.table.separatorStyle = UITableViewCellSeparatorStyleNone;
        //创建并设置空白视图对象
        UIView *header = [[UIView alloc] initWithFrame:CGRectMake(0.0f, 0.0f
        - self.table.bounds.size.height,[MenuConfiguration menuWidth],self.
        table.bounds.size.height)];
        header.backgroundColor = [UIColor color:[MenuConfiguration itemsColor]
        withAlpha:[MenuConfiguration menuAlpha]];
        [self.table addSubview:header];
    }
    return self;
}
```

菜单的显示与隐藏，需要使用 show、hide、bounceAnimationDuration、addFooter、removeFooter 和 onBackgroundTap:方法实现。其中，show 方法实现菜单的显示。程序代码如下：

```
- (void)show
{
    [self addSubview:self.table];
    if (!self.table.tableFooterView) {
        [self addFooter];
    }
    //实现显示菜单
    [UIView animateWithDuration:[MenuConfiguration animationDuration]
    animations:^{
        self.layer.backgroundColor = [UIColor color:[MenuConfiguration
        mainColor] withAlpha:[MenuConfiguration backgroundAlpha]].CGColor;
                                                       //设置图层的背景颜色
        self.table.frame = endFrame;                   //设置框架
        self.table.contentOffset = CGPointMake(0, [MenuConfiguration
        bounceOffset]);
    } completion:^(BOOL finished) {
        [UIView animateWithDuration:[self bounceAnimationDuration]
        animations:^{
            self.table.contentOffset = CGPointMake(0, 0);//设置内容的偏移量
        }];
    }];
}
```

hide 方法实现菜单的隐藏。程序代码如下：

```
- (void)hide
{
    [UIView animateWithDuration:[self bounceAnimationDuration] animations:^{
        self.table.contentOffset = CGPointMake(0, [MenuConfiguration
        bounceOffset]);
    } completion:^(BOOL finished) {
```

```
    [UIView animateWithDuration:[MenuConfiguration animationDuration]
    animations:^{
        self.layer.backgroundColor = [UIColor color:[MenuConfiguration
        mainColor] withAlpha:0.0].CGColor;              //设置背景颜色
        self.table.frame = startFrame;
    } completion:^(BOOL finished) {
        MenuCell *cell = (MenuCell *)[self.table cellForRowAtIndexPath:
        currentIndexPath];                              //实例化
        [cell setSelected:NO withCompletionBlock:^{
          }];
         currentIndexPath = nil;
         //移除
        [self removeFooter];
        [self.table removeFromSuperview];
        [self removeFromSuperview];
    }];
  }];
}
```

tableView: didSelectRowAtIndexPath:方法实现表视图的行的选择。程序代码如下：

```
-   (void)tableView:(UITableView  *)tableView  didSelectRowAtIndexPath:
(NSIndexPath *)indexPath
{
    currentIndexPath = indexPath;           //设置当前的索引
    MenuCell *cell = (MenuCell *)[tableView cellForRowAtIndexPath:
    indexPath];
    [cell setSelected:YES withCompletionBlock:^{
       [self.menuDelegate didSelectItemAtIndex:indexPath.row];
                                            //调用 didSelectItemAtIndex:方法
    }];
}
```

（17）创建一个基于 UIControl 类的 MenuButton 类。

（18）打开 MenuButton.h 文件，编写代码，实现头文件、属性和方法的声明。程序代码如下：

```
#import <UIKit/UIKit.h>
#import "MenuConfiguration.h"                   //头文件
@interface MenuButton : UIControl
//属性
@property (nonatomic, unsafe_unretained) BOOL isActive;
……
@property (nonatomic, strong) UIImageView *arrow;
- (UIImageView *)defaultGradient;               //方法
@end
```

（19）打开 MenuButton.m 文件，编写代码，实现菜单的按钮效果。使用的方法如表 8-9 所示。

表 8-9　MenuButton.m 文件中方法总结

方　　法	功　　能
initWithFrame:	初始化菜单上的按钮（此按钮用于打开和关闭菜单）
defaultGradient	获取默认的渐变
layoutSubviews	布局
beginTrackingWithTouch:withEvent:	开始触摸

续表

方法	功能
continueTrackingWithTouch:withEvent:	继续触摸
endTrackingWithTouch:withEvent:	结束触摸
cancelTrackingWithEvent:	取消触摸
drawRect:	绘制一个图形
newSpotlightGradient	获取新的渐变
setSpotlightGradientRef:	设置新的渐变

要实现菜单的按钮效果，需要 3 部分组成：设置菜单按钮、绘制触摸时的菜单效果以及实现触摸。其中设置菜单按钮，需要使用 initWithFrame:、defaultGradient、layoutSubviews、newSpotlightGradient 和 setSpotlightGradientRef:方法实现。initWithFrame:方法实现菜单上按钮的初始化。程序代码如下：

```
- (id)initWithFrame:(CGRect)frame
{
    self = [super initWithFrame:frame];
    if (self) {
        if ([self defaultGradient]) {
        } else {
            [self setSpotlightCenter:CGPointMake(frame.size.width/2, frame.
            size.height*(-1)+10)];
            [self setBackgroundColor:[UIColor clearColor]];  //设置背景颜色
            [self setSpotlightStartRadius:0];
            [self setSpotlightEndRadius:frame.size.width/2];
        }
        frame.origin.y -= 2.0;
        //设置标题
        self.title = [[UILabel alloc] initWithFrame:frame];
        self.title.textAlignment = NSTextAlignmentCenter;     //设置对齐方法
        self.title.backgroundColor = [UIColor clearColor];
        self.title.textColor = [UIColor blackColor];           //设置文本颜色
        self.title.font = [UIFont boldSystemFontOfSize:20.0];
        [self addSubview:self.title];
        //设置图片
        self.arrow = [[UIImageView alloc] initWithImage:[MenuConfiguration
        arrowImage]];
        [self addSubview:self.arrow];
    }
    return self;
}
```

layoutSubviews 方法对子视图进行重新布局。程序代码如下：

```
- (void)layoutSubviews
{
    [self.title sizeToFit];
    self.title.center = CGPointMake(self.frame.size.width/2, (self.frame.
    size.height-2.0)/2);//设置中心位置
```

```objc
    self.arrow.center = CGPointMake(CGRectGetMaxX(self.title.frame) +
    [MenuConfiguration arrowPadding], self.frame.size.height / 2);
}
```

要实现触摸时菜单的效果，需要使用 drawRect:方法实现。程序代码如下：

```objc
- (void)drawRect:(CGRect)rect
{
    CGContextRef context = UIGraphicsGetCurrentContext();      //创建图像上下文
    CGGradientRef gradient = self.spotlightGradientRef;
    float radius = self.spotlightEndRadius;
    float startRadius = self.spotlightStartRadius;
    CGContextDrawRadialGradient (context, gradient, self.spotlightCenter,
    startRadius, self.spotlightCenter, radius, kCGGradientDrawsAfterEnd
    Location);                                                  //绘制
}
```

实现触摸，需要使用 beginTrackingWithTouch:withEvent:、continueTrackingWithTouch:withEvent:、endTrackingWithTouch:withEvent:和 cancelTrackingWithEvent:方法共同实现。程序代码如下：

```objc
//开始触摸
- (BOOL)beginTrackingWithTouch:(UITouch *)touch withEvent:(UIEvent *)event
{
    self.isActive = !self.isActive;
    CGGradientRef defaultGradientRef = [[self class] newSpotlightGradient];
                                                    //创建渐变对象
    [self setSpotlightGradientRef:defaultGradientRef];
    CGGradientRelease(defaultGradientRef);          //释放渐变对象
    return YES;
}
//继续触摸
- (BOOL)continueTrackingWithTouch:(UITouch *)touch withEvent:(UIEvent *)event
{
    return YES;
}
//结束触摸
- (void)endTrackingWithTouch:(UITouch *)touch withEvent:(UIEvent *)event
{
    self.spotlightGradientRef = nil;
}
//取消触摸
- (void)cancelTrackingWithEvent:(UIEvent *)event
{
    self.spotlightGradientRef = nil;
}
```

（20）创建一个基于 UIView 类的 Navigation 类。

（21）打开 Navigation.h 文件，编写代码，实现头文件、协议、属性和方法的声明。程序代码如下：

```
#import <UIKit/UIKit.h>
//头文件
#import "MenuTable.h"
#import "MenuButton.h"
//协议
@protocol NavigationMenuDelegate <NSObject>
- (void)didSelectItemAtIndex:(NSUInteger)index;
@end
@interface Navigation : UIView<MenuDelegate>
//属性
@property (nonatomic, weak) id <NavigationMenuDelegate> delegate;
……
@property (nonatomic, strong) UIView *menuContainer;
//方法
- (id)initWithFrame:(CGRect)frame title:(NSString *)title;
- (void)displayMenuInView:(UIView *)view;
@end
```

(22)打开 Navigation.m 文件,编写代码,实现具有下拉菜单的导航栏效果。使用的方法如表 8-10 所示。

表 8-10　Navigation.m文件中方法总结

方　　法	功　　能
initWithFrame:title:	初始化导航栏
displayMenuInView:	显示视图中的菜单
onHandleMenuTap:	轻拍导航栏上标题的附近
onShowMenu	显示菜单
onHideMenu	隐藏菜单
rotateArrow:	旋转导航栏上的三角形图片
didSelectItemAtIndex:	选择条目
didBackgroundTap	轻拍背景

其中,initWithFrame:title:方法对导航栏进行初始化。程序代码如下:

```
- (id)initWithFrame:(CGRect)frame title:(NSString *)title
{
    self = [super initWithFrame:frame];
    if (self) {
        frame.origin.y += 1.0;
        self.menuButton = [[MenuButton alloc] initWithFrame:frame];
        self.menuButton.title.text = title;
        [self.menuButton addTarget:self action:@selector(onHandleMenuTap:)
             forControlEvents:UIControlEventTouchUpInside];           //添加动作
        [self addSubview:self.menuButton];
    }
    return self;
}
```

onHandleMenuTap:方法实现在导航栏标题的附近轻拍。程序代码如下:

```
- (void)onHandleMenuTap:(id)sender
{
    if (self.menuButton.isActive) {
        [self onShowMenu];         //显示
    } else {
        [self onHideMenu];         //隐藏
    }
}
```

rotateArrow:方法实现导航栏上三角形图片的旋转动画。程序代码如下:

```
- (void)rotateArrow:(float)degrees
{
    [UIView animateWithDuration:[MenuConfiguration animationDuration]
    delay:0 options:UIViewAnimationOptionAllowUserInteraction animations:^{
        self.menuButton.arrow.layer.transform = CATransform3DMakeRotation
        (degrees, 0, 0, 1);        //旋转
    } completion:NULL];
}
```

（23）打开 Main.storyboard 文件，添加 NavigationController 导航控制器到画布中，将此控件关联的根视图设置为 View Controller 视图控制器。

（24）在 ViewController.h 文件中，进行头文件的声明以及协议的遵守。

（25）打开 ViewController.m 文件，实现显示一个具有菜单的导航栏，单击菜单中的任意一项，背景颜色会发生相应的改变。程序代码如下：

```
//视图加载，将创建的具有菜单的导航栏对象添加到当前视图中
- (void)viewDidLoad
{
    [super viewDidLoad];
    // Do any additional setup after loading the view, typically from a nib.
    if (self.navigationItem) {
        CGRect frame = CGRectMake(0.0, 0.0, 200.0, self.navigationController.
        navigationBar.bounds.size.height);          //设置框架
        Navigation *menu = [[Navigation alloc] initWithFrame:frame title:@"
        菜单"];                                      //创建导航对象
        [menu displayMenuInView:self.view];
        menu.items = @[@"Cyan Color", @"Red Color", @"Black Color", @"Orange
        Color", @"Blue Color", @"Green Color"];    //菜单中的菜单项
        menu.delegate = self;
        self.navigationItem.titleView = menu;
    }
    self.view.backgroundColor=[UIColor cyanColor];
}
//选择菜单项中的任意一项后，实现的功能
- (void)didSelectItemAtIndex:(NSUInteger)index
{
    //判断索引是否为 0
    if (index==0) {
        self.view.backgroundColor=[UIColor cyanColor];
    }else if (index==1){                            //判断索引是否为 1
```

```
            self.view.backgroundColor=[UIColor redColor];
        }else if (index==2){                           //判断索引是否为 2
            self.view.backgroundColor=[UIColor blackColor];
        }else if (index==3){                           //判断索引是否为 3
            self.view.backgroundColor=[UIColor orangeColor];
        }else if (index==4){                           //判断索引是否为 4
            self.view.backgroundColor=[UIColor blueColor];
        }else{
            self.view.backgroundColor=[UIColor greenColor];
        }
}
```

【代码解析】

由于本实例中的代码和方法非常的多，为了方便读者的阅读，笔者绘制了一些执行流程图如图 8.11～8.13 所示。其中，单击运行按钮后一直到在模拟器上显示具有导航栏的界面，需要使用 viewDidLoad、initWithFrame:title:、initWithFrame:、defaultGradient、arrowImage、displayMenuInView:、layoutSubviews、arrowPadding 和 drawRect:方法共同实现。它们的执行流程如图 8.11 所示。

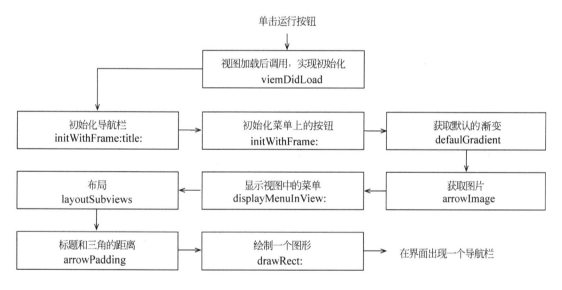

图 8.11　实例 52 程序执行流程 1

这时，在界面的导航栏是不显示下拉菜单的。如果想要显示下拉菜单，需要轻拍导航栏中间的标题或按钮附近位置，将其打开，需要使用 beginTrackingWithTouch:withEvent:、newSpotlightGradient、setSpotlightGradientRef:、endTrackingWithTouch:withEvent:、onHandleMenuTap:、onShowMenu、initWithFrame:items:、mainColor、color:withAlpha:、itemCellHeight、menuWidth、itemsColor、menuAlpha、rotateArrow:、animationDuration、show、addFooter、menuWidth、numberOfSectionsInTableView:、tableView: numberOfRowsInSection:、tableView:heightForRowAtIndexPath:、backgroundAlpha、bounceOffset、layoutSubviews、arrowPadding、tableView: cellForRowAtIndexPath:、initWithStyle:reuseIdentifier:、itemTextColor、selectionColor、initWithFrame:andColor:、drawRect:（MenuButton.m）、drawRect:（CellSelection.m）和 bounceAnimationDuration 方法共同实现。它们的执行流程如图 8.12 所示。

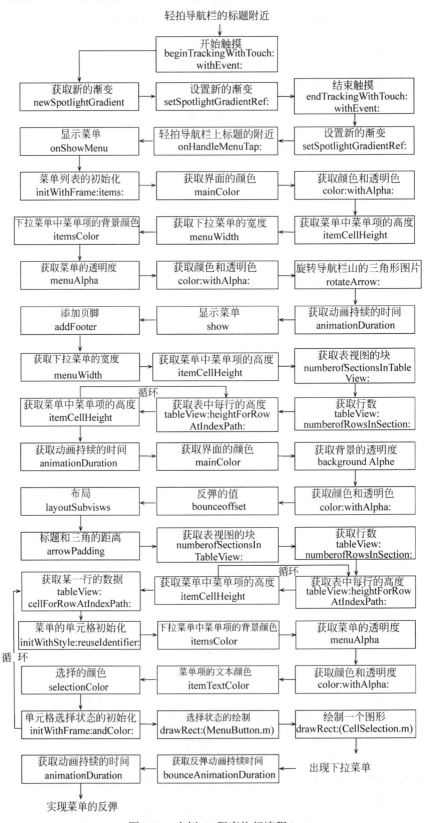

图 8.12 实例 52 程序执行流程 2

第 8 章 导航栏

如果想要将打开的下拉菜单进行关闭，可以有 3 种方法：第一种是轻拍导航栏中间的标题或按钮；第二种是轻拍背景实现关闭；最后一种是在下拉菜单中进行菜单项的选择。以下以最后一种关闭菜单的方式为例，实现程序的执行流程如图 8.13 所示。

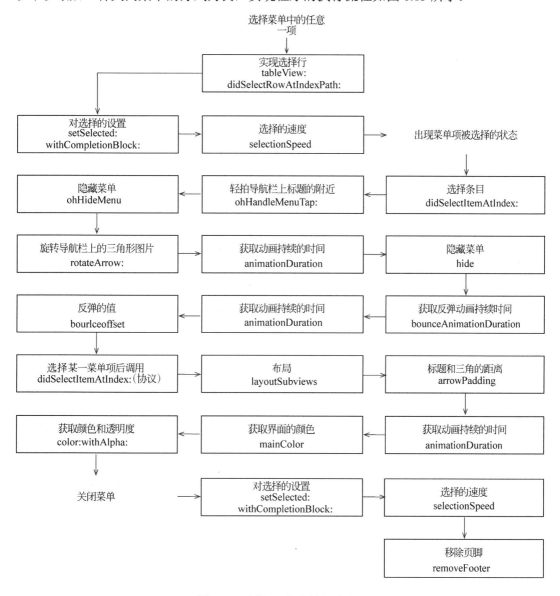

图 8.13 实例 52 程序执行流程 3

实例 53 具有页面控件的导航栏

【实例描述】

本实例实现的功能是为导航栏添加一个自制的页面控件。当用户滑动导航栏上的页面控件，就会出现相应的内容。运行效果如图 8.14 所示。

图 8.14 实例 53 运行效果

【实现过程】

（1）创建一个项目，命名为"具有页面控件的导航栏"。

（2）创建一个基于 UIView 的 Navigation 类。

（3）打开 Navigation.h 文件，编写代码，实现协议、属性和方法的声明。程序代码如下：

```
#import <UIKit/UIKit.h>
//协议
@protocol NavigationDelegate
- (void)titleViewChange:(int)page;
@end
@interface Navigation : UIView
//属性
@property(weak)NSObject<NavigationDelegate>* delegate;
……
@property(strong)UIScrollView* scrollView;
@property(strong)NSArray* items;
//方法
-(void)scrollForward;
-(void)scrollBackward;
- (id)initWithFrame:(CGRect)frame titleItems:(NSArray*)titleItems;
@end
```

（4）打开 Navigation.m 文件，编写代码，实现具有页面控件的导航栏。使用的方法如表 8-11 所示。

第 8 章 导航栏

表 8-11 Navigation.m 文件中方法总结

方　　法	功　　能
initWithFrame: titleItems:	导航栏的初始化
scrollForward	向前滚动
scrollBackward	向后滚动
drawRect:	绘制页面控件

要实现具有页面控件的导航栏，需要实现两个部分：绘制具有页面控件的导航栏以及实现滚动。其中，绘制具有页面控件的导航栏，需要使用 initWithFrame: titleItems:和 drawRect:方法实现。initWithFrame: titleItems:方法对框架以及标题条目进行初始化。程序代码如下：

```
- (id)initWithFrame:(CGRect)frame titleItems:(NSArray*)titleItems
{
    self = [super initWithFrame:frame];
    if (self) {
        [self setOpaque:NO];
        currentPage = 0;
        items = titleItems;
        pageColor = [UIColor lightGrayColor];
        currentPageColor = [UIColor grayColor];
        //创建并设置滚动视图对象
        scrollView = [[UIScrollView alloc] initWithFrame:frame];
        [scrollView setContentSize:CGSizeMake([items count]*CGRectGet
        Width(frame),40)];
        [scrollView setBackgroundColor:[UIColor clearColor]];//设置背景颜色
        [scrollView setPagingEnabled:YES];
        [scrollView setShowsHorizontalScrollIndicator:NO];
                                            //不显示水平的滚动指示器
        [scrollView setBounces:NO];
        [scrollView setScrollEnabled:NO];
        for (int i = 0; i < [items count]; i++) {
            //创建并设置按钮对象
            UIButton* button = [[UIButton alloc] initWithFrame:CGRectMake
            (i*CGRectGetWidth(scrollView.frame), 0, 120, 40)];
            [button setBackgroundColor:[UIColor clearColor]];//设置背景颜色
            //设置标题的背景颜色
            [button setTitleColor:[UIColor blackColor] forState:UIControl
            StateNormal];[button.titleLabel setFont:[UIFont fontWithName:
            @"HelveticaNeue" size:20]];
            [button.titleLabel setAdjustsFontSizeToFitWidth:YES];
            [button.titleLabel setMinimumScaleFactor:3];
            [button.titleLabel setTextAlignment:NSTextAlignmentCenter];
                                                    //设置对齐方式
            [button setTitle:[titleItems objectAtIndex:i] forState:UIControl
            StateNormal];
            [button addTarget:self action:@selector(scrollForward) forControl
            Events: UIControlEventTouchUpInside];      //添加动作
            //创建并设置手势对象
            UISwipeGestureRecognizer *swipe = [[UISwipeGestureRecognizer
            alloc] initWithTarget:self action:@selector(scrollForward)];
```

· 329 ·

```
            swipe.direction = UISwipeGestureRecognizerDirectionLeft;
                                                    //设置滑动手势的方向
            [button addGestureRecognizer:swipe];
            UISwipeGestureRecognizer *leftswipe = [[UISwipeGesture
            Recognizer alloc] initWithTarget:self action:@selector
            (scrollBackward)];
            leftswipe.direction = UISwipeGestureRecognizerDirectionRight;
            [button addGestureRecognizer:leftswipe];    //添加手势识别器
            [scrollView addSubview:button];
        }
        [self addSubview:scrollView];
        [self setNeedsDisplay];
    }
    return self;
}
```

drawRect:方法实现对页面控件的绘制。程序代码如下：

```
- (void)drawRect:(CGRect)rect
{
    CGContextRef context = UIGraphicsGetCurrentContext();
                                                    //获取当前的图像上下文
    CGContextSaveGState(context);                   //保存
    CGContextSetAllowsAntialiasing(context, TRUE);
    CGFloat diameter = 4.0f;
    CGFloat space = 12.0f;
    CGFloat dotsWidth = [items count] * diameter + MAX(0, [self.items count]
    - 1) * space;
    CGFloat x = CGRectGetMidX(self.bounds) - dotsWidth / 2;
    CGFloat y = CGRectGetHeight( self.bounds)- diameter - 4;
    //绘制点
    for (int i = 0 ; i < [self.items count] ; i++)
    {
        CGRect dotRect = CGRectMake(x, y, diameter, diameter) ;
        //判断是否是当前页
        if (i == currentPage)
        {
            CGContextSetFillColorWithColor(context, currentPageColor.CGColor) ;
                                                    //设置颜色
            //填充指定矩形中的椭圆
            CGContextFillEllipseInRect(context, CGRectInset(dotRect, -0.5f,
            -0.5f)) ;
        }
        else
        {
            CGContextSetFillColorWithColor(context, pageColor.CGColor) ;
            //填充指定矩形中的椭圆
            CGContextFillEllipseInRect(context, CGRectInset(dotRect, -0.5f,
            -0.5f)) ;
        }
        x += diameter + space;
    }
    CGContextRestoreGState(context);
    CGContextSetFillColorWithColor(context, [UIColor clearColor].CGColor);
```

```
                                                                //填充颜色
    CGContextFillRect(context, self.bounds);
}
```

scrollForward 方法实现向前滚动。程序代码如下：

```
-(void)scrollForward {
    int nextPage =  (currentPage + 1) % [self.items count];
    //判断是否实现了 titleViewChange:方法
    if ([delegate respondsToSelector:@selector(titleViewChange:)]) {
        [delegate titleViewChange:nextPage];
    }
    CGFloat pageWidth = scrollView.bounds.size.width;   //获取滚动视图的宽度
    currentPage = nextPage;
    //设置内容视图的偏移量
    [scrollView setContentOffset:CGPointMake(nextPage*pageWidth,0) animated:YES];
    [self setNeedsDisplay];
}
```

scrollBackward 方法实现向后滚动。程序代码如下：

```
-(void)scrollBackward {
    int nextPage =  (currentPage - 1);
    if (nextPage == -1) nextPage = [self.items count]-1;
    //判断是否实现了 titleViewChange:方法
    if ([delegate respondsToSelector:@selector(titleViewChange:)]) {
        [delegate titleViewChange:nextPage];
    }
    CGFloat pageWidth = scrollView.bounds.size.width;
    currentPage = nextPage;
    //设置内容视图的偏移量
    [scrollView setContentOffset:CGPointMake(nextPage*pageWidth, 0) animated:YES];
    [self setNeedsDisplay];
}
```

（5）打开 Main.storyboard 文件，添加 Navigation Controller 导航控制器到画布中。将此控制器关联的根视图设置为 View Controller 视图控制器。将 View Controller 视图控制器的设计界面的 Class 设置为 Navigation。

（6）在 ViewController.h 文件中，实现头文件和遵守协议的声明。

（7）打开 ViewControlller.m 文件，编写代码，实现在界面显示页面控件的导航栏，并且可以实现当导航栏的标题改变后，在标签中显示相应内容。程序代码如下：

```
- (void)viewDidLoad
{
    [super viewDidLoad];
    // Do any additional setup after loading the view, typically from a nib.
    NSArray* titles = @[@"Page 1",@"Page 2",@"Page 3",@"Page 4"];
    //创建并设置具有页面控件的导航栏对象
    Navigation *view = [[Navigation alloc] initWithFrame:CGRectMake(0, 0, 120, 40) titleItems:titles];
    view.delegate = self;
    [view setCurrentPageColor:[UIColor redColor]];          //设置当前页的颜色
    [view setPageColor:[UIColor lightGrayColor]];
```

```
    self.navigationItem.titleView = view;
    //创建并设置标签栏对象
    UILabel* testLabel = [[UILabel alloc] initWithFrame:CGRectMake(0, 0, 320,
480)];
    [testLabel setTextAlignment:NSTextAlignmentCenter];//设置对齐方式
    testLabel.font=[UIFont fontWithName:@"Verdana" size:30];
    [testLabel setTag:1];                                  //设置 tag 值
    [testLabel setText:@"第 1 页"];
    [self.view addSubview:testLabel];
}
//当导航栏的标题改变后,实现在标签中显示相应内容
-(void)titleViewChange:(int)page {
    [UIView beginAnimations:@"" context:nil];
    //设置过渡动画
    [UIView setAnimationTransition:UIViewAnimationTransitionFlipFromLeft
    forView:self.view cache:YES];
    UILabel* testLabel = (UILabel*)[self.view viewWithTag:1];
    if (page == 0) {
        [testLabel setText:@"第 1 页"];                       //设置文本内容
    } else {
        [testLabel setText:[NSString stringWithFormat:@"第%d页",page+1]];
    }
    [UIView commitAnimations];
}
```

【代码解析】

本实例关键功能是页面控件的选择实现。下面就是这个知识点的详细讲解。

如果想要实现页面控件的选择功能,必须要获取当前的页面,在本实例中使用 currentpPage 表示,使用当前的页面和 drawRect:方法中的 i 进行比较(i 是一个增量,用来控制分段控件中点的个数)。比较 i 是否等于 currentPage,如果相等就绘制被选择的状态。代码如下:

```
if (i == currentPage)
{
    //如果 i 是当前的页
    CGContextSetFillColorWithColor(context, currentPageColor.CGColor) ;
    CGContextFillEllipseInRect(context, CGRectInset(dotRect, -0.5f, -0.5f)) ;
}
else
{
    //如果 i 不是当前的页
    CGContextSetFillColorWithColor(context, pageColor.CGColor) ;
    CGContextFillEllipseInRect(context, CGRectInset(dotRect, -0.5f, -0.5f)) ;
}
```

实例 54　包含多个按钮的导航栏

【实现描述】

本实例实现的功能是为导航栏添加多个按钮,当按钮很多时,可以实现滚动效果。单击按钮,会出现相应的内容。其中,用于显示相应内容的是标签。运行效果如图 8.15 所示。

第 8 章 导航栏

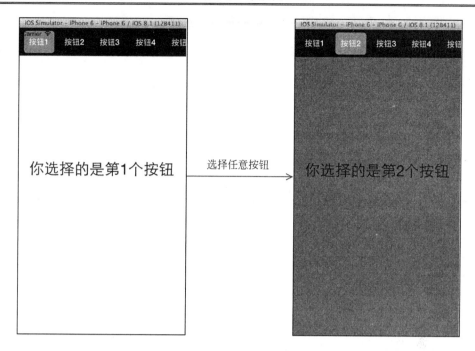

图 8.15 实例 54 运行效果

【实现过程】
（1）创建一个项目，命名为"包含多个按钮的导航栏"。
（2）创建一个基于 UIScrollView 类的 Navigation 类。
（3）打开 Navigation.h 文件，编写代码，实现协议、实例变量、对象、属性和方法的声明。程序代码如下：

```
#import <UIKit/UIKit.h>
@class Navigation;
//协议
@protocol NavigationDelegate <NSObject>
- (NSInteger) numberOfMenuItems;
@optional
- (NSString *) scrollView:(Navigation *)scrollView titleForMenuIndex:
(NSInteger) index;
- (void) scrollView:(Navigation *)scrollView menuItemSelectedAtIndex:
(NSInteger) index;
@end
@interface Navigation : UIScrollView{
    //实例变量
    NSInteger _selectedButton;
    float _horizontalPadding;
    ……
    float _btnCornerRadius;
    //对象
    UIColor *_btnColor;
    ……
    UIColor *_btnSelectedColor;
    __weak id <NavigationDelegate> _menuDelegate;
}
//属性
@property (nonatomic, weak) IBOutlet id <NavigationDelegate> menuDelegate;
```

```
@property (nonatomic) float horizontalPadding;
......
@property (nonatomic, strong) UIColor *btnSelectedColor;
//方法
- (void)selectedButton:(id)sender;
- (void)setSelectedIndex:(NSInteger)index;
@end
```

（4）打开 Navigation.m 文件，编写代码，实现包含多个按钮的导航栏效果。使用的方法如表 8-12 所示。

表 8-12　Navigation.m文件中方法总结

方　　法	功　　能
selectedButton:	选择按钮
setSelectedIndex:	设置选择的索引
selectedIndex	获取选择的按钮
initWithCoder:	初始化实例化对象
initWithFrame:	包含多个按钮导航栏的初始化
configure	设置
layoutSubviews	布局

要实现包含多个按钮的导航栏效果，需要分为两个部分：对包含多个按钮的导航栏的绘制以及实现选择功能。这里需要讲解几个重要的方法（其他方法请读者参考源代码）。其中，要实现绘制包含多个按钮的导航栏，需要使用 setSelectedIndex:、selectedIndex、initWithCoder:、initWithFrame:、configure 和 layoutSubviews 方法共同实现。Configure 方法对导航栏的一些属性内容进行设置。程序代码如下：

```
- (void)configure {
    self.showsHorizontalScrollIndicator = NO;
    self.canCancelContentTouches = YES;
    self.clipsToBounds = YES;
    self.backgroundColor = [UIColor blackColor];           //设置背景颜色
    _btnHorizontalPadding = 10.0f;
    _horizontalPadding = 10.0f;
    _verticalPadding = 5.0f;
    _btnSpacing = 10.0f;
    _btnCornerRadius = 5.0f;
    _btnFont = [UIFont boldSystemFontOfSize:15.0f];         //设置字体
    _btnFontColor = [UIColor whiteColor];
    _btnColor = [UIColor clearColor];
    _btnSelectedFontColor = [UIColor whiteColor];           //设置按钮选择的字体颜色
    _btnSelectedColor = [UIColor orangeColor];
    _btnShadowColor = [UIColor darkGrayColor];
}
```

layoutSubviews 方法实现对子视图的布局。程序代码如下：

```
- (void)layoutSubviews {
    [[self subviews] makeObjectsPerformSelector:@selector(removeFromSuperview)];
    int xPos = _horizontalPadding;
    //判断是否实现了 numberOfMenuItems 方法
    if ([_menuDelegate respondsToSelector:@selector(numberOfMenuItems)]) {
        for (int i=0; i<[_menuDelegate numberOfMenuItems]; i++) {
```

```objc
        NSString *menuItem = [_menuDelegate scrollView:self titleFor
        MenuIndex:i];
        //创建并设置按钮对象
        UIButton *customButton = [UIButton buttonWithType:UIButton
        TypeCustom];
        [customButton setTitle:menuItem forState:UIControlStateNormal];
                                                            //设置标题
        customButton.titleLabel.font = _btnFont;
        customButton.titleLabel.textAlignment = NSTextAlignmentCenter;
        customButton.titleLabel.textColor = _btnFontColor;
        customButton.backgroundColor = _btnColor;      //设置背景颜色
        customButton.clipsToBounds = YES;
        customButton.layer.cornerRadius = _btnCornerRadius;
        customButton.tag = i;                          //设置 tag
        [customButton addTarget:self action:@selector(selectedButton:)
         forControlEvents:UIControlEventTouchUpInside];  //添加动作
        NSMutableParagraphStyle *style = [[NSParagraphStyle defaultParag
        raphStyle] mutableCopy];
        [style setLineBreakMode:NSLineBreakByClipping]; //设置换行模式
        NSDictionary *a=[NSDictionary dictionaryWithObjectsAndKeys:
        customButton.titleLabel.font,NSFontAttributeName,style,
        NSParagraphStyleAttributeName, nil];
        //获取文本绘制所占据的矩形空间
        float buttonWidth=[menuItem boundingRectWithSize:CGSizeMake(9999,
        customButton.titleLabel.frame.size.height) options:0 attributes:a
        context:nil].size.width;
        customButton.frame = CGRectMake(xPos,_verticalPadding,(buttonWidth +
        (2*_btnHorizontalPadding)),self.frame.size.height-(2*_vertical
        Padding));                                     //设置框架
        xPos += customButton.frame.size.width + _btnSpacing;
        if (_selectedButton == i) {
            //设置选择的按钮
            customButton.backgroundColor = _btnSelectedColor;
                                                        //设置背景颜色
            customButton.titleLabel.textColor = _btnSelectedFontColor;
            customButton.titleLabel.shadowColor = _btnShadowColor;
                                                        //设置阴影的颜色
            customButton.titleLabel.shadowOffset = CGSizeMake(1, 1);
        }
        [self addSubview:customButton];                 //添加视图对象
        if (i == [_menuDelegate numberOfMenuItems]-1) {
            xPos += _horizontalPadding-_btnSpacing;
        }
    }
}
self.contentSize = CGSizeMake(xPos,self.frame.size.height);
}
```

要实现导航栏中按钮的选择,需要使用 selectedButton:方法。程序代码如下:

```objc
- (void) selectedButton:(id)sender {
    NSInteger index = ((UIButton *)sender).tag;    //获取 tag 值
    _selectedButton = index;
    [self setNeedsLayout];
    //判断是否实现了 scrollView:menuItemSelectedAtIndex:方法
    if ([_menuDelegate respondsToSelector:@selector(scrollView:menuItem
```

```
SelectedAtIndex:)]) {
    [_menuDelegate scrollView:self menuItemSelectedAtIndex:index];
  }
}
```

（5）打开 ViewController.h 文件，编写代码，实现头文件、对象和插座变量的声明。程序代码如下：

```
#import <UIKit/UIKit.h>
#import "Navigation.h"                    //头文件
@interface ViewController : UIViewController<NavigationDelegate>{
                                          //遵守的协议
    NSArray *array;                       //对象
    IBOutlet  UILabel *label;             //插座变量
}
@end
```

（6）打开 Main.storyboard 文件，对 View Controller 视图控制器的设计界面进行设计，效果如图 8.16 所示。

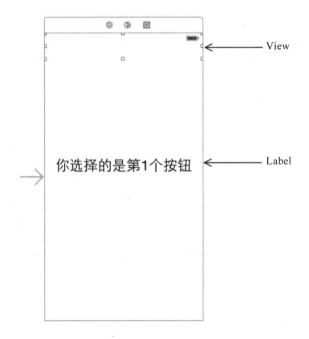

图 8.16　实例 54View Controller 视图控制器的设计界面

需要添加的视图、控件以及对它们的设置如表 8-13 所示。

表 8-13　实例 54 视图、控件设置

视图、控件	属 性 设 置	其　　他
View	Title：⌒融智技术堡院⌒	Class：Navigation
Label	Text：你选择的是第1个按钮 Font：System 29.0 Alignment：居中对齐	与插座变量label关联

（7）打开 ViewController.m 文件。编写代码，实现为导航栏的多个按钮添加标题，并且单击按钮后，会在标签中显示相应的内容。程序代码如下：

```objectivec
- (void)viewDidLoad
{
    [super viewDidLoad];
    // Do any additional setup after loading the view, typically from a nib.
    array= [NSArray arrayWithObjects:@"你选择的是第1个按钮",@"你选择的是第2个按钮",@"你选择的是第3个按钮",@"你选择的是第4个按钮",@"你选择的是第5个按钮,",@"你选择的是第6个按钮",@"你选择的是第7个按钮",@"你选择的是第8个按钮",@"你选择的是第9个按钮", nil];           //创建数组
}
//获取数组中对象的个数
- (NSInteger) numberOfMenuItems {
    return [array count];
}
//为按钮添加标题
- (NSString *) scrollView:(Navigation *)scrollView titleForMenuIndex:(NSInteger) index {
    return [NSString stringWithFormat:@"按钮%d",index+1];
}
//实现单击某一按钮后的动作
- (void) scrollView:(Navigation *)scrollView menuItemSelectedAtIndex:(NSInteger) index {
    self.view.backgroundColor = [UIColor colorWithRed:arc4random()%200/255.0 green:arc4random()%150/255.0 blue:arc4random()%100/255.0 alpha:1];
                                                                //随机产生背景色
    [label setText:[NSString stringWithFormat:@"%@",[array objectAtIndex:index]]];
}
```

【代码解析】

本实例关键功能是按钮选择后的状态改变以及背景颜色的随机显示。下面依次讲解这两个知识点。

1．按钮选择后的状态

为了将选择的按钮和没有选择的按钮区分开，可以在每次选择按钮后，都调用 layoutSubviews 方法，判断在导航栏上显示的按钮，哪一个是被选择的按钮，如果某一个是被选择的按钮，就实现如下代码：

```objectivec
if (_selectedButton == i) {
    customButton.backgroundColor = _btnSelectedColor;
    customButton.titleLabel.textColor = _btnSelectedFontColor;
    customButton.titleLabel.shadowColor = _btnShadowColor;
    customButton.titleLabel.shadowOffset = CGSizeMake(1, 1);
}
```

2．背景颜色的随机显示

当按下导航栏上的任意一个按钮后，视图的背景颜色都是不一样的，它是随机的颜色。要实现颜色的随机显示，需要使用 arc4random(void)函数实现，其语法形式如下：

```
arc4random(void);
```

在代码中，使用了arc4random(void)函数为背景产生随机颜色，代码如下：

```
self.view.backgroundColor = [UIColor colorWithRed:arc4random()%200/255.0
green:arc4random()%150/255.0 blue:arc4random()%100/255.0 alpha:1];
```

实例 55　导航栏的颜色调节

【实例描述】

本实例实现的功能是对导航栏的颜色进行调节。当用户想要调节导航栏颜色时，单击Edit按钮，在出现的视图中，滑动滑块控件。运行效果如图8.17所示。

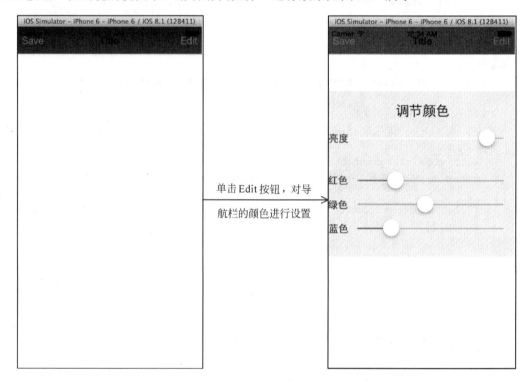

图 8.17　实例 55 运行效果

【实现过程】

当用户滑动亮度滑块时，滑块的值就会作为控制导航栏亮暗的元素。具体的实现步骤如下。

（1）创建一个项目，命名为"导航栏的颜色调节"。

（2）创建一个基于UINavigationBar类的NavigationBarTint类。

（3）在NavigationBarTint.h文件中实现属性的声明。

（4）打开NavigationBarTint.m文件，编写代码，实现导航栏的颜色控件。使用的方法如表8-14所示。

表8-14 NavigationBarTint.m文件中方法总结

方 法	功 能
initWithFrame:	导航栏的初始化
setColourAdjustFactor:	设置颜色的调节因子
adjustColourBrightness:withFactor:	使用因子调节颜色的亮暗
drawRcct:	绘制

这里需要讲解几个重要的方法（其他方法请读者参考源代码）。其中，setColourAdjustFactor：方法用来设置颜色的调整因子。程序代码如下：

```
-(void)setColourAdjustFactor:(float)colourAdjustFactor{
    _colourAdjustFactor = colourAdjustFactor;
    [self setNeedsDisplay];
}
```

adjustColourBrightness:withFactor:方法实现使用因子调节颜色的亮暗。程序代码如下：

```
- (UIColor *)adjustColourBrightness:(UIColor *)color withFactor:(double)factor
{
    CGColorRef cgColour = [color CGColor];
    CGFloat *oldComponents = (CGFloat *)CGColorGetComponents(cgColour);
    CGFloat newComponents[4];
    int numComponents = CGColorGetNumberOfComponents(cgColour);
                                                    //获取颜色组成部分的个数
    switch (numComponents)
    {
        case 2:
        {
            //设置新的颜色分量
            newComponents[0] = oldComponents[0]*factor;
            newComponents[1] = oldComponents[0]*factor;
            newComponents[2] = oldComponents[0]*factor;
            newComponents[3] = oldComponents[1];
            break;
        }
        case 4:
        {
            //设置新的颜色分量
            newComponents[0] = oldComponents[0]*factor;
            newComponents[1] = oldComponents[1]*factor;
            newComponents[2] = oldComponents[2]*factor;
            newComponents[3] = oldComponents[3];
            break;
        }
    }
    CGColorSpaceRef colorSpace = CGColorSpaceCreateDeviceRGB();
    //创建色彩空间
    CGColorRef newColor = CGColorCreate(colorSpace, newComponents);
    CGColorSpaceRelease(colorSpace);                //释放色彩空间
    UIColor *retColor = [UIColor colorWithCGColor:newColor];//创建颜色对象
    CGColorRelease(newColor);
```

```
    return retColor;
}
```

drawRect:方法实现渐变颜色的绘制。程序代码如下:

```
- (void)drawRect:(CGRect)rect {
    [super drawRect:rect];
    if (self.tintColor == nil){
        NSLog(@"UINavigationBarAdjustableTint: Didn't do anything because
        there is no tint colour set. Please set the tintColor property.");
                                                                //输出
        return;
    }
    //实例化颜色对象
    UIColor *bottomColor = self.tintColor;
    UIColor *topColor = [self adjustColourBrightness:bottomColor withFactor:
    self.colourAdjustFactor];
    CGContextRef context = UIGraphicsGetCurrentContext();    //创建图形上下文
    CGFloat locations[2] = { 0.0, 1.0 };
    CGColorSpaceRef myColorspace = CGColorSpaceCreateDeviceRGB();
                                                            //创建色彩空间
    NSArray *colours = [NSArray arrayWithObjects:(__bridge id)topColor.
    CGColor, (__bridge id) bottomColor.CGColor, nil];
    CGGradientRef gradient = CGGradientCreateWithColors(myColorspace,
    (__bridge CFArrayRef)colours, locations);               //创建渐变对象
    CGContextDrawLinearGradient(context, gradient, CGPointMake(0, 0),
    CGPointMake(0,self.frame.size.height), 0);              //填充渐变色
    CGGradientRelease(gradient);
    CGColorSpaceRelease(myColorspace);
}
```

（5）打开 ViewController.h 文件，编写代码，实现头文件、插座变量和动作的声明。程序代码如下：

```
#import <UIKit/UIKit.h>
//头文件
#import "NavigationBarTint.h"
@interface ViewController : UIViewController{
    //插座变量
    IBOutlet NavigationBarTint *navi;
    IBOutlet UISlider *slider;
    IBOutlet UISlider *redslider;
    IBOutlet UISlider *greenslider;
    IBOutlet UISlider *blueslider;
    IBOutlet UIView *vv;
}
//动作
- (IBAction)edit:(id)sender;
- (IBAction)save:(id)sender;
- (IBAction)lightChange:(id)sender;
- (IBAction)colorChange:(id)sender;
@end
```

（6）打开 Main.storyboard 文件，对 View Controller 视图控制器的设计界面进行设计，效果如图 8.18 所示。

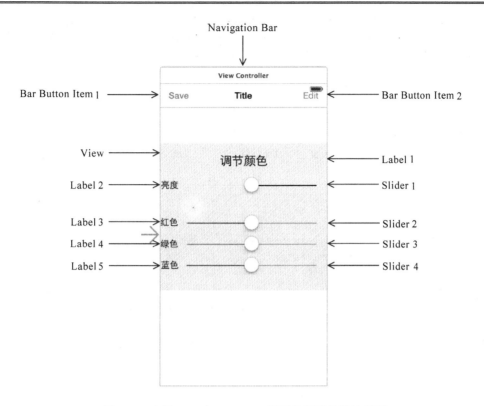

图 8.18 实例 55View Controller 视图控制器的设计界面

需要添加的视图、控件以及对它们的设置如表 8-15 所示。

表 8-15 实例 55 视图、控件设置

视图、控件	属 性 设 置	其 他
Navigation Bar		Class：NavigationBarTint 与插座变量navi关联
Bar Button Item1	Identifier：Save Tint：绿色	与动作save:关联
Bar Button Item2	Identifier：Edit Tint：绿色	与动作edit:关联
Slider1	Max Track Tint：黑色	与插座变量slider关联 与动作lightChange:关联
Slider2	Min Track Tint：红色	与插座变量redslider关联 与动作colorChange:关联
Slider3	Min Track Tint：绿色	与插座变量greenslider关联 与动作colorChange:关联
Slider4	Min Track Tint：蓝色	与插座变量blueslider关联 与动作colorChange:关联
View	Background：浅粉色	与插座变量vv关联
Label1	Text：调节颜色 Font：System 23.0 Alignment：居中对齐	

视图、控件	属性设置	其他
Label2	Text：亮度	
Label3	Text：红色	
Label4	Text：绿色	
Label5	Text：蓝色	

（7）打开 ViewController.m 文件，编写代码，实现通过滑块对导航栏上的颜色进行控制。程序代码如下：

```
- (void)viewDidLoad
{
    [super viewDidLoad];
    // Do any additional setup after loading the view, typically from a nib.
    float initialRed = 150;
    float initialGreen = 22;
    float initialBlue = 18;
    navi.tintColor = [UIColor colorWithRed:initialRed/255.0f green:
    initialGreen/255.0f blue:initialBlue/255.0f alpha:1.0f];
                                                //设置导航栏的填充颜色
    navi.colourAdjustFactor=0.6;                //设置颜色调整的因子
}
//以动画的形式显示放有滑块控件的视图
- (IBAction)edit:(id)sender {
    [UIView beginAnimations:@"" context:nil];
    vv.frame=CGRectMake(0, 106, 320, 278);      //设置框架
    [UIView commitAnimations];
}
//以动画的形式退出放有滑块控件的视图
- (IBAction)save:(id)sender {
    [UIView beginAnimations:@"" context:nil];
    vv.frame=CGRectMake(0, 500, 320, 278);      //设置框架
    [UIView commitAnimations];
}
//调节导航栏颜色的亮暗
- (IBAction)lightChange:(id)sender {
    navi.colourAdjustFactor=slider.value;
}
//调节导航栏的颜色
- (IBAction)colorChange:(id)sender {
    navi.tintColor = [UIColor colorWithRed:redslider.value green:
    greenslider.value blue:blueslider.value alpha:1.0f];//设置导航栏的颜色
}
```

【代码解析】

本实例关键功能是实现颜色的改变和亮度的调节。下面依次讲解这两个知识点。

1．颜色的改变

如表想要改变导航栏的背景颜色，需要使用 UINavigationBar 类的 tintColor 属性，其语法形式如下：

```
@property(nonatomic, retain) UIColor *tintColor;
```

在代码中，使用了 tintColor 属性对导航栏的背景颜色进行改变。代码如下：

```
navi.tintColor = [UIColor colorWithRed:redslider.value green:greenslider.
value blue:blueslider.value alpha:1.0f];
```

2. 亮度的调节

如果要对导航栏的亮暗进行调节,需要使用调整因子对新的分量进行设置,代码如下:

```
case 2:
{
    newComponents[0] = oldComponents[0]*factor;
    newComponents[1] = oldComponents[0]*factor;
    newComponents[2] = oldComponents[0]*factor;
    newComponents[3] = oldComponents[1];
    break;
}
case 4:
{
    newComponents[0] = oldComponents[0]*factor;
    newComponents[1] = oldComponents[1]*factor;
    newComponents[2] = oldComponents[2]*factor;
    newComponents[3] = oldComponents[3];
    break;
}
```

实例 56 滚动的导航栏

【实例描述】

本实例实现的功能是让导航栏进行滚动。当选择导航栏中的某一个条目时,指示器就会执行选择的条目(作为一个提醒用户的操作),并且条目的标题会发生改变,背景颜色也会出现相应的改变。运行效果如图 8.19 所示。

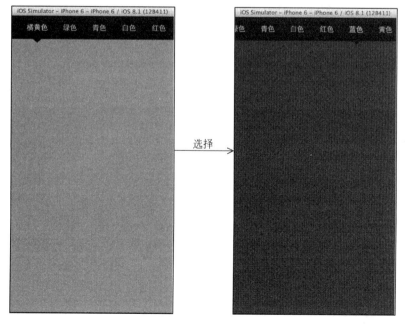

图 8.19 实例 56 运行效果

【实现过程】

（1）创建一个项目，命名为"可以滚动的导航栏"。

（2）创建一个基于 UIView 类的 Indicator 类。

（3）在 Indicator.h 文件中编写代码，实现属性的声明。

（4）打开 Indicator.m 文件，编写代码，实现指示器的绘制。其中，drawRect:方法实现的就是绘制指示器。程序代码如下：

```
- (void)drawRect:(CGRect)rect
{
    CGContextRef context = UIGraphicsGetCurrentContext();
    CGContextClearRect(context, rect);
    CGContextBeginPath(context);
    //绘制3个点
    CGContextMoveToPoint (context, CGRectGetMinX(rect), CGRectGetMinY(rect));
    CGContextAddLineToPoint(context, CGRectGetMidX(rect), CGRectGetMaxY(rect));
    CGContextAddLineToPoint(context, CGRectGetMaxX(rect), CGRectGetMinY(rect));
    CGContextClosePath(context);
    CGContextSetFillColorWithColor(context, self.color.CGColor);
    CGContextFillPath(context);         //绘制
}
```

（5）创建一个基于 UIView 类的 NavigationBar 类。

（6）打开 Navigation.h 文件，编写代码，实现协议、属性和方法的声明。程序代码如下：

```
#import <UIKit/UIKit.h>
@class NavigationBar ;
//协议
@protocol NavigationBarDelegate <NSObject>
-                           (void)itemAtIndex:(NSUInteger)index
didSelectInPagesContainerTopBar:(NavigationBar *)bar;
@end
@interface NavigationBar : UIView
//属性
@property (strong, nonatomic) UIImage *backgroundImage;
……
@property (strong, nonatomic) UIScrollView *scrollView;
@property (strong, nonatomic) NSArray *itemViews;
//方法
- (CGPoint)centerForSelectedItemAtIndex:(NSUInteger)index;
- (CGPoint)contentOffsetForSelectedItemAtIndex:(NSUInteger)index;
- (void)layoutItemViews;
@end
```

（7）打开 NavigationBar.m 文件，编写代码，实现导航栏的绘制。使用的方法如表 8-16 所示。

表 8-16 NavigationBar.m文件中方法总结

方法	功能
initWithFrame:	导航栏的初始化
centerForSelectedItemAtIndex:	获取选择条目的中心
contentOffsetForSelectedItemAtIndex:	获取选择条目的内容偏移量
setItemTitleColor:	设置条目的标题颜色

续表

方　　法	功　　能
setBackgroundImage:	设置背景视图
setItemTitles:	设置条目的标题
setFont:	设置字体
addItemView	添加条目视图
layoutItemViews	布局条目视图
layoutSubviews	布局子视图
backgroundImageView	获取背景图片视图

这里需要讲解几个重要的方法（其他方法请读者参考源代码）。其中，initWithFrame:方法实现对导航栏的初始化。程序代码如下：

```
- (id)initWithFrame:(CGRect)frame
{
    self = [super initWithFrame:frame];
    if (self) {
        self.scrollView = [[UIScrollView alloc] initWithFrame:self.bounds];
        //创建滚动视图对象
        self.scrollView.autoresizingMask = UIViewAutoresizingFlexibleWidth
        UIViewAutoresizingFlexibleHeight;
        self.scrollView.showsHorizontalScrollIndicator = NO;
        [self addSubview:self.scrollView];
        self.font = [UIFont systemFontOfSize:14];          //设置字体
        self.itemTitleColor = [UIColor whiteColor];        //设置条目标题颜色
    }
    return self;
}
```

setItemTitles:方法对条目的标题进行设置。程序代码如下：

```
- (void)setItemTitles:(NSArray *)itemTitles
{
    if (_itemTitles != itemTitles) {
        _itemTitles = itemTitles;
        NSMutableArray *mutableItemViews = [NSMutableArray arrayWithCapacity:
        itemTitles.count];
        //创建按钮
        for (NSUInteger i = 0; i < itemTitles.count; i++) {
            UIButton *itemView = [self addItemView];        //添加条目视图
            [itemView setTitle:itemTitles[i] forState:UIControlStateNormal];
            [mutableItemViews addObject:itemView];          //添加对象
        }
        self.itemViews = [NSArray arrayWithArray:mutableItemViews];
        [self layoutItemViews];
    }
}
```

addItemView方法实现对条目视图的添加。程序代码如下：

```
- (UIButton *)addItemView
{
    CGRect frame = CGRectMake(0., 0., DAPagesContainerTopBarItemViewWidth,
    CGRectGetHeight(self.frame));
    UIButton *itemView = [[UIButton alloc] initWithFrame:frame];
                                                        //创建按钮对象
```

```
[itemView addTarget:self action:@selector(itemViewTapped:) forControlEvents:
UIControlEventTouchUpInside];                                    //添加方法
itemView.titleLabel.font = self.font;
[itemView setTitleColor:self.itemTitleColor forState:UIControlState
 Normal];
[self.scrollView addSubview:itemView];                           //添加视图对象
return itemView;
}
```

layoutItemViews 方法对条目视图进行布局。程序代码如下：

```
- (void)layoutItemViews
{
    CGFloat x = DAPagesContainerTopBarItemsOffset;
    //对条目视图进行布局
    for (NSUInteger i = 0; i < self.itemViews.count; i++) {
        CGFloat width = [self.itemTitles[i] sizeWithFont:self.font].width;
        UIView *itemView = self.itemViews[i];
        itemView.frame = CGRectMake(x, 0., width, CGRectGetHeight(self.
        frame));                                                //设置框架
        x += width + DAPagesContainerTopBarItemsOffset;
    }
    //设置内容视图的尺寸
    self.scrollView.contentSize = CGSizeMake(x, CGRectGetHeight(self.
    scrollView.frame));CGRect frame = self.scrollView.frame;
    //判断获取的宽度是否大于 x
    if (CGRectGetWidth(self.frame) > x) {
        frame.origin.x = (CGRectGetWidth(self.frame) - x) / 2.;
        frame.size.width = x;
    } else {
        frame.origin.x = 0.;
        frame.size.width = CGRectGetWidth(self.frame);
    }
    self.scrollView.frame = frame;                              //设置滚动视图的框架
}
```

（8）创建一个基于 UIViewController 类的 Container 类。

（9）打开 Container.h 文件，编写代码，实现头文件、属性和方法的声明。程序代码如下：

```
#import <UIKit/UIKit.h>
//头文件
#import "NavigationBar.h"
#import "Indicator.h"
@interface Container : UIViewController<NavigationBarDelegate,UIScrollView
Delegate>   //遵守的协议
//属性
@property (strong, nonatomic) NSArray *viewControllers;
……
@property (readonly, assign, nonatomic) CGFloat scrollHeight;
//方法
- (void)setSelectedIndex:(NSUInteger)selectedIndex animated:(BOOL)animated;
- (void)updateLayoutForNewOrientation:(UIInterfaceOrientation)orientation;
@end
```

（10）打开 Container.m 文件，编写代码，绘制导航控制器。使用的方法如表 8-17 所示。

表 8-17 Container.m文件中方法总结

方　　法	功　　能
init	初始化
initWithCoder:	初始化实例化对象
setUp	设置属性
viewDidLoad	视图加载后调用，实现初始化
setSelectedIndex: animated:	设置选择的索引动画
updateLayoutForNewOrientation:	为新位置更新布局
setPageIndicatorViewSize:	设置指示器视图的尺寸
setPageItemsTitleColor:	设置条目的标题的填充颜色
setSelectedIndex:	设置选择的索引
setSelectedPageItemTitleColor:	设置选择条目的标题颜色
setTopBarBackgroundColor:	设置顶部导航栏的背景颜色
setTopBarHeight:	设置顶部导航栏的高度
setTopBarItemLabelsFont:	设置顶部导航栏的条目的标签字体
setViewControllers:	设置视图控制器
setPageIndicatorImage:	设置指示器的图像
layoutSubviews	布局子视图
pageIndicatorCenterY	获取指示器中心的y值
pageIndicatorView	获取指示器
scrollHeight	获取导航视图的高度
scrollWidth	获取导航视图的宽度
startObservingContentOffsetForScrollView:	设置滚动视图显示区域的偏移量
stopObservingContentOffset	停止滚动视图显示区域的偏移量
itemAtIndex: didSelectInPagesContainerTopBar:	选择条目时调用
scrollViewDidEndDecelerating:	用户拖动，之后继续滚动调用
scrollViewDidEndDragging: willDecelerate:	用户拖动并停止调用
scrollViewDidEndScrollingAnimation:	滚动视图滚动并伴有动画，当动画结束后调用
scrollViewDidScroll:	滚动视图滚动并没用动画，调用
observeValueForKeyPath: ofObject: change: context:	条目和指示器的移动

要绘制导航控制器，需要实现两个部分：导航控制器的创建以及实现导航栏的滑动选择。其中，创建导航控制器需要使用 init、initWithCoder:、setUp、viewDidLoad、setSelectedIndex: animated:、setPageIndicatorViewSize:、setPageItemsTitleColor:、setSelectedIndex:、setSelectedPageItemTitleColor:、setTopBarBackgroundColor:、setTopBarHeight:、setTopBarItemLabelsFont:、setViewControllers:、setPageIndicatorImage:、layoutSubviews、pageIndicatorCenterY、pageIndicatorView、scrollHeight、scrollWidth、startObservingContentOffsetForScrollView:和 observeValueForKeyPath: ofObject: change: context:方法实现。setUp方法对导航控制器的属性进行设置。程序代码如下：

```
- (void)setUp
{
    //设置导航栏、指示器和条目的属性
    _topBarHeight = 44.;
    _topBarBackgroundColor = [UIColor colorWithWhite:0.1 alpha:1.];
    //背景颜色
```

```objectivec
_topBarItemLabelsFont = [UIFont systemFontOfSize:12];        //字体
_pageIndicatorViewSize = CGSizeMake(22., 9.);
self.pageItemsTitleColor = [UIColor lightGrayColor];
self.selectedPageItemTitleColor = [UIColor whiteColor];      //标题颜色
}
```

viewDidLoad 方法在视图加载后调用,实现初始化。程序代码如下

```objectivec
- (void)viewDidLoad
{
    [super viewDidLoad];
    self.shouldObserveContentOffset = YES;
    self.scrollView = [[UIScrollView alloc] initWithFrame:CGRectMake
    (0.,self.topBarHeight,CGRectGetWidth(self.view.frame),CGRectGetHeight
    (self.view.frame) - self.topBarHeight)];     //实例化对象
    self.scrollView.autoresizingMask = UIViewAutoresizingFlexibleWidth |
    UIViewAutoresizingFlexibleHeight;
    self.scrollView.delegate = self;
    self.scrollView.pagingEnabled = YES;
    self.scrollView.showsHorizontalScrollIndicator = NO;
                                          //不显示水平的滚动指示器
    [self.view addSubview:self.scrollView];
    [self startObservingContentOffsetForScrollView:self.scrollView];
    //创建并设置导航栏
    self.topBar = [[NavigationBar alloc] initWithFrame:CGRectMake(0.,0.,
    CGRectGetWidth(self.view.frame),self.topBarHeight)];    //实例化对象
    self.topBar.autoresizingMask = UIViewAutoresizingFlexibleBottomMargin
    | UIViewAutoresizingFlexibleWidth;
    self.topBar.itemTitleColor = self.pageItemsTitleColor; //设置标题的颜色
    self.topBar.delegate = self;
    [self.view addSubview:self.topBar];
    self.topBar.backgroundColor = self.topBarBackgroundColor;//设置背景颜色
}
```

setViewControllers:方法对导航控制器控制的视图控制器进行设置。程序代码如下:

```objectivec
- (void)setViewControllers:(NSArray *)viewControllers
{
    if (_viewControllers != viewControllers) {
        _viewControllers = viewControllers;
        self.topBar.itemTitles = [viewControllers valueForKey:@"title"];
        for (UIViewController *viewController in viewControllers) {
            [viewController willMoveToParentViewController:self];
                                          //删除视图控制器
            viewController.view.frame = CGRectMake(0., 0., CGRectGetWidth
            (self.scrollView.frame), self.scrollHeight);
                                          //设置视图控制器中视图的框架
            [self.scrollView addSubview:viewController.view];
            [viewController didMoveToParentViewController:self];
        }
        [self layoutSubviews];
        self.selectedIndex = 0;
        self.pageIndicatorView.center = CGPointMake([self.topBar centerFor
        SelectedItemAtIndex:self.selectedIndex].x,[self pageIndicatorCenterY]);
                                          //设置指示器视图的中心点
    }
}
```

layoutSubviews 方法对子视图进行重新布局。程序代码如下:

```objc
- (void)layoutSubviews
{
    self.topBar.frame = CGRectMake(0., 0., CGRectGetWidth(self.view.
    bounds), self.topBarHeight);
    CGFloat x = 0.;
    for (UIViewController *viewController in self.viewControllers) {
        viewController.view.frame = CGRectMake(x, 0, CGRectGetWidth(self.
        scrollView.frame), self.scrollHeight);          //设置框架
        x += CGRectGetWidth(self.scrollView.frame);
    }
    //设置滚动视图
    self.scrollView.contentSize = CGSizeMake(x, self.scrollHeight);
    [self.scrollView setContentOffset:CGPointMake(self.selectedIndex *
    self.scrollWidth, 0.) animated:YES];               //设置内容视图的偏移量
    //设置指示器视图
    self.pageIndicatorView.center = CGPointMake([self.topBar centerFor
    SelectedItemAtIndex:self.selectedIndex].x, [self pageIndicatorCenterY]);
    //设置导航栏的滚动视图
    self.topBar.scrollView.contentOffset = [self.topBar contentOffset
    ForSelectedItemAtIndex:self.selectedIndex];
    self.scrollView.userInteractionEnabled = YES;
}
```

pageIndicatorView 方法获取指示器的视图。程序代码如下:

```objc
- (UIView *)pageIndicatorView
{
    if (!_pageIndicatorView) {
        if (self.pageIndicatorImage) {
            _pageIndicatorView = [[UIImageView alloc] initWithImage:self.page
            IndicatorImage];//实例化对象
        } else {
            _pageIndicatorView = [[Indicator alloc] initWithFrame:CGRect
            Make(0.,44,self.pageIndicatorViewSize.width,self.pageIndicator
            ViewSize.height)];
            [(Indicator *)_pageIndicatorView setColor:self.topBarBackground
            Color];                                    //设置颜色
        }
        [self.view addSubview:self.pageIndicatorView]; //添加视图对象
    }
    return _pageIndicatorView;
}
```

observeValueForKeyPath: ofObject: change: context:方法实现对导航栏以及指示器的滚动进行设置。程序代码如下:

```objc
- (void)observeValueForKeyPath:(NSString *)keyPath ofObject:(id)object
change:(NSDictionary *)change
context:(void *)context{
    CGFloat oldX = self.selectedIndex * CGRectGetWidth(self.scrollView.
    frame);
    //判断 oldX 是否不等于滚动视图中内容视图的偏移量的 x 值以及 shouldObserve
    ContentOffset 的值
    if (oldX != self.scrollView.contentOffset.x && self.shouldObserve
    ContentOffset) {
        BOOL scrollingTowards = (self.scrollView.contentOffset.x > oldX);
        //设置布尔值
        NSInteger targetIndex = (scrollingTowards) ? self.selectedIndex + 1 :
```

```
            self.selectedIndex - 1;
        CGFloat ratio = (self.scrollView.contentOffset.x - oldX) / CGRect
        GetWidth(self.scrollView.frame);
        CGFloat previousItemContentOffsetX = [self.topBar contentOffset
        ForSelectedItemAtIndex:self.selectedIndex].x;
        CGFloat nextItemContentOffsetX = [self.topBar contentOffset
        ForSelectedItemAtIndex:targetIndex].x;
        CGFloat previousItemPageIndicatorX = [self.topBar centerForSelected
        ItemAtIndex:self.selectedIndex].x;
        CGFloat nextItemPageIndicatorX = [self.topBar centerForSelected
        ItemAtIndex:targetIndex].x;
    //实现导航栏以及指示器的滚动
        //判断 scrollingTowards 是否为 YES
        if (scrollingTowards) {
            self.topBar.scrollView.contentOffset = CGPointMake(previousItem
            ContentOffsetX +(nextItemContentOffsetX - previousItemContent
            OffsetX) * ratio , 0.);          //设置内容视图的偏移量
            self.pageIndicatorView.center = CGPointMake(previousItemPage
            IndicatorX +(nextItemPageIndicatorX - previousItemPageIndicatorX)
            * ratio,[self pageIndicatorCenterY]);
        } else {
            self.topBar.scrollView.contentOffset = CGPointMake(previous
            ItemContentOffsetX - (nextItemContentOffsetX - previousItem
            Content OffsetX) * ratio , 0.);
            //设置中心位置
            self.pageIndicatorView.center = CGPointMake(previousItemPage
            IndicatorX - (nextItemPageIndicatorX - previousItemPage
            IndicatorX) * ratio, [self pageIndicatorCenterY]);
        }
    }
}
```

如果想要实现导航栏的滚动选择，需要使用 itemAtIndex: didSelectInPagesContainerTopBar：、scrollViewDidEndDecelerating：、scrollViewDidEndDragging：、willDecelerate：、scrollViewDidEndScrollingAnimation: 和 scrollViewDidScroll: 方法实现。itemAtIndex: didSelectInPagesContainerTopBar:方法实现导航栏条目的选择。程序代码如下：

```
- (void)itemAtIndex:(NSUInteger)index didSelectInPagesContainerTopBar:
(Container *)bar
{
    [self setSelectedIndex:index animated:YES];
}
```

（11）打开 ViewController.h 文件，编写代码，实现头文件和属性的声明。程序代码如下：

```
#import <UIKit/UIKit.h>
#import "Container.h"
@interface ViewController : UIViewController
@property (strong, nonatomic) Container *pagesContainer;         //属性
@end
```

（12）打开 ViewController.m 文件，编写代码，实现导航控制器以及视图控制器对象的创建。程序代码如下：

```
- (void)viewDidLoad
{
    [super viewDidLoad];
    // Do any additional setup after loading the view, typically from a nib.
```

```
//创建并设置导航控制器对象
self.pagesContainer = [[Container alloc] init];
[self.pagesContainer willMoveToParentViewController:self];
self.pagesContainer.view.frame = self.view.bounds;          //设置框架
self.pagesContainer.view.autoresizingMask = UIViewAutoresizingFlexible
Width | UIViewAutoresizingFlexibleHeight;
[self.view addSubview:self.pagesContainer.view];            //添加视图对象
[self.pagesContainer didMoveToParentViewController:self];
UIViewController *orange = [[UIViewController alloc] init];
orange.view.backgroundColor=[UIColor orangeColor];          //设置背景颜色
orange.title = @"橘黄色";                                    //设置标题
……
UIViewController *yellow= [[UIViewController alloc] init];
yellow.view.backgroundColor=[UIColor yellowColor];
yellow.title = @"黄色";
self.pagesContainer.viewControllers = @[ orange, green,cyan ,white,red,
blue,yellow];
}
```

【代码解析】

本实例关键功能是导航栏和指示器的滚动。下面依次讲解这两个知识点。

1. 导航栏的滚动

要想实现导航栏的滚动，需要使用 UIScrollView 的 contentOffset 属性，其语法形式如下：

```
@property(nonatomic) CGPoint contentOffset;
```

在代码中，使用了 contentOffset 属性让导航栏进行了滚动。代码如下：

```
self.topBar.scrollView.contentOffset = CGPointMake(previousItemContent
OffsetX +(nextItemContentOffsetX - previousItemContentOffsetX) * ratio ,
0.);
```

2. 指示器的滚动

要实现指示器的滚动，需要使用 UIView 的 center 属性，其语法形式如下：

```
@property(nonatomic) CGPoint center;
```

在代码中，使用了 center 属性让指示器进行了滚动。代码如下：

```
self.pageIndicatorView.center = CGPointMake(previousItemPageIndicatorX
+(nextItemPageIndicatorX - previousItemPageIndicatorX) * ratio,[self page
IndicatorCenterY]);
```

实例 57 具有导航记录的导航栏

【实例描述】

在浏览器中或其他的地方，用户进行浏览后，会出现一个浏览的历史记录，以便用户可以看到自己浏览过的地方。或者是不小心关闭后，可以通过历史记录很快地找到此网页或者其他。本实例就是要实现这么一个功能，在导航栏上显示导航记录（历史记录），可以让用户看到自己浏览过的内容。运行效果如图 8.20 所示。

图 8.20　实例 57 运行效果

【实现过程】

（1）创建一个项目，命名为"具有导航记录的导航栏"。

（2）添加图片 1.png、2.png 和 3.png 到创建项目的 Supporting Files 文件夹中。

（3）创建一个基于 NSObject 类的 FolderStyle 类。

（4）打开 FolderStyle.h 文件，编写代码，实现导航栏的风格枚举及定义。程序代码如下：

```
#import <Foundation/Foundation.h>
enum {
    DTFolderBarStyleNormal = 0,
    DTFolderBarStyleFixedLeftHome,
};
typedef NSInteger DTFolderBarStyle;
@interface FolderStyle : NSObject
@end
```

（5）创建一个基于 NSObject 类的 FolderConfig 类。

（6）打开 FolderConfig.h 文件，编写代码，实现宏定义。程序代码如下：

```
#import <Foundation/Foundation.h>
#define kFolderItemIcon              @"3.png"
#define kFolderItemTextColor         [UIColor whiteColor]
#define kAddFolderDuration           0.2f
#define kDeleteFolderDuration        0.2f
#define kBarBlackgroundImage         @"1.png"
#define kFolderNameLimitLength       10
#define kItemBackgroundImage         @"2.png"
@interface FolderConfig : NSObject
@end
```

（7）创建一个基于 UIView 类的 FolderItem 类。

（8）打开 FolderItem.h 文件，编写代码，实现头文件、宏定义、实例变量、属性和方法的声明。程序代码如下：

```
#import <UIKit/UIKit.h>
#import "FolderConfig.h"                    //头文件
//宏定义
#define kBackgroundImageViewTag 1
#define kTitleLabelTag          2
#define kIconImageViewTag       3
@interface FolderItem : UIView{
    //实例变量
    id      _targer;
    SEL     _action;
}
//属性
@property (nonatomic, retain) UIImage *backgroundImage;
@property (nonatomic, retain) UIImage *highlightedImage;
......
@property (nonatomic, readonly) UILabel *textLable;
//方法
+ (id)itemWithFolderName:(NSString *)folderName targer:(id)targer action:(SEL)action;
+ (id)itemWithImage:(UIImage *)iconImage targer:(id)targer action:(SEL)action;
- (id)initWithFolderName:(NSString *)folderName targer:(id)targer action:(SEL)action;
- (id)initWithImage:(UIImage *)iconImage targer:(id)targer action:(SEL)
```

```
action;
@end
```

（9）打开 FolderItem.m 文件，编写代码，实例导航记录条目的绘制。使用的方法如表 8-18 所示。

表 8-18 FolderItem.m文件中方法总结

方　　法	功　　能
itemWithFolderName: targer: action:	使用导航记录的名称、对象和动作（方法、选择器）获取导航记录的条目
itemWithImage: targer: action:	使用导航记录的图像、对象和动作（方法、选择器）获取导航记录的条目
initWithFolderName: targer: action:	使用导航记录的名称、对象和动作（方法、选择器）进行初始化
initWithImage: targer: action:	使用导航记录的图像、对象和动作（方法、选择器）进行初始化
setViewWithFolderName:	使用导航记录名称设置导航记录条目的视图
setViewWithImage:	使用图像设置导航记录条目的视图
tapItem:	轻拍导航记录条目
title	获取标签的文本
setBackgroundImage:	设置背景图像
backgroundImage	获取背景图像
textLable	获取标签

要实现导航记录的绘制，需要使用 itemWithFolderName: targer: action:、itemWithImage: targer: action:、initWithFolderName: targer: action:、initWithImage: targer: action:、initWithImage: targer: action:、initWithImage: targer: action:、setViewWithFolderName:、setViewWithImage:、title、setBackgroundImage:、backgroundImage 和 textLable 方法实现。这里需要讲解几个重要的方法（其他方法请读者参考源代码）。其中，initWithFolderName: targer: action:方法实现使用导航记录的名称、对象和动作（方法、选择器）进行初始化。程序代码如下：

```
- (id)initWithFolderName:(NSString *)folderName targer:(id)targer action:
(SEL)action
{
   CGFloat fontSize = [UIFont systemFontSize];            //获取标准字体的大小
   CGFloat width = folderName.length * fontSize - 10.0f;  //获取宽度
   CGFloat height = UIInterfaceOrientationIsLandscape([[UIApplication
sharedApplication] statusBarOrientation]) ? 32.0f : 44.0f;
   self = [super initWithFrame:CGRectMake(0.0f, 0.0f, width, height)];
                                                         //初始化 self
   if (self == nil) return nil;
   [self setUserInteractionEnabled:YES];
   [self setAutoresizingMask:UIViewAutoresizingFlexibleHeight];
   [self setAutoresizesSubviews:YES];                    //设置自动调整
   _targer = targer;
   _action = action;
   [self setViewWithFolderName:folderName];
   UITapGestureRecognizer *tapGesture = [[UITapGestureRecognizer alloc]
   initWithTarget:self action:@selector(tapItem:)];//创建轻拍的手势识别器对象
   [self addGestureRecognizer:tapGesture];
   return self;
}
```

setViewWithFolderName:方法实现使用导航记录名称设置导航记录条目的视图。程序代码如下：

```objc
- (void)setViewWithFolderName:(NSString *)folderName
{
    UIEdgeInsets edgeInsets = UIEdgeInsetsMake(20.0f, 1.0f, 20.0f, 44.0f);
    //创建并设置背景的图像视图对象
    UIImage *bgImage = [[UIImage imageNamed:kItemBackgroundImage] resizable
    ImageWithCapInsets:edgeInsets];
    UIImageView *backgroundImageView = [[UIImageView alloc] initWithImage:
    bgImage];
    [backgroundImageView setFrame:self.frame];         //设置图像视图对象的框架
    [backgroundImageView setTag:kBackgroundImageViewTag];
    [backgroundImageView setContentMode:UIViewContentModeScaleToFill];
    [backgroundImageView setAutoresizingMask:UIViewAutoresizingFlexible
    Height];                                           //设置自适应
    if (folderName.length > kFolderNameLimitLength) {
        folderName = [folderName substringWithRange:NSMakeRange(0, kFolder
        NameLimitLength - 3)];
        folderName = [folderName stringByAppendingString:@"..."];
    }
    //设置并创建标签对象
    UILabel *titleLabel = [[UILabel alloc] initWithFrame:CGRectZero];
    [titleLabel setText:folderName];                   //设置文本内容
    [titleLabel sizeToFit];
    [titleLabel setBackgroundColor:[UIColor clearColor]];   //设置背景颜色
    [titleLabel setTag:kTitleLabelTag];
    [titleLabel setAutoresizingMask:UIViewAutoresizingFlexibleTopMargin |
    UIViewAutoresizingFlexibleBottomMargin | UIViewAutoresizingFlexible
    LeftMargin | UIViewAutoresizingFlexibleRightMargin];   //设置自适应
    CGRect bgFrame = backgroundImageView.frame;
    bgFrame.size.width = titleLabel.frame.size.width + 44.0f;
    [backgroundImageView setFrame:bgFrame];            //设置框架
    [self setFrame:bgFrame];
    [titleLabel setCenter:self.center];                //设置中心位置
    [self addSubview:backgroundImageView];
    [self addSubview:titleLabel];
}
```

在本文件中除了绘制导航记录外，还为导航记录添加了轻拍的手势。手势实现的方法是 tapItem:方法。程序代码如下：

```objc
- (IBAction)tapItem:(UITapGestureRecognizer *)sender
{
    [[UIApplication sharedApplication] sendAction:_action to:_targer from:
    self forEvent:nil];
}
```

（10）创建一个基于 UIView 类的 FolderBar 类。

（11）打开 FolderBar.h 文件，编写代码，实现头文件、宏定义、实例变量、对象、属性和方法等的声明。程序代码如下：

```objc
#import <UIKit/UIKit.h>
//头文件
#import "FolderItem.h"
#import "FolderStyle.h"
#import "FolderConfig.h"
```

```
#define kBackgroundViewTag 2
……
#define kButtonType UIButtonTypeCustom
typedef void (^DTAnimationsBlock) (void);
typedef void (^DTCompletionBlock) (BOOL finshed);
@interface FolderBar : UIView{
    //实例变量
    DTFolderBarStyle _style;
    //对象
    NSMutableArray *_folderItems; // Saved folderItem array
    FolderItem *_leftItem;
    UIButton *_actionButton;
}
//属性
@property (nonatomic, assign ,readonly) DTFolderBarStyle style;
……
@property (nonatomic, retain) FolderItem *leftItem;
//方法
+ (id)folderBarWithFrame:(CGRect)frame;
+ (id)folderBarWithFrame:(CGRect)frame style:(DTFolderBarStyle)style;
- (void)setFolderItems:(NSArray *)folderItems animated:(BOOL)animated;
@end
```

（12）打开 FolderBar.m 文件，编写代码，实现导航栏的绘制以及导航记录的删除。使用的方法如表 8-19 所示。

表 8-19　FolderBar.m 文件中方法总结

方　　法	功　　能
folderBarWithFrame:	使用框架创建导航栏对象并获取此对象
folderBarWithFrame: style:	使用框架和风格创建导航栏对象并获取此对象
initWithFrame:	使用框架初始化
initWithFrame:style:	使用框架和风格初始化
setFolderItems:	设置导航记录条目
setFolderItems:animated:	设置导航记录条目以及动画
folderItems	获取导航记录条目
addFolderItem:animated:	添加导航记录条目
deleteFolderItem:animated:	删除导航记录条目
style	获取风格
setBackgroundImage:	设置背景图像
backgroundImage	获取背景图像

要实现导航栏的绘制，需要使用 folderBarWithFrame:、folderBarWithFrame: style:、initWithFrame:、initWithFrame:style:、setFolderItems:、setFolderItems:、setFolderItems:animated:、folderItems、style、setBackgroundImage:和 backgroundImage 方法共同实现。其中，initWithFrame:style:方法实现的功能是使用框架和风格对导航栏进行初始化。程序代码如下：

```
-(id)initWithFrame:(CGRect)frame style:(DTFolderBarStyle)style
{
    self = [super initWithFrame:frame];
    if (self == nil) return nil;
    [self setClipsToBounds:YES];
```

```objc
[self setBackgroundColor:[UIColor whiteColor]];          //设置背景颜色
_style = style;
_folderItems = [NSMutableArray new];                     //创建新的可变数组
UIViewAutoresizing autoresizing = UIViewAutoresizingFlexibleWidth |
UIViewAutoresizingFlexibleHeight;
//背景对象的设置
UIImageView *backgroundView = [[UIImageView alloc] initWithFrame:
self.bounds];
[backgroundView setImage:[UIImage imageNamed:kBarBlackgroundImage]];
                                                         //设置图像视图显示的图像
[backgroundView setTag:kBackgroundViewTag];              //设置 tag 值
[backgroundView setContentMode:UIViewContentModeScaleToFill];
[backgroundView setAutoresizingMask:autoresizing];       //设置自适应
CGRect scrollViewFrame = self.bounds;
if (_style == DTFolderBarStyleFixedLeftHome ) {
    //设置 DTFolderBarStyleFixedLeftHome 风格的导航栏
    scrollViewFrame.origin.x += 22.0f;
    scrollViewFrame.size.width -= 22.0f;
}
//滚动视图对象的设置
UIScrollView *scrollView = [[UIScrollView alloc] initWithFrame:
scrollViewFrame];
[scrollView setTag:kScrollViewTag];                      //设置 tag 值
[scrollView setShowsHorizontalScrollIndicator:NO];
[scrollView setShowsVerticalScrollIndicator:NO];
[scrollView setContentOffset:CGPointMake(0, 0)];//设置内容视图的偏移量
[scrollView setAutoresizingMask:autoresizing];
CGRect folderItemViewFrame = self.bounds;
folderItemViewFrame.size.width = 0;
UIView *folderItemView = [[UIView alloc] initWithFrame:folderItem
ViewFrame];
[folderItemView setBackgroundColor:[UIColor clearColor]];//设置背景颜色
[folderItemView setClipsToBounds:YES];
[folderItemView setTag:kFolderItemViewTag];
[folderItemView setAutoresizingMask:UIViewAutoresizingFlexibleHeight];
                                                         //设置自适应
[self addSubview:backgroundView];                        //添加视图对象
[self addSubview:scrollView];
if (_actionButton != nil) {
    [self addSubview:_actionButton];                     //添加视图对象
}
[scrollView addSubview:folderItemView];
return self;
}
```

如果要想实现导航记录条目的添加,需要使用 addFolderItem:animated:方法,程序代码如下:

```objc
- (void)addFolderItem:(FolderItem *)folderItem animated:(BOOL)animated
{
    NSTimeInterval duration = animated ? kAddFolderDuration : 0.0f;
    UIView *folderItemView = (UIView *)[self viewWithTag:kFolderItem
    ViewTag];                                            //实例化对象
    CGRect folderItemViewFrame = folderItemView.frame;
                                                         //获取 folderItemView 的框架
    CGFloat nextItemPosition = folderItemViewFrame.size.width - 22.0f;
    folderItemViewFrame.size.width += folderItem.frame.size.width;
    //判断
```

```
    if (nextItemPosition < 0.0f) {
        nextItemPosition = 0.0f;
    }
    CGRect folderItemFrame = folderItem.frame;              //获取框架
    folderItemFrame.origin.x = nextItemPosition;            //设置 x 的位置
    [folderItem setFrame:folderItemFrame];                  //设置框架
    CGFloat folderItemViewWidth = nextItemPosition + folderItemFrame.size.
    width;
    folderItemViewFrame.size.width = folderItemViewWidth;   //设置宽度
    DTAnimationsBlock animations = ^{
        [folderItemView insertSubview:folderItem atIndex:0];
        [folderItemView setFrame:folderItemViewFrame];
                                        //为导航记录条目视图设置框架
    };
    [UIView animateWithDuration:duration animations:animations];
    UIScrollView *scrollView = (UIScrollView *)[self viewWithTag:kScroll
    ViewTag];
    [scrollView setContentSize:CGSizeMake((folderItemViewWidth + 44.0f),
    0)]//设置内容视图的尺寸
    if (folderItemViewWidth > scrollView.frame.size.width) {
        CGPoint offset = scrollView.contentOffset;
                                        //获取滚动视图显示区域的偏移量
        offset.x = folderItemViewWidth - scrollView.frame.size.width +
        44.0f;
        [scrollView setContentOffset:offset animated:YES];
    }
}
```

如果要想实现导航记录条目的删除,需要使用 deleteFolderItem:animated:方法。程序代码如下:

```
- (void)deleteFolderItem:(FolderItem *)folderItem animated:(BOOL)animated
{
    NSTimeInterval duration = animated ? kDeleteFolderDuration : 0.0f;
    UIView *folderItemView = (UIView *)[self viewWithTag:kFolderItem
    ViewTag];                                               //实例化对象
    CGRect folderItemViewFrame = folderItemView.frame;
                                        //获取 folderItemView 的框架
    folderItemViewFrame.size.width -= ( folderItem.frame.size.width -
    22.0f );
    UIScrollView *scrollView = (UIScrollView *)[self viewWithTag:kScroll
    ViewTag];
    CGSize contenSize = scrollView.contentSize;//获取内容视图的尺寸
    contenSize.width = folderItemViewFrame.size.width + 44.0f;
                                        //实现删除
    DTAnimationsBlock animations = ^{
        [folderItemView setFrame:folderItemViewFrame];
        [scrollView setContentSize:contenSize];         //设置内容视图的尺寸
    };
    DTCompletionBlock completion = ^(BOOL finsh){
        [folderItem removeFromSuperview];               //移除视图
    };
    [UIView animateWithDuration:duration animations:animations completion:
    completion];
}
```

(13)创建一个基于 UINavigationController 类的 Navigation 类。

(14)打开 Navigation.h 文件,编写代码,实现头文件、宏定义、实例变量、对象、属性和方法等的声明。程序代码如下：

```
#import <UIKit/UIKit.h>
//头文件
#import "FolderBar.h"
#import "FolderItem.h"
//宏定义
#define kUIApplicationChangeStatusBarFrameDuration 0.355
typedef void (^DTAnimationsBlock) (void);
typedef void (^DTCompletionBlock) (BOOL finshed);
@interface Navigation : UINavigationController{
    //对象
    UIView *blockView;
    FolderBar *_folderBar;
    DTFolderBarStyle _folderStyle;                //实例变量
}
//属性
@property (nonatomic, readonly) FolderBar *folderBar;
@property (nonatomic, getter = isFolderBarHidden) BOOL folderBarHidden;
//方法
+ (instancetype)navigationWithRootViewController:(UIViewController *)rootViewController;
+ (instancetype)navigationWithRootViewController:(UIViewController *)rootViewController folderStyle:(DTFolderBarStyle)folderStyle;
- (void)setFolderBarHidden:(BOOL)folderBarHidden animated:(BOOL)animated;
@end
```

（15）打开 Navigation.m 文件，编写代码，实现导航控制器的设置。使用的方法如表 8-20 所示。

表 8-20　Navigation.m文件中方法总结

方　　法	功　　能
navigationWithRootViewController:	使用根视图控制器创建导航控制器对象并获取此对象
navigationWithRootViewController:folderStyle:	使用根视图控制器和风格创建导航控制器对象并获取此对象
initWithRootViewController:folderStyle:	初始化导航控制器
viewDidLoad	视图加载后调用，实现初始化
viewWillAppear:	视图即将可见时调用
viewWillDisappear:	视图即将被驳回时调用
pushViewController:animated:	入栈操作
folderBar	获取导航栏的风格
setFolderBarHidden:	设置是否隐藏导航栏
setFolderBarHidden:animated:	设置导航栏的隐藏以及动画
isFolderBarHidden	获取是否隐藏导航栏
tapFolderItem:	轻拍导航记录条目
FolderBarFrameChange:	改变导航栏的框架

要创建导航控制器，需要使用 navigationWithRootViewController:、navigationWithRootViewController:folderStyle:、initWithRootViewController:folderStyle:、viewDidLoad、viewWillAppear:、viewWillDisappear:、folderBar 和 FolderBarFrameChange:方法共同实现。其中，initWithRootViewController: folderStyle:方法实现使用根视图控制器和导航栏风格对导航控制器进行初始化。程序代码如下：

```objc
- (id)initWithRootViewController:(UIViewController *)rootViewController
folderStyle:(DTFolderBarStyle)folderStyle
{
    _folderStyle = folderStyle;
    self = [super initWithRootViewController:rootViewController];
    if (self == nil) return nil;
    [self.view setAutoresizesSubviews:YES];         //设置自动调整
    [self.navigationBar setTintColor:[UIColor blackColor]];
    return self;
}
```

viewDidLoad 方法在视图加载后调用,实现导航栏的创建。程序代码如下:

```objc
- (void)viewDidLoad
{
    [super viewDidLoad];
    // Do any additional setup after loading the view.
    if (_folderBar == nil)
        _folderBar = [FolderBar folderBarWithFrame:self.navigationBar.frame
            style:_folderStyle];
    [self.view addSubview:_folderBar];
}
```

FolderBarFrameChange:方法实现导航栏框架的改变。程序代码如下:

```objc
- (void)FolderBarFrameChange:(NSNotification *)sender
{
    CGRect folderBarFrame = [_folderBar frame];
    folderBarFrame.origin.y = 20.0f;                //设置导航栏 y 的位置
    [UIView animateWithDuration:kUIApplicationChangeStatusBarFrameDuration
    animations:^(){
        [_folderBar setFrame:folderBarFrame];       //设置导航栏的框架
    }];
}
```

要实现视图控制器的控制,需要使用 pushViewController:animated:方法实现。程序代码如下:

```objc
- (void)pushViewController:(UIViewController *)viewController animated:
(BOOL)animated
{
    [super pushViewController:viewController animated:animated];
    if (_folderBar == nil)
        _folderBar = [FolderBar folderBarWithFrame:self.navigationBar.frame
            style:_folderStyle];                    //实例化对象
    NSMutableArray *folderItems = [NSMutableArray arrayWithArray:
    _folderBar.folderItems];
    FolderItem *folderItem = nil;
    //判断视图控制器的标题是否为空
    if (viewController.title == nil) {
        folderItem = [FolderItem itemWithImage:[UIImage imageNamed:kFolder
        ItemIcon] targer:self action:@selector(tapFolderItem:)];//实例化对象
        //设置风格
        if (_folderStyle == DTFolderBarStyleFixedLeftHome ) {
            [_folderBar setLeftItem:folderItem];
        } else {
            [folderItems addObject:folderItem];
            [_folderBar setFolderItems:folderItems animated:YES];
        }
```

```
    } else {
        folderItem = [FolderItem itemWithFolderName:viewController.title
        targer:self action:@selector(tapFolderItem:)];        //实例化对象
        [folderItem.textLable setTextColor:kFolderItemTextColor];
                                                              //设置文本颜色
        [folderItems addObject:folderItem];
        [_folderBar setFolderItems:folderItems animated:YES];
    }
}
```

要实现导航栏的显示和隐藏，需要使用 setFolderBarHidden:、setFolderBarHidden:animated:和 isFolderBarHidden 方法。setFolderBarHidden:animated:方法实现对导航栏的隐藏及动画设置。程序代码如下：

```
- (void)setFolderBarHidden:(BOOL)folderBarHidden animated:(BOOL)animated
{
    //判断 animated 是否为 YES
    if (animated) {
        DTAnimationsBlock animations = ^{
            CGRect folderBarFrame = _folderBar.frame;
                                             //获取_folderBar 对象的框架
            CGFloat width = _folderBar.frame.size.width;
                                             //获取_folderBar 对象的宽度
            if (folderBarHidden) {
                folderBarFrame.origin.x -= width;
                                             //设置 folderBarFrame 对象的 x 值
            } else {
                [_folderBar setHidden:folderBarHidden];
                folderBarFrame.origin.x += width;
                                             //设置 folderBarFrame 对象的 x 值
            }
            [_folderBar setFrame:folderBarFrame];
        };
        DTCompletionBlock completion = ^(BOOL finished){
            if(folderBarHidden)[self setFolderBarHidden:folderBarHidden];
        };
        [UIView animateWithDuration:UINavigationControllerHideShowBarDuration
         animations:animations completion:completion];        //动画
        return;
    }
    [self setFolderBarHidden:folderBarHidden];
}
```

要实现导航记录条目的轻拍，需要使用 tapFolderItem:方法。程序代码如下：

```
- (IBAction)tapFolderItem:(FolderItem *)sender
{
    NSMutableArray *folderItems = [NSMutableArray arrayWithArray:_
    folderBar.folderItems];
    NSUInteger index = [folderItems indexOfObject:sender];
    NSRange range = NSMakeRange(index + 1, folderItems.count - (index + 1));
                                                              //设置范围
    [folderItems removeObjectsInRange:range];                 //删除
    [_folderBar setFolderItems:folderItems animated:YES];
    UIViewController *viewComtroller = [self.viewControllers objectAtIndex:
    index];
    [self popToViewController:viewComtroller animated:YES];
}
```

（16）创建一个基于 UITableViewController 类的 TableViewController 类。

（17）打开 Navigation.h 文件，编写代码。实现头文件、实例变量、对象和方法等的声明。程序代码如下：

```
#import <UIKit/UIKit.h>
#import "Navigation.h"
@interface TableViewController : UITableViewController{
    NSInteger _folderNumber;
}
+ (id)demoWithFolderNumber:(NSInteger)folderNumber;
@end
```

（18）打开 TableViewController.m 文件，编写代码，实现表视图控制器的初始化、表内容的填充，以及实现响应选择的行。使用的方法如表 8-21 所示。

表 8-21　TableViewController.m 文件中方法总结

方　　法	功　　能
demoWithFolderNumber:	使用导航记录数创建表视图控制器并获取
initWithFolderNumer:	表视图控制器的初始化
numberOfSectionsInTableView:	获取表视图的块
tableView:numberOfRowsInSection:	获取行数
tableView:cellForRowAtIndexPath:	获取某一行的数据
tableView:didSelectRowAtIndexPath:	实现响应选择的行

其中，实现表视图控制器的初始化，需要使用 demoWithFolderNumber: 和 initWithFolderNumer: 方法实现。initWithFolderNumer: 方法实现表视图控制器的初始化。程序代码如下：

```
- (id)initWithFolderNumer:(NSInteger)folderNumber
{
    self = [super initWithStyle:UITableViewStylePlain];
    if (self == nil ) return nil;
    _folderNumber = folderNumber;
    //判断_folderNumber 是否等于 0
    if (_folderNumber == 0) {
        return self;
    }
    NSString *title = [NSString stringWithFormat:@"历史记录 %d", _folderNumber];
    [self setTitle:title];
    return self;
}
```

numberOfSectionsInTableView:、tableView:numberOfRowsInSection: 和 tableView:cellForRowAtIndexPath: 方法实现表的内容填充。程序代码如下：

```
//获取表视图的块
- (NSInteger)numberOfSectionsInTableView:(UITableView *)tableView
{
    return 1;
}
//获取行数
- (NSInteger)tableView:(UITableView *)tableView numberOfRowsInSection:(NSInteger)section
```

```
{
    return 3;
}
//获取某一行的数据
- (UITableViewCell *)tableView:(UITableView *)tableView cellForRowAtIndexPath:
(NSIndexPath *)indexPath
{
    static NSString *CellIdentifier = @"Cell";
    UITableViewCell *cell = [tableView dequeueReusableCellWithIdentifier:
    CellIdentifier];
    if (cell == nil) {
        //实例化对象
        cell = [[UITableViewCell alloc] initWithStyle:UITableView
        CellStyleDefault reuseIdentifier:CellIdentifier];
        [cell setAccessoryType:UITableViewCellAccessoryDisclosureIndicator];
                                                            //设置附件类型
    }
    if (indexPath.row == 0) {
        [cell.textLabel setText:@"Push next folder"];       //设置文本内容
    }
    if (indexPath.row == 1) {
        [cell.textLabel setText:@"Puth and hide folderBar"];//设置文本内容
    }
    if (indexPath.row == 2) {
        [cell.textLabel setText:@"Back"];                   //设置文本内容
        [cell setAccessoryType:UITableViewCellAccessoryNone];//设置附件类型
    }
    return cell;
}
```

要实现响应选择的行，需要使用 tableView:didSelectRowAtIndexPath:方法。程序代码如下：

```
- (void)tableView:(UITableView *)tableView didSelectRowAtIndexPath: (NSIndexPath *)indexPath
{
    [tableView deselectRowAtIndexPath:indexPath animated:YES];
    Navigation *navigation = (Navigation *)self.navigationController;
    //实例化对象
    TableViewController*demo = [TableViewController demoWithFolderNumber:_
    folderNumber + 1];
    //判断选择的行
    if (indexPath.row == 0) {
        [self.navigationController pushViewController:demo animated:YES];
    }
    if (indexPath.row == 1) {
        [navigation setFolderBarHidden:YES animated:YES];
        [navigation pushViewController:demo animated:YES];
    }
    if (indexPath.row == 2) {
        [self dismissViewControllerAnimated:YES completion:nil];
    }
}
```

（19）打开 ViewController.m 文件，编写代码，实现表视图的创建和填充，以及响应选择的行。使用的方法如表 8-22 所示。

表 8-22 ViewController.m文件中方法总结

方　　法	功　　能
viewDidload	视图加载后调用，实现初始化
numberOfSectionsInTableView:	获取表视图的块
tableView:numberOfRowsInSection:	获取行数
tableView:cellForRowAtIndexPath:	获取某一行的数据
tableView:didSelectRowAtIndexPath:	实现响应选择的行

表视图的创建及填充需要使用 viewDidload、numberOfSectionsInTableView:、tableView:numberOfRowsInSection:和 tableView:cellForRowAtIndexPath:方法实现。程序代码如下：

```objectivec
- (void)viewDidLoad
{
    [super viewDidLoad];
    // Do any additional setup after loading the view, typically from a nib.
    UITableView *table = [[UITableView alloc] initWithFrame:self.view.bounds style:UITableViewStylePlain];       //实例化对象
    [table setDataSource:self];
    [table setDelegate:self];
    [self setView:table];
}
//获取表视图的块
- (NSInteger)numberOfSectionsInTableView:(UITableView *)tableView
{
    return 1;
}
//获取行数
- (NSInteger)tableView:(UITableView *)tableView numberOfRowsInSection:(NSInteger)section
{
    return 2;
}
//获取某一行的数据
- (UITableViewCell *)tableView:(UITableView *)tableView cellForRowAtIndexPath:(NSIndexPath *)indexPath
{
    static NSString *CellIdentifier = @"Cell";
    UITableViewCell *cell = [tableView dequeueReusableCellWithIdentifier:CellIdentifier];
    if (cell == nil) {
        cell = [[UITableViewCell alloc] initWithStyle:UITableViewCellStyleDefault reuseIdentifier:CellIdentifier];
        [cell setAccessoryType:UITableViewCellAccessoryDisclosureIndicator];
                                              //设置附件类型
    }
    if (indexPath.row == 0) {
        [cell.textLabel setText:@"Normal Style"];        //设置文本内容
    }
    if (indexPath.row == 1) {
        [cell.textLabel setText:@"Fixed Left Home Style"];
    }
    return cell;
}
```

要实现响应选择的行，需要使用 tableView:didSelectRowAtIndexPath:方法。程序代码如下：

```
- (void)tableView:(UITableView *)tableView didSelectRowAtIndexPath:(NSIndex
Path *)indexPath
{
    [tableView deselectRowAtIndexPath:indexPath animated:YES];
    DTFolderBarStyle style;
    //判断是否 row 的索引是否为 0
    if (indexPath.row == 0) {
        style = DTFolderBarStyleNormal;
    }
    if (indexPath.row == 1) {
        style = DTFolderBarStyleFixedLeftHome;
    }
    TableViewController *demo = [TableViewController demoWithFolderNumber:0];
                                                    //实例化对象
    [demo setModalTransitionStyle:UIModalTransitionStyleFlipHorizontal];
    Navigation *navigation = [Navigation navigationWithRootViewController:demo
    folderStyle:style];
    [self presentViewController:navigation animated:YES completion:nil];
                                                    //以动画的形式显示视图控制器
}
```

（20）打开 Main.storyboard 文件，添加 Navigation Controller 导航控制器到画布中，将此控制器关联的根视图改为 View Controller 视图控制器。将 Navigation Controller 导航控制器的 Class 设置为 Navigation。

【代码解析】

由于本实例中的代码和方法非常的多，为了方便读者的阅读，笔者绘制了一些执行流程图如图 8.21～8.24 所示。其中，单击运行按钮后一直到在模拟器上显示具有导航栏的界面，需要使用 viewDidLoad（Navigation.m）、folderBarWithFrame:style:、initWithFrame:style:、viewWillAppear:、viewDidload（ViewController.m）、numberOfSectionsInTableView（ViewController.m）、tableView:numberOfRowsInSection:（ViewController.m）和 tableView:cellForRowAtIndexPath:（ViewController.m）方法共同实现。它们的执行流程如图 8.21 所示。

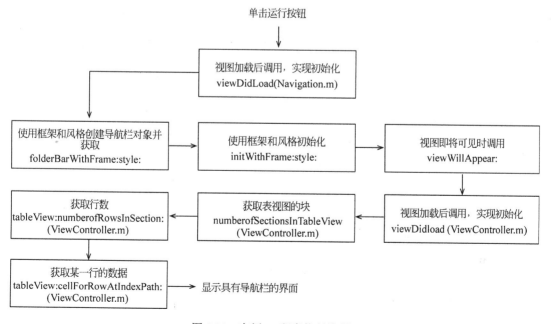

图 8.21 实例 57 程序执行流程 1

如果想要显示不同风格的导航栏，可以选择表视图中的内容，以下以选择内容为 Normal Style 的行为例，实现的执行流程如图 8.22 所示。

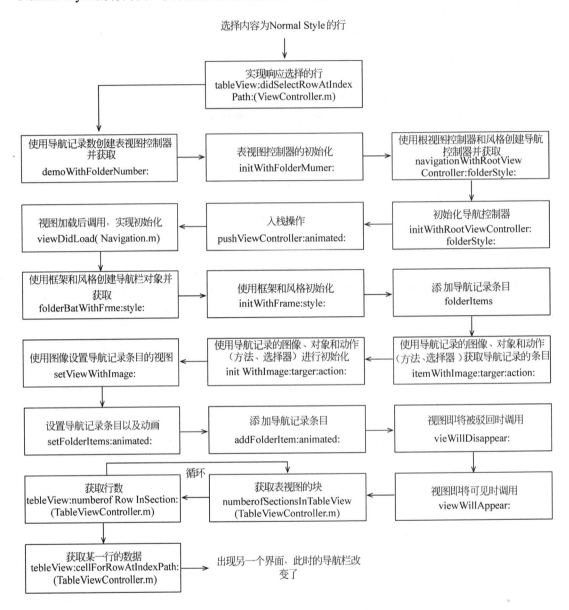

图 8.22　实例 57 程序执行流程 2

本实例的重点就是在导航栏中显示导航记录，需要使用 tableView:didSelectRowAtIndexPath:（TableViewController.m）、demoWithFolderNumber:、initWithFolderNumer:、pushViewController:animated:、folderItems、itemWithFolderName:targer: action:、initWithFolderName:targer: action:、setViewWithFolderName:、textLable、setFolderItems:animated:、addFolderItem:animated:、numberOfSectionsInTableView（TableViewController.m）和 tableView:numberOfRowsInSection:（TableViewController.m）方法共同实现。它们的执行流程如图 8.23 所示。

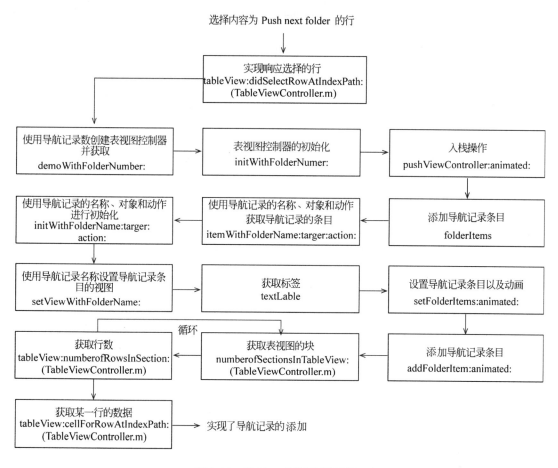

图 8.23 实例 57 程序执行流程 3

在本实例中添加的导航记录条目也是可以删除的，用户只需轻拍条目，就可以将它之后的导航栏删除，需要使用 tapItem:、tapFolderItem:、folderItems、setFolderItems:animated: 和 deleteFolderItem:animated:方法共同实现。它们的执行流程如图 8.24 所示。

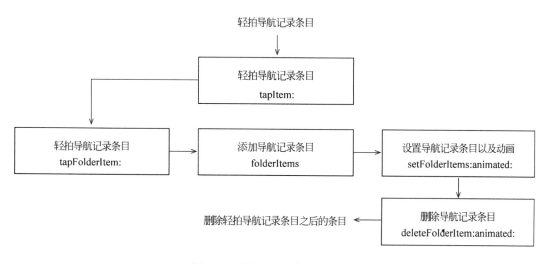

图 8.24 实例 57 程序执行流程 4

第 9 章 标 签 栏

在 iPhone 或者 iPad 中，除了导航栏可以实现视图切换外，标签栏也可以实现视图切换的功能。它的优势在于可以使用户的选择明确。当用户选择标签栏上的条目时，相应的视图就会出现。本章将主要讲解标签栏的相关实例。

实例 58　右上角带有数字的标签栏

【实例描述】

在 iPhone 中，为了显示软件更新或者来电的个数等以提醒用户，在此应用程序的右上角会出现一些带文本的红色图标。在标签栏中也是不例外的。本实例实现的功能就是在标签栏的右上角添加一个具有数字的红色图标。运行效果如图 9.1 所示。

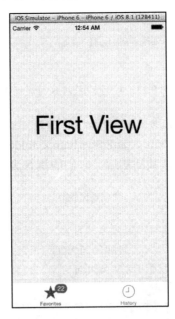

图 9.1　实例 58 运行效果

【实现过程】

（1）创建一个项目，命名为"右上角带有数字的标签栏"。

（2）打开 Main.storyboard 文件，删除 View Controller 视图控制器，添加 Tab Bar Controller 标签栏控制器，对标签栏控件器的设计界面进行设计，效果如图 9.2 所示。

第 9 章　标签栏

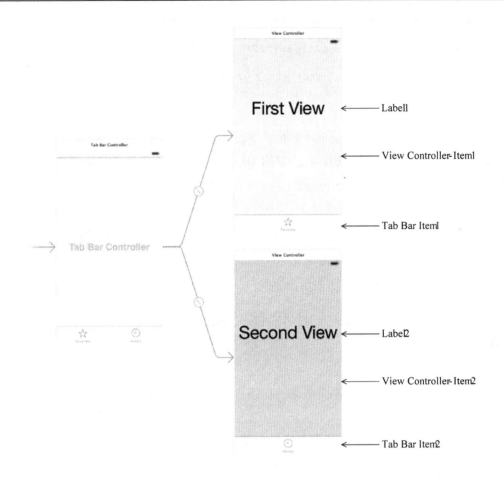

图 9.2　标签栏控件器的设计界面

需要添加的视图、控件以及对它们的设置如表 9-1 所示。

表 9-1　实例 58 视图、控件设置

视图、控件	属 性 设 置	其 他
View Controller-Item1 的设计界面	Background：浅蓝色	
Tab Bar Item1	Identifier：Favorites	Class：ViewController
Label1	Text：First View Font：System 52.0 Alignment：居中	
View Controller-Item2 的设计界面	Background：粉色	
Tab Bar Item2	Identifier：History	
Label2	Text：Second View Font：System 52.0 Alignment：居中	

（3）打开 ViewController.m 文件，编写代码，实现在标签栏的右上方出现数字。程序代码如下：

```
- (void)viewDidLoad
{
    self.tabBarItem.badgeValue=@"22";              //在控件的右上角显示"22"
    [super viewDidLoad];
    // Do any additional setup after loading the view, typically from a nib.
}
```

【代码解析】

本实例关键功能是右上角数字的添加。下面就是这个知识点的详细讲解。

要想实现右上角数字的添加，需要使用 UITabBarItem 的 badgeValue 属性，其语法形式如下：

```
@property(nonatomic, copy) NSString *badgeValue;
```

在代码中，使用 badgeValue 属性在标签栏条目的右上方添加了一个数字，代码如下：

```
self.tabBarGradientLayer.colors=gradientColors;
```

实例 59　具有渐变效果的标签栏

【实现描述】

本实例实现的是一个具有渐变效果的标签栏，从而使标签栏更为突出和别致。运行效果如图 9.3 所示。

图 9.3　实例 59 运行效果

【实现过程】

在标签栏控件的图层上再插入一个具有渐变的图层。具体的实现步骤如下。

（1）创建一个项目，命名为"具有渐变效果的标签栏"。

（2）创建一个基于 UITabBar 类的 TabBar 类。

（3）打开 TabBar.h 文件，编写代码，实现属性和方法的声明。程序代码如下：

```
#import <UIKit/UIKit.h>
@interface TabBar : UITabBar
//属性
@property (nonatomic, assign) BOOL gradientVerticalFlag;
@property (nonatomic, strong) CAGradientLayer *tabBarGradientLayer;
//方法
-(void)setTabBarGradientTintColors:(NSArray*)barTintGradientColors verticalFlag:(BOOL)flag;
@end
```

（4）打开 TabBar.m 文件，编程代码，实现为标签栏添加渐变色。首先，使用 setTabBarGradientTintColors:verticalFlag:方法实现对标签栏填充渐变颜色和是否实现垂直填充进行设置。程序代码如下：

```
- (void)setTabBarGradientTintColors:(NSArray *)barTintGradientColors verticalFlag:(BOOL)flag
{
    self.gradientVerticalFlag = flag;
    if (!self.tabBarGradientLayer) {
        //对渐变图层对象进行创建和设置
        self.tabBarGradientLayer = [CAGradientLayer layer];
        self.tabBarGradientLayer.opacity = 0.5;                 //设置可见度
        self.tabBarGradientLayer.frame = self.bounds;           //设置框架
        self.tabBarGradientLayer.colors = barTintGradientColors;
        if(flag)
        {
            self.tabBarGradientLayer.startPoint = CGPointMake(0.0, 0.5);
                                                                //设置开始点
            self.tabBarGradientLayer.endPoint = CGPointMake(1.0, 0.5);
                                                                //设置结束点
        }
    }
    //遍历，并为 gradientColors 数组添加对象
    NSMutableArray *gradientColors = [NSMutableArray array];
    //遍历
    for (id Color in barTintGradientColors) {
        if([Color isKindOfClass:[UIColor class]])
            [gradientColors addObject:(id)[Color CGColor]];    //添加对象
        else
            [gradientColors addObject:Color];
    }
    self.tabBarGradientLayer.colors=gradientColors;             //设置渐变颜色组
}
```

在 drawRect:方法中实现为标签栏的图层插入具有渐变的图层。程序代码如下：

```
- (void)drawRect:(CGRect)rect
{
    [self.layer insertSublayer:self.tabBarGradientLayer atIndex:1];
                //插入
}
```

（5）打开 ViewController.h 文件，编写代码，实现头文件、插座变量和对象的声明。程序代码如下：

```
#import <UIKit/UIKit.h>
#import "TabBar.h"
@interface ViewController : UIViewController{
    IBOutlet TabBar *tabbar1;
    IBOutlet TabBar *tabbar2;                              //插座变量
    NSArray *color;
}
@end
```

（6）打开 Main.storyboard 文件，对 View Controller 视图控制器的设计界面进行设计，效果如图 9.4 所示。

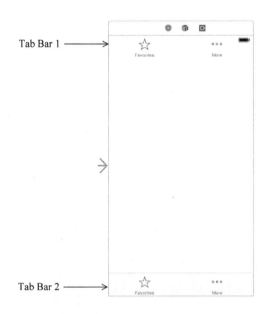

图 9.4　实例 59View Controller 视图控制器的设计界面

需要添加的视图、控件以及对它们的设置如表 9-2 所示。

表 9-2　实例 59 视图、控件设置

视图、控件	属 性 设 置	其　　他
Tab Bar1		Class：TabBar 与插座变量 tabbar1 关联
Tab Bar2		Class：TabBar 与插座变量 tabbar1 关联

（7）打开 ViewController.m 文件，编写代码，在 viewDidLoad 方法中实现具有渐变效果的标签栏对象的设置。程序代码如下：

```
- (void)viewDidLoad
{
    [super viewDidLoad];
    // Do any additional setup after loading the view, typically from a nib.
    //设置 tabbar1 标签栏
```

```
    color=[NSArray arrayWithObjects:[UIColor cyanColor],[UIColor orange
Color],[UIColor greenColor], nil];
    [tabbar1 setTabBarGradientTintColors:color verticalFlag:NO];
    //设置 tabbar2 标签栏
    color=[NSArray arrayWithObjects:[UIColor cyanColor],[UIColor brown
Color], nil];
    [tabbar2 setTabBarGradientTintColors:color verticalFlag:YES];
}
```

【代码解析】

本实例关键功能是渐变颜色的添加。下面就是这个知识点的详细讲解。要实现渐变颜色的添加，需要实现以下 3 步。

1. 创建并设置渐变图层

如果想要添加渐变层，首先要创建一个渐变图层，在本实例中代码如下：

```
self.tabBarGradientLayer = [CAGradientLayer layer];
……
self.tabBarGradientLayer.colors = barTintGradientColors;
```

2. 设置渐变图层的颜色组

要对渐变图层的颜色组进行设置，需要使用 CAGradientLayer 的 colors 属性，其语法形式如下：

```
@property(copy) NSArray *colors;
```

在代码中，使用了 colors 属性对渐变图层的颜色组进行了设置，代码如下：

```
self.tabBarGradientLayer.colors=gradientColors;
```

3. 将渐变颜色图层添加到标签栏的图层上

要想将渐变颜色图层添加到标签栏的图层上，需要使用 CALayer 的 insertSublayer:atIndex:方法，其语法形式如下：

```
- (void)insertSublayer:(CALayer *)aLayer atIndex:(unsigned)index;
```

其中，(CALayer *)aLayer 表示要插入的图层；(unsigned)index 表示索引。在代码中，使用了 insertSublayer:atIndex:方法将渐变图层添加到了标签栏的图层上，代码如下：

```
[self.layer insertSublayer:self.tabBarGradientLayer atIndex:1];
```

其中，self.tabBarGradientLayer 表示插入的图层；1 表示索引。

实例 60　中间凸起的标签栏

【实现描述】

在大多数情况下，用户看到的标签栏是一个矩形的样子。这种形式的标签栏往往缺乏了自己的特点和相应重点强调的部分。本实例就在这个方面做了改进，实现了一个中间凸起的标签栏。运行效果如图 9.5 所示。

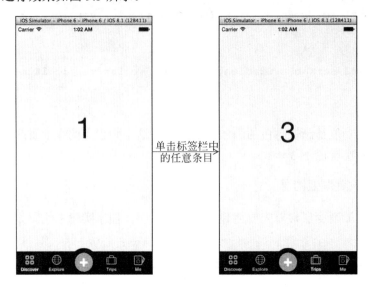

图 9.5 实例 60 运行效果

【实现过程】

（1）创建一个项目，命名为"中间凸起的标签栏"。

（2）添加图片 1.png、2.png、3.png、4.png、5.png 和 6.png 到创建项目的 Supporting Files 文件夹中。

（3）创建一个基于 UIView 类的 TabBar 类。

（4）打开 TabBar.h 文件，编写代码，实现协议和属性的声明。程序代码如下：

```
#import <UIKit/UIKit.h>
//协议
@protocol tabbarDelegate <NSObject>
-(void)touchBtnAtIndex:(NSInteger)index;
@end
@class tabbarView;
@interface TabBar : UIView
//属性
@property(nonatomic,strong) UIImageView *tabbarView;
@property(nonatomic,strong) UIImageView *tabbarViewCenter;
……
@property(nonatomic,weak) id<tabbarDelegate> delegate;
@end
```

（5）打开 TabBar.m 文件。编写代码，实现对标签栏的绘制以及选择条目。使用的方法如表 9-3 所示。

表 9-3 TabBar.m文件中的方法总结

方　　法	功　　能
initWithFrame:	初始化标签栏
layoutView	布局子视图
layoutBtn	布局按钮
btn1Click:	单击按钮，实现条目的选择

这里需要讲解几个重要的方法（其他方法请读者参考源代码）。要绘制标签栏，需要使用 initWithFrame:、layoutView 和 layoutBtn 方法实现。其中，layoutView 方法实现对标签栏用到的视图进行布局。程序代码如下：

```
-(void)layoutView
{
    //创建并设置图像视图对象，作为标签栏
    _tabbarView = [[UIImageView alloc]initWithImage:[UIImage image Named:@"1.png"]];
    [_tabbarView setFrame:CGRectMake(0, 9, _tabbarView.bounds.size.width, 51)];                       //设置框架
    [_tabbarView setUserInteractionEnabled:YES];
    //创建并设置图像视图对象，作为中间凸起标签栏的背景
    _tabbarViewCenter=[[UIImageView alloc]initWithImage:[UIImage image Named:@"5.png"]];
    _tabbarViewCenter.center=CGPointMake(self.center.x,self. Bounds.size.height/2.0);                 //设置中心位置
    [_tabbarViewCenter setUserInteractionEnabled:YES];
    //创建并设置按钮对象，作为中间的标签栏图标
    _button_center = [UIButton buttonWithType:UIButtonTypeCustom];
    _button_center.adjustsImageWhenHighlighted = YES;
    //设置背景图像
[_button_centersetBackgroundImage:[UIImageimageNamed:@"6.png"]forState:UIControlStateNormal];
    [_button_center setFrame:CGRectMake(0, 0, 46, 46)];
                            //设置框架
    _button_center.center =CGPointMake(_tabbarViewCenter. bounds.size. width/ 2.0, _tabbarViewCenter.bounds.size.height/2.0 + 5) ;
    //添加对象
    [_tabbarViewCenter addSubview:_button_center];
    [self addSubview:_tabbarView];
    [self addSubview:_tabbarViewCenter];
    [self layoutBtn];
}
```

layoutBtn 方法实现对按钮的布局，这些按钮会实现标签栏中被选中条目的状态。程序代码如下：

```
-(void)layoutBtn
{
    //创建并设置按钮对象
    _button_1 = [UIButton buttonWithType:UIButtonTypeCustom];
    [_button_1 setFrame:CGRectMake(0, 0, 64, 60)];           //设置框架
    [_button_1 setTag:101];                                  //设置tag值
    [_button_1 addTarget:self action:@selector(btn1Click:) forControlEvents:UIControlEventTouchUpInside];
    _button_2 = [UIButton buttonWithType:UIButtonTypeCustom];
    [_button_2 setFrame:CGRectMake(65, 0, 64, 60)];          //设置框架
    [_button_2 setTag:102];
    [_button_2 addTarget:self action:@selector(btn1Click:) forControlEvents:UIControlEventTouchUpInside];
    _button_3 = [UIButton buttonWithType:UIButtonTypeCustom];
    [_button_3 setFrame:CGRectMake(202, 0, 64, 60)];
    [_button_3 setTag:103];                                  //设置tag值
    [_button_3 addTarget:self action:@selector(btn1Click:) forControlEvents:UIControlEventTouchUpInside];
    _button_4 = [UIButton buttonWithType:UIButtonTypeCustom];
```

```objc
    [_button_4 setFrame:CGRectMake(267, 0, 64, 60)];
    [_button_4 setTag:104];                                          //设置tag值
    [_button_4 addTarget:self action:@selector(btn1Click:) forControl
Events:UIControlEventTouchUpInside];
    //添加视图对象
    [_tabbarView addSubview:_button_1];
    [_tabbarView addSubview:_button_2];
    [_tabbarView addSubview:_button_3];
    [_tabbarView addSubview:_button_4];
}
```

btn1Click:方法是单击按钮后调用的方法,实现条目的选择。程序代码如下:

```objc
-(void)btn1Click:(id)sender
{
    UIButton *btn = (UIButton *)sender;              //实例化对象btn
    switch (btn.tag) {
        case 101:
        {
            [_tabbarView setImage:[UIImage imageNamed:@"1.png"]]; //设置图像
            [self.delegate touchBtnAtIndex:0];      //调用touchBtnAtIndex:方法
            break;
        }
        case 102:
        {
            [_tabbarView setImage:[UIImage imageNamed:@"2.png"]];
            [self.delegate touchBtnAtIndex:1];      //调用touchBtnAtIndex:方法
            break;
        }
        case 103:
            [_tabbarView setImage:[UIImage imageNamed:@"3.png"]];
            [self.delegate touchBtnAtIndex:2];      //调用touchBtnAtIndex:方法
            break;
        case 104:
            [_tabbarView setImage:[UIImage imageNamed:@"4.png"]];
            [self.delegate touchBtnAtIndex:3];      //调用touchBtnAtIndex:方法
            break;
        default:
            break;
    }
}
```

(6) 创建3个基于UIViewController类的SecondViewController、ThreeViewController和FourViewController类。

(7) 创建一个基于UITabBarController类的AATabBarController类。

(8) 打开AATabBarController.h文件,编写代码,实现头文件和属性的声明。程序代码如下:

```objc
#import <UIKit/UIKit.h>
//头文件
#import "TabBar.h"
#import "ViewController.h"
#import "SecondViewController.h"
#import "ThreeViewController.h"
#import "FourViewController.h"
@interface AATabBarController : UITabBarController<tabbarDelegate>
//属性
@property(nonatomic,strong) TabBar *tabbar;
```

```
@property(nonatomic,strong) NSArray *arrayViewcontrollers;
@end
```

(9)打开 AATabBarController.m 文件,编写代码,实现标签栏控制器的设置。使用的方法如表 9-4 所示。

表 9-4　AATabbarController.m文件中的方法总结

方　　法	功　　能
viewDidLoad	视图加载后调用,实现初始化
getViewcontrollers	得到视图控制器
touchBtnAtIndex:	实现视图切换

其中,viewDidLoad 方法实现对标签栏对象的创建以及设置等功能。程序代码如下:

```
- (void)viewDidLoad
{
  [super viewDidLoad];
  // Do any additional setup after loading the view.
  CGFloat orginHeight = self.view.frame.size.height- 60;
  _tabbar = [[TabBar alloc]initWithFrame:CGRectMake(0, orginHeight, 320, 60)];                                                      //实例化对象
  _tabbar.delegate = self;                          //设置委托
  [self.view addSubview:_tabbar];
  _arrayViewcontrollers = [self getViewcontrollers];
  [self touchBtnAtIndex:0];
}
```

getViewcontrollers 方法实现获取视图控制器。程序代码如下:

```
-(NSArray *) getViewcontrollers
{
  NSArray* tabBarItems = nil;
  //实例化视图控制器对象
  ViewController *first=[self.storyboard instantiateViewController
  WithIdentifier:@"first"];
  SecondViewController*second=[self.storyboardinstantiateView Controller
  WithIdentifier:@"second"];
  ThreeViewController *three=[self.storyboard instantiateViewController
  WithIdentifier:@"three"];
  FourViewController *four=[self.storyboard instantiateViewController
  WithIdentifier:@"four"];
  //将视图控制器放入到数组中
  tabBarItems = [NSArray arrayWithObjects:
           [NSDictionary dictionaryWithObjectsAndKeys:first, @"view
           Controller", nil],
           [NSDictionarydictionaryWithObjectsAndKeys: second, @"view
           Controller", nil],
           [NSDictionary dictionaryWithObjectsAndKeys:three, @"view
           Controller", nil],
           [NSDictionary dictionaryWithObjectsAndKeys:four, @"view
           Controller", nil],
           nil];
  return tabBarItems;
}
```

touchBtnAtIndex:方法实现单击按钮后,通过索引切换视图。程序代码如下:

```
-(void)touchBtnAtIndex:(NSInteger)index
{
```

```
    UIView* currentView = [self.view viewWithTag:SELECTED _VIEW_
    CONTROLLER_TAG];
    [currentView removeFromSuperview];                           //删除移动视图
    //获取索引对应的视图控制器
    NSDictionary* data = [_arrayViewcontrollers objectAtIndex:index];
    UIViewController *viewController = data[@"viewController"];
    viewController.view.frame = CGRectMake(0,0,self.view.frame.size.width,
    self.view.frame.size.height- 50);
    [self.view insertSubview:viewController.view belowSubview:_tabbar];
        //插入视图
}
```

（10）打开 Main.storyboard 文件，对画布进行设计。效果如图 9.6 所示。

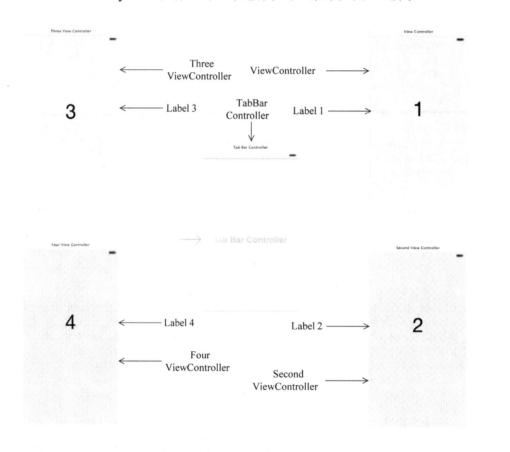

图 9.6　画布效果

需要添加的视图、控件以及对它们的设置如表 9-5 所示。

表 9-5　实例 60 视图、控件设置

视图、控件	属性设置	其　他
Tab Bar Controller		Class：AATabBarController
View Controller	Background：浅紫色	Class：ViewController
Label1	Text：1 Font：System 66.0 Alignment：居中	
Second View Controller	Background：浅绿色	Class：SecondViewController

续表

视图、控件	属 性 设 置	其 他
Label2	Text：2 Font：System 66.0 Alignment：居中	
Three View Controller	Background：浅蓝色	Class：ThreeViewController
Label3	Text：3 Font：System 66.0 Alignment：居中	
Four View Controller	Background：橘黄色	Class：FourViewController
Label4	Text：4 Font：System 66.0 Alignment：居中	

【代码解析】

本实例关键功能是标签栏条目的选择和对应的视图的切换。下面依次讲解这两个知识点。

1．标签栏条目的选择

在标签栏中，为了使用户可以清楚地看到自己选择的条目，需要使用特殊的标记。在本实例中，选择某一条目后，就会触发 btn1Click:方法。在此方法中，使用 setImage 属性将选择条目的背景颜色进行改变。代码如下：

```
switch (btn.tag) {
    case 101:
    {
        [_tabbarView setImage:[UIImage imageNamed:@"1.png"]];
    ......
    default:
        break;
}
```

2．对应的视图的切换

btn1Click:方法除了可以对背景的颜色进行改变外，还可以调用 touchBtnAtIndex:方法，这个方法实现视图切换的功能，程序会根据 touchBtnAtIndex:方法后面的索引，调用 touchBtnAtIndex:方法。通过如下代码实现视图的切换：

```
NSDictionary* data = [_arrayViewcontrollers objectAtIndex:index];
UIViewController *viewController = data[@"viewController"];
viewController.view.frame = CGRectMake(0,0,self.view.frame.size.width,
self.view.frame.size.height- 50);
[self.view insertSubview:viewController.view belowSubview:_tabbar];
            //插入视图
```

实例 61　标签栏控制器实现的视图切换效果

【实例描述】

标签栏最重要的功能就是实现视图切换，但是在视图切换时，没有任何的动画效果。

所以本实例实现的功能就是在视图切换时添加动画效果。运行效果如图 9.7 所示。

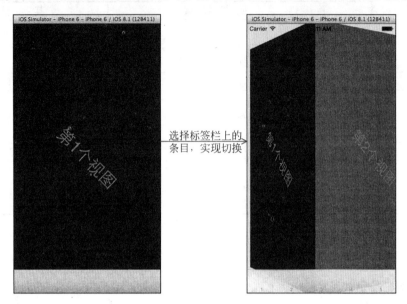

图 9.7　实例 61 运行效果

【实现过程】

当用户选择标签栏上的一个标签时，相应的视图就会出现，这时，就会出现动画效果。具体的实现步骤如下。

（1）创建一个项目，命名为"标签栏控制器实现的视图切换效果"。

（2）创建一个基于 UITabBarController 类的 TabBarViewController 类。

（3）打开 TabBarViewController.h 文件，编写代码，实现数据结构的定义和属性的声明。程序代码如下：

```
#import <UIKit/UIKit.h>
//数据结构的定义
typedef enum {
    CubeTabBarControllerAnimationNone = -1,
    CubeTabBarControllerAnimationOutside = 0,
    CubeTabBarControllerAnimationInside
} CubeTabBarControllerAnimation;
@interface TabBarViewController : UITabBarController
//属性
@property(nonatomic, assign) CubeTabBarControllerAnimation animation;
@property(nonatomic, strong) UIColor *backgroundColor;
```

（4）打开 TabBarViewController.m 文件，编写代码，在视图的切换过程中实现动画效果。需要在 setSelectedViewController:方法中实现。程序代码如下：

```
- (void)setSelectedViewController:(UIViewController *)next
{
    //判断 animation 是否为 CubeTabBarControllerAnimationNone
    if (self.animation == CubeTabBarControllerAnimationNone) {
        [super setSelectedViewController:next];    //设置选择的视图控制器
        return;
    }
```

```objc
//判断next是否为self.selectedViewController
if (next == self.selectedViewController) {
    return;
}
self.view.userInteractionEnabled = NO;
NSUInteger nextIndex = [self.viewControllers indexOfObject:next];
UIViewController *current = self.selectedViewController;
next.view.frame = current.view.frame;                       //设置框架
CGFloat halfWidth = current.view.bounds.size.width / 2.0;
CGFloat duration = 0.7;
CGFloat perspective = -1.0/1000.0;
UIView *superView = current.view.superview;                 //实例化对象
CATransformLayer *transformLayer = [[CATransformLayer alloc] init];
transformLayer.frame = current.view.layer.bounds;
[current.view removeFromSuperview];                         //删除
//添加视图对象
[transformLayer addSublayer:current.view.layer];
[transformLayer addSublayer:next.view.layer];
[superView.layer addSublayer:transformLayer];
UIColor *originalBackgroundColor = superView.backgroundColor;
superView.backgroundColor = self.backgroundColor;           //设置背景颜色
//显示事务，实现3D效果的旋转切换
[CATransaction begin];
[CATransaction setDisableActions:YES];
CATransform3D transform = CATransform3DIdentity;            //变化
//设置动画效果
switch (self.animation) {
    case CubeTabBarControllerAnimationOutside:
        transform = CATransform3DTranslate(transform, 0, 0, -halfWidth);
                                                            //移动
        transform = CATransform3DRotate(transform, (nextIndex >
        self.selectedIndex) ? M_PI_2 : -M_PI_2, 0, 1, 0);   //旋转
        transform = CATransform3DTranslate(transform, 0, 0, halfWidth);
        break;
    case CubeTabBarControllerAnimationInside:
        transform = CATransform3DTranslate(transform, 0, 0, halfWidth);
        transform = CATransform3DRotate(transform, (nextIndex >
        self.selectedIndex) ? -M_PI_2 : M_PI_2, 0, 1, 0);
        transform = CATransform3DTranslate(transform, 0, 0, -halfWidth);
                                                            //移动
        break;
    default:
        break;
}
next.view.layer.transform = transform;                      //设置图层的改变
[CATransaction commit];
[CATransaction begin];
//动画结束时调用的块
[CATransaction setCompletionBlock:^(void) {
    [next.view.layer removeFromSuperlayer];                 //移除指定的图层
    next.view.layer.transform = CATransform3DIdentity;
    [current.view.layer removeFromSuperlayer];
    superView.backgroundColor = originalBackgroundColor;    //背景颜色
    [superView addSubview:current.view];                    //添加视图对象
    [transformLayer removeFromSuperlayer];
    [super setSelectedViewController:next];                 //设置选择的视图控制器
    self.view.userInteractionEnabled = YES;
}];
```

```
    CABasicAnimation *transformAnimation = [CABasicAnimation animation
    WithKeyPath:@"transform"];
    transform = CATransform3DIdentity;
    transform.m34 = perspective;                    //透视效果
    transformAnimation.fromValue = [NSValue value WithCATransform3D: trans
    form];                                           //动画的开始值
    transform = CATransform3DIdentity;
    transform.m34 = perspective;
    switch (self.animation) {
        //设置动画效果
        case CubeTabBarControllerAnimationOutside:
            transform = CATransform3DTranslate(transform, 0, 0,
            -halfWidth);
            transform = CATransform3DRotate(transform, (nextIndex >
            self.selectedIndex) ? -M_PI_2 : M_PI_2, 0, 1, 0);    //旋转
            transform = CATransform3DTranslate(transform, 0, 0, halfWidth);
            break;
        case CubeTabBarControllerAnimationInside:
            transform = CATransform3DTranslate(transform, 0, 0,
            halfWidth);                              //移动
            transform = CATransform3DRotate(transform, (nextIndex >
            self.selectedIndex) ? M_PI_2 : -M_PI_2, 0, 1, 0);
            transform = CATransform3DTranslate(transform, 0, 0,
            -halfWidth);
            break;
        default:
            break;
    }
    transformAnimation.toValue = [NSValue valueWithCATrans form3D: trans
    form];                                           //动画的结束值
    transformAnimation.duration = duration;          //设置动画所需时间
    [transformLayer addAnimation:transformAnimation forKey:@"rotate"];
    transformLayer.transform = transform;
    [CATransaction commit];
}
```

（5）打开 AppDelegate.h 文件，编写代码，实现头文件和属性的声明。程序代码如下：

```
#import <UIKit/UIKit.h>
#import "TabBarViewController.h"
@interface AppDelegate : UIResponder <UIApplicationDelegate, UITabBar
ControllerDelegate>
//属性
@property(nonatomic, strong) TabBarViewController *cubeTabBarController;
@property(nonatomic, strong) NSArray *colorArray;
@property (strong, nonatomic) UIWindow *window;
@end
```

（6）打开 AppDelegate.m 文件，编写代码。在 application:didFinishLaunchingWithOptions: 方法中实现视图控制器对象的创建，以及将创建的视图控制器添加到标签栏中，实现视图的动画切换。程序代码如下：

```
- (BOOL)application:(UIApplication *)application didFinishLaunching With
Options:(NSDictionary *)launchOptions
{
    self.window = [[UIWindow alloc] initWithFrame:[[UIScreen mainScreen]
    bounds]];
    [self.window makeKeyAndVisible];
    self.colorArray = [[NSArray alloc] initWithObjects:[UIColor redColor],
```

```objc
        [UIColor orangeColor], [UIColor yellowColor], [UIColor greenColor],
        [UIColor blueColor], [UIColor purpleColor], nil];         //实例化对象
    NSMutableArray *array = [[NSMutableArray alloc] init];
    //创建并设置视图控制器对象和标签栏条目对象
    for (int ii = 0; ii < 5; ii++) {
        UIViewController *vc = [[UIViewController alloc] initWithNibName:
        nil bundle:nil];
        vc.view.backgroundColor = [UIColor colorWithWhite:(ii+1) *0.2
        alpha:1.0];//设置背景
        UILabel *label = [[UILabel alloc] init];
        label.font = [UIFont systemFontOfSize:36];              //设置字体
        label.text = [NSString stringWithFormat:@"第%d 个视图", ii+1];
        label.textColor=[UIColor orangeColor];                  //设置字体颜色
        [label sizeToFit];
        vc.tabBarItem = [[UITabBarItem alloc] initWithTitle:[NSString
        stringWithFormat:@"%d", ii+1] image:nil tag:0];
                                            //标签栏条目对象的创建
        [vc.view addSubview:label];
        label.center = vc.view.center;
        label.transform = CGAffineTransformMakeRotation(M_PI_4);   //旋转
        [array addObject:vc];
    }
    self.cubeTabBarController = [[TabBarViewController alloc] initWith
    NibName:nil bundle:nil];              //创建
    self.cubeTabBarController.delegate = self;
    self.cubeTabBarController.backgroundColor = [self.colorArray object
    AtIndex:2];                           //背景颜色的设置
    self.cubeTabBarController.viewControllers = array;
    self.window.rootViewController = self.cubeTabBarController;
                                            //设置根视图
    return YES;
}
```

【代码解析】

本实例关键功能是视图切换时的立方体动画。下面就是这个知识点的详细讲解。

在本实例中立方体的动画效果是通过显示事务实现的（显示事务是实现显示动画的一种方式）。在显示事务中一般通过 CATransaction 的 begin 和 commit 方法实现动画。要实现立方体的效果，需要 CATransform3DTranslate、CATransform3DRotate 和 CATransform3DTranslate 函数修改 CATransform3D 的数据结构。程序代码如下：

```objc
switch (self.animation) {
  case CubeTabBarControllerAnimationOutside:
    transform = CATransform3DTranslate(transform, 0, 0, -halfWidth);
    transform = CATransform3DRotate(transform, (nextIndex > self.
    selectedIndex) ? M_PI_2 : -M_PI_2, 0, 1, 0);
    transform = CATransform3DTranslate(transform, 0, 0, halfWidth);
    break;
  case CubeTabBarControllerAnimationInside:
    transform = CATransform3DTranslate(transform, 0, 0, halfWidth);
    transform = CATransform3DRotate(transform, (nextIndex > self.
    selectedIndex) ? -M_PI_2 : M_PI_2, 0, 1, 0);
    transform = CATransform3DTranslate(transform, 0, 0, -halfWidth);
    break;
  default:
    break;
}
```

实例 62 具有动画效果的标签栏

【实例描述】

本实例实现的功能是一个具有动画效果的标签栏，它可以使用户清楚地看到当前选择的条目。运行效果如图 9.8 所示。

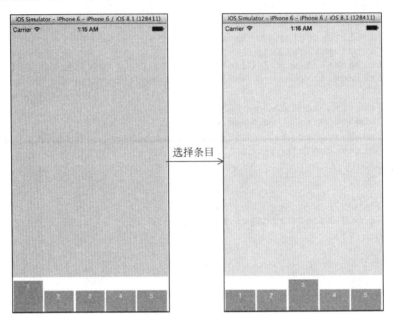

图 9.8 实例 62 运行效果

【实现过程】

当用户选择标签栏上的一个标签时，相应的视图就会出现，并且选择的条目会伸长，之前选择的条目会变短。具体的实现步骤如下。

（1）创建一个项目，命名为"具有动画效果的标签栏"。

（2）添加图片 1.png 和 2.png 到创建项目的 Supporting Files 文件夹中。

（3）创建一个基于 UIView 类的 Tabbar 类。

（4）打开 Tabbar.h 文件，编写代码，实现协议、对象、属性和方法的声明。程序代码如下：

```
#import <UIKit/UIKit.h>
//协议
@protocol TabbarDelegate<NSObject>
-(void)PlaylistBtnClick;
-(void)ArtistBtnClick;
-(void)AlbumBtnClick;
-(void)AllSongsBtnClick;
-(void)MoreBtnClick;
@end
@interface Tabbar : UIView{
    //对象
```

```
    UIButton *PlaylistBtn;
    UIButton *ArtistBtn;
    UIButton *AlbumBtn;
    UIButton *AllSongsBtn;
    UIButton *MoreBtn;
}
//属性
@property(nonatomic,strong)id <TabbarDelegate> delegate;
@property(nonatomic,strong) UIButton *PlaylistBtn;
……
@property(nonatomic,strong) UIButton *MoreBtn;
-(void)buttonClickAction:(id)sender;                        //方法
@end
```

（5）打开 Tabbar.m 文件，编写代码，实现具有动画的标签栏的绘制。使用的方法如表 9-6 所示。

表 9-6　Tabbar.m 文件中方法总结

方　　法	功　　能
initWithFrame:	具有动画效果标签栏的初始化
checkSelectedBtn:	检测选择的按钮
callButtonAction:	调用按钮的动作
buttonClickAction:	单击按钮，被选择的按钮弹出

要实现具有动画效果的标签栏的绘制需要实现两个部分：绘制标签栏；实现标签栏的动画效果。这里需要讲解几个重要的方法（其他方法请读者参考源代码）。要绘制标签栏，需要使用 initWithFrame:方法实现，程序代码如下：

```
- (id)initWithFrame:(CGRect)frame
{
    CGRect frame1=CGRectMake(frame.origin.x, frame.origin.y, 320, 165);
    self = [super initWithFrame:frame1];
    if (self) {
        [self setBackgroundColor:[UIColor whiteColor]];     //设置背景颜色
        PlaylistBtn=[UIButton buttonWithType:UIButtonTypeCustom];
        [PlaylistBtn setFrame:CGRectMake(3, 7, 60, 71)];    //设置框架
        [PlaylistBtn setSelected:YES];
        [PlaylistBtn setTag:1];
        [PlaylistBtn.titleLabel setFont:[UIFont systemFontOfSize:12]];
                                                            //设置字体
        [PlaylistBtn setContentVerticalAlignment:UIControlContentVerticalAlignmentTop];
        PlaylistBtn.contentEdgeInsets = UIEdgeInsetsMake(4, 0, 0, 0);
        [PlaylistBtn setBackgroundImage:[UIImage imageNamed:@"2.png"]
            forState:UIControlStateNormal];                 //设置背景图像
        [PlaylistBtn setBackgroundImage:[UIImage imageNamed:@"2.png"]
            ForState:UIControlStateSelected];               //设置背景图像
        [PlaylistBtn setTitle:@"1" forState:UIControlStateNormal];
        [PlaylistBtn addTarget:self action:@selector(buttonClickAction:)
            forControlEvents:UIControlEventTouchUpInside];  //添加动作
        [self addSubview:PlaylistBtn];
        ……
        MoreBtn=[UIButton buttonWithType:UIButtonTypeCustom];
        [MoreBtn setFrame:CGRectMake(256, 27, 60, 71)];
        [MoreBtn setTag:5];                                 //设置 tag 值
        [MoreBtn.titleLabel setFont:[UIFont systemFontOfSize:12]];
```

```objc
    [MoreBtn setContentVerticalAlignment:UIControlContent Vertical
    AlignmentTop];
    MoreBtn.contentEdgeInsets = UIEdgeInsetsMake(4, 0, 0, 0);
    [MoreBtn setBackgroundImage:[UIImage imageNamed:@"2.png"]
    forState: UIControlStateNormal];                        //设置背景图像
    [MoreBtn setBackgroundImage:[UIImage imageNamed:@"2.png"]
    forState:UIControlStateSelected];
    [MoreBtn setTitle:@"5" forState:UIControlStateNormal];  //设置标题
    [MoreBtn addTarget:self action:@selector(buttonClickAction:)
    forControlEvents:UIControlEventTouchUpInside];
    [self addSubview:MoreBtn];
    }
    return self;
}
```

要实现标签栏的动画效果，需要使用 buttonClickAction:、callButtonAction: 和 checkSelectedBtn:方法共同实现。其中，buttonClickAction:方法实现单击按钮，被选择的按钮以动画形式弹出，与没有选择的按钮区分。程序代码如下：

```objc
-(void)buttonClickAction:(id)sender{
    UIButton *btn=(UIButton *)sender;
    CGRect rec=btn.frame;                        //获取btn的框架
    //判断
    if (!btn.selected) {
        //弹出
        [UIView animateWithDuration:0.3 animations:^{
            //设置框架
            [btn setFrame:CGRectMake(rec.origin.x, rec.origin.y-20, rec.
            size. width, rec.size.height)];
        } completion:^(BOOL finished) {
        }];
        btn.selected=YES;
        [self checkSelectedBtn:btn];
        [self  callButtonAction:btn];
    }
}
```

checkSelectedBtn:方法实现对选择按钮的检测，就是判断在 initFrame:方法中创建的按钮对象是否是被选择状态的，如果不是，就将按钮以动画形式弹回，和没有选择的按钮会在一起。程序代码如下：

```objc
-(void)checkSelectedBtn:(UIButton *)sender{
    int buttonTag=sender.tag;                        //获取tag值
    if (PlaylistBtn.selected && PlaylistBtn.tag!=buttonTag) {
        CGRect TempFrame=PlaylistBtn.frame;      //获取PlaylistBtn的框架
        //弹回
        [UIView animateWithDuration:0.3 animations:^{
            [PlaylistBtn setFrame:CGRectMake(TempFrame.origin.x, TempFrame.
            origin.y+20, TempFrame.size.width, TempFrame.size.height)];
            //设置框架
        }completion:^(BOOL finished) {
        }];
        PlaylistBtn.selected=NO;
    }
    ……
    else if (MoreBtn.selected &&MoreBtn.tag!=buttonTag) {
```

```
        CGRect TempFrame=MoreBtn.frame;              //获取 MoreBtn 的框架
        [UIView animateWithDuration:0.3 animations:^{
            [MoreBtn setFrame:CGRectMake(TempFrame.origin.x, TempFrame.
            origin.y+20, TempFrame.size.width, TempFrame.size.height)];
            //设置框架
        }completion:^(BOOL finished) {
        }];
        MoreBtn.selected=NO;
    }
}
```

callButtonAction:方法实现调用被选择按钮的动作。程序代码如下：

```
-(void)callButtonAction:(UIButton *)sender{
    int value=sender.tag;
    //判断 value 值是否为 1
    if (value==1) {
        [self.delegate PlaylistBtnClick];
    }
    //判断 value 值是否为 2
    if (value==2) {
        [self.delegate ArtistBtnClick];
    }
    //判断 value 值是否为 3
    if (value==3) {
        [self.delegate AlbumBtnClick];
    }
    //判断 value 值是否为 4
    if (value==4) {
        [self.delegate AllSongsBtnClick];
    }
    //判断 value 值是否为 5
    if (value==5) {
        [self.delegate MoreBtnClick];
    }
}
```

（6）打开 ViewController.h 文件，编写代码，实现头文件和对象的声明。程序代码如下：

```
#import <UIKit/UIKit.h>
#import "Tabbar.h"
@interface ViewController : UIViewController<TabbarDelegate>{
    Tabbar *tabbar;
}
```

（7）打开 ViewController.m 文件，编写代码，实现具有动画效果的标签栏对象的创建与设置，以及被选择按钮的动作。程序代码如下：

```
- (void)viewDidLoad
{
    [super viewDidLoad];
    // Do any additional setup after loading the view, typically from a nib.
    tabbar=[[Tabbar alloc]initWithFrame:CGRectMake(0, 500, 0, 0)];
        //实例化对象
```

```
        tabbar.delegate=self;
        [self.view addSubview:tabbar];                          //添加对象
        self.view.backgroundColor=[UIColor cyanColor];
}
-(void)PlaylistBtnClick{
        self.view.backgroundColor=[UIColor cyanColor];          //设置视图的背景
}
-(void)ArtistBtnClick{
        self.view.backgroundColor=[UIColor orangeColor];        //设置视图的背景
}
-(void)AlbumBtnClick{
        self.view.backgroundColor=[UIColor yellowColor];        //设置视图的背景
}
-(void)AllSongsBtnClick{
        self.view.backgroundColor=[UIColor colorWithRed:0.5 green:0.5 blue:
        0.5 alpha:0.9];
}
-(void)MoreBtnClick{
        self.view.backgroundColor=[UIColor colorWithRed:0.7 green:0.3 blue:
        0.5 alpha:0.9];
}
```

【代码解析】

本实例关键功能是标签栏条目的弹出和弹回。下面就是这个知识点的详细讲解。

在本实例中，标签栏的条目是由按钮实现的。当用户单击按钮后，就会执行 buttonClickAction:方法，在此方法中实现了标签栏条目的弹出，以及对按钮的检测和动作的调用。首先，条目的弹出代码如下：

```
[UIView animateWithDuration:0.3 animations:^{
        [btn setFrame:CGRectMake(rec.origin.x, rec.origin.y-20, rec.
        size.width, rec.size.height)];
    } completion:^(BOOL finished) {
}];
```

按钮的检测，需要调用 checkSelectedBtn:方法，在此方法中实现按钮的弹回，以 PlaylistBtn 按钮弹回为例，代码如下：

```
[UIView animateWithDuration:0.3 animations:^{
        [PlaylistBtn setFrame:CGRectMake(TempFrame.origin.x, TempFrame.
        origin.y+20, TempFrame.size.width, TempFrame.size.height)];
    }completion:^(BOOL finished) {
}];
```

实例 63 滚动的标签栏

【实例描述】

当标签栏的条目很多的时候，在标签栏中是放不下的，它会出现一个 more 的条目，将剩余的条目放到里面。为了摆脱这种效果。本实例实现了一个可以滚动的标签栏，用于显示数据的是标签。运行效果如图 9.9 所示。

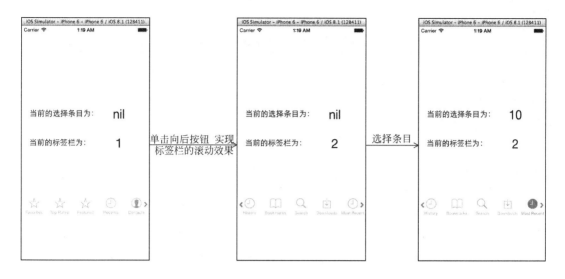

图 9.9　实例 63 运行效果

【实现过程】

标签栏是在滚动视图的基础上实现的。具体的实现步骤如下。

（1）创建一个项目，命名为"滚动的标签栏"。

（2）添加图片 1.png 和 2.png 到创建项目的 Supporting Files 文件夹中。

（3）创建一个基于 UIScrollView 类的 TabBar 类。

（4）打开 TabBar.h 文件，编写代码，实现协议、对象、属性和方法的声明。程序代码如下：

```
#import <UIKit/UIKit.h>
@class TabBar;
//协议
@protocol TabBarDelegate <NSObject>
- (void)infiniTabBar:(TabBar *)tabBar didScrollToTabBarWithTag:(int)tag;
- (void)infiniTabBar:(TabBar *)tabBar didSelectItemWithTag:(int)tag;
@end
@interface TabBar : UIScrollView<UIScrollViewDelegate, UITabBarDelegate>{
    //对象
    NSMutableArray *tabBars;
    UIButton * btnnext;
    UIButton * btnprev;
}
//属性
@property (nonatomic, assign) id <TabBarDelegate> infiniTabBarDelegate;
@property (nonatomic, retain) NSMutableArray *tabBars;
//方法
- (id)initWithItems:(NSArray *)items;
- (int)currentTabBarTag;
- (BOOL)scrollToTabBarWithTag:(int)tag animated:(BOOL)animated;
@end
```

（5）打开 TabBar.m 文件，编写代码，实现一个会滚动的标签栏。使用的方法如表 9-7 所示。

表 9-7　TabBar.m文件中方法总结

方　　法	功　　能
initWithItems:	标签栏的初始化
currentTabBarTag	当前标签栏的标签值
scrollToTabBarWithTag:animated:	使用标签值和动画实现标签栏的滚动
scrollViewDidEndDecelerating:	用户拖动，之后继续滚动
scrollViewDidEndScrollingAnimation:	在代码中滚动，并且带有动画
tabBar:didSelectItem:	选择标签栏上的条目
scrollToPreviousTabBar	标签栏的向前滚动
scrollToNextTabBar	标签栏的向后滚动

这里需要讲解几个重要的方法（其他方法请读者参考源代码）。要实现一个滚动的标签栏，首先应创建标签栏，需要使用 initWithItems:方法实现。程序代码如下：

```
- (id)initWithItems:(NSArray *)items {
    self = [super initWithFrame:CGRectMake(0.0, 411.0, 320.0, 49.0)];
    if (self) {
        self.pagingEnabled = YES;
        self.delegate = self;
        self.tabBars = [[NSMutableArray alloc] init];
        //实例化对象
        float x = 0.0;
        double ceil_items = ceil(items.count / 5.0);
        //遍历
        for (double d = 0; d < ceil(items.count / 5.0); d ++) {
            UITabBar *tabBar = [[UITabBar alloc] initWithFrame:CGRectMake(x,
                0.0, 320.0, 49.0)];//实例化
            tabBar.delegate = self;                            //设置委托
            int len = 0;
            //添加标签栏的条目
            for (int i = d * 5; i < d * 5 + 5; i ++)
                if (i < items.count)
                    len ++;
            tabBar.items = [items objectsAtIndexes:[NSIndexSet indexSet
            WithIndexesInRange:NSMakeRange(d * 5, len)]];   //设置条目
            [self addSubview:tabBar];
            [self.tabBars addObject:tabBar];                   //添加对象
            x += 320.0;
            //对向前向后滚动的按钮进行设置
            if (d == 0) {
                //创建并设置向后滚动的按钮对象
                btnnext = [UIButton buttonWithType:UIButtonTypeCustom];
                [btnnext setFrame:CGRectMake(x-15.0, 15.0, 10.0, 14.0)];
                                                              //设置框架
                [btnnext setBackgroundImage:[UIImage imageNamed:@"1.png"]
                forState:UIControlStateNormal];              //设置背景图像
                [btnnext addTarget:self action:@selector (scrollToNextTab
                Bar) forControlEvents:UIControlEventTouchUpInside];
                                                              //添加动作
                [self addSubview:btnnext];
                //创建并设置向前滚动的按钮对象
```

```
            btnprev = [UIButton buttonWithType:UIButtonTypeCustom];
            [btnprev setFrame:CGRectMake(-10.0, 15.0, 10.0, 14.0)];
                                                        //设置框架
            [btnprev setBackgroundImage:[UIImage imageNamed:@"2.png"]
            forState:UIControlStateNormal];
            [btnprev addTarget:self action:@selector(scrollToPrevious
            TabBar) forControlEvents:UIControlEventTouchUpInside];
                                                        //添加动作
            [self addSubview:btnprev];
        }
        else if ((ceil_items-1)==d) {
            btnprev = [UIButton buttonWithType:UIButtonTypeCustom];
            [btnprev setFrame:CGRectMake(x-315.0, 15.0, 10.0, 14.0)];
                                                        //设置框架
            [btnprev setBackgroundImage:[UIImage imageNamed:@"2.png"]
            forState:UIControlStateNormal];
            [btnprev addTarget:self action:@selector(scrollToPrevious
            TabBar) forControlEvents:UIControlEventTouchUpInside];
                                                        //添加动作
            [self addSubview:btnprev];
            btnnext = [UIButton buttonWithType:UIButtonTypeCustom];
            [btnnext setFrame:CGRectMake(x-335.0, 15.0, 10.0, 14.0)];
            [btnnext setBackgroundImage:[UIImage imageNamed:@"1.png"]
            forState:UIControlStateNormal];
            [btnnext addTarget:self action:@selector(scrollToNextTab
            Bar) forControlEvents:UIControlEvent Touch UpInside];;
                                                        //添加对象
            [self addSubview:btnnext];
        }
        else {
            btnnext = [UIButton buttonWithType:UIButtonTypeCustom];
            [btnnext setFrame:CGRectMake(x-15.0, 15.0, 10.0, 14.0)];
            [btnnext setBackgroundImage:[UIImage imageNamed:@"1.png"]
            forState:UIControlStateNormal]           //设置背景图像
            [btnnext addTarget:self action:@selector(scrollToNextTab
            Bar) forControlEvents:UIControlEventTouchUpInside];;
            [self addSubview:btnnext];
            btnprev = [UIButton buttonWithType:UIButtonTypeCustom];
            [btnprev setFrame:CGRectMake(x-315.0, 15.0, 10.0, 14.0)];
                                                        //设置框架
            [btnprev setBackgroundImage:[UIImage imageNamed:@"2.png"]
            forState:UIControlStateNormal];
            [btnprev addTarget:self action:@selector(scrollToPrevious
            TabBar) forControlEvents:UIControlEventTouchUpInside];
            [self addSubview:btnprev];
        }
    }
    self.contentSize = CGSizeMake(x, 49.0);
    //设置内容视图的尺寸
    }
    return self;
}
```

在本实例中滚动标签栏除了可以使用手指拖动的形式进行滚动外，还可以单击按钮进行滚动，要实现此效果，需要使用 scrollToTabBarWithTag:、scrollToPreviousTabBar 和 scrollToNextTabBar 方法。其中，scrollToTabBarWithTag:方法通过标签值和动画实现标签栏的滚动效果。程序代码如下：

```objc
- (BOOL)scrollToTabBarWithTag:(int)tag animated:(BOOL)animated {
    int countbar = [self.tabBars count];
    //遍历
    for (UITabBar *tabBar in self.tabBars)
        if ([self.tabBars indexOfObject:tabBar] == tag) {
            UITabBar *tabBar = [self.tabBars objectAtIndex:tag];
                                            //实例化对象
            [self scrollRectToVisible:tabBar.frame animated:animated];
                                            //滚动查看内容
            if (animated == NO)
                [self scrollViewDidEndDecelerating:self];
            if (tag == 0) {
                [btnprev setFrame:CGRectMake(tabBar.frame.origin.x-10.0,
                15.0, 10.0, 14.0)];         //设置框架
            }
            else if ((countbar-1)==tag) {
                [btnnext setFrame:CGRectMake(tabBar.frame.origin.x-15.0,
                15.0, 10.0, 14.0)];
                [btnprev setFrame:CGRectMake(tabBar.frame.origin.x+5.0,
                15.0, 10.0, 14.0)];         //设置框架
            }
            else {
                [btnprev setFrame:CGRectMake(tabBar.frame.origin.x+5.0,
                15.0, 10.0, 14.0)];         //设置框架
                [btnnextsetFrame:CGRectMake((tabBar.frame.origin.x*(tag+
                1)) - 15.0, 15.0, 10.0, 14.0)];
            }
            return YES;
        }
    return NO;
}
```

scrollToPreviousTabBar 和 scrollToNextTabBar 方法实现了单击按钮后的滚动效果。程序代码如下：

```objc
//单击向前滚动的按钮实现向前滚动
- (void)scrollToPreviousTabBar {
    [self scrollToTabBarWithTag:self.currentTabBarTag - 1 animated:YES];
}
//单击向后滚动的按钮实现向后滚动
- (void)scrollToNextTabBar {
    [self scrollToTabBarWithTag:self.currentTabBarTag + 1 animated:YES];
}
```

在本实例中还实现了选择标签栏条目后，在界面显示标签栏的标签。要实现此功能，需要使用 tabBar: didSelectItem:方法。程序代码如下：

```objc
- (void)tabBar:(UITabBar *)cTabBar didSelectItem:(UITabBarItem *)item {
    for (UITabBar *tabBar in self.tabBars)
        if (tabBar != cTabBar)
            //如果标签栏不是当前的标签栏，则取消选择的状态
            tabBar.selectedItem = nil;
    [infiniTabBarDelegate infiniTabBar:self didSelectItemWithTag: item.
    tag];
}
```

（6）打开 ViewController.h 文件，编写代码，实现头文件、插座变量和属性的声明。程序代码如下：

```
#import <UIKit/UIKit.h>
#import "TabBar.h"
@interface ViewController : UIViewController<TabBarDelegate>{
    IBOutlet UILabel *dLabel;                              //插座变量
    IBOutlet UILabel *fLabel;
}
@property (nonatomic, retain) TabBar *tabBar;
@end
```

（7）打开 Main.storyboard 文件，对 View Controller 视图控制器的设计界面进行设计，效果如图 9.10 所示。

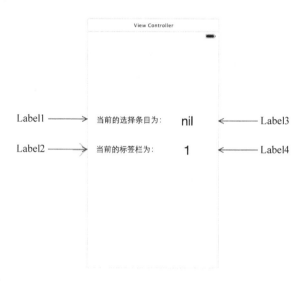

图 9.10　实例 63View Controller 视图控制器的设计界面

需要添加的视图、控件以及对它们的设置如表 9-8 所示。

表 9-8　实例 63 视图、控件设置

视图、控件	属 性 设 置	其　　他
Label1	Text：当前的选择条目为： Font：System 19.0 Alignment：居中	
Label2	Text：nil Font：System 31.0 Alignment：居中	
Label3	Text：当前的标签栏为： Font：System 19.0 Alignment：居中	与插座变量 dLabel 关联
Label4	Text：1 Font：System 31.0 Alignment：居中	与插座变量 fLabel 关联

（8）打开 ViewController.m 文件，编写代码，对具有滚动效果的标签栏对象进行创建和设置，以及实现当标签栏上的选择条目和当前标签栏的标签值发生改变时，显示在界面上。在 viewDidLoad 方法中实现具有滚动效果的标签栏对象的创建和设置。程序代码如下：

```objc
- (void)viewDidLoad
{
    [super viewDidLoad];
    // Do any additional setup after loading the view, typically from a nib.
    //创建标签栏的条目对象
    UITabBarItem *favorites = [[UITabBarItem alloc] initWithTabBarSystemItem:UITabBarSystemItemFavorites tag:0];  //创建标签条目favorites
    UITabBarItem *topRated = [[UITabBarItem alloc] initWithTabBarSystemItem:UITabBarSystemItemTopRated tag:1];
    UITabBarItem *featured = [[UITabBarItem alloc] initWithTabBarSystemItem:UITabBarSystemItemFeatured tag:2];   //创建标签条目featured
    UITabBarItem *recents = [[UITabBarItem alloc] initWithTabBarSystemItem:UITabBarSystemItemRecents tag:3];    //创建标签条目recents
    UITabBarItem *contacts = [[UITabBarItem alloc] initWithTabBarSystemItem:UITabBarSystemItemContacts tag:4];   //创建标签条目contacts
    UITabBarItem *history = [[UITabBarItem alloc] initWithTabBarSystemItem:UITabBarSystemItemHistory tag:5];    //创建标签条目history
    UITabBarItem *bookmarks = [[UITabBarItem alloc] initWithTabBarSystemItem:UITabBarSystemItemBookmarks tag:6];
    UITabBarItem *search = [[UITabBarItem alloc] initWithTabBarSystemItem:UITabBarSystemItemSearch tag:7];     //创建标签条目search
    UITabBarItem *downloads = [[UITabBarItem alloc] initWithTabBarSystemItem:UITabBarSystemItemDownloadstag:8];
    UITabBarItem *mostRecent = [[UITabBarItem alloc] initWithTabBarSystemItem:UITabBar SystemItem MostRecent tag:9];
    UITabBarItem *mostViewed = [[UITabBarItem alloc] initWithTabBarSystemItem:UITabBarSystemItemMore tag:10];    //创建标签条目mostViewed
    //创建并设置标签栏
    self.tabBar = [[TabBar alloc] initWithItems:[NSArray arrayWithObjects:favorites,topRated,featured,recents,contacts,history,bookmarks,search,downloads,mostRecent,mostViewed, nil]];
    self.tabBar.showsHorizontalScrollIndicator = NO;
    self.tabBar.infiniTabBarDelegate = self;
    self.tabBar.bounces = NO;
    [self.view addSubview:self.tabBar];                //添加视图对象
}
```

infiniTabBar:didScrollToTabBarWithTag:和 infiniTabBar:didSelectItemWithTag:方法实现当标签栏上的选择条目和当前标签栏的标签值发生改变时，显示在界面上。程序代码如下：

```objc
- (void)infiniTabBar:(TabBar *)tabBar didScrollToTabBarWithTag:(int)tag {
    fLabel.text = [NSString stringWithFormat:@"%d", tag + 1];
}
- (void)infiniTabBar:(TabBar *)tabBar didSelectItemWithTag:(int)tag {
    dLabel.text = [NSString stringWithFormat:@"%d", tag + 1];
}
```

【代码解析】

本实例关键功能是标签栏的滚动。下面就是这个知识点的详细讲解。

在本实例中，标签栏的滚动分为了两种，一种是使用手指拖动标签栏进行滚动的方法，这种滚动是滚动视图自身提供的。另一种是单击按钮后，通过标签值进行滚动。要使用标签栏进行滚动，需要使用 UIScrollView 的 scrollRectToVisible:animated:方法实现。其语法形式如下：

```
- (void)scrollRectToVisible:(CGRect)rect animated:(BOOL)animated;
```

其中，(CGRect)rect 表示可见区域；(BOOL)animated 表示是否需要滚动动画。在本实例中，单击按钮进行滚动，首先要获取当前标签栏的标签值，然后通过标签值创建一个标签栏。代码如下：

```
UITabBar *tabBar = [self.tabBars objectAtIndex:tag];
```

最后通过 scrollRectToVisible:animated:方法，将标签栏移动到新创建标签框架（指定框架）的可视区域。程序代码如下：

```
[self scrollRectToVisible:tabBar.frame animated:animated];
```

其中，**tabBar.frame** 表示可见区域，animated 表示需要动画。

第 10 章　菜　　单

在 iPhone 或者 iPad 中，除导航栏、标签栏之外，最常用于界面切换的就是菜单。菜单是一系列的选项列表，用户可以通过这些选择列表进行选择，从而实现对应的操作。本章主要讲解关键菜单的相关实例。

实例 64　立方体菜单

【实例描述】

本实例实现的功能是显示一个立方体菜单，当用户滑动菜单时，就会看到立方体的效果。运行效果如图 10.1 所示。

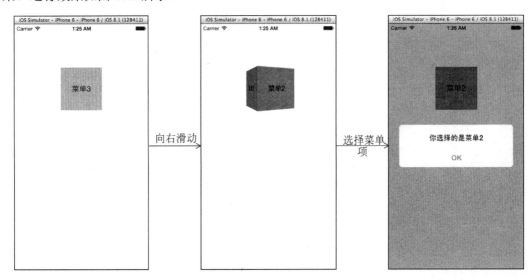

图 10.1　实例 64 运行效果

【实现过程】

当用户向左或者向右滑动菜单时，就会出现立方体动画效果，从而产生此菜单是立方体的感觉。具体的实现步骤如下。

（1）创建一个项目，命名为"立方体菜单"。

（2）创建一个基于 UIView 类的 CubeMenu 类。

（3）打开 CubeMenu.h 文件，编写代码，实现实例变量和属性的声明。程序代码如下：

```
#import <UIKit/UIKit.h>
@interface CubeMenu : UIView{
```

第 10 章 菜单

```
    int i;
}
//属性
@property (strong, nonatomic) UIButton *fw;
@property (strong, nonatomic) UIButton *sw;
@property (strong, nonatomic) UIButton *tw;
@end
```

（4）打开 CubeMenu.m 文件，编写代码，实现立方体的菜单的绘制以及菜单项的选择。使用的方法如表 10-1 所示。

表 10-1　CubeMenu.m 文件中方法总结

方　　法	功　　能
initWithFrame:	立方体菜单的初始化
selectMenuItemLeft	向左滑动
selectMenuItemRight	向右滑动
buttonClicked:	单击按钮实现选择

这里需要讲解几个重要的方法（其他方法请读者参考源代码）。其中，绘制立方体菜单需要使用 initWithFrame:、selectMenuItemLeft 和 selectMenuItemRight 方法实现。initWithFrame:方法实现立方体按钮的初始化。程序代码如下：

```
- (id)initWithFrame:(CGRect)frame
{
    self = [super initWithFrame:frame];
    if (self)
    {
        //实现3个按钮对象的创建和设置
        self.fw = [[UIButton alloc] initWithFrame:CGRectMake(0.0, 0.0,
        frame.size.width, frame.size.height)];
        [self.fw setShowsTouchWhenHighlighted:YES];   //设置按钮在触摸时高亮
        [self.fw setBackgroundColor:[UIColor cyanColor]];
        [self.fw setTitle:@"菜单1" forState: UIControlStateNormal];
        [self.fw setTitleColor:[UIColor blackColor] forState:UIControl
        StateNormal];                                 //设置按钮的标题颜色
        [self.fw setTag:100];
        [self.fw addTarget:self action:@selector(buttonClicked:) for
        Control Events: UIControlEventTouchUpInside];
        [self addSubview:self.fw];                    //添加视图对象
        self.sw = [[UIButton alloc] initWithFrame:CGRectMake(0.0, 0.0,
        frame.size.width, frame.size.height)];
        [self.sw setShowsTouchWhenHighlighted:YES];
        [self.sw setBackgroundColor:[UIColor orangeColor]]; //设置背景颜色
        [self.sw setTitle:@"菜单2" forState: UIControlStateNormal];
        [self.sw setTitleColor:[UIColor blackColor] forState: UIControl
        StateNormal];
        [self.sw setTag:101];
        [self.sw addTarget:self action:@selector(buttonClicked:)for Control
        Events: UIControlEventTouchUpInside];         //添加对象
        [self addSubview:self.sw];
        ……
        //向左滑动的手势
        UISwipeGestureRecognizer *grl = [[UISwipeGestureRecognizer alloc]
        init];
        [grl setDirection:UISwipeGestureRecognizerDirectionLeft];
```

```objc
                                                    //设置方向
    [grl addTarget:self action:@selector(selectMenuItemLeft)];
    [self addGestureRecognizer:grl];
    //向右滑动的手势
    UISwipeGestureRecognizer *grr = [[UISwipeGestureRecognizer alloc]
init];
    [grr setDirection:UISwipeGestureRecognizerDirectionRight];
    [grr addTarget:self action:@selector(selectMenuItemRight)];
                                                    //添加动作
    [self addGestureRecognizer:grr];
    i = 0;
    }
    return self;
}
```

selectMenuItemLeft 方法通过滑动的手势实现向左滑动。程序代码如下：

```objc
- (void) selectMenuItemLeft
{
    CATransition *animation = [CATransition animation];
    animation.duration = 0.8;                       //设置动画所需时间
    animation.type = @"cube";
    animation.subtype = kCATransitionFromRight;     //设置方向
    //找到最上层的 View
    UIButton *lastView = (UIButton *)[self.subviews lastObject];
    //判断 lastView 的 tag 值是否为 100
    if (lastView.tag==100)
        [self bringSubviewToFront:self.tw];
        else if (lastView.tag==101)                 //判断 lastView 的 tag 值是否为 101
        [self bringSubviewToFront:self.fw];
        else if (lastView.tag==102)                 //判断 lastView 的 tag 值是否为 102
        [self bringSubviewToFront:self.sw];
    [self.layer addAnimation:animation forKey:@"animation"];    //添加动作
}
```

selectMenuItemRight 方法通过滑动的手势实现向右滑动。程序代码如下：

```objc
- (void) selectMenuItemRight
{
    CATransition *animation = [CATransition animation];
    animation.duration = 0.8;
    animation.type = @"cube";                       //设置动画类型
    animation.subtype = kCATransitionFromLeft;
    //找到最上层的 View
    UIButton *lastView = (UIButton *)[self.subviews lastObject];
    //判断 lastView 的 tag 值是否为 100
    if (lastView.tag==100)
        [self bringSubviewToFront:self.sw];
        else if (lastView.tag==101)                 //判断 lastView 的 tag 值是否为 101
        [self bringSubviewToFront:self.tw];
        else if (lastView.tag==102)                 //判断 lastView 的 tag 值是否为 102
        [self bringSubviewToFront:self.fw];
    [self.layer addAnimation:animation forKey:@"animation"];
}
```

绘制好菜单后，需要实现菜单的选择，使用 buttonClicked:方法。它的功能是选择菜单后，弹出相应的警告视图。程序代码如下：

```objc
- (void) buttonClicked:(id) sender
```

```
{
    UIButton *button = (UIButton *)sender;
    //判断 button 的 tag 值是否为 100
    if (button.tag==100) {
        UIAlertView *aa=[[UIAlertView alloc]initWithTitle:@"你选择的是菜单
        1" message:nil delegate:nil cancelButtonTitle:@"OK" otherButton
        Titles: nil];                              //创建警告视图
        [aa show];
    }else if (button.tag==101) {          //判断 button 的 tag 值是否为 101
        UIAlertView *aa=[[UIAlertView alloc]initWithTitle:@"你选择的是菜单
        2" message:nil delegate:nil cancelButtonTitle: @"OK"other Button
        Titles: nil];
        [aa show];
    }else{
        UIAlertView *aa=[[UIAlertView alloc]initWithTitle:@"你选择的是菜单
        3" message:nil delegate:nil cancelButtonTitle:@"OK" otherButton
        Titles: nil];
        [aa show];
    }
}
```

（5）打开 AppDelegate.m 文件，编写代码，实现头文件的声明，以及立方体菜单在界面上的显示。程序代码如下：

```
#import "AppDelegate.h"
#import "CubeMenu.h"
@implementation AppDelegate
- (BOOL)application:(UIApplication *)application didFinishLaunching With
Options:(NSDictionary *)launchOptions
{
    self.window = [[UIWindow alloc] initWithFrame:[[UIScreen mainScreen]
    bounds]];
    self.window.backgroundColor = [UIColor whiteColor];      //设置背景颜色
    CubeMenu *cubeMenu = [[CubeMenu alloc] initWithFrame:CGRectMake(110.0,
    100.0, 100.0, 100.0)];
    [self.window addSubview:cubeMenu];                       //添加视图对象
    [self.window makeKeyAndVisible];
    return YES;
}
```

【代码解析】

本实例关键功能是实现立方体菜单的效果。下面就是这个知识点的详细讲解。

立方体菜单的效果的实现使用了 CATransition 的私有动画效果 @"cube"，代码如下：

```
animation.type = @"cube";
```

使用了 UIView 的 bringSubviewToFront:方法显示了指定的视图（bringSubviewToFront:方法的功能是移动指定的视图，让它显示在其他子视图的前面）。其语法如下：

```
- (void)bringSubviewToFront:(UIView *)view;
```

其中，(UIView *)view 表示指定的视图。在代码中，使用了 bringSubviewToFront:方法将指定的视图进行了显示，代码如下：

```
if (lastView.tag==100)
    [self bringSubviewToFront:self.tw];
else if (lastView.tag==101)
    [self bringSubviewToFront:self.fw];
```

```
else if (lastView.tag==102)
    [self bringSubviewToFront:self.sw];
```

实例 65　仿 Windows 8 的 Metro 风格

【实例描述】

在 Windows 8 操作系统中，Windows 操作系统的风格已经完全改变成了 Metro UI。本实例实现的功能是在 iOS 操作系统下模仿 Window 8 的 Metro 风格。运行效果如图 10.2 所示。

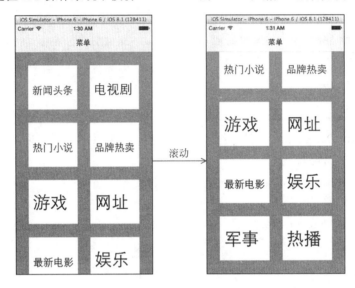

图 10.2　实例 65 运行效果

【实现过程】

（1）创建一个项目，命名为"仿 Windows 8 的 Metro 风格"。
（2）添加"菜单内容.plist"文件到创建项目的 Supporting Files 文件夹中。
（3）创建一个基于 UIControl 类的 MenuItem 类。
（4）打开 MenuItem.h 文件，编写代码，实现方法的声明。程序代码如下：

```
#import <UIKit/UIKit.h>
@interface MenuItem : UIControl
-(void)addTitle:(NSString *)title;                                //方法
@end
```

（5）打开 MenuItem.m 文件，编写代码，实现对菜单项的初始化以及方法的声明。程序代码如下：

```
//初始化
- (id)initWithFrame:(CGRect)frame
{
    self = [super initWithFrame:frame];
    if (self) {
        [self setAutoresizesSubviews:YES];
    }
    return self;
}
```

```
//添加标题
-(void)addTitle:(NSString *)title {
    self.backgroundColor = [UIColor whiteColor];
    CGRect parentFrame = self.frame;                              //获取框架
    CGFloat margin = 10.0;
    CGRect titleFrame = CGRectMake(margin, 0.0, parentFrame.size.width -
    (margin *2), parentFrame.size.height);
    UILabel *titleLabel = [[UILabel alloc] initWithFrame:titleFrame];
      //实例化对象
    titleLabel.text = title;
    titleLabel.backgroundColor = [UIColor clearColor];        //设置背景颜色
    titleLabel.font = [UIFont fontWithName:@"HelveticaNeue-Light" size:
    40];
    titleLabel.adjustsFontSizeToFitWidth = YES;
    titleLabel.contentMode = UIViewContentModeScaleAspectFit; //设置显示模式
    titleLabel.autoresizingMask = UIViewAutoresizingFlexibleWidth|UIView
    AutoresizingFlexibleHeight|UIViewAutoresizingFlexibleBottomMargin|UI
    ViewAutoresizingFlexibleRightMargin;                           //自适应
    [self addSubview:titleLabel];
}
```

（6）创建一个基于 UIScrollView 类的 Menu 类。

（7）打开 Menu.h 文件，编写代码，实现头文件、实例变量和方法的声明。程序代码如下：

```
#import <UIKit/UIKit.h>
#import "MenuItem.h"
@interface Menu : UIScrollView{
    //实例变量
    int _columns;
    int _columnInc;
    CGFloat _marginSize;
    CGFloat _gutterSize;
    CGFloat _rowHeight;
    CGFloat _xOffset;
    CGFloat _yOffset;
}
//方法
-(id) initWithColumns:(int)col marginSize:(CGFloat)margin gutterSize:
(CGFloat)gutter rowHeight:(CGFloat)height;
-(MenuItem *) createMenuItem;
@end
```

（8）打开 Menu.m 文件，编写代码，实现仿 Windows 8 的 metro 风格。使用的方法如表 10-2 所示。

表 10-2　Menu.m文件中方法总结

方　　法	功　　能
init	初始化
initWithColumns:gutterSize:rowHeight:	使用列数、沟槽的尺寸和行的高度等进行初始化
createMenuItem	创建菜单项

其中，init 方法实现对仿 Window 8 的 metro 风格进行初始化。程序代码如下：

```
-(id)init {
    CGRect windowSize = [[UIScreen mainScreen] applicationFrame];
    self = [super initWithFrame:windowSize];
```

```
    if (self) {
        self.scrollEnabled = YES;
        self.userInteractionEnabled = YES;
        self.showsVerticalScrollIndicator = NO;         //不显示垂直方向的滚动条
    }
    return self;
}
```

initWithColumns:gutterSize:rowHeight:方法实现使用列数、视图与视图之间的距离、行的高度等进行初始化。程序代码如下：

```
-(id)initWithColumns:(int)col    marginSize:(CGFloat)margin    gutterSize:
(CGFloat) gutter rowHeight:(CGFloat)height{
    self = [self init];
    if (self) {
        //设置一些属性
        _columns = col;
        _marginSize = margin;
        _gutterSize = gutter;
        _rowHeight = height;
        _xOffset = gutter;
        _yOffset = gutter;
    }
    return self;
}
```

createMenuItem 方法实现对菜单项的创建。程序代码如下：

```
-(MenuItem *) createMenuItem {
    CGFloat adjustedMargin = (_marginSize * (_columns - 1) / _columns);
    CGFloat menuWidth = (self.frame.size.width - (_gutterSize * 2));
    CGFloat itemWidth = (menuWidth / _columns) - adjustedMargin;
    CGRect itemFrame = CGRectMake(_xOffset, _yOffset, itemWidth, _rowHeight);
                                                                    //获取框架
    MenuItem *item = [[MenuItem alloc] initWithFrame:itemFrame];
    _columnInc++;
    //判断_columnInc 是否大于等于_columns
    if (_columnInc >= _columns) {
        _columnInc = 0;
        _yOffset = _yOffset + _rowHeight + _marginSize;
        _xOffset = _gutterSize;
    } else {
        _xOffset = _xOffset + _marginSize + itemWidth;
        //设置内容视图的尺寸
        self.contentSize = CGSizeMake(self.contentSize.width, _yOffset +
        _marginSize + _rowHeight);
    }
    return item;
}
```

（9）打开 Main.storyboard 文件，添加 Navigation Controller 导航控制器到画布中，将此控制器的根视图设为 View Controller。将 Navigation Item 下的 Title 属性设置为"菜单"。

（10）打开 ViewController.h 文件，编写代码，实现头文件、属性和方法的声明。程序代码如下：

```
#import <UIKit/UIKit.h>
#import "Menu.h"
@interface ViewController : UIViewController
@property (nonatomic, strong) Menu *menuView;                       //属性
```

```
-(void)populateMenu;                                              //方法
@end
```

（11）打开 ViewController.m 文件，编写代码，实现在界面显示仿 Windows 8 的 Metro 风格。程序代码如下：

```
- (void)viewDidLoad
{
    [super viewDidLoad];
    _menuView = nil;
    _menuView = [[Menu alloc] initWithColumns:2 marginSize:30 gutterSize:30
    rowHeight:100];                                              //实例化
    _menuView.backgroundColor = [UIColor grayColor];   //设置背景颜色
    self.view = _menuView;
    [self populateMenu];
}
//填充菜单
-(void)populateMenu {
    NSString *path = [[NSBundle mainBundle] pathForResource:@"菜单内容"
    ofType:@"plist"];
    NSArray *books = [NSArray arrayWithContentsOfFile:path];
    //实现填充
    for (NSString *bookTitle in books) {
        MenuItem *menuItem = [_menuView createMenuItem];   //创建菜单项
        [menuItem addTitle:bookTitle];                       //添加标题
        [self.view addSubview:menuItem];
    }
}
@end
```

【代码解析】

本实例的关键功能是菜单项的布局。下面就是这个知识点的详细讲解。

在菜单的填充中会多次调用 createMenuItem 方法，它不仅用来对菜单项进行创建，还要对创建的菜单项进行布局。布局使用的是将_columns（列数）和_columnInc 进行比较。如果_columnInc 小于_columns，进行横排；如果_columnInc 大于_columns，就进行竖排。代码如下：

```
if (_columnInc >= _columns) {
    _columnInc = 0;
    _yOffset = _yOffset + _rowHeight + _marginSize;
    _xOffset = _gutterSize;
} else {
    _xOffset = _xOffset + _marginSize + itemWidth;
    self.contentSize = CGSizeMake(self.contentSize.width, _yOffset +
_marginSize + _rowHeight);
}
```

实例 66　下 拉 菜 单

【实例描述】

为了给 iPhone 或者 iPad 腾出更多的空间，苹果公司将菜单设计成下拉菜单。本实例的功能就是实现一个下拉菜单的功能。运行效果如图 10.3 所示。

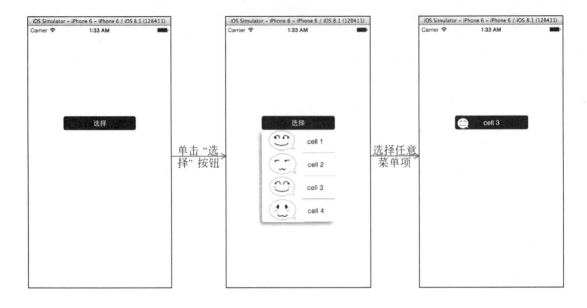

图 10.3 实例 66 运行效果

【实现过程】

（1）创建一个项目，命名为"下拉菜单"。

（2）添加图片 1.png、2.png、3.png 和 4.png 到创建项目的 Supporting Files 文件夹中。

（3）创建一个基于 UIView 类的 Menu 类。

（4）打开 Menu.h 文件，编写代码，实现协议、对象、属性和方法的声明。程序代码如下：

```
#import <UIKit/UIKit.h>
@class Menu;
//协议
@protocol MenuDelegate
- (void) niDropDownDelegateMethod: (Menu *) sender;
@end
@interface Menu : UIView <UITableViewDelegate, UITableViewDataSource>{
    NSString *animationDirection;                          //对象
    UIImageView *imgView;                                  //对象
}
//属性
@property(nonatomic, strong) UITableView *table;
@property(nonatomic, strong) UIButton *btnSender;
……
@property (nonatomic, retain) NSString *animationDirection;
//方法
-(void)hideDropDown:(UIButton *)b;
- (id)showDropDown:(UIButton *)b :(CGFloat *)height :(NSArray *)array :(NSArray *)imgArr :(NSString *)direction;
@end
```

（5）打开 Menu.m 文件，编写代码，实现下拉菜单的效果以及选择。使用的方法如表 10-3 所示。

表 10-3 Menu.m文件中方法总结

方法	功能
showDropDown::::	显示下拉菜单
hideDropDown:	隐藏下拉菜单
numberOfSectionsInTableView:	获取表视图的块
tableView:heightForRowAtIndexPath:	获取表中每行的高度
tableView:numberOfRowsInSection:	获取行数
tableView:cellForRowAtIndexPath:	获取某一行的数据
tableView:didSelectRowAtIndexPath:	响应选择的行

其中，要实现绘制下拉菜单的效果，需要使用 showDropDown::::、number Of SectionsInTableView:和 tableView:heightForRowAtIndexPath:、tableView:numberOf RowsIn Section:、tableView:cellForRowAtIndexPath:方法实现。用 showDropDown::::方法实现显示一个下拉菜单。程序代码如下：

```objc
- (id)showDropDown:(UIButton *)b :(CGFloat *)height :(NSArray *)arr :(NSArray *)imgArr :(NSString *)direction {
    btnSender = b;
    animationDirection = direction;
    self.table = (UITableView *)[super init];
    if (self) {
        CGRect btn = b.frame;                              //获取框架
        self.list = [NSArray arrayWithArray:arr];
        self.imageList = [NSArray arrayWithArray:imgArr];
        //判断direction是否等于"up"
        if ([direction isEqualToString:@"up"]) {
            self.frame = CGRectMake(btn.origin.x, btn.origin.y, btn.size.width, 0);
            self.layer.shadowOffset = CGSizeMake(-5, -5);
        }else if ([direction isEqualToString:@"down"]) {
                                              //判断direction是否等于"down"
            self.frame = CGRectMake(btn.origin.x, btn.origin.y+btn.size.height, btn.size.width, 0);
            self.layer.shadowOffset = CGSizeMake(-5, 5);
        }
        self.layer.masksToBounds = NO;
        self.layer.cornerRadius = 8;                       //设置圆角半径
        self.layer.shadowRadius = 5;
        self.layer.shadowOpacity = 0.5;                    //设置阴影的不透明度
        //表视图对象的创建和设置
        table = [[UITableView alloc] initWithFrame:CGRectMake(0, 0, btn.size.width, 0)];
        table.delegate = self;                             //设置委托
        table.dataSource = self;
        table.layer.cornerRadius = 5;
        table.separatorStyle = UITableViewCellSeparatorStyleSingleLine;
                                                           //设置分割线的风格
        table.separatorColor = [UIColor blueColor];        //设置分隔线的颜色
        //动画的实现
        [UIView beginAnimations:nil context:nil];
        [UIView setAnimationDuration:0.5];                 //设置动画所需时间
         //判断direction是否等于"up"
        if ([direction isEqualToString:@"up"]) {
            //向上显示一个下拉菜单
```

```objc
        self.frame = CGRectMake(btn.origin.x, btn.origin.y-*height,
        btn.size.width, *height);
    } else if([direction isEqualToString:@"down"]) {
                                            //判断direction是否等于"down"
        //向下显示一个下拉菜单
        self.frame = CGRectMake(btn.origin.x, btn.origin.y+btn.size.
        height, btn.size.width, *height);
    }
    table.frame = CGRectMake(0, 0, btn.size.width, *height);  //设置框架
    [UIView commitAnimations];
    [b.superview addSubview:self];                          //添加视图对象
    [self addSubview:table];
    }
    return self;
}
```

numberOfSectionsInTableView:、tableView:heightForRowAtIndexPath:、tableView:numberOfRowsInSection:和 tableView:cellForRowAtIndexPath:方法实现对下拉菜单的内容填充以及设置。程序代码如下：

```objc
//获取表视图的块
- (NSInteger)numberOfSectionsInTableView:(UITableView *)tableView {
    return 1;
}
//获取表中每行的高度
-(CGFloat)tableView:(UITableView*)tableView heightForRowAtIndexPath: (NSIndexPath *)indexPath {
    return 50;
}
//获取行数
- (NSInteger)tableView:(UITableView*)tableView numberOfRowsInSection: (NSInteger)section {
    return [self.list count];
}
//获取某一行的数据
- (UITableViewCell *)tableView:(UITableView *)tableView cellForRowAtIndexPath: (NSIndexPath *)indexPath {
    static NSString *CellIdentifier = @"Cell";
    UITableViewCell*cell=[tableViewdequeueReusableCellWithIdentifier:
    CellIdentifier];
    //判断cell是否为空
    if (cell == nil) {
        cell = [[UITableViewCell alloc] initWithStyle: UITableView
        CellStyleDefault reuseIdentifier:CellIdentifier];
        cell.textLabel.font = [UIFont systemFontOfSize:15];     //设置字体
        cell.textLabel.textAlignment = NSTextAlignmentCenter;  //设置对齐方式
    }
    //填充内容
    if ([self.imageList count] == [self.list count]) {
        cell.textLabel.text =[list objectAtIndex:indexPath.row]; //设置文本
        cell.imageView.image = [imageList objectAtIndex:indexPath.row];
                                                              //设置图像
    } else if ([self.imageList count] > [self.list count]) {
        cell.textLabel.text =[list objectAtIndex:indexPath.row];
        if (indexPath.row < [imageList count]) {
            cell.imageView.image = [imageList objectAtIndex:indexPath.row];
                                                              //设置图像
```

```
    }
} else if ([self.imageList count] < [self.list count]) {
    cell.textLabel.text =[list objectAtIndex:indexPath.row];    //设置文本
    if (indexPath.row < [imageList count]) {
        cell.imageView.image = [imageList objectAtIndex:indexPath.row];
    }
}
cell.textLabel.textColor = [UIColor blackColor];              //设置文本颜色
UIView * v = [[UIView alloc] init];
v.backgroundColor = [UIColor grayColor];                      //设置背景颜色
cell.selectedBackgroundView = v;
return cell;
}
```

要实现菜单中响应选择的行，需要使用 tableView:didSelectRowAtIndexPath:方法。程序代码如下：

```
- (void)tableView:(UITableView *)tableView didSelectRowAtIndex Path:
(NSIndexPath *)indexPath {
    [self hideDropDown:btnSender];
    //设置按钮的标题
    UITableViewCell *c = [tableView cellForRowAtIndexPath:indexPath];
    [btnSender setTitle:c.textLabel.text forState:UIControlStateNormal];
                                        //设置标题
    for (UIView *subview in btnSender.subviews) {
        if ([subview isKindOfClass:[UIImageView class]]) {
            [subview removeFromSuperview];               //删除
        }
    }
    //为按钮添加图片
    imgView.image = c.imageView.image;                   //设置图像
    imgView = [[UIImageView alloc] initWithImage:c.imageView.image];
    imgView.frame = CGRectMake(5, 5, 25, 25);
    [btnSender addSubview:imgView];                      //添加视图对象
    [self.delegate niDropDownDelegateMethod:self];
}
```

hideDropDown:方法实现下拉菜单的隐藏。程序代码如下：

```
-(void)hideDropDown:(UIButton *)b {
    CGRect btn = b.frame;
    //动画的实现
    [UIView beginAnimations:nil context:nil];
    [UIView setAnimationDuration:0.5];                   //设置动画所需时间
    if ([animationDirection isEqualToString:@"up"]) {
        //隐藏向上的下拉菜单
        self.frame = CGRectMake(btn.origin.x, btn.origin.y, btn.size.width, 0);
    }else if ([animationDirection isEqualToString:@"down"]) {
        //隐藏向下的下拉菜单
        self.frame = CGRectMake(btn.origin.x, btn.origin.y+btn.size.height,
        btn.size.width, 0);
    }
    table.frame = CGRectMake(0, 0, btn.size.width, 0);   //设置框架
    [UIView commitAnimations];
}
```

（6）打开 ViewController.h 文件，编写代码，实现头文件、插座变量、对象以及动作的声明。程序代码如下：

```
#import <UIKit/UIKit.h>
#import "Menu.h"
@interface ViewController : UIViewController<MenuDelegate>{
    IBOutlet UIButton *btnSelect;                        //插座变量
    Menu *dropDown;
}
- (IBAction)selectClicked:(id)sender;                    //动作
@end
```

（7）打开 Main.storyboard 文件，在设计界面中，添加一个 Button 按钮控件，将此按钮和插座变量 btnSelect 以及动作 selectClicked:方法关联。将 Text 属性设置为"选择"，将 Text Color 属性设置为白色，将 Background 属性设置为黑色。

（8）打开 ViewController.m 文件，编写代码，实现单击按钮后，下拉菜单的显示或隐藏。程序代码如下：

```
- (void)viewDidLoad
{
    [super viewDidLoad];
    // Do any additional setup after loading the view, typically from a nib.
    btnSelect.layer.borderWidth = 1;                     //设置边框宽度
    btnSelect.layer.cornerRadius = 5;                    //设置圆角半径
}
//实现单击按钮后显示或者隐藏下拉菜单
- (IBAction)selectClicked:(id)sender {
    NSArray * arr = [[NSArray alloc] init];              //实例化对象
    arr = [NSArray arrayWithObjects:@"cell 1", @"cell 2", @"cell 3", @"cell 4", @"cell 5", @"cell 6", @"Cell 7", @"cell 8",nil];
    NSArray * arrImage = [[NSArray alloc] init];
    arrImage = [NSArray arrayWithObjects:[UIImage imageNamed:@"1.png"], [UIImage imageNamed:@"2.png"], [UIImage imageNamed:@"3.png"], [UIImage imageNamed:@"4.png"], [UIImage imageNamed:@"1.png"], [UIImage imageNamed:@"2.png"], [UIImage imageNamed:@"3.png"], nil];
    //判断 dropDown 是否为空
    if(dropDown == nil) {
        CGFloat f = 200;
        dropDown=[[Menu alloc]showDropDown:sender:&f:arr:arrImage: @"down"];
            //创建
        dropDown.delegate = self;
    }
    else {
        [dropDown hideDropDown:sender];
        dropDown = nil;                                  //将 dropDown 赋值为空
    }
}
- (void) niDropDownDelegateMethod: (Menu *) sender {
    dropDown = nil;                                      //将 dropDown 赋值为空
}
```

【代码解析】

本实例的关键功能是选择菜单项后按钮标题和图片的更新。下面依次讲解这两个知识点。

1. 按钮标题的更新

要实现选择菜单项后，按钮标题的更新。需要使用 UIButton 的 setTitle:forState:方法。其语法形式如下：

```
- (void)setTitle:(NSString *)title forState:(UIControlState)state;
```

其中，(NSString *)title 表示在指定的状态下使用的标题；(UIControlState)state 表示指定的状态，按钮的状态如表 10-4 所示。

表 10-4 按钮的状态

状　态	功　能
UIControlStateNormal	正常的状态
UIControlStateHighlighted	高亮的状态
UIControlStateDisabled	禁用的状态
UIControlStateSelected	选中的状态
UIControlStateApplication	应用的状态
UIControlStateReserved	保留的状态

在代码中，就使用了 setTitle:forState:方法实现了按钮的标题更新。代码如下：

```
[btnSender setTitle:c.textLabel.text forState:UIControlStateNormal];
```

其中，c.textLabel.text 表示在指定的状态下使用的标题，这个参数是通过 UITableView 的 cellForRowAtIndexPath:方法获取的。其语法形式如下：

```
- (UITableViewCell *)cellForRowAtIndexPath:(NSIndexPath *)indexPath;
```

其中，(NSIndexPath *)indexPath 表示指定行的索引。本实例中的代码如下：

```
UITableViewCell *c = [tableView cellForRowAtIndexPath:indexPath];
```

其中，indexPath 表示指定行的索引。

2. 更新按钮的图片

要更新图片，需要创建一个图像视图，图像视图就是通过 UITableView 的 cellForRowAtIndexPath:方法获取行的图像，最后添加到按钮中。代码如下：

```
imgView.image = c.imageView.image;
imgView = [[UIImageView alloc] initWithImage:c.imageView.image];
imgView.frame = CGRectMake(5, 5, 25, 25);
[btnSender addSubview:imgView];
```

实例 67　浮动的菜单

【实例描述】

本实例实现的功能是显示一个浮动的菜单。单击界面的按钮，此菜单就会展开，当用户再一次单击此按钮，菜单就会关闭。并且，单击菜单中的任意一个菜单项，导航栏的颜色也会发生相应的改变。运行效果如图 10.4 所示。

图 10.4　实例 67 运行效果

【实现过程】

（1）创建一个项目，命名为"浮动的菜单"。

（2）添加图片 1.png、2.png、3.png、4.png、5.png、6.png 和 7.png 到创建项目的 Supporting Files 文件夹中。

（3）创建一个基于 UIView 类的 Menu 类。

（4）打开 Menu.h 文件，编写代码，实现数据结构的定义、对象、属性、方法、协议以及全局变量的声明。程序代码如下：

```
#import <UIKit/UIKit.h>
//数据结构的定义
typedef enum {
    kQBKOverlayMenuViewPositionDefault,
    kQBKOverlayMenuViewPositionTop,
    kQBKOverlayMenuViewPositionBottom
} MenuViewPosition;
struct MenuViewOffset {
    CGFloat bottomOffset;
    CGFloat topOffset;
};
typedef struct MenuViewOffset MenuViewOffset;
@protocol MenuViewDelegate;
@interface Meun : UIView{
    //对象
    UIButton *_mainButton;
    UIImageView *_mainBackgroundImageView;
}
//属性
@property (nonatomic, weak, readonly) id<MenuViewDelegate> delegate;
@property (nonatomic, weak) UIView *parentView;
……
@property (nonatomic, strong, readonly) NSMutableArray *additionalButtons;
//方法
-   (id)initWithDelegate:(id  <MenuViewDelegate>)delegate  position:(Menu
```

```
ViewPosition)position offset:(MenuViewOffset)offset;
......
- (CGRect)createFoldedContentViewFrameForPosition:(MenuViewPosition) po
sition;
- (CGRect)createUnfoldedContentViewFrameForPosition:(MenuViewPosition) po
sition;
@end
//协议
@protocol MenuViewDelegate <NSObject>
@optional
- (void)overlayMenuView:(Meun *)overlayMenuView didActivateAdditional Bu
ttonWithIndex:(NSInteger)index;
@end
//全局变量
extern NSString *QBKOverlayMenuDidPerformUnfoldActionNotification;
extern NSString *QBKOverlayMenuDidPerformFoldActionNotification;
extern NSString *QBKOverlayMenuDidActivateAdditionalButtonNotification;
```

(5) 打开 Meun.m 文件，编写代码，实现浮动菜单的效果以及菜单项的选择。使用的方法如表 10-5 所示。

表 10-5　Meun.m 文件中方法总结

方　　法	功　　能
initWithDelegate:position:offset:	使用委托、位置和偏移量初始化
initWithDelegate:position:	使用委托和位置初始化
initWithFrame:	使用框架初始化
setParentView:	设置父视图
mainButtonPressed	主按钮的按下
additionalButtonPressed:	其他按钮的按下
unfoldWithAnimationDuration:	实现展开
foldWithAnimationDuration:	实现折叠
addButtonWithImage:	添加按钮
setupMainButton	创建主按钮
setupContentView	创建内容视图
createFoldedContentViewFrameForPosition:	创建折叠的内容视图框架
createUnfoldedContentViewFrameForPosition:	创建展开的内容视图框架
createFoldedMainFrameForPosition:	创建折叠的主框架
createUnfoldedMainFrameForPosition:	创建展开的主框架

这里需要讲解几个重要的方法（其他方法请读者参考源代码）。要实现浮动效果的绘制，需要 initWithDelegate:position:offset:、initWithDelegate:position:、initWithFrame:、setParentView:、addButtonWithImage:、setupMainButton、setupContentView、createFoldedContentViewFrameForPosition: 和 createFoldedMainFrameForPosition:方法。其中，initWithDelegate:position:offset:方法实现使用委托、位置和偏移量对浮动的菜单进行初始化。程序代码如下：

```
- (id)initWithDelegate:(id <MenuViewDelegate>)delegate position:(MenuView
Position) position offset:(MenuViewOffset)offset
{
    if (delegate && [delegate conformsToProtocol: @protocol(Menu View
```

```
Delegate)]) {
        _delegate = delegate;                              //设置委托
        _position = position;                              //设置位置
        _offset = offset;
        _unfolded = NO;
        return (self = [self initWithFrame:CGRectZero]);
    }
    return nil;
}
```

setupMainButton 方法实现创建一个主按钮，用于打开或关闭菜单项。程序代码如下：

```
- (void)setupMainButton
{
    CGFloat x = (44 - 28) / 2;
    CGFloat y = (44 - 28) / 2 - 1;
    _mainButton = [[UIButton alloc] initWithFrame:CGRectMake(x, y, 28, 28)];
                                                                //实例化
    //设置背景图像
    [_mainButton setBackgroundImage:[UIImage imageNamed:@"6.png"] for
State:UIControlStateNormal];
    [_mainButton setAutoresizingMask:UIViewAutoresizing FlexibleLeft Mar
gin];
    [_mainButton addTarget:self action:@selector(mainButtonPressed) for
ControlEvents:UIControlEventTouchUpInside];                 //添加动作
}
```

addButtonWithImage: 方法实现添加作为菜单项的按钮，并放置在一个视图中，这个视图被称为内容视图。程序代码如下：

```
- (void)addButtonWithImage:(UIImage *)image index:(NSInteger)index
{
    if (!_additionalButtons) _additionalButtons = [[NSMutableArray alloc] init];
    CGFloat tamHuecoBoton = (320 - (44 + 5 * 2 + 5* 2 )) /4 ;
    CGFloat posXBotonCentrado = (tamHuecoBoton - 22) / 2;
    CGFloat posYBotonCentrado = (([_contentView bounds].size.height -22) /
2) - 1 ;
    CGFloat posXBoton = (4 - ([_additionalButtons count] + 1)) *
tamHuecoBoton + posXBotonCentrado;
    //按钮对象的创建和设置
    UIButton *newButton = [[UIButton alloc] initWithFrame: CGRectMake
(posXBoton, posYBotonCentrado, 22, 22)];                    //实例化
    [newButton setBackgroundImage:image forState:UIControlStateNormal];
                                                            //设置背景图像
    [newButton setAutoresizingMask:UIViewAutoresizingNone];
    [newButton addTarget:self action:@selector(additionalButtonPressed:)
forControlEvents:UIControlEventTouchUpInside];              //添加动作
    [_additionalButtons insertObject:newButton atIndex:index];
    [_contentView addSubview:newButton];
```

}
```

setupContentView 方法实现创建一个内容视图,用于放置作为菜单项的按钮。程序代码如下:

```
- (void)setupContentView
{
    _contentView = [[UIView alloc] initWithFrame:[self createFoldedContent
    ViewFrameForPosition:_position]];                        //实例化
    [_contentView setClipsToBounds:YES];
    [_contentView setAutoresizingMask:UIViewAutoresizingFlexibleWidth];
}
```

setParentView:方法实现创建一个父视图,在这个视图中放置浮动菜单(即主按钮和放置菜单项的内容视图)。程序代码如下:

```
- (void)setParentView:(UIView *)view
{
    _parentView = view;
    CGRect frame = [self createFoldedMainFrameForPosition:_position];
    [self setFrame:frame];                                   //设置框架
    [self setClipsToBounds:YES];
    [self setupMainButton];
    [self addSubview:_mainButton];                           //添加视图对象
    _mainBackgroundImageView = [[UIImageView alloc] initWithImage:[UIImage
    imageNamed:@"7.png"]];
    [self addSubview:_mainBackgroundImageView];
    [self sendSubviewToBack:_mainBackgroundImageView];
    [_mainBackgroundImageView setAlpha:0.0f];                //设置透明度
    [self setupContentView];
    [self addSubview:_contentView];
    [view addSubview:self];
}
```

createFoldedMainFrameForPosition:方法实现针对位置后期折叠的主框架进行创建。程序代码如下:

```
- (CGRect)createFoldedMainFrameForPosition:(MenuViewPosition)position
{
    CGRect frame;
    switch (_position) {
        case kQBKOverlayMenuViewPositionBottom:
            frame = CGRectMake([_parentView bounds].size.width - (44 + 5),
            [_parentView bounds].size.height - (44 + 10 + _offset.bottom
            Offset), 44, 44 );                               //设置frame
            break;
        case kQBKOverlayMenuViewPositionTop:
            frame = CGRectMake([_parentView bounds].size.width - (44 + 10),
            10 + _offset.topOffset, 44, 44 );                //设置frame
            break;
        default:
```
```

```objc
            frame = CGRectMake([_parentView bounds].size.width - (44 + 5 ),
            [_parentView bounds].size.height - (44 + 10 + _offset.bottom
            Offset), 44, 44 );                          //设置 frame
            break;
        }
    return frame;
}
```

createUnfoldedMainFrameForPosition:方法实现针对位置的后期展开的主框架的创建。程序代码如下:

```objc
- (CGRect)createUnfoldedMainFrameForPosition:(MenuViewPosition)position
{
    CGRect frame;
    switch (_position) {
        case kQBKOverlayMenuViewPositionBottom:
            frame = CGRectMake(5 , [[self superview] bounds].size.height -
            (44+10+_offset.bottomOffset),[[selfsuperview]bounds].size.
            width - 5 * 2, 44 );                //设置 frame
            break;
        case kQBKOverlayMenuViewPositionTop:
            frame = CGRectMake(5 , 10+ _offset.topOffset, [[self superview]
            bounds].size.width - 5 * 2, 44 );           //设置 frame
            break;
        default:
            frame = CGRectMake(5 , [[self superview] bounds].size.height -
            (44+10+_offset.bottomOffset),[[selfsuperview]bounds] .size.
            width - 5 * 2, 44 );                //设置 frame
            break;
    }
    return frame;
}
```

以上就实现了一个折叠的浮动菜单的效果,如果想要将此菜单项展开或者将展开的菜单项再一次折叠,需要使用 mainButtonPressed、unfoldWithAnimationDuration:和 foldWithAnimationDuration:方法实现。mainButtonPressed 方法实现按下主按钮后,展开或者折叠浮动菜单的菜单项。程序代码如下:

```objc
- (void)mainButtonPressed
{
    if (!_unfolded) {
        [self unfoldWithAnimationDuration:0.5];         //展开
    } else {
        [self foldWithAnimationDuration:0.2];           //折叠
    }
}
```

unfoldWithAnimationDuration:方法实现浮动菜单项的展开。程序代码如下:

```objc
- (void)unfoldWithAnimationDuration:(float)duration
{
    [UIView animateWithDuration:duration animations:^{
        CGRect newFrame = [self createUnfoldedMainFrame ForPosition:
        _position];
        [self setFrame:newFrame];                            //设置框架
        [_mainBackgroundImageView setAlpha:.9f];             //设置透明度
        CGAffineTransform xform = CGAffineTransformMakeRotation(-M_PI_2);
        [_mainButton setTransform:xform];                    //设置改变
        _unfolded = YES;
    } completion:^(BOOL finished) {
        //设置背景图像
        [_mainButton setBackgroundImage:[UIImage imageNamed:@"5.png"]
         forState:UIControlStateNormal];
    }];
}
```

foldWithAnimationDuration:方法实现浮动菜单项的折叠。程序代码如下:

```objc
- (void)foldWithAnimationDuration:(float)duration
{
    [UIView animateWithDuration:duration animations:^{
        CGRect newFrame = [self createFoldedMainFrameForPosition:_position];
        [self setFrame:newFrame];                            //设置框架
        [_contentViewsetFrame:[selfcreateFoldedContentViewFrameForPosition:
        _position]];
        [_mainBackgroundImageView setAlpha:0.0f];            //设置透明度
        CGAffineTransform xform = CGAffineTransformMakeRotation(0);
        [_mainButton setTransform:xform];
        _unfolded = NO;
    } completion:^(BOOL finished) {
        //设置背景图像
        [_mainButton setBackgroundImage:[UIImage imageNamed:@"6.png"]
         forState: UIControlStateNormal];
    }];
}
```

如果要实现菜单项的选择,需要使用 additionalButtonPressed:方法。程序代码如下:

```objc
- (void)additionalButtonPressed:(id)sender
{
    [_delegate overlayMenuView:self didActivateAdditionalButton
     WithIndex:[_additionalButtonsindexOfObject:sender]];
}
```

(6)打开 ViewController.h 文件,编写代码,实现头文件、对象和插座变量的声明。程序代码如下:

```objc
#import <UIKit/UIKit.h>
#import "Meun.h"
@interface ViewController : UIViewController<MenuViewDelegate>{
    Meun *_qbkOverlayMenu;
    IBOutlet UINavigationBar *bar;                           //插座变量
}
@end
```

（7）打开 Main.storyboard 文件，对 View Controller 视图控制器的设计界面进行设计，效果如图 10.5 所示。

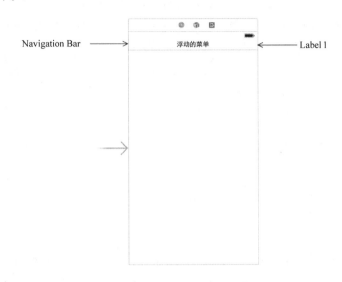

图 10.5　实例 67View Controller 视图控制器的设计界面

需要添加的视图、控件以及对它们的设置如表 10-6 所示。

表 10-6　实例 67 视图、控件设置

视图、控件	属 性 设 置	其 他
Label	Text：浮动的菜单 Font：Body Alignment：居中	
Navigation Bar		与插座变量 bar 关联

（8）打开 ViewController.m 文件，编写代码，实现浮动菜单对象的创建，并按下相应的菜单项实现导航栏的颜色改变。程序代码如下：

```
- (void)viewDidLoad
{
    [super viewDidLoad];
    // Do any additional setup after loading the view, typically from a nib.
    [super viewDidLoad];
    MenuViewOffset offset;
    offset.bottomOffset = 44;
    //创建并设置菜单对象
    _qbkOverlayMenu = [[Meun alloc] initWithDelegate:self position:
kQBKOverlayMenuViewPositionBottom offset:offset];
    [_qbkOverlayMenu setParentView:[self view]];
    //添加按钮
    [_qbkOverlayMenu addButtonWithImage:[UIImage imageNamed:@"1.png"] index:0];
    [_qbkOverlayMenu addButtonWithImage:[UIImage imageNamed:@"2.png"] index:1];
```

```
    [_qbkOverlayMenu    addButtonWithImage:[UIImage    imageNamed:@"3.png"]
    index:2];
    [_qbkOverlayMenu    addButtonWithImage:[UIImage    imageNamed:@"4.png"]
    index:3];
}
//实现选择菜单项的响应
-  (void)overlayMenuView:(Meun  *)overlayMenuView  didActivateAdditional
ButtonWithIndex:(NSInteger)index
{
    //判断索引值是否为 0
    if (index==0) {
        bar.backgroundColor=[UIColor redColor];
    }else if (index==1){                              //判断索引值是否为 1
        bar.backgroundColor=[UIColor blueColor];
    }else if (index==2){                              //判断索引值是否为 2
        bar.backgroundColor=[UIColor orangeColor];
    }else{
        bar.backgroundColor=[UIColor greenColor];
    }
}
```

【代码解析】

本实例关键功能是菜单项的展开与折叠。下面就是这个知识点的详细讲解。

要实现菜单项的展开和折叠，最主要的方式是获取内容视图和主视图的框架。以菜单项的折叠为例，首先在 createFoldedContentViewFrameForPosition: 和 createFoldedMainFrameForPosition:方法中获取了框架。其中，在 createFoldedContentViewFrameForPosition:方法中获取框架的代码如下：

```
CGRect frame = CGRectMake([self bounds].origin.x, [self bounds].origin.y,
0, 44 );
return frame;
```

然后，在 foldWithAnimationDuration:方法中实现菜单项的折叠，其中，就要调用 createFoldedContentViewFrameForPosition:和 createFoldedMainFrameForPosition:方法，对视图本身和内容视图的框架进行重新的设置。代码如下：

```
CGRect newFrame = [self createFoldedMainFrameForPosition:_position];
[self setFrame:newFrame];
[_contentView   setFrame:[self  createFoldedContentViewFrameForPosition:
_position]];
```

实例 68　具有按钮的菜单

【实例描述】

本实例实现的功能是显示一个具有按钮的菜单。当用户单击导航栏左边的按钮，就会在界面的左边显示一个具有按钮的菜单，用户选择菜单中的任意一项，会改变导航栏的背景颜色。当用户单击导航栏右边的按钮，就会在界面的右边显示一个具有按钮的菜单，用户选择菜单中的任意一项，会改变界面的背景颜色。运行效果如图 10.6 所示。

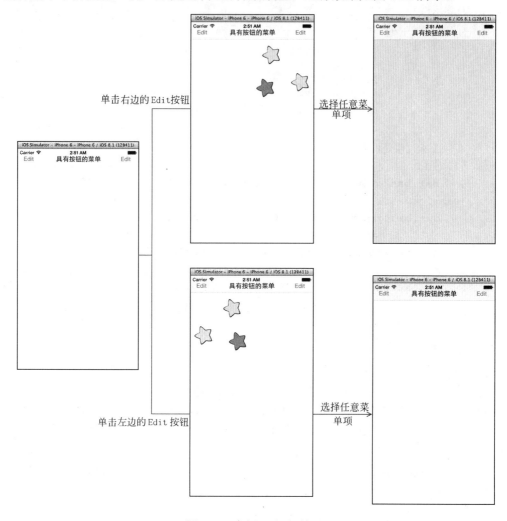

图 10.6　实例 68 运行效果

【实现过程】

（1）创建一个项目，命名为"具有按钮的菜单"。

（2）添加图片 1.png、2.png 和 3.png 到创建项目的 Supporting Files 文件夹中。

（3）创建一个基于 UIView 类的 Menu 类。

（4）打开 Menu.h 文件，编写代码，实现头文件、协议、数据结构的定义、实例变量、属性和方法的声明。程序代码如下：

```
#import <UIKit/UIKit.h>
#import <objc/runtime.h>                                //头文件
@class Menu;
//协议
@protocol MenuDelegate <NSObject>
@optional
-(void)buttonMenu:(Menu*)buttonMenu    titleForTappedButton:(NSString *)title;
@end
```

```
//数据结构的定义
typedef NS_ENUM(NSInteger, TJLBarButtonMenuSide) {
    TJLBarButtonMenuRightTop,
    TJLBarButtonMenuLeftTop
};
typedef void (^TJLButtonTappedBlock)(Menu *buttonView, NSString *title);
@interface Menu : UIView{
    TJLButtonTappedBlock buttonTappedBlock;                    //实例变量
}
//属性
@property(strong, nonatomic) NSMutableArray *constraintsArray;
……
@property(nonatomic) NSLayoutAttribute finalButtonConstant;
@property(nonatomic) NSLayoutAttribute finalButtonConstant1;
//方法
- (instancetype)initWithViewController:(UIViewController *)viewController images:(NSArray *)images buttonTitles:(NSArray *)titles position:(TJLBarButtonMenuSide)position;    //初始化
- (instancetype)initWithViewController:(UIViewController *)viewController delegate:(id<MenuDelegate>)delegate images:(NSArray *)images buttonTitles:(NSArray *)titles position:(TJLBarButtonMenuSide)position;
- (void)show;
- (void)setButtonTappedBlock:(TJLButtonTappedBlock)block;
@end
```

(5)打开 Menu.m 文件,编写代码,实现具有按钮的菜单的绘制以及轻拍菜单项关闭菜单等。使用的方法如表 10-7 所示。

表 10-7　Menu.m 文件中方法总结

方　　法	功　　能
initWithViewController:delegate:images:buttonTitles:position:	使用视图控制器、协议、图像、按钮标题和位置对菜单初始化
setupPositions:	设置位置
initWithViewController:images:buttonTitles:position:	使用视图控制器、图像、按钮标题和位置对菜单初始化
buttonTapped:	轻拍按钮后隐藏菜单
setButtonTappedBlock:	设置按钮轻拍的块
show	显示
hide:	隐藏

这里需要讲解几个重要的方法(其他方法请读者参考源代码)。其中,要绘制具有按钮的菜单需要使用 initWithViewController:delegate:images:buttonTitles:position:、setupPositions:、initWithViewController: images:buttonTitles:position:、show 和 hide:方法。initWithViewController:delegate:images:buttonTitles: position:方法实现使用视图控制器、协议、图像、按钮标题和位置对菜单初始化。程序代码如下:

```
- (instancetype)initWithViewController:(UIViewController *)viewController delegate:(id)delegate images:(NSArray *)images buttonTitles:(NSArray *) titles position:(TJLBarButtonMenuSide)position {
self = [super init];
//判断 sel 是否为空
    if(!self) {
        return nil;
    }
```

```objc
    [self setupPositions:position];
    self.parentView = viewController;
    if(delegate) self.delegate = delegate;                    //设置委托
    self.translatesAutoresizingMaskIntoConstraints = NO;
    //添加约束
    [self addConstraints:@[ [NSLayoutConstraint constraintWithItem:self attribute:NSLayoutAttributeWidth relatedBy:NSLayoutRelationEqual toItem:nil attribute:NSLayoutAttributeNotAnAttribute multiplier:1.0 constant:150], [NSLayoutConstraint constraintWithItem:self attribute:NSLayoutAttributeHeight relatedBy:NSLayoutRelationEqual toItem:self attribute:NSLayoutAttributeWidth multiplier:1.0 constant:0]]];
    self.backgroundColor = [UIColor clearColor];              //设置背景颜色
    self.buttonArray = [NSMutableArray new];
    //创建并设置按钮对象
    self.constraintsArray = [NSMutableArray new];
    for(NSUInteger i = 0; i < images.count; i++) {
        UIButton *b = [UIButton buttonWithType:UIButtonTypeCustom];//实例化
        b.translatesAutoresizingMaskIntoConstraints = NO;
        [b setImage:images[i] forState:UIControlStateNormal];     //设置图像
        b.hidden = YES;                                           //隐藏按钮
        [b addTarget:self action:@selector(buttonTapped:) forControlEvents:UIControlEventTouchUpInside];
        objc_setAssociatedObject(b, &key, titles[i], OBJC_ASSOCIATION_ASSIGN);
        self.buttonArray[i] = b;
        [self addSubview:b];                                      //添加视图对象
        //创建约束
        NSLayoutConstraint *first = [NSLayoutConstraint constraintWithItem:b attribute:NSLayoutAttributeTop relatedBy:NSLayoutRelationEqual toItem: self attribute:NSLayoutAttributeTop multiplier:1.0 constant:-40];
        NSLayoutConstraint *second = [NSLayoutConstraint constraintWithItem:b attribute:self.rightLeftPosition relatedBy:NSLayoutRelationEqual toItem:self attribute:self.rightLeftPosition multiplier:1.0 constant: self.initialButtonConstant];
        [self.constraintsArray addObject:@[first, second]];
        [self addConstraints:@[first, second]];                   //添加一个约束
        [b addConstraints:@[ [NSLayoutConstraint constraintWithItem: b attribute:NSLayoutAttributeWidth relatedBy:NSLayoutRelationLessThanOrEqual toItem:nil attribute:NSLayoutAttributeNotAnAttribute multiplier:1.0 constant:50], [NSLayoutConstraint constraintWithItem:b attribute: NSLayoutAttributeHeight relatedBy:NSLayoutRelationLessThanOrEqual toItem: b attribute:NSLayoutAttributeWidth multiplier:1.0 constant:0]]];
    }
    return self;
}
```

setupPositions:方法实现对按钮及菜单项位置的设置。程序代码如下:

```objc
- (void)setupPositions:(TJLBarButtonMenuSide)position {
    switch(position) {
        //右边的菜单
        case TJLBarButtonMenuRightTop:
            self.rightLeftPosition = NSLayoutAttributeRight;  //设置左右的位置
            self.initialButtonConstant = -10;
            self.finalButtonConstant = NSLayoutAttributeLeft;
            self.finalButtonConstant1 = 15;
            break;
```

```objc
        //左边的菜单
        case TJLBarButtonMenuLeftTop:
            self.rightLeftPosition = NSLayoutAttributeLeft;
            self.initialButtonConstant = 10;
            self.finalButtonConstant = NSLayoutAttributeRight;
            self.finalButtonConstant1 = -15;
            break;
        default:
            break;
    }
}
```

show方法实现具有按钮的菜单的显示。程序代码如下：

```objc
- (void)show {
    [self.parentView.view addSubview:self];
    [self layoutSubviews];
    CGFloat position = (self.parentView.navigationController) ? 0 : 44;
    //添加约束
    [self.parentView.view addConstraints:@[[NSLayoutConstraint constraint
WithItem:self attribute:NSLayoutAttributeTop relatedBy: NSLayout
RelationEqual toItem:self.parentView.view attribute:NSLayout Attribute
Top multiplier:1.0 constant:position],[NSLayoutConstraint constraint
WithItem:self attribute:self.rightLeftPosition relatedBy:NSLayout
RelationEqual toItem:self.parentView.view attribute:self.right
LeftPosition multiplier:1.0 constant:0]]];
    [self.parentView.view layoutSubviews];
    //创建约束
    self.firstConstraints = [@[@[[NSLayoutConstraint constraintWithItem:
self.buttonArray[0] attribute:self.finalButtonConstant relatedBy:NSLayout
RelationEqual toItem:self attribute:self.finalButtonConstant multiplier: 1.0
constant:self.finalButtonConstant1], [NSLayoutConstraint constraint
WithItem:self.buttonArray[0] attribute:NSLayoutAttributeTop relatedBy:
NSLayoutRelationEqual toItem:self attribute:NSLayoutAttributeTop
multiplier:1.0 constant:15]],@[[NSLayoutConstraint constraintWithItem:
self.buttonArray[1] attribute:self.rightLeftPosition relatedBy:
NSLayoutRelationEqual toItem:self attribute:self.rightLeftPosition
multiplier:1.0 constant:self.initialButtonConstant], [NSLayoutConstraint
constraintWithItem:self.buttonArray[1] attribute:NSLayoutAttributeBottom
relatedBy:NSLayoutRelationEqual toItem:self attribute:NSLayoutAttribute
Bottom multiplier:1.0 constant:-15]]] mutableCopy];
    //判断数组buttonArray中元素的个数是否大于2
    if(self.buttonArray.count > 2) {
        self.firstConstraints[2]=@[[NSLayoutConstraintconstraintWithItem:
        self.buttonArray[2] attribute:self.finalButtonConstant relatedBy:
        NSLayoutRelationEqual toItem:self attribute:self. finalButton
        Constant multiplier:1.0 constant:0], [NSLayoutConstraint constraint
        WithItem:self.buttonArray[2] attribute:NSLayoutAttributeBottom relatedBy:
        NSLayoutRelationEqual toItem:self attribute:NSLayoutAttributeBottom
        multiplier:1.0 constant:0]];
    }
    //动画
    [UIView animateWithDuration:.15 delay:0 options:UIViewAnimationOption
CurveEaseOut animations:^{
        [self.buttonArray[0] setHidden:NO];                    //不隐藏
        [self removeConstraints:self.constraintsArray[0]];    //移除约束
        [self addConstraints:self.firstConstraints[0]];       //添加约束
        [self layoutIfNeeded];
    } completion:^(BOOL finished) {
        [UIView animateWithDuration:.15 delay:0 options:UIViewAnimation
```

```objc
        OptionCurveEaseOut animations:^{
            [self.buttonArray[1] setHidden:NO];
            [self removeConstraints:self.constraintsArray[1]];   //移除约束
            [self addConstraints:self.firstConstraints[1]];      //添加约束
            [self layoutIfNeeded];
        } completion:^(BOOL complete) {
            [UIView animateWithDuration:.15 delay:0 options:UIViewAnimation
            OptionCurveEaseOut animations:^{
                if(self.buttonArray.count > 2) {
                    [self.buttonArray[2] setHidden:NO];
                    [self removeConstraints:self.constraintsArray[2]];//移除约束
                    [self addConstraints:self.firstConstraints[2]];//添加约束
                    [self layoutIfNeeded];
                }
            } completion:nil];
        }];
    }];
}
```

hide:方法实现将显示的菜单进行隐藏。程序代码如下：

```objc
- (void)hide:(NSString *)buttonTitle {
    [UIView animateWithDuration:.15 delay:0 options:UIViewAnimationOption
    CurveEaseOut animations:^{
        [self removeConstraints:self.firstConstraints[0]];    //移除约束
        [self addConstraints:self.constraintsArray[0]];       //添加约束
        [self layoutIfNeeded];
    } completion:^(BOOL finished) {
        [self.buttonArray[0] setHidden:YES];
        [UIView animateWithDuration:.15 delay:0 options:UIViewAnimationOption
        CurveEaseOut animations:^{
            [self removeConstraints:self.firstConstraints[1]];   //移除约束
            [self addConstraints:self.constraintsArray[1]];      //添加约束
            [self layoutIfNeeded];
        }completion:^(BOOL complete) {
            [self.buttonArray[1] setHidden:YES];                 //移除按钮
            [UIView animateWithDuration:.15 delay:0 options:UIViewAnimationOption
            CurveEaseOut animations:^{
                if(self.buttonArray.count > 2) {
                    [self removeConstraints:self.firstConstraints[2]];//移除约束
                    [self addConstraints:self.constraintsArray[2]];//添加约束
                    [self layoutIfNeeded];
                }
            }completion:^(BOOL final) {
                if(buttonTappedBlock) buttonTappedBlock(self, buttonTitle);
                if([self.delegate respondsToSelector:@selector(buttonMenu:title
                ForTappedButton:)])
                    [self.delegate buttonMenu:self titleForTappedButton: buttonTitle];
                [self removeFromSuperview];                      //移除视图
            }];
        }];
    }];
}
```

要实现菜单项的功能，需要使用 buttonTapped:方法，它的功能是在轻拍按钮后隐藏菜单。程序代码如下：

```objc
- (void)buttonTapped:(UIButton *)sender {
    NSString *title = objc_getAssociatedObject(sender, &key);
```

```
        [self hide:title];
}
```

（6）打开 ViewController.h 文件，编写代码，实现头文件、对象、插座变量以及动作的声明。程序代码如下：

```
#import <UIKit/UIKit.h>
#import "Menu.h"
@interface ViewController : UIViewController{
    Menu *barMenu;
    IBOutlet UINavigationBar *bar;                  //插座变量
}
//动作
- (IBAction)showright:(id)sender;
- (IBAction)showleft:(id)sender;
@end
```

（7）打开 Main.storyboard 文件，对 View Controller 视图控制器的设计界面进行设计，效果如图 10.7 所示。

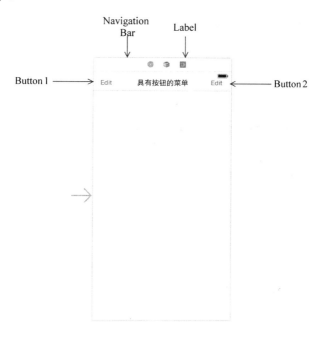

图 10.7　实例 68View Controller 视图控制器的设计界面

需要添加的视图、控件以及对它们的设置如表 10-8 所示。

表 10-8　实例 68 视图、控件设置

视图、控件	属 性 设 置	其 他
Label	Text：具有按钮的菜单 Alignment：居中	
Navigation Bar		与插座变量 bar 关联
Button1	Title：Edit	与动作 showleft:关联
Button2	Title：Edit	与动作 showright:关联

（8）打开 ViewController.m 文件，编写代码，实现在单击导航栏左右两边的 Edit 按钮

后出现菜单，并且在选择菜单项后，不仅可以关闭菜单，还可以对导航栏以及界面背景的颜色进行改变。程序代码如下：

```objc
//单击右边的Edit按钮（菜单项可以设置界面的背景颜色）
- (IBAction)showright:(id)sender {
    NSArray *images = @[[UIImage imageNamed:@"1.png"],[UIImage imageNamed:
@"2.png"],[UIImage imageNamed:@"3.png"]];              //实例化数组对象
    barMenu = [[Menu alloc]initWithViewController:self images:images button
Titles:@[@"yellow",@"gray" ,@"orange",] position:TJLBar ButtonMenu RightTop];
    //实现界面背景颜色的设置
    [barMenu setButtonTappedBlock:^(Menu *buttonView, NSString *title) {
        //判断title是否与"yellow"相同
        if ([title isEqualToString:@"yellow"]) {
            self.view.backgroundColor=[UIColor yellowColor];    //设置背景颜色
        }else if ([title isEqualToString:@"gray"]){      //判断title 是否与
"gray"相同
            self.view.backgroundColor=[UIColor grayColor];
        }else{
            self.view.backgroundColor=[UIColor orangeColor];
        }
    }];
    [barMenu show];
}
//单击左边的按钮（菜单项可以设置导航栏的背景颜色）
- (IBAction)showleft:(id)sender {
    NSArray *images = @[[UIImage imageNamed:@"1.png"],[UIImage imageNamed:
@"2.png"],[UIImage imageNamed:@"3.png"]];              //实例化数组对象
    barMenu = [[Menu alloc]initWithViewController:self images:imagesbutton
Titles:@[@"yellow",@"gray" ,@"orange",] position:TJLBarButtonMenuLeftTop];
    //设置导航栏的背景颜色
    [barMenu setButtonTappedBlock:^(Menu *buttonView, NSString *title) {
        //判断title是否与"yellow"相同
        if ([title isEqualToString:@"yellow"]) {
            bar.backgroundColor=[UIColor yellowColor];
        }else if ([title isEqualToString:@"gray"]){   //判断title是否与"gray"相同
            bar.backgroundColor=[UIColor grayColor];
        }else{
            bar.backgroundColor=[UIColor orangeColor];       //设置背景颜色
        }
    }];
    [barMenu show];
}
```

【代码解析】

本实例关键功能是自动布局。下面就是这个知识点的详细讲解。

要实现自动布局，需要使用到约束。因为自动布局是一直基于约束的、描述性的系统布局。要使用约束，首先要创建使用 NSLayoutConstraint 的 constraint WithItem: attribute: relatedBy:toItem:attribute: multiplier:constant:方法，其语法形式如下：

```
+ (id)constraintWithItem:(id)view1    attribute:(NSLayoutAttribute)attr1
relatedBy:(NSLayoutRelation)relation               toItem:(id)view2
attribute:(NSLayoutAttribute)attr2    multiplier:(CGFloat)multiplier
constant:(CGFloat)c;
```

其中，参数说明如下。

❏ (id)view1 表示要约束的对象。

第 10 章 菜单

- (NSLayoutAttribute)attr1 表示对象的布局属性。
- (NSLayoutRelation)relation 表示布局关系。
- (id)view2 表示参照对象。
- (NSLayoutAttribute)attr2 表示参照对象的布局属性。
- (CGFloat)multiplier 表示乘数。
- (CGFloat)c 表示常数。

在代码中，使用了 constraintWithItem:attribute:relatedBy:toItem:attribute:multiplier: constant:方法进行了自动布局，代码如下：

```
  NSLayoutConstraint *first = [NSLayoutConstraint constraintWithItem:b
attribute:NSLayoutAttributeTop
relatedBy:NSLayoutRelationEqual toItem:self attribute:NSLayoutAttribute
Top multiplier:1.0 constant:-40];
```

其中，参数说明如下。
- b 表示要约束的对象。
- NSLayoutAttributeTop 表示对象的布局属性。
- NSLayoutRelationEqual n 表示布局关系。
- self 表示参照对象。
- NSLayoutAttributeTop 表示参照对象的布局属性。
- 1.0 表示乘数。
- -40 表示常数。

创建好自动约束后，需要使用 UIView 的 addConstraint:或 addConstraints:方法，实现将创建的约束添加到指定的视图中。其中 addConstraint:方法实现将指定的一个约束添加到指定的视图中。其语法形式如下：

```
- (void)addConstraint:(NSLayoutConstraint *)constraint;
```

其中，(NSLayoutConstraint *)constraint 表示指定的一个约束。addConstraints:方法实现将指定的多个个约束添加到指定的视图中。其语法形式如下：

```
- (void)addConstraints:(NSArray *)constraints;
```

其中，(NSArray *)constraints 表示一个数组对象，用来保存多个约束。在代码中，使用了 addConstraints:方法将创建的约束添加到视图中，代码如下：

```
[self addConstraints:@[first, second]];
```

实例 69 仿 Tumblr iOS App 菜单

【实例描述】

本实例实现的功能是显示一个仿 Tumblr iOS App 的菜单。其中，当用户单击界面显示的按钮，就会以动画的形式出现菜单。选择菜单中的任意一项，或者单击背景，菜单就会以动画的形式隐藏。运行效果如图 10.8 所示。

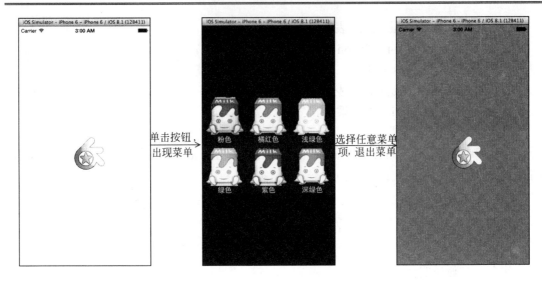

图 10.8　实例 69 运行效果

【实现过程】

（1）创建一个项目，命名为"仿 Tumblr iOS App 菜单"。

（2）添加图片 1.png、2.png、3.png、4.png、5.png、6.png、7.png 和 8.png 到创建项目的 Supporting Files 文件夹中。

（3）创建一个基于 UIButton 类的 MenuItem 类。

（4）打开 MenuItem.h 文件，编写代码，实现宏定义、新类型的定义、对象、属性以及方法的声明。程序代码如下：

```
#import <UIKit/UIKit.h>
//宏定义
#define CHTumblrMenuViewImageHeight 90
#define CHTumblrMenuViewTitleHeight 20
typedef void (^CHTumblrMenuViewSelectedBlock)(void);          //新类型的定义
@interface MenuItem : UIButton{
    //属性
    UIImageView *iconView_;
    UILabel *titleLabel_;
}
@property(nonatomic,copy)CHTumblrMenuViewSelectedBlock selectedBlock;    //属性
-(id)initWithTitle:(NSString*)title andIcon:(UIImage*)icon andSelectedBlock:
(CHTumblrMenuViewSelectedBlock)block;                          //方法
@end
```

（5）打开 MenuItem.m 文件，编写代码，实现对菜单项的绘制。使用的方法有 initWithTitle: andIcon:andSelectedBlock:和 setFrame:。其中，initWithTitle: andIcon:andSelectedBlock:方法实现对菜单项的初始化。程序代码如下：

```
-(id)initWithTitle:(NSString*)title   andIcon:(UIImage*)icon   andSelected
Block:(CHTumblrMenuViewSelectedBlock)block{
    self = [super init];
    if (self) {
```

```
      //图像视图对象的创建和设置
      iconView_ = [UIImageView new];
      iconView_.image = icon;                              //设置图像
   //标签对象的创建和设置
      titleLabel_ = [UILabel new];
      titleLabel_.textAlignment = NSTextAlignmentCenter;  //设置对齐方式
      titleLabel_.backgroundColor = [UIColor clearColor]; //设置背景颜色
      titleLabel_.textColor = [UIColor whiteColor];       //设置文本颜色
      titleLabel_.text = title;
      _selectedBlock = block;
      [self addSubview:iconView_];
      [self addSubview:titleLabel_];
   }
   return self;
}
```

setFrame:方法实现对菜单项框架的设置。程序代码如下：

```
- (void)setFrame:(CGRect)frame
{
   [super setFrame:frame];
   iconView_.frame = CGRectMake(0, 0, CHTumblrMenuViewImageHeight,
CHTumblrMenuViewImageHeight);                             //设置框架
   titleLabel_.frame = CGRectMake(0, CHTumblrMenuViewImageHeight,
CHTumblrMenuViewImageHeight, CHTumblrMenuViewTitleHeight);
}
```

（6）创建一个基于 UIView 类的 Menu 类。

（7）打开 Menu.h 文件，编写代码，实现头文件、宏定义、对象、属性和方法的声明。程序代码如下：

```
#import <UIKit/UIKit.h>
#import "MenuItem.h"                                      //头文件
//宏定义
#define CHTumblrMenuViewTag 1999
……
#define TumblrBlue [UIColor colorWithRed:45/255.0f green:68/255.0f blue:94/255.0f alpha:1.0]
@interface Menu : UIView<UIGestureRecognizerDelegate>{
   //对象
   UIImageView *backgroundView_;
   NSMutableArray *buttons_;
}
@property (nonatomic, readonly)UIImageView *backgroundImgView;//属性
//方法
- (void)addMenuItemWithTitle:(NSString*)title andIcon:(UIImage*)icon and
SelectedBlock:
(CHTumblrMenuViewSelectedBlock)block;
- (void)show;
@end
```

（8）打开 Menu.m 文件，编写代码，实现 Tumblr iOS App 菜单。使用的方法如表 10-9 所示。

表 10-9 Menu.m文件中方法总结

方　　法	功　　能
initWithFrame:	菜单的初始化
addMenuItemWithTitle:andIcon:andSelectedBlock:	添加菜单项
frameForButtonAtIndex:	获取框架
layoutSubviews	布局子视图
dismiss:	隐藏
buttonTapped:	轻拍按钮
riseAnimation	弹出动画
dropAnimation	弹回动画
animationDidStart:	动画开始时调用，判断动画的关键字，并执行相应的功能
show	显示

要实现 Tumblr iOS App 菜单，需要实现两个部分：显示菜单和菜单的关闭。其中，实现菜单的显示，需要使用 initWithFrame:、addMenuItemWithTitle:andIcon:andSelectedBlock:、frameForButtonAtIndex:、layoutSubviews、riseAnimation 和 show 方法。initWithFrame:方法实现对菜单的初始化。程序代码如下：

```objc
- (id)initWithFrame:(CGRect)frame
{
    self = [super initWithFrame:frame];
    if (self) {
        UITapGestureRecognizer *ges = [[UITapGestureRecognizer alloc] initWithTarget:self action:@selector(dismiss:)];//创建轻拍手势识别器对象
        ges.delegate = self;                       //设置委托
        [self addGestureRecognizer:ges];
        self.backgroundColor = [UIColor clearColor];//设置背景颜色
        backgroundView_ = [[UIImageView alloc] initWithFrame:self.bounds];
        backgroundView_.backgroundColor = TumblrBlue;
        backgroundView_.autoresizingMask = UIViewAutoresizing Flexible Height | UIViewAutoresizingFlexibleWidth;
        [self addSubview:backgroundView_];         //添加视图对象
        buttons_ = [[NSMutableArray alloc] initWithCapacity:6];
    }
    return self;
}
```

addMenuItemWithTitle:andIcon:andSelectedBlock:方法实现为菜单添加菜单项的功能。程序代码如下：

```objc
- (void)addMenuItemWithTitle:(NSString*)title andIcon:(UIImage*)icon andSelectedBlock:
(CHTumblrMenuViewSelectedBlock)block
{
    MenuItem *button = [[MenuItem alloc] initWithTitle:title andIcon:icon andSelectedBlock:block];
    //添加动作
    [buttonaddTarget:self action:@selector(buttonTapped:)forControlEvents:UIControlEventTouchUpInside];
    [self addSubview:button];
    [buttons_ addObject:button];
}
```

frameForButtonAtIndex:方法实现框架的获取。程序代码如下：

```
- (CGRect)frameForButtonAtIndex:(NSUInteger)index
{
    NSUInteger columnCount = 3;
    NSUInteger columnIndex = index % columnCount;           //获取列的索引
    NSUInteger rowCount=buttons_.count/columnCount+(buttons_.count%columnCount>0?1:0);
    NSUInteger rowIndex = index / columnCount;              //获取行的索引
    CGFloat itemHeight = (CHTumblrMenuViewImageHeight + CHTumblrMenuViewTitleHeight) * rowCount + (rowCount > 1?(rowCount - 1) * CHTumblrMenuViewHorizontalMargin:0);
    CGFloat offsetY = (self.bounds.size.height - itemHeight) / 2.0;
    CGFloat verticalPadding = (self.bounds.size.width - CHTumblrMenuViewHorizontalMargin * 2 - CHTumblrMenuViewImageHeight * 3) / 2.0;
    CGFloat offsetX = CHTumblrMenuViewHorizontalMargin;
    offsetX += (CHTumblrMenuViewImageHeight+ verticalPadding) * columnIndex;
    offsetY += (CHTumblrMenuViewImageHeight + CHTumblrMenuViewTitleHeight + CHTumblrMenuViewVerticalPadding) * rowIndex;
    return CGRectMake(offsetX, offsetY, CHTumblrMenuViewImageHeight, (CHTumblrMenuViewImageHeight+CHTumblrMenuViewTitleHeight));      //返回
}
```

layoutSubviews 方法实现子视图的布局，即实现菜单项的布局。程序代码如下：

```
- (void)layoutSubviews
{
    [super layoutSubviews];
    for (NSUInteger i = 0; i < buttons_.count; i++) {
        MenuItem *button = buttons_[i];
        button.frame = [self frameForButtonAtIndex:i];      //设置框架
    }
}
```

riseAnimation 方法实现菜单的显示动画，即弹出动画。程序代码如下：

```
- (void)riseAnimation
{
    NSUInteger columnCount = 3;
    NSUIntegerrowCount=buttons_.count/columnCount+(buttons_.count %columnCount>0?1:0);
    for (NSUInteger index = 0; index < buttons_.count; index++) {
        MenuItem *button = buttons_[index];
        button.layer.opacity = 0;                           //设置透明度
        CGRect frame = [self frameForButtonAtIndex:index];  //获取框架
        NSUInteger rowIndex = index / columnCount;
        NSUInteger columnIndex = index % columnCount;
        CGPoint fromPosition = CGPointMake(frame.origin.x + CHTumblr MenuViewImageHeight /2.0,frame.origin.y+(rowCount - rowIndex + 2)*200 + (CHTumblrMenuViewImageHeight + CHTumblrMenuViewTitleHeight) / 2.0);
                                                            //获取开始的位置
        CGPoint toPosition = CGPointMake(frame.origin.x + CHTumblrMenuViewImageHeight / 2.0,frame.origin.y+(CHTumblrMenuViewImageHeight + CHTumblrMenuViewTitleHeight) / 2.0);
        double delayInSeconds = rowIndex * columnCount * CHTumblrMenuViewAnimationInterval;
        //判断 columnIndex
        if (!columnIndex) {
            delayInSeconds += CHTumblrMenuViewAnimationInterval;
```

```objectivec
    else if(columnIndex == 2) {
        delayInSeconds += CHTumblrMenuViewAnimationInterval * 2;
    }
    //动画效果
    CABasicAnimation *positionAnimation;
    positionAnimation = [CABasicAnimation animationWithKeyPath:@"position"];
    positionAnimation.fromValue = [NSValue valueWithCGPoint:fromPosition];
                                    //开始值
    positionAnimation.toValue = [NSValue valueWithCGPoint:toPosition];
                                    //结束值
    positionAnimation.timingFunction = [CAMediaTimingFunction functionWithControlPoints:0.45f :1.2f :0.75f :1.0f];
    positionAnimation.duration = CHTumblrMenuViewAnimationTime;  //所需时间
    positionAnimation.beginTime = [button.layer convertTime:CACurrentMediaTime() fromLayer:nil] + delayInSeconds;
    [positionAnimation setValue:[NSNumber numberWithUnsignedInteger:index] forKey:CHTumblrMenuViewRriseAnimationID];
    positionAnimation.delegate = self;
    [button.layer addAnimation:positionAnimation forKey:@"riseAnimation"];
                                    //添加动画
    }
}
```

animationDidStart:方法在动画开始时调用，判断动画的关键字，并执行相应的功能。程序代码如下：

```objectivec
- (void)animationDidStart:(CAAnimation *)anim
{
    NSUInteger columnCount = 3;
    //判断
    if([anim valueForKey:CHTumblrMenuViewRriseAnimationID]) {
        NSUInteger index = [[anim valueForKey: CHTumblrMenuViewRriseAnimationID] unsignedIntegerValue];
        UIView *view = buttons_[index];
        CGRect frame = [self frameForButtonAtIndex:index];  //获取框架
        CGPoint toPosition = CGPointMake(frame.origin.x + CHTumblrMenuViewImageHeight / 2.0,frame.origin.y + (CHTumblrMenuViewImageHeight + CHTumblrMenuViewTitleHeight) / 2.0);            //开始的位置
        CGFloat toAlpha = 1.0;
        view.layer.position = toPosition;                   //设置位置
        view.layer.opacity = toAlpha;                       //设置透明度
    }
    else if([anim valueForKey:CHTumblrMenuViewDismissAnimationID]) {
        NSUInteger index = [[anim valueForKey:CHTumblrMenuViewDismissAnimationID] unsignedIntegerValue];
        NSUInteger rowIndex = index / columnCount;
        UIView *view = buttons_[index];                     //实例化按钮
        CGRect frame = [self frameForButtonAtIndex:index];
        CGPoint toPosition = CGPointMake(frame.origin.x + CHTumblrMenuViewImageHeight / 2.0,frame.origin.y - (rowIndex + 2)*200 + (CHTumblrMenuViewImageHeight + CHTumblrMenuViewTitleHeight) / 2.0);
                                                            //获取结束的位置
        view.layer.position = toPosition;                   //设置位置
    }
}
```

show方法实现菜单的显示。程序代码如下：

```objc
- (void)show
{
    UIViewController *appRootViewController;
    UIWindow *window;
    window = [UIApplication sharedApplication].keyWindow;
    appRootViewController = window.rootViewController;
    UIViewController *topViewController = appRootViewController;
    //判断topViewController.presentedViewController是否不为空
    while (topViewController.presentedViewController != nil)
    {
        topViewController = topViewController.presentedViewController;
    }
    if ([topViewController.view viewWithTag:CHTumblrMenuViewTag]) {
        [[topViewController.view viewWithTag:CHTumblrMenuViewTag] removeFromSuperview];//移除
    }
    self.frame = topViewController.view.bounds;
    [topViewController.view addSubview:self];              //添加视图对象
    [self riseAnimation];
}
```

要实现菜单的隐藏,需要使用 dismiss:、buttonTapped:、dropAnimation 方法以及 frameForButtonAtIndex:和 animationDidStart:方法。dismiss:方法实现菜单的隐藏。程序代码如下:

```objc
- (void)dismiss:(id)sender
{
    [self dropAnimation];                            //调用 dropAnimation 方法
    double delayInSeconds = CHTumblrMenuViewAnimationTime + CHTumblrMenuViewAnimationInterval * (buttons_.count + 1);
    dispatch_time_t popTime = dispatch_time(DISPATCH_TIME_NOW, (int64_t)(delayInSeconds * NSEC_PER_SEC));
    //在指定的时间执行删除视图的操作
    dispatch_after(popTime, dispatch_get_main_queue(), ^(void){
        [self removeFromSuperview];                  //移除
    });
}
```

buttonTapped:方法实现按钮的轻拍并响应。程序代码如下:

```objc
- (void)buttonTapped:(MenuItem*)btn
{
    [self dismiss:nil];                              //调用 dismiss:方法
    double delayInSeconds = CHTumblrMenuViewAnimationTime + CHTumblrMenuViewAnimationInterval * (buttons_.count + 1);
    dispatch_time_t popTime = dispatch_time(DISPATCH_TIME_NOW, (int64_t)(delayInSeconds * NSEC_PER_SEC));
    dispatch_after(popTime, dispatch_get_main_queue(), ^(void){
        //延后执行 selectedBlock()
        btn.selectedBlock();
    });
}
```

dropAnimation 方法实现菜单的隐藏动画,即弹回动画。程序代码如下:

```objc
- (void)dropAnimation
{
```

```objectivec
    NSUInteger columnCount = 3;
    for (NSUInteger index = 0; index < buttons_.count; index++) {
        MenuItem *button = buttons_[index];                     //实例化对象
        CGRect frame = [self frameForButtonAtIndex:index];
        NSUInteger rowIndex = index / columnCount;
        NSUInteger columnIndex = index % columnCount;
        CGPoint toPosition = CGPointMake(frame.origin.x + CHTumblrMenuView
        ImageHeight / 2.0,frame.origin.y - (rowIndex + 2)*200 + (CHTumblr
        MenuViewImageHeight + CHTumblrMenuViewTitleHeight) / 2.0);
                                                                //获取结束位置
        CGPoint fromPosition = CGPointMake(frame.origin.x + CHTumblrMenu
        ViewImageHeight / 2.0,frame.origin.y+(CHTumblrMenuViewImageHeight +
        CHTumblrMenuViewTitleHeight) / 2.0);
        double delayInSeconds = rowIndex * columnCount * CHTumblrMenu View
        AnimationInterval;
        //判断
        if (!columnIndex) {
            delayInSeconds += CHTumblrMenuViewAnimationInterval;
        }
        else if(columnIndex == 2) {
            delayInSeconds += CHTumblrMenuViewAnimationInterval * 2;
        }
        //动画实现
        CABasicAnimation *positionAnimation;
        positionAnimation = [CABasicAnimation animationWithKeyPath: @"position"];
        positionAnimation.fromValue = [NSValue valueWithCGPoint:fromPosition];
                //设置开始值
        positionAnimation.toValue = [NSValue valueWithCGPoint:toPosition];
                //设置结束值
        positionAnimation.timingFunction = [CAMediaTimingFunction function
        WithControlPoints:0.3 :0.5f :1.0f :1.0f];
        positionAnimation.duration = CHTumblrMenuViewAnimationTime;
                //设置所需时间
        positionAnimation.beginTime = [button.layer convertTime: CACurrent
        MediaTime() fromLayer:nil] + delayInSeconds;
        [positionAnimation setValue:[NSNumber numberWithUnsignedInteger:
        index] forKey:CHTumblrMenuViewDismissAnimationID];      //设置值
        positionAnimation.delegate = self;                      //设置委托
        [button.layer addAnimation:positionAnimation forKey:@"rise Animation"];
    }
}
```

（9）打开 ViewController.m 文件，编写代码，实现按钮的初始化，以及单击按钮后菜单的显示。程序代码如下：

```objectivec
- (void)viewDidLoad
{
    [super viewDidLoad];
    // Do any additional setup after loading the view, typically from a nib.
    //创建并设置按钮对象
    UIButton *button = [UIButton buttonWithType:UIButtonTypeCustom];
                        //实例化
    [button setImage:[UIImage imageNamed:@"7.png"] forState:UIControl
    StateHighlighted];          //设置图像
    [button setImage:[UIImage imageNamed:@"8.png"] forState:UIControlState
    Normal];                    //设置图像
    button.frame = CGRectMake((self.view.bounds.size.width - 59)/2.0, (self.
```

```objc
    view.bounds.size.height - 48)/2.0, 80, 80);          //设置框架
    [button addTarget:self action:@selector(showMenu) forControlEvents:
    UIControlEventTouchUpInside];                        //添加动作
    [self.view addSubview:button];
}
//显示菜单
- (void)showMenu
{
    //创建并设置菜单对象
    Menu *menuView = [[Menu alloc] init];
    [menuView addMenuItemWithTitle:@"粉色" andIcon:[UIImage imageNamed:
    @"1.png"] andSelectedBlock:^{
        self.view.backgroundColor=[UIColor colorWithHue:1.0 saturation:0.5
        brightness:1 alpha:1.0];            //背景颜色
    }];
    [menuView addMenuItemWithTitle:@"橘红色" andIcon:[UIImage imageNamed:
    @"2.png"] andSelectedBlock:^{
        self.view.backgroundColor=[UIColor colorWithHue:0.1 saturation:
        0.3 brightness:1 alpha:1.0];        //背景颜色
    }];
    [menuView addMenuItemWithTitle:@"浅绿色" andIcon:[UIImage imageNamed:
     @"3.png"] andSelectedBlock:^{
        //背景颜色
        self.view.backgroundColor=[UIColor colorWithHue:0.3 saturation:
        0.3 brightness:1 alpha:1.0];
    }];
    [menuView addMenuItemWithTitle:@"绿色" andIcon:[UIImage imageNamed:
    @"4.png"] andSelectedBlock:^{
        self.view.backgroundColor=[UIColor greenColor];  //背景颜色
    }];
    [menuView addMenuItemWithTitle:@"紫色" andIcon:[UIImage imageNamed:
    @"5.png"] andSelectedBlock:^{
        self.view.backgroundColor=[UIColor colorWithHue:0.7 saturation:0.3
        brightness:1 alpha:1.0];            //背景颜色
    }];
    [menuView addMenuItemWithTitle:@"深绿色" andIcon:[UIImage imageNamed:
    @"6.png"] andSelectedBlock:^{
        self.view.backgroundColor=[UIColor colorWithHue:0.5 saturation:0.7
        brightness:0.5 alpha:1.0];          //背景颜色
    }];
    //显示
    [menuView show];
}
```

【代码解析】

由于本实例中的代码和方法非常的多,为了方便读者的阅读,笔者绘制了一些执行流程图如图 10.9~10.10 所示。其中,单击运行按钮后一直到在模拟器上显示菜单,使用了 viewDidLoad、showMenu、initWithFrame:、addMenuItemWithTitle:andIcon:andSelectedBlock:、initWithTitle:andIcon:andSelectedBlock:、setFrame:、show、riseAnimation、frameForButtonAtIndex:、layoutSubviews 和 animationDidStart:方法共同实现。它们的执行流程如图 10.9 所示。

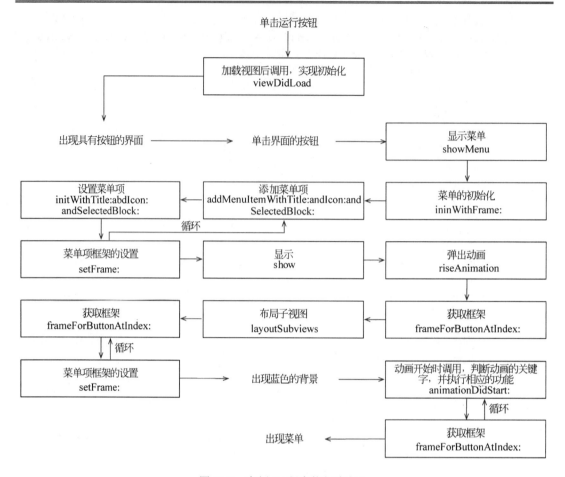

图 10.9　实例 69 程序执行流程 1

退出菜单有两种方法,一种是轻拍菜单项以外的部分;另一种是选择任意的菜单项。以选择任意的菜单项为例,使用了 buttonTapped:、dismiss:、dropAnimation、frameForButtonAtIndex:和 animationDidStart:方法共同实现。它们的执行流程如图 10.10 所示。

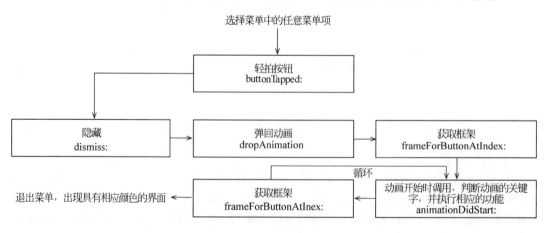

图 10.10　实例 69 程序执行流程 2

实例 70 边栏菜单

【实例描述】

菜单的形式是千变万化的,本实例的功能是实现一个边栏菜单的效果。当用户使用滑动的手势滑动界面,就会显示出边栏菜单,在此菜单中可以实现视图的切换。运行效果如图 10.11 所示。

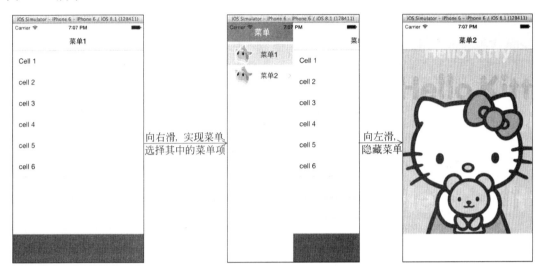

图 10.11 实例 70 运行效果

【实现过程】

(1)创建一个项目,命名为"边栏菜单"。
(2)添加图片 1.png 和 2.jpg 到创建项目的 Supporting Files 文件夹中。
(3)创建一个基于 UIView 类的 Menu 类。
(4)打开 Menu.h 文件,编写代码,实现协议、属性和方法的声明。程序代码如下:

```
#import <UIKit/UIKit.h>
@class Menu;
//协议
@protocol LeftViewDelegate <NSObject>
- (void)LeftView:(Menu *)view selectedIndex:(NSInteger)integer;
@end
@interface Menu : UIView<UITableViewDataSource,UITableViewDelegate>
//属性
@property (nonatomic, weak)   id<LeftViewDelegate>delegate;
@property (nonatomic, strong) NSArray *viewControuers;
- (id)initWithArray:(NSArray *)array;                //方法
@end
```

(5)打开 Menu.m 文件,编写代码,实现边栏菜单的绘制以及被选菜单项的响应。使用的方法如表 10-10 所示。

表 10-10　Menu.m 文件中方法总结

方　　法	功　　能
initWithArray:	菜单的初始化
tableView:numberOfRowsInSection:	获取行数
tableView:cellForRowAtIndexPath:	获取某一行的数据
tableView:didSelectRowAtIndexPath:	选择行的响应

这里需要讲解几个重要的方法（其他方法请读者参考源代码）。如果要实现菜单的绘制，需要使用 initWithArray:、tableView:numberOfRowsInSection: 和 tableView:cellForRowAtIndexPath:方法。其中，initWithArray:方法实现对菜单的初始化。程序代码如下：

```
- (id)initWithArray:(NSArray *)array
{
    self = [super initWithFrame:CGRectMake(0, 0, 170, 568)];
    if (self) {
        self.backgroundColor=[UIColor colorWithRed:239.0/255 green:250.0/
        255 blue:233.0/255 alpha:1];
        self.viewControllers=array;
        //标签对象的创建和设置
        UILabel *titleLabel_=[[UILabel alloc] initWithFrame:CGRectMake(0, 0,
        170, 44)];
        titleLabel_.backgroundColor=[UIColor grayColor];       //设置背景颜色
        titleLabel_.layer.shadowColor=[UIColor blackColor].CGColor;//设置阴影颜色
        titleLabel_.layer.shadowOffset=CGSizeMake(-2, -5);
        titleLabel_.layer.shadowOpacity=0.5;                   //设置阴影的透明度
        titleLabel_.text=@"菜单";
        titleLabel_.font=[UIFont boldSystemFontOfSize:20];
        titleLabel_.textAlignment=NSTextAlignmentCenter;       //设置对齐方式
        titleLabel_.textColor=[UIColor whiteColor];
        [self addSubview:titleLabel_];
        //表视图对象的创建和设置
        UITableView *tableView_=[[UITableView alloc] initWithFrame:CGRect
        Make(0, 50, 170, 435)];
        tableView_.backgroundColor=[UIColor clearColor];       //设置背景颜色
        tableView_.separatorStyle=UITableViewCellSeparatorStyleNone;
        tableView_.rowHeight=50;
        tableView_.delegate=self;                              //设置委托
        tableView_.dataSource=self;
        [self addSubview:tableView_];
        NSIndexPath *indexPath_=[NSIndexPath indexPathForRow:0 inSection:0];
        [tableView_ selectRowAtIndexPath:indexPath_ animated:YES scrollPosition:
        UITableViewScrollPositionNone];        //根据指定的索引路径选择行
    }
    return self;
}
```

tableView:numberOfRowsInSection: 和 tableView:cellForRowAtIndexPath:方法实现对表视图的内容填充。程序代码如下：

```
//获取行数
- (NSInteger)tableView:(UITableView *)tableView numberOfRowsInSection:
(NSInteger)section{
    return self.viewControllers.count;
}
//获取某一行的数据
```

```
-(UITableViewCell*)tableView:(UITableView*)tableViewcellForRowAtIndexPath:
(NSIndexPath *)indexPath{
    static NSString *cellString_=@"tableCell";
    UITableViewCell*cell=[tableViewdequeueReusableCellWithIdentifier:cellString_];
    //判断 cell 是否为空
    if (cell==nil) {
        cell=[[UITableViewCell alloc] initWithStyle:UITableViewCellStyleSubtitle
        reuseIdentifier:cellString_];
        cell.accessoryType=UITableViewCellAccessoryDisclosureIndicator;
         //设置附件类型
    }
    UIViewController *viewController_=self.viewControllers[indexPath.row];
    cell.imageView.image=viewController_.tabBarItem.image;       //设置图像
    cell.textLabel.text=viewController_.tabBarItem.title;
    return cell;
}
```

如果想要实现选择菜单项的响应，需要使用 tableView:didSelectRowAtIndexPath:方法实现。程序代码如下：

```
- (void)tableView:(UITableView *)tableView didSelectRow AtIndexPath: (NS In
dexPath *)indexPath{
    if ([self.delegate respondsToSelector:@selector (LeftView:selectedIndex:)]) {
        [self.delegate LeftView:self selectedIndex:indexPath.row];
        //调用 LeftView:selectedIndex:方法
    }
}
```

（6）创建一个基于 UIViewController 类的 FirstViewController 类。

（7）打开 FirstViewController.h 文件，编写代码，实现对象的声明，以及要遵守协议的声明。程序代码如下：

```
#import <UIKit/UIKit.h>
@interface FirstViewController : UIViewController <UITableViewDelegate,
UITableViewDataSource>{
    NSArray *array;
}
@end
```

（8）打开 FirstViewController.m 文件，编写代码，实现视图控制器的设置。使用的方法如表 10-11 所示。

表 10-11 FirstViewController.m文件中方法总结

方　　法	功　　能
initWithNibName:bundle:	初始化视图控制器
viewDidLoad	视图加载后调用，实现初始化
tableView:numberOfRowsInSection:	获取行数
tableView:cellForRowAtIndexPath:	获取某一行的数据

initWithNibName:bundle:方法实现对视图控制器的初始化。程序代码如下：

```
- (id)initWithNibName:(NSString *)nibNameOrNil bundle:(NSBundle *)nibBundleOrNil
{
    self = [super initWithNibName:nibNameOrNil bundle:nibBundleOrNil];
    if (self) {
        self.title=@"菜单1";
```

```
        self.tabBarItem.image=[UIImage imageNamed:@"1.png"];    //设置图像
    }
    return self;
}
```

（9）创建一个基于 UIViewController 类的 SecondViewController 类。

（10）打开 SecondViewController.m 文件，编写代码，实现初始化视图控制器以及对视图控制器的设置。程序代码如下：

```
//初始化
- (id)initWithNibName:(NSString *)nibNameOrNil bundle:(NSBundle *)nibBundleOrNil
{
    self = [super initWithNibName:nibNameOrNil bundle:nibBundleOrNil];
    if (self) {
        self.title=@"菜单2";
        self.tabBarItem.image=[UIImage imageNamed:@"1.png"];    //设置图像
    }
    return self;
}
//视图加载后调用
- (void)viewDidLoad
{
    [super viewDidLoad];
    // Do any additional setup after loading the view.
    [super viewDidLoad];
    self.view.backgroundColor=[UIColor whiteColor];             //设置背景颜色
    UIImageView *imageView_=[[UIImageView alloc] initWithFrame: CGRect
    Make(0, 0, 320, [UIScreen mainScreen].bounds.size.height-45)];
    imageView_.contentMode=UIViewContentModeScaleAspectFit;    //设置图像显示模式
    imageView_.image=[UIImage imageNamed:@"2.jpg"];
    [self.view addSubview:imageView_];
    imageView_=nil;
}
```

（11）创建一个基于 UITabBarController 类的 TabBarViewController 类。

（12）打开 TabBarViewController.h 文件，编写代码，实现头文件和属性的声明。程序代码如下：

```
#import <UIKit/UIKit.h>
//头文件
#import "FirstViewController.h"
#import "SecondViewController.h"
#import "Menu.h"
@interface TabBarViewController : UITabBarController<LeftViewDelegate>
//属性
@property (nonatomic, strong) UIView *transitionView;
@end
```

（13）打开 TabBarViewController.m 文件，编写代码，实现标签栏控制器的设置。使用的方法如表 10-12 所示。

表 10-12　TabBarViewController.m文件中方法总结

方　　法	功　　能
viewDidLoad	视图加载后调用，实现初始化
swipeGesterClick:	实现滑动的手势
LeftView:selectedIndex:	实现菜单中的选择索引

其中，viewDidLoad 方法对标签栏控制器的设计界面进行初始化设置。程序代码如下：

```objc
- (void)viewDidLoad
{
    [super viewDidLoad];
    for (UIView *view_ in self.view.subviews) {
        if ([view_ isKindOfClass:[UITabBar class]]) {
            [view_ removeFromSuperview];            //移除指定视图对象
        }
        else{
            self.transitionView=view_;
        }
    }
    self.transitionView.frame=CGRectMake(0, 0, 320, 568);    //设置框架
    //左划
    UISwipeGestureRecognizer *swipeLeftGesture_=[[UISwipeGestureRecognizer alloc]initWithTarget:
    self action:@selector(swipeGesterClick:)];
    swipeLeftGesture_.direction=UISwipeGestureRecognizerDirectionLeft;//设置滑动方向
    [self.transitionView addGestureRecognizer:swipeLeftGesture_];
    swipeLeftGesture_=nil;
    //右划
    UISwipeGestureRecognizer *swipeRightGesture_=[[UISwipeGestureRecognizer alloc]initWithTarget:
    self action:@selector(swipeGesterClick:)];
    [self.transitionView addGestureRecognizer:swipeRightGesture_];
                                                    //添加手势识别器对象
    swipeRightGesture_=nil;
    FirstViewController *oneViewController_=[[FirstViewController alloc] init];
    UINavigationController *oneNav_=[[UINavigationController alloc] init
    WithRootViewController:oneViewController_];//实例化导航控制器对象
    oneViewController_=nil;
    SecondViewController *twoViewController_=[[SecondViewController alloc]
    init];                                          //实例化视图控制器对象
    UINavigationController *twoNav_=[[UINavigationController alloc] init
    WithRootViewController:twoViewController_];
    twoViewController_=nil;
    self.viewControllers=@[oneNav_, twoNav_];
    oneNav_=nil;
    twoNav_=nil;
    //菜单对象的创建与设置
    Menu *menu=[[Menu alloc] initWithArray:self.viewControllers];
    menu.delegate=self;                             //设置委托
    [self.view insertSubview:menu atIndex:0];
    menu=nil;
}
```

swipeGesterClick:方法实现滑动的手势。向右滑动时，出现菜单；向左滑动时隐藏菜单。程序代码如下：

```objc
- (void)swipeGesterClick:(UISwipeGestureRecognizer *)swipeGesture{
    float x=self.transitionView.frame.origin.x;
    if (swipeGesture.direction==UISwipeGestureRecognizerDirectionLeft && x>0) {
        x=0;
        [UIView beginAnimations:nil context:nil];
        self.transitionView.frame=CGRectMake(x, 0, 320, 568);    //设置框架
```

```
        [UIView commitAnimations];
    }
    else if (swipeGesture.direction==UISwipeGestureRecognizer Direction
Right && x==0) {
        x=160;
        [UIView beginAnimations:nil context:nil];
        self.transitionView.frame=CGRectMake(x, 0, 320, 568);    //设置框架
        [UIView commitAnimations];
    }
}
```

【代码解析】

本实例关键功能是边栏菜单的显示隐藏和菜单项的响应。下面依次讲解这两个知识点。

1. 边栏菜单的显示隐藏

要实现边栏菜单的显示和隐藏，需要使用 UIView 的 frame 属性。程序代码如下：

```
if (swipeGesture.direction==UISwipeGestureRecognizerDirectionLeft && x>0)
{
    x=0;
    [UIView beginAnimations:nil context:nil];
    self.transitionView.frame=CGRectMake(x, 0, 320, 568);
    [UIView commitAnimations];
}
else if (swipeGesture.direction==UISwipeGestureRecognizerDirectionRight && x==0) {
    x=160;
    [UIView beginAnimations:nil context:nil];
    self.transitionView.frame=CGRectMake(x, 0, 320, 568);
    [UIView commitAnimations];
}
```

2. 菜单项的响应

要实现选择菜单项后，出现对应的视图控制器，需要使用 UITabBarController 的 selectedIndex 属性。它的功能是显示 viewControllers 中对应索引的视图控制器。其语法形式如下：

```
@property(nonatomic) NSUInteger selectedIndex;
```

在代码中，使用了 selectedIndex 属性，显示了通过选择菜单项后对应出现的视图控制器。代码如下：

```
self.selectedIndex=integer;
```

实例 71　九宫格菜单

【实例描述】

本实例实现的功能是在界面显示一个九宫格菜单。当用户选择其中的某一菜单项时，会弹出相应的警告视图。运行效果如图 10.12 所示。

第 10 章　菜单

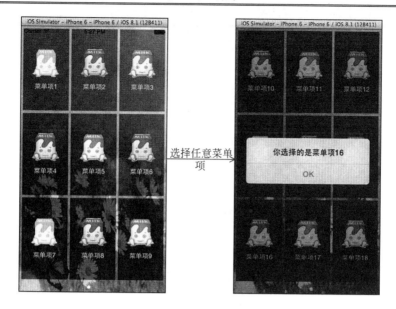

图 10.12　实例 71 运行效果

【实现过程】

（1）创建一个项目，命名为"九宫格菜单"。

（2）添加图片 1.png、2.png、3.png、4.png 和 5.png 到创建项目的 Supporting Files 文件夹中。

（3）创建一个基于 NSObject 类的 Data 类。

（4）打开 Data.h 文件，编写代码，实现属性和方法的声明。程序代码如下：

```
#import <Foundation/Foundation.h>
@interface Data : NSObject
//属性
@property (nonatomic, retain) NSString *imageName;
@property (nonatomic, retain) NSString *labelName;
- (id)initWith:(NSString *)image label:(NSString *)label;    //方法
@end
```

（5）打开 Data.m 文件，编写代码，实现声明方法的定义，即对数据的初始化。程序代码如下：

```
- (id)initWith:(NSString *)image label:(NSString *)label
{
    self = [super init];
    if (self) {
        _imageName =image;                                //设置图像
        _labelName =label;                                //设置标签
    }
    return self;
}
@end
```

（6）创建一个基于 UIControl 类的 MenuItem 类。

（7）打开 MenuItem.h 文件，编写代码，实现头文件、宏定义、属性和方法的声明。程序代码如下：

```
#import <UIKit/UIKit.h>
#import "Data.h"                                                    //头文件
//宏定义
#define kMarginScale 0.05
#define kLabelHeightScale 0.2
#define kLabelGap 2
@interface MenuItem : UIControl<UIGestureRecognizerDelegate>
//属性
@property (nonatomic, retain) Data *data;
@property (nonatomic, retain) UIView *mainView;
@property (nonatomic, retain) UIImageView *image;
@property (nonatomic, retain) UILabel *label;
//方法
- (id)initWithFrame:(CGRect)frame name:(Data *)gridData;
- (void)setupFrame;
-(void)getBlack;
@end
```

（8）打开 MenuItem.m 文件，编写代码，实现对菜单项的绘制以及菜单项的选择。使用的方法如表 10-13 所示。

表 10-13　MenuItem.m 文件中方法总结

方　　法	功　　能
initWithFrame:name:	菜单项的初始化
didTap:	实现轻拍
getBlack	设置背景
setupUI	设置菜单项
setupFrame	设置菜单项的框架

这里需要讲解几个重要的方法（其他方法请读者参考源代码）。要绘制菜单项需要使用 initWithFrame:name:、setupUI 和 setupFrame 方法。其中，initWithFrame:name:方法实现对菜单项的初始化。程序代码如下：

```
- (id)initWithFrame:(CGRect)frame name:(Data *)gridData
{
    self = [super initWithFrame:frame];
    if (self)
    {
        _data = gridData;
        UITapGestureRecognizer *tapGesture = [[UITapGestureRecognizer alloc]
        initWithTarget:self action:@selector(didTap:)];        //实例化轻拍手势
        [tapGesture setDelegate:self];
        [self addGestureRecognizer:tapGesture];
        self.backgroundColor = [UIColor colorWithPatternImage:[UIImage
        imageNamed:@"5.png"]];
        [self setupUI];
    }
    return self;
}
```

setupUI 方法实现对菜单项的设置。程序代码如下：

```
- (void)setupUI
{
    self.clipsToBounds = YES;
    //空白视图对象的创建与设置
```

```
    UIView *v = [[UIView alloc]init];
    self.mainView = v;
    [self addSubview:self.mainView];
    //图像视图对象的创建和设置
    UIImageView *image = [[UIImageView alloc]initWithImage:[UIImage
    imageNamed:_data.imageName]];
    self.image = image;
    [v addSubview:self.image];
    //标签对象的创建和设置
    UILabel *label = [[UILabel alloc]init];
    label.text = _data.labelName;
    label.textAlignment = NSTextAlignmentCenter;
    label.textColor = [UIColor whiteColor];              //设置文本颜色
    label.backgroundColor = [UIColor clearColor];        //设置背景颜色
    label.adjustsFontSizeToFitWidth = YES;
    if (UI_USER_INTERFACE_IDIOM() == UIUserInterfaceIdiomPhone)
    {
        label.font = [UIFont systemFontOfSize:14];       //设置字体
    }
    self.label = label;
    [v addSubview:self.label];
}
```

setupFrame 方法实现菜单项框架的设置。程序代码如下:

```
- (void)setupFrame
{
    float width = CGRectGetWidth(self.frame);            //获取宽度
    float height = CGRectGetHeight(self.frame);          //获取高度
    float newWidth = width > height ? height : width;
    float margin = newWidth * kMarginScale;
    float labelGap = kLabelGap * margin;
    float labelHeight = newWidth * kLabelHeightScale;    //为 labelHeight 赋值
    float iconWidth = newWidth - 2 * margin;
    float imagWidth = newWidth * 0.6;
    CGRect rect = CGRectMake(0, 0, newWidth, newWidth);
    _mainView.frame = rect;                              //设置框架
    _mainView.center = CGPointMake(width * 0.5, height * 0.5);   //设置中心位置
    _image.frame = CGRectMake(0, 0, imagWidth, imagWidth);
    _image.center = CGPointMake(newWidth * 0.5, margin + imagWidth * 0.5);
    _label.frame = CGRectMake(0, 0, iconWidth, labelHeight);
    _label.center = CGPointMake(newWidth*0.5, margin + imagWidth+labelGap+
    labelHeight * 0.5);
}
```

(9) 创建一个基于 UIView 类的 Menu 类。

(10) 打开 Menu.h 文件,编写代码,实现头文件、宏定义、协议、对象、实例变量、属性以及方法的声明。程序代码如下:

```
#import <UIKit/UIKit.h>
//头文件
#import "MenuItem.h"
@class Data;
//宏定义
#define  kPageControlHeight   ((UI_USER_INTERFACE_IDIOM() == UIUserInter
faceIdiomPhone)?20:40)
……
#define kGapScale 0.05
```

```objc
//协议
@protocol MenuDelegate <NSObject>
- (void)touchAction:(NSString *)sender;
@end
@interface Menu : UIView<UIScrollViewDelegate>{
    //对象
    UIScrollView *_scrollView;
    UIPageControl *_pageControl;
    NSMutableArray* _dataSource;
    NSMutableArray * _elementSource;
    //实例变量
    NSInteger _row;
    NSInteger _column;
    BOOL _isPage;
}
//属性
@property (nonatomic, assign, readonly) NSInteger row;
@property (nonatomic, assign, readonly) NSInteger column;
@property (nonatomic, assign) id<MenuDelegate> delegate;
@property (nonatomic, retain) UIImageView *backgroundImageView;
//方法
- (id)initWithFrame:(CGRect)frame withSource:(NSMutableArray *)array withRowNum:(NSInteger)row withColumnNum:(NSInteger)column isPage Control:(BOOL)isPage;
- (void)setupFrame;
@end
```

（11）打开 Menu.m 文件，编写代码，实现九宫格菜单的效果。使用的方法如表 10-14 所示。

表 10-14　Menu.m 文件中方法总结

方　　法	功　　能
initWithFrame:withSource:withRowNum:withColumnNum:isPageControl:	九宫格菜单的初始化
loadUI	加载界面
setupFrame	菜单框架的设置
setScrollView:withData:isPage:page:	设置滚动视图
setupUI:	创建菜单项
btnDown:	按下按钮的响应
scrollViewDidEndDecelerating:	用户拖动，之后继续滚动
changePage:	改变当前的页

要实现九宫格菜单的效果，需要完成两个部分：绘制菜单；实现菜单项的选择。其中，绘制菜单需要使用 initWithFrame:withSource:withRowNum:withColumnNum:isPageControl:、loadUI、scrollViewDidEnd Decelerating: 以及 changePage: 方法实现。其中，initWithFrame:withSource:withRowNum:withColumnNum:isPageControl: 方法实现九宫格菜单的初始化。程序代码如下：

```objc
- (id)initWithFrame:(CGRect)frame withSource:(NSMutableArray *)array withRowNum:(NSInteger)row withColumnNum:(NSInteger)column isPageControl:(BOOL)isPage
{
    self = [super initWithFrame:frame];
    if (self)
```

```objc
    {
        _elementSource = [[NSMutableArray alloc]initWithCapacity:0];
        //实例化对象_elementSource
        _dataSource = [[NSMutableArray alloc]initWithArray:array];
        //实例化对象_dataSource
        _row = row;
        _column = column;
        _isPage = isPage;
        [self loadUI];
    }
    return self;
}
```

loadUI 方法对界面进行加载，实际上就是进行初始化。程序代码如下：

```objc
- (void)loadUI
{
    //创建并设置图像视图对象
    UIImageView *v = [[UIImageView alloc]initWithImage:[UIImage imageNamed:@"4.jpg"]];
    v.frame = self.bounds;                                      //设置框架
    self.backgroundImageView = v;
    [self addSubview:self.backgroundImageView];                 //添加视图对象
    if (_dataSource.count <= 0 || self.column <= 0|| self.row <= 0)
    {
        UIAlertView *alert = [[UIAlertView alloc]initWithTitle:@"sorry"
        message:@"the data for the grid is wrong! " delegate:nil cancelButton
        Title:@"OK" otherButtonTitles:nil, nil];                //创建警告视图
        [alert show];
        return;
    }
    //创建并设置滚动视图对象
    _scrollView = [[UIScrollView alloc]init];
    _scrollView.scrollEnabled = YES;
    _scrollView.delegate = self;                                //设置委托
    _scrollView.showsVerticalScrollIndicator = NO;
    _scrollView.showsHorizontalScrollIndicator = NO;
    _scrollView.frame = self.frame;                             //设置框架
    _scrollView.backgroundColor = [UIColor clearColor];
    [self addSubview:_scrollView];
    [self setupUI:_dataSource];
    [self setupFrame];
}
```

setupFrame 方法实现对菜单框架的设置。程序代码如下：

```objc
- (void)setupFrame
{
    ......
    if (totalRow > self.row)
    {
        NSInteger num = (totalRow % self.row) ? (totalRow / self.row + 1) :
        (totalRow / self.row);
        if (_isPage)
        {
            //判断 num 是否大于
            if (num > 1)
            {
                _scrollView.pagingEnabled = YES;
                CGSize size = CGSizeMake(CGRectGetWidth(self.frame) * num,
```

```objc
                    CGRectGetHeight(self.frame));
            [_scrollView setContentSize:size];      //设置内容视图的尺寸
            if (!_pageControl)
            {
                _pageControl = [[UIPageControl alloc]init];     //实例化页面空间
            }
            //创建并设置页面控件对象
            _pageControl.frame =CGRectMake(0, _scrollView.frame.size.
            height-kPageControlHeight, kViewWidth, kPageControlHeight);
                        //设置框架
            _pageControl.numberOfPages = num;
            _pageControl.currentPage = 0;
            [_pageControl addTarget:self action:@selector(changePage:)
            forControlEvents:UIControlEventValueChanged];       //添加动作
            [self addSubview:_pageControl];
        }
        for (int i = 0; i < num; i++)
        {
            //实例化对象
            NSRange rang = NSMakeRange(i * self.row * self.column, ((i !=
            num-1 )? (self.row * self.column): (dataCount- (num-1)* self.
            row*self.column)));
            NSMutableIndexSet *_index = [NSMutableIndexSet indexSetWith
            IndexesInRange:rang];
            NSArray *arr = [_elementSource objectsAtIndexes:_index];
            float height = CGRectGetHeight(self.frame);     //获取高度
            float width = CGRectGetWidth(self.frame);       //获取宽度
            newRect.origin.x = i * width;
            newRect.origin.y = CGRectGetMinY(self.frame);   //获取y值
            newRect.size.width = width;
            newRect.size.height = height - kPageControlHeight;//设置高度
            [self setScrollView:newRect withData:(NSMutableArray *)arr
            isPage:YES page:i];
        }
     ……
    }
    else
    {
        [self setScrollView:self.frame withData:_elementSource isPage:NO
page:0];
    }
}
```

setScrollView:isPage:page:方法实现对滚动视图的设置。程序代码如下：

```objc
-(void)setScrollView:(CGRect)newRect     withData:(NSMutableArray  *)arr
isPage:(BOOL)isPage page:(NSInteger)page
{
    NSInteger elementCount = arr.count;
    float width = CGRectGetWidth(newRect);                      //获取宽度
    float height = CGRectGetHeight(newRect);                    //获取高度
    //为变量赋值
    float column_d = width / self.column;
    float row_d = height / self.row;
    float margin = (column_d > row_d ? row_d : column_d) * kGapScale;
    float icon_width = (width - (self.column + 1) * margin)/self.column;
    float icon_height = (height- (self.row + 1) * margin)/self.row;
```

```
            float column_distance = icon_width + margin;
            float row_distance = icon_height + margin;
            //遍历
            for (NSInteger i = 0;i < elementCount;)
            {
                CGFloat y = margin + icon_height * 0.5 + (i / self.column) * row_distance;
                //遍历
                for (NSInteger col = 0; (col < self.column) && (i < elementCount); col++)
                {
                    MenuItem *ele = [arr objectAtIndex:i++];
                    CGFloat x = margin + icon_width * 0.5 + col * column_distance +
                    (isPage?page * width:0);
                    CGRect itemViewFrame = CGRectZero;
                    itemViewFrame.size = CGSizeMake(icon_width, icon_height);   //设置尺寸
                    ele.frame = itemViewFrame;
                    ele.center = CGPointMake(x, y);                             //设置中心位置
                    [ele setupFrame];
                }
            }
        }
```

setupUI:方法实现对菜单项的创建。程序代码如下：

```
- (void)setupUI:(NSMutableArray *)array
{
    for (int i = 0; i < array.count; i++)
    {
    //创建并设置菜单项对象
        MenuItem *ele = [[MenuItem alloc]initWithFrame:CGRectZero name:
        [array objectAtIndex:i]];
        [ele addTarget:self action:@selector(btnDown:) forControlEvents:
        UIControlEventTouchUpInside];                                //添加动作
        [_elementSource addObject:ele];
        [_scrollView addSubview:ele];
    }
}
```

changePage:方法实现改变页数的功能。程序代码如下：

```
- (void)changePage:(id)sender
{
    CGRect frame = _scrollView.frame;
    frame.origin.x = frame.size.width * _pageControl.currentPage;
    frame.origin.y = 0;
    [_scrollView scrollRectToVisible:frame animated:YES];   //滚动查看内容
}
```

菜单项的选择，需要使用btnDown:方法实现。程序代码如下：

```
- (void)btnDown:(MenuItem *)ele
{
    //判断是否实现了touchAction:方法
    if ([self.delegate respondsToSelector:@selector(touchAction:)]) {
        [self.delegate touchAction:ele.data.labelName];
    }
}
```

（12）打开ViewController.h文件，编写代码，实现头文件和属性的声明。程序代码如下：

```
#import <UIKit/UIKit.h>
```

```
#import "Menu.h"
@interface ViewController : UIViewController<MenuDelegate>
@property (nonatomic, retain)Menu *scroll;
@end
```

(13)打开 ViewController.m 文件，编写代码，实现九宫格对象的创建和设置，以及在选择某一项后弹出警告视图。程序代码如下：

```
- (void)viewDidLoad
{
    [super viewDidLoad];
    NSMutableArray *array = [[NSMutableArray alloc]initWithCapacity:12];
                                                    //实例化数组对象
    //图像的设置
    for (int i = 0; i < 18; i++)
    {
        NSString *image = nil;
        if (i % 2) {
            image = @"1.png";
        }
        else if (i % 3) {
            image = @"2.png";
        }
        else
        {
            image = @"3.png";
        }
        Data *data = [[Data alloc]initWith:image label:[NSString stringWithFormat:@"菜单项%i",i+1]];
        [array addObject:data];                     //添加对象
    }
    CGRect newRect = self.view.bounds;
    scroll = [[Menu alloc]initWithFrame:newRect withSource:array withRowNum:3 withColumnNum:3 isPageControl:YES];  //实例化对象
    scroll.delegate = self;                         //设置委托
    [self.view addSubview:scroll];
}
//选择某一项后弹出警告视图
- (void)touchAction:(NSString *)sender
{
    NSString *st=[NSString stringWithFormat:@"你选择的是%@",sender];
    UIAlertView *alert=[[UIAlertView alloc]initWithTitle:st message:nil delegate:nil cancelButtonTitle:@"OK" otherButtonTitles: nil];
                                                    //创建警告视图
    [alert show];
}
```

【代码解析】

由于本实例中的代码和方法非常的多，为了方便读者的阅读，笔者绘制了一些执行流程图如图 10.13 和图 10.14 所示。其中，单击运行按钮后一直到在模拟器上显示九宫格菜单，使用了 viewDidLoad、initWith:label:、initWithFrame:withSource:withRowNum:withColumnNum:isPageControl:、loadUI（Menu.m）、setupUI:（Menu.m）、initWithFrame:name:、setupUI（MenuItem.m）、setScrollView:withData:isPage:page:和 setupFrame（MenuItem.m）方法共同实现。它们的执行流程如图 10.13 所示。

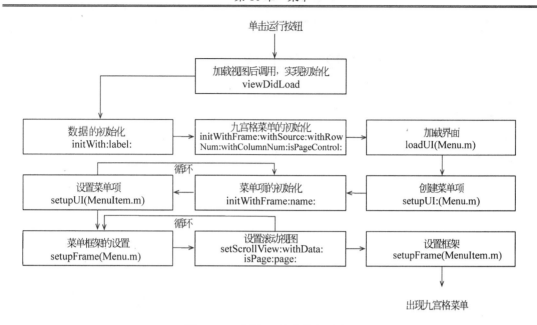

图 10.13　实例 71 程序执行流程 1

在本实例中菜单是可以滑动的，从而进行菜单页面的切换。实现切换的方式有两种，一种是滑动页面，使用的方法是 scrollViewDidEndDecelerating:方法；另一种是轻拍页面控件，使用的方法是 changePage:方法。它们的执行流程如图 10.14 所示。菜单除了可以进行切换外，菜单中的菜单项也是可以进行选择的，使用的方法有：didTap:、btnDown:、touchAction:和 getBlack。它们的执行流程如图 10.14 所示。

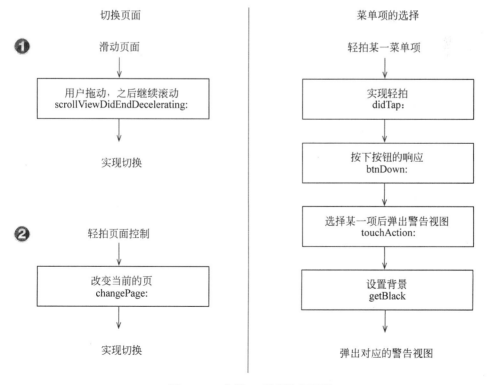

图 10.14　实例 71 程序执行流程 2

实例 72　侧面弹出式菜单

【实例描述】
　　本实例的功能是实现一个侧面弹出式的菜单。当用户单击"显示/隐藏"按钮，就会以弹出的方法显示一个侧面菜单，选择菜单中的任意一项时，就会出现一个警告视图。当用户再一次单击"显示/隐藏"按钮，菜单就会以动画形式隐藏。运行效果如图所示。

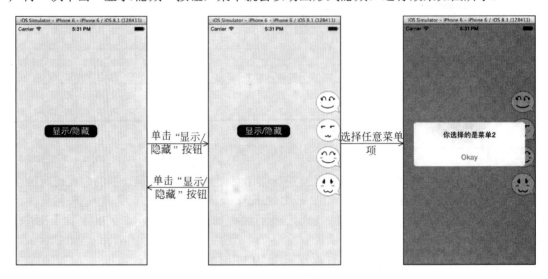

图 10.15　实例 72 运行效果

【实现过程】
　　（1）创建一个项目，命名为"侧面弹出式菜单"。
　　（2）添加图片 1.png、2.png、3.png 和 4.png 到创建项目的 Supporting Files 文件夹中。
　　（3）创建一个 UIView 类的 Menu 类。
　　（4）打开 Menu.h 文件，编写代码，实现宏定义、新数据类型以及数据类型的定义、属性和方法的声明。程序代码如下：

```
#import <UIKit/UIKit.h>
#define kAnimationDelay 0.08                                    //宏定义
typedef CGFloat (^EasingFunction)(CGFloat, CGFloat, CGFloat, CGFloat);
//定义新的数据类型
//数据类型的定义
typedef enum {
    HMSideMenuPositionLeft,
    HMSideMenuPositionRight,
    HMSideMenuPositionTop,
    HMSideMenuPositionBottom
} HMSideMenuPosition;
@interface Menu : UIView
//属性
@property (nonatomic, assign, readonly) BOOL isOpen;
@property (nonatomic, assign) CGFloat itemSpacing;
……
```

```
@property (nonatomic, assign) CGFloat menuHeight;
//方法
- (id)initWithItems:(NSArray *)items;
- (void)open;
- (void)close;
@end
```

(5)打开 Menu.m 文件,编写代码,实现侧面弹出式菜单的绘制。使用的方法如表 10-15 所示。

表 10-15　Menu.m文件中方法总结

方　　法	功　　能
initWithItems:	菜单的初始化
setItems:	设置菜单项
open	打开菜单
close	关闭菜单
showItem:	显示菜单项
hideItem:	隐藏菜单项
pointInside:withEvent:	判断触摸的点是否在视图中
layoutSubviews	布局子视图
animateLayer:withKeyPath:to:	动画的实现
变量	功能
static EasingFunction easeOutElastic = ^CGFloat(CGFloat t, CGFloat b, CGFloat c, CGFloat d)	获取弹出菜单的弹性值,即振幅

这里需要讲解几个重要的方法(其他方法请读者参考源代码)。其中,initWithItems:方法实现菜单的初始化。程序代码如下:

```
- (id)initWithItems:(NSArray *)items {
    self = [super initWithFrame:CGRectZero];
    if (self) {
        self.items = items;                          //设置条目
        _animationDuration = 1.3f;
        _menuPosition = HMSideMenuPositionRight;
    }
    return self;
}
```

setItems:方法实现对菜单项的设置,即为菜单添加菜单项。程序代码如下:

```
- (void)setItems:(NSArray *)items {
    for (UIView *item in items) {
        item.layer.opacity = 0;                      //设置透明度
        [item removeFromSuperview];                  //移除
    }
    _items = items;
    for (UIView *item in items) {
        [self addSubview:item];
    }
}
```

showItem:方法实现菜单项的显示。程序代码如下:

```
- (void)showItem:(UIView *)item {
    [NSObject cancelPreviousPerformRequestsWithTarget:item.layer];
```

```objc
    item.layer.opacity = 1.0f;                              //设置透明度
    CGPoint position = item.layer.position;
    position.x += self.menuPosition == HMSideMenuPositionRight ? -self.menuWidth : self.menuWidth;
    [self animateLayer:item.layer withKeyPath:@"position.x" to:position.x];
    item.layer.position = position;                         //设置位置
}
```

hideItem:方法实现菜单项的隐藏。程序代码如下：

```objc
- (void)hideItem:(UIView *)item {
    CGPoint position = item.layer.position;
    position.x += self.menuPosition == HMSideMenuPositionRight ? self.menuWidth : -self.menuWidth;
    [self animateLayer:item.layer withKeyPath:@"position.x" to:position.x];
    item.layer.position = position;                         //设置位置
    [item.layer performSelector:@selector(setOpacity:) withObject:[NSNumber numberWithFloat:0.0f] afterDelay:0.5];  //经过0.5秒后执行setOpacity
}
```

layoutSubviews 方法对子视图进行布局，即对菜单以及菜单中的菜单项进行布局。程序代码如下：

```objc
- (void)layoutSubviews {
    [super layoutSubviews];
    //为变量赋值
    self.menuWidth = 0;
    self.menuHeight = 0;
    CGFloat __block biggestHeight = 0;
    //枚举对象指向一个给定的块
    [self.items enumerateObjectsUsingBlock:^(UIView *item, NSUInteger idx, BOOL *stop) {
        self.menuWidth = MAX(item.frame.size.width, self.menuWidth);
        biggestHeight = MAX(item.frame.size.height, biggestHeight);
    }];
    //为变量赋值
    self.menuHeight = (biggestHeight*self.items.count)+(self.itemSpacing * (self.items.count - 1));
    CGFloat x = 0;
    CGFloat y = 0;
    CGFloat itemInitialX = 0;
    x = self.menuPosition == HMSideMenuPositionRight ? self.superview.frame.size.width : 0 - self.menuWidth;
    y = (self.superview.frame.size.height / 2) - (self.menuHeight / 2);
    itemInitialX = self.menuWidth / 2;
    self.frame = CGRectMake(x, y, self.menuWidth, self.menuHeight);;
    //设置框架
    //枚举
    [self.items enumerateObjectsUsingBlock:^(UIView *item, NSUInteger idx, BOOL *stop) {
        [item setCenter:CGPointMake(itemInitialX, (idx * biggestHeight)+(idx * self.itemSpacing) + (biggestHeight / 2))];   //设置中心位置
    }];
}
```

animateLayer:withKeyPath:to:方法实现了菜单的弹出收回效果。程序代码如下：

```objc
- (void)animateLayer:(CALayer *)layer withKeyPath:(NSString *)keyPath to:(CGFloat)endValue {
    CGFloat startValue = [[layer valueForKeyPath:keyPath] floatValue];
    CAKeyframeAnimation *animation = [CAKeyframeAnimation animation
```

```
            WithKeyPath:keyPath];
//动画速度的设置
animation.timingFunction = [CAMediaTimingFunction functionWithName:
 kCAMediaTimingFunctionLinear];
animation.fillMode = kCAFillModeForwards;      //当动画结束后,layer会一直
保持着动画最后的状态
animation.removedOnCompletion = NO;
animation.duration = self.animationDuration;       //动画的时间
CGFloat steps = 100;
NSMutableArray *values = [NSMutableArray arrayWithCapacity:steps];
CGFloat delta = endValue - startValue;
EasingFunction function = easeOutElastic;
for (CGFloat t = 0; t < steps; t++) {
    [values addObject:@(function(animation.duration * (t / steps), start
     Value, delta, animation.duration))];        //添加对象
}
animation.values = values;             //设置动画采用的值的数组
[layer addAnimation:animation forKey:nil];       //添加动画
}
```

static EasingFunction easeOutElastic = ^CGFloat(CGFloat t, CGFloat b, CGFloat c, CGFloat d)变量获取弹出菜单的弹性值即振幅。程序代码如下:

```
static EasingFunction easeOutElastic = ^CGFloat(CGFloat t, CGFloat b,
CGFloat c, CGFloat d) {
    CGFloat amplitude = 5;
    CGFloat period = 0.6;
    CGFloat s = 0;
    //判断t是否等于0
    if (t == 0) {
        return b;             //返回b的值
    }
    else if ((t /= d) == 1) {    //判断t/=d是否等于1
        return b + c;           //返回b+c的值
    }
    if (!period) {
        period = d * .3;
    }
    //判断amplitude是否小于abc(c)
    if (amplitude < abs(c)) {
        amplitude = c;
        s = period / 4;
    }
    else {
        s = period / (2 * M_PI) * sin(c / amplitude);
    }
    return (amplitude * pow(2,-10*t)*sin((t*d-s)*(2 * M_PI)/period)+c+b);
};
```

(6) 创建一个基于UIView类的分类Action。

(7) 打开Action.h文件,编写代码,实现头文件以及方法的声明。

```
#import <UIKit/UIKit.h>
#import <objc/runtime.h>
@interface UIView (Action)
- (void)setMenuActionWithBlock:(void (^)(void))block;            //方法
@end
```

（8）打开 Action.m 文件，编写代码，在此文件中，代码分为动作的设置以及实现轻拍。其中 setMenuActionWithBlock:方法实现使用块对菜单的动作进行设置。程序代码如下：

```
- (void)setMenuActionWithBlock:(void (^)(void))block {
    UITapGestureRecognizer *gesture = objc_getAssociatedObject(self, &kActionHandlerTapGestureKey);
    //判断 gesture 是否为空
    if (!gesture) {
        gesture = [[UITapGestureRecognizer alloc] initWithTarget:self
        action:@selector(handleActionForTapGesture:)];   //创建对象 gesture
        [self addGestureRecognizer:gesture];
        objc_setAssociatedObject(self, &kActionHandlerTapGestureKey, gesture,
        OBJC_ASSOCIATION_RETAIN);                        //关联对象
    }
    objc_setAssociatedObject(self, &kActionHandlerTapBlockKey, block,
    OBJC_ASSOCIATION_COPY);
}
```

handleActionForTapGesture:方法实现轻拍的动作。程序代码如下：

```
- (void)handleActionForTapGesture:(UITapGestureRecognizer *)gesture {
    //判断 gesture 的状态是否为 UIGestureRecognizerStateRecognized
    if (gesture.state == UIGestureRecognizerStateRecognized) {
        void(^action)(void) = objc_getAssociatedObject(self, &kActionHandlerTapBlockKey);
        if (action) {
            action();
        }
    }
}
```

（9）打开 ViewController.h 文件，编写代码，实现头文件、插座变量、属性以及动作的声明。程序代码如下：

```
#import <UIKit/UIKit.h>
//头文件
#import "Menu.h"
#import "UIView+Action.h"
@interface ViewController : UIViewController{
    IBOutlet UIButton *button;                           //插座变量
}
@property (nonatomic, strong)Menu *menu;                 //属性
- (IBAction)toggleMenu:(id)sender;                       //动作
@end
```

（10）打开 Main.storyboard 文件，对 View Controller 视图控制器的设计界面进行设计。

效果如图 10.16 所示。

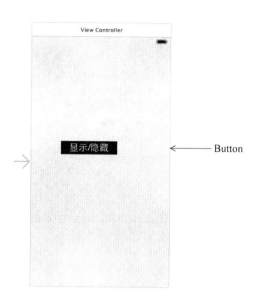

图 10.16 实例 72View Controller 视图控制器的设计界面

需要添加的视图、控件以及对它们的设置如表 10-16 所示。

表 10-16 实例 72 视图、控件设置

视图、控件	属 性 设 置	其 他
设置界面	Background：浅灰色	
Button	Title：显示/隐藏 Font：System 21.0 Text Color：白色 Background：黑色	与插座变量 button 关联 与动作 toggleMenu:关联

（11）打开 ViewController.m 文件，编写代码，实现单击按钮显示或隐藏菜单。程序代码如下：

```
- (void)viewDidLoad
{
    [super viewDidLoad];
    // Do any additional setup after loading the view, typically from a nib.
    button.layer.cornerRadius=10;                    //设置圆角半径
    //创建菜单项
    UIView *one = [[UIView alloc] initWithFrame:CGRectMake(0, 0, 60, 60)];
    [one setMenuActionWithBlock:^{
        UIAlertView *alertView = [[UIAlertView alloc] initWithTitle:nil
            message:@"你选择的是菜单 1"delegate:nil cancelButtonTitle:@"Okay"
            otherButtonTitles:nil, nil];                //创建警告视图
        [alertView show];
    }];
    UIImageView *oneIcon = [[UIImageView alloc] initWithFrame:CGRectMake(0,
    0, 60, 60)];
    [oneIcon setImage:[UIImage imageNamed:@"1.png"]];        //设置图像
```

```objc
    [one addSubview:oneIcon];
    ……
    UIView *four = [[UIView alloc] initWithFrame:CGRectMake(0, 0, 60, 60)];
    [four setMenuActionWithBlock:^{
        UIAlertView *alertView = [[UIAlertView alloc] initWithTitle:nil
            message:@"你选择的是菜单 4"delegate:nil cancelButtonTitle:@"Okay"
            otherButtonTitles:nil, nil];                   //创建警告视图
        [alertView show];
    }];
    UIImageView *fourIcon = [[UIImageView alloc] initWithFrame: CGRectMake
    (0, 0, 60, 60)];
    [fourIcon setImage:[UIImage imageNamed:@"4.png"]]; //设置图像
    [four addSubview:fourIcon];
    //创建并设置菜单
    self.menu = [[Menu alloc] initWithItems:@[one, two, three, four]];
    [self.menu setItemSpacing:5.0f];
    [self.view addSubview:self.menu];                  //添加视图对象
}
//实现菜单的显示和隐藏
- (IBAction)toggleMenu:(id)sender {
    if (self.menu.isOpen)
        [self.menu close];                             //关闭菜单
    else
        [self.menu open];                              //打开菜单
}
```

【代码解析】

本实例关键功能是菜单的弹出收回。下面就是这个知识点的详细讲解。

要实现弹出收回的动画效果，需要使用 CAKeyframeAnimation 关键帧动画。在这类中使用 value 属性指定动画的各个阶段，从而实现弹出弹回动画效果。其语法形式如下：

```objc
@property(copy) NSArray *values;
```

在代码中，使用了 value 属性将实现菜单的弹出收回，代码如下：

```objc
NSMutableArray *values = [NSMutableArray arrayWithCapacity:steps];
CGFloat delta = endValue - startValue;
EasingFunction function = easeOutElastic;
for (CGFloat t = 0; t < steps; t++) {
    [values addObject:@(function(animation.duration * (t / steps), start
Value, delta, animation.duration))];
}
animation.values = values;
```

实例 73　分享菜单

【实例描述】

在 iPhone 或者 iPad，分享是不可避免的。为了让用户可以快速、便捷地实现分享功能，苹果公司提供了专有的分享菜单。本实例的功能就是以自定义的形式，显示一个与众不同

的分享菜单。当用户长按界面就会出现此菜单,当用户选择菜单中的任意一项,就会弹出相应的警告实现。运行效果如图 10.17 所示。

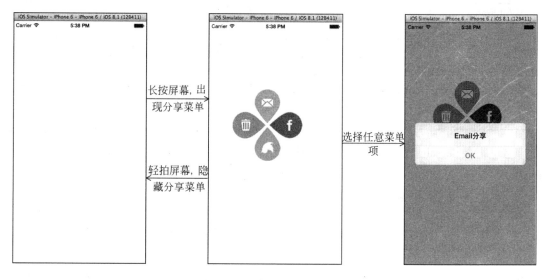

图 10.17　实例 73 运行效果

【实现过程】

当用户长按界面就会出现分享菜单,当用户选择菜单中的任意一项,就会弹出相应的警告实现。具体的实现步骤如下。

（1）创建一个项目,命名为"分享菜单"。

（2）添加图片 1.png、2.png、3.png 和 4.png 到创建项目的 Supporting Files 文件夹中。

（3）创建一个基于 UIButton 类的 MenuItem 类。

（4）打开 MenuItem.h 文件,编写代码,实现宏定义、数据类型的定义以及属性和方法的声明。程序代码如下：

```
#import <UIKit/UIKit.h>
#define DegreesToRadians(degrees) (degrees * M_PI / 180.f)    //宏定义
//数据类型的定义
typedef enum {
    FAFancyButtonFadeIn,
    FAFancyButtonFadeOut
} MenuItemtate;
@interface MenuItem : UIButton
//属性
@property (nonatomic) CGFloat degree;
@property (nonatomic) MenuItemtate state;
//方法
- (void)show;
- (void)hide;
@end
```

（5）打开 MenuItem.m 文件,编写代码,实现菜单项的动画效果。使用的方法如表 10-17 所示。

表 10-17 MenuItem.m 文件中方法总结

方法	功能
initWithFrame:	菜单项的初始化
show	显示菜单项
hide	隐藏菜单项
animationDidStop:finished:	过渡动画完成后调用
fadeInAnimation	淡入动画
fadeOutAnimation	淡出动画
rotateAnimationFromDegree: toDegree:delegate:	旋转动画

这里需要讲解几个重要的方法（其他方法请读者参考源代码）。Animation DidStop: finished:方法在过渡动画完成后调用，通过菜单的状态实现。程序代码如下：

```
- (void)animationDidStop:(CAAnimation *)anim finished:(BOOL)flag{
   CABasicAnimation *animation = (CABasicAnimation *)anim; //实例化对象
   if ([animation.keyPath isEqualToString:@"transform.scale"] && flag){
      switch (self.state) {
         case FAFancyButtonFadeIn:
            [self.layer addAnimation:[self rotateAnimationFromDegree:0
            toDegree:self.degree delegate:nil] forKey:@"FancyButton Rota
            tion"];                                    //添加动画
            self.transform = CGAffineTransformMakeRotation(Degrees ToRadians
            (self.degree));
            break;
         case FAFancyButtonFadeOut:
            self.hidden = YES;                         //设置隐藏
            break;
         default:
            break;
      }
   }else if ([animation.keyPath isEqualToString:@"transform.rotation.z"] && flag){
      [self.layer addAnimation:[self fadeOutAnimation] forKey:@"Fancy
      ButtonFadeOut"];                                 //添加动画
      self.transform=CGAffineTransformMakeRotation(DegreesToRadians
      (self.degree));                                  //旋转
   }
}
```

fadeInAnimation 方法实现淡入动画效果。程序代码代码如下：

```
- (CABasicAnimation *)fadeInAnimation{
   self.state = FAFancyButtonFadeIn;
   CABasicAnimation *scaleAnimation = [CABasicAnimation animationWithKey
   Path:@"transform.scale"];
   scaleAnimation.fromValue = [NSValue valueWithCATransform3D:
   CATransform3DMakeScale(.01, .01, .1)];              //设置开始值
   scaleAnimation.toValue = [NSValue valueWithCATransform3D: CATransform
   3DMakeScale(1, 1, 1)];
   scaleAnimation.duration = 0.2;
   scaleAnimation.delegate = self;                     //设置委托
   return scaleAnimation;
}
```

rotateAnimationFromDegree:方法实现旋转动画效果。程序代码如下：

```
- (CABasicAnimation *)rotateAnimationFromDegree:(CGFloat)from toDegree:
(CGFloat)to delegate:(id)delegate{
   CABasicAnimation* rotationAnimation;
   rotationAnimation = [CABasicAnimation animationWithKeyPath:@"transform.
   rotation.z"];
   rotationAnimation.fromValue = [NSNumber numberWithFloat: DegreesTo
   Radians(from)];
   rotationAnimation.toValue = [NSNumber numberWithFloat: DegreesTo
   Radians(to)];
   rotationAnimation.duration = 0.3f;                    //设置时间间隔
   rotationAnimation.delegate = delegate;
   rotationAnimation.removedOnCompletion = NO;
   rotationAnimation.fillMode = kCAFillModeForwards;
   return rotationAnimation;
}
```

（6）创建一个基于 UIView 类的 Menu 类。

（7）打开 Menu.h 文件，编写代码，实现头文件、协议、属性以及方法的声明。程序代码如下：

```
#import <UIKit/UIKit.h>
//头文件
#import "MenuItem.h"
@class Menu;
//协议
@protocol MenuDelegate <NSObject>
- (void)fancyMenu:(Menu *)menu didSelectedButton AtIndex: (NSUInteger)index;
@end
@interface Menu : UIView
//属性
@property (nonatomic, assign) id<MenuDelegate> delegate;
@property (nonatomic, strong) NSArray *buttonImages;
@property (nonatomic) BOOL onScreen;
//方法
- (void)show;
- (void)hide;
@end
```

（8）打开 Menu.m 文件，编写代码，实现分享菜单的绘制、菜单的显示和隐藏以及菜单项的选择。使用的方法如表 10-18 所示。

表 10-18　Menu.m文件中方法总结

方　　法	功　　能
addButtons	添加菜单项
handleLongPress:	长按
handleTap:	轻拍
addGestureRecognizerForView:	添加手势
willMoveToSuperview:	通知父视图改变到特定的父视图
buttonPressed:	菜单项的选择
showButton:	显示菜单项
hideButton:	隐藏菜单项
hide	隐藏菜单
show	显示菜单
setButtonImages:	设置菜单项的图像

菜单的绘制需要使用 addButtons 和 setButtonImages:方法实现。其中，addButtons 方法实现菜单项的添加。程序代码如下：

```
- (void)addButtons{
    self.frame = CGRectMake(100, 100, ((UIImage *)[self.buttonImages lastO
bject]).size.height * 2, ((UIImage *)[self.buttonImages lastObject]).
size.height * 2);                              //设置框架
    //判断 subviews 的个数是否大于 0
    if (self.subviews.count > 0)
        [self.subviews makeObjectsPerformSelector:@selector(removeFromSuperview)];
    NSInteger i = 0;
    CGFloat degree = 360.f/self.buttonImages.count;
    //添加菜单项
    for (UIImage *image in self.buttonImages){
        MenuItem *fancyButton = [[MenuItem alloc] initWithFrame: CGRect Make
(self.frame.size.width/2 - image.size.width/2, 0, image. size.
    width, image.size.height)];
        [fancyButton setBackgroundImage:image forState:UIControlState Normal];
                                              //设置背景图像
        fancyButton.degree = i*degree;
        fancyButton.hidden = YES;             //移除按钮
        fancyButton.tag = i + 292;            //设置 tag 值
        [fancyButton addTarget:self action:@selector(buttonPressed:) for
ControlEvents:UIControlEventTouchUpInside];
        [self addSubview:fancyButton];
        i++;
    }
}
```

实现长按屏幕菜单的显示或轻拍屏幕菜单的隐藏，需要使用 handleLongPress:、handleTap:、addGestureRecognizerForView:、willMoveToSuperview:、showButton:、hideButton:、hide 和 show 方法。handleLongPress:方法实现长按屏幕显示菜单的功能。

```
- (void)handleLongPress:(UILongPressGestureRecognizer *)sender{
    if (self.onScreen)
        return;
    UIView *superView = [sender view];
    CGPoint pressedPoint = [sender locationInView:superView];  //获取位置
    CGPoint newCenter = pressedPoint;
    if ((pressedPoint.x - self.frame.size.width/2) < 0){
        newCenter.x = self.frame.size.width/2;  //设置 newCenter 的 x 的值
    }
    if ((pressedPoint.x + self.frame.size.width/2) > superView. frame.
 size.width){
        newCenter.x = superView.frame.size.width - self.frame.size.width/2;
                                              //设置 newCenter 的 x 的值
    }
    if ((pressedPoint.y - self.frame.size.height/2) <0){
        newCenter.y = self.frame.size.height/2;  //设置 newCenter 的 y 的值
    }
    if ((pressedPoint.y + self.frame.size.height/2) > superView.frame.size.
height){
        newCenter.y = superView.frame.size.height - self.frame.size.
    height/2;                                 //设置 newCenter 的 y 的值
    }
    self.center = newCenter;
    [self show];                              //调用显示菜单的方法
}
```

addGestureRecognizerForView:方法实现将手势添加到视图中。程序代码如下：

```objc
- (void)addGestureRecognizerForView:(UIView *)view{
    UILongPressGestureRecognizer *longPress = [[UILongPressGestureReco
gnizer alloc] initWithTarget:self action:@selector(handleLongPress:)];
                                                   //创建长按的手势识别器
    [view addGestureRecognizer:longPress];
    UITapGestureRecognizer *tap = [[UITapGestureRecognizer alloc] initWith
Target:self action:@selector(handleTap:)];         //创建轻拍的手势识别器
    [view addGestureRecognizer:tap];
}
```

show 方法实现菜单的显示。程序代码如下：

```objc
- (void)show{
    self.onScreen = YES;
    float delay = 0.f;
//遍历
    for (MenuItem *button in self.subviews){
        [self performSelector:@selector(showButton:) withObject:button afterDelay:delay];
        delay += 0.05;
    }
}
```

hide 方法实现菜单的隐藏。程序代码如下：

```objc
- (void)hide{
    //遍历
    for (MenuItem *button in self.subviews){
        [button hide];
    }
    self.onScreen = NO;
}
```

buttonPressed:方法实现分享菜单中菜单项的选择。程序代码如下：

```objc
- (void)buttonPressed:(MenuItem *)button{
if (self.delegate){
    //判断是否实现了 fancyMenu:didSelectedButtonAtIndex:方法
        if([self.delegate respondsToSelector:@selector(fancyMenu:didSelected
        ButtonAtIndex:)]){
            [self.delegate fancyMenu:self didSelectedButtonAtIndex: button.
            tag - 292];
        }
    }
}
```

（9）打开 ViewController.h 文件，编写代码，实现头文件、遵守协议和属性的声明。程序代码如下：

```objc
#import <UIKit/UIKit.h>
#import "Menu.h"
@interface ViewController : UIViewController<MenuDelegate>
@property (nonatomic, strong) Menu *menu;
@end
```

（10）打开 ViewController.m 文件，编写代码。在此文件中，代码实现：菜单对象的创建和设置；选择菜单项后弹出相应的警告视图。其中，在 viewDidLoad 方法中实现菜单对象的创建和设置。程序代码如下：

```objc
- (void)viewDidLoad
{
    [super viewDidLoad];
    // Do any additional setup after loading the view, typically from a nib.
    NSArray *images = @[[UIImage imageNamed:@"1.png"],[UIImage imageNamed:
@"4.png"],[UIImage imageNamed:@"3.png"],[UIImage imageNamed:@"2.png"]];
    self.menu = [[Menu alloc] init];                    //实例化对象
    self.menu.delegate = self;                          //设置委托
    self.menu.buttonImages = images;
    [self.view addSubview:self.menu];
}
```

fancyMenu:didSelectedButtonAtIndex:方法实现选择菜单项后，弹出相应的警告视图。程序代码如下：

```objc
- (void)fancyMenu:(Menu *)menu didSelected ButtonAtIndex: (NSUInteger)index{
    //判断 index 是否为 0
    if (index==0) {
        UIAlertView *alert=[[UIAlertView alloc]initWithTitle:@"Email 分享"
        message:nil delegate:nil cancelButtonTitle:@"OK" otherButtonTitles: nil];
        [alert show];
    }else if (index==1) {                               //判断 index 是否为 1
        UIAlertView *alert=[[UIAlertView alloc]initWithTitle:@"Facebook 分
享" message:nil delegate:nil cancelButtonTitle:@"OK" otherButtonTitles: nil];
        [alert show];
    }else if (index==2) {                               //判断 index 是否为 2
        UIAlertView *alert=[[UIAlertView alloc]initWithTitle:@"Twitter 分享"
        message:nil delegate:nil cancelButtonTitle:@"OK" otherButtonTitles: nil];
        [alert show];
    }else{
        UIAlertView *alert=[[UIAlertView alloc]initWithTitle:@"不分享"
        message:nil delegate:nil cancelButtonTitle:@"OK" otherButtonTitles: nil];
        [alert show];
    }
}
```

【代码解析】

本实例关键功能是菜单的显示和隐藏时的动画效果。下面就是这个知识点的详细讲解。

以菜单显示的动画效果为例。菜单出现，首先是淡出的过渡动画，使用 fadeInAnimation 方法实现。在此过渡动画中，设置了菜单项的状态。当过渡动画结束后，要实现旋转的功能，需要重写 CAAnimation 的 animationDidStop:finished:方法。其语法形式如下：

```objc
- (void)animationDidStop:(CAAnimation *)theAnimation finished:(BOOL)flag;
```

在本代码中，就是重写了 animationDidStop:finished:方法，实现了在过渡动画结束后的旋转动画。代码如下：

```objc
- (void)animationDidStop:(CAAnimation *)anim finished:(BOOL)flag{
    CABasicAnimation *animation = (CABasicAnimation *)anim;
    if ([animation.keyPath isEqualToString:@"transform.scale"] && flag){
        switch (self.state) {
            ......
            self.transform = CGAffineTransformMakeRotation(DegreesToRadians
(self.degree));
        }
```

实例 74　扇 形 菜 单

【实现描述】

本实例实现的功能是一个扇形菜单的显示。当用户单击界面的按钮，扇形菜单就会展开，让用户进行选择。当用户单击后，改变界面的背景颜色。当用户再次单击此按钮，菜单就会隐藏。运行效果如图 10.18 所示。

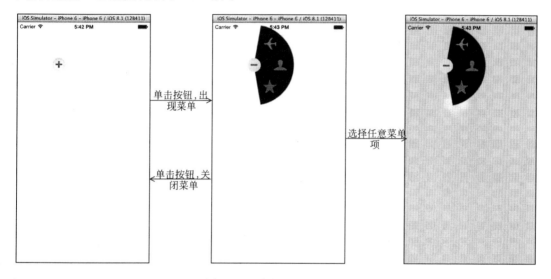

图 10.18　实例 74 运行效果

【实现过程】

当用户单击界面的按钮，判断菜单是否是展开状态，如果不是，则展开菜单。当用户再次单击此按钮，菜单就会隐藏。在菜单中选择任意菜单项，界面的背景颜色就会发生改变。具体的实现步骤如下。

（1）创建一个项目，命名为"扇形菜单"。

（2）添加图片 1.png、2.png、3.png、4.png、5.png 和 6.png 到创建项目的 Supporting Files 文件夹中。

（3）创建一个基于 NSObject 类的 MenuItem 类。

（4）打开 MenuItem.h 文件，编写代码，实现属性和方法的声明。程序代码如下：

```
#import <Foundation/Foundation.h>
@interface MenuItem : NSObject
//属性
@property (nonatomic) SEL action;
@property (nonatomic, assign) id target;
@property (nonatomic, strong) UIImage* normalImage;
//方法
- (id)initWithNormalImage:(UIImage *)normalImage target:(id)target action:
(SEL)action;
@end
```

（5）打开 MenuItem.m 文件，编写代码，实现方法的定义及初始化。程序代码如下：

```
- (id)initWithNormalImage:(UIImage *)normalImage target:(id)target action:
(SEL)action {
    self = [super init];
    if (self) {
        //对变量进行赋值
        _action = action;
        _target = target;
        _normalImage = normalImage;
    }
    return self;
}
```

（6）创建一个基于 UIView 类的 Menu 类。

（7）打开 Menu.h 文件，编写代码，实现头文件、对象、属性以及方法的声明。程序代码如下：

```
#import <UIKit/UIKit.h>
#import "MenuItem.h"                                    //头文件
@interface Menu : UIView{
    //对象
    NSArray* _items;
    NSMutableArray* _leavesLayers;
    NSMutableArray* _imagesLayers;
    UITapGestureRecognizer* _tapGestureRecognizer;
}
//属性
@property (nonatomic, strong) UIButton* wheelButton;
@property (nonatomic, assign, getter = isOn) BOOL on;
//方法
- (id)initWithItems:(NSArray*)items;
- (void)setOn:(BOOL)on animated:(BOOL)animated;
……
- (void)fold;
@end
```

（8）打开 Menu.m 文件，编写代码，实现扇形菜单的绘制、打开与隐藏以及菜单项的选择。使用的方法如表 10-19 所示。

表 10-19　Menu.m文件中方法总结

方　　法	功　　能
initWithItems:	菜单的初始化
layoutSubviews	布局
toggleWithAnimation:	切换
setOn:	设置控制菜单按钮的状态
setImages:	设置图像
addImages	添加图像
addLeaves	填充菜单项的背景
wheelButtonAction:	主按钮的动画
tapAction:	轻拍的动作
expand	展开
fold	折叠
static inline CGFloat DegreesToRadians(CGFloat inValue)	获取度数

这里需要讲解几个重要的方法（其他方法请读者参考源代码）。菜单的绘制需要使用 initWithItems:、layoutSubviews、setImages:、addImages 和 addLeaves 方法实现。其中，initWithItems:方法实现扇形菜单的初始化。程序代码如下：

```
- (id)initWithItems:(NSArray*)items {
    self = [super initWithFrame:CGRectMake(0.0f, 0.0f, 199.0f, 199.0f)];
    if (self) {
        _items = items;
        _on = NO;
        //轻拍手势的创建和设置
        _tapGestureRecognizer = [[UITapGestureRecognizer alloc] initWith
        Target:self action:@selector(tapAction:)];
        _tapGestureRecognizer.delegate = self;                //设置委托
        [self addGestureRecognizer:_tapGestureRecognizer];
        //添加手势识别器对象
        _leavesLayers = [NSMutableArray new];
        _imagesLayers = [NSMutableArray new];
        [self addLeaves];
        [self addImages];
        //按钮对象的创建和设置，此按钮用于打开或隐藏菜单
        UIImage* image = [UIImage imageNamed:@"2.png"];
        _wheelButton = [[UIButton alloc] init];
        [_wheelButton setImage:image forState:UIControlStateNormal];//设置图像
        [self addSubview:_wheelButton];
        [_wheelButton addTarget:self action:@selector(wheelButtonAction:)
        forControlEvents:UIControlEventTouchUpInside];        //添加动作
        [self setExclusiveTouch:NO];
        [self setBackgroundColor:[UIColor clearColor]];
    }
    return self;
}
```

addImages 方法实现为菜单项添加图像的功能。程序代码如下：

```
- (void)addImages {
    [_imagesLayers makeObjectsPerformSelector:@selector(removeFromSuperlayer)];
    [_imagesLayers removeAllObjects];                //删除数组中的所有对象
    //遍历所有的图像
    [_items enumerateObjectsUsingBlock:^(MenuItem* obj, NSUInteger idx,
    BOOL *stop) {
        UIImage* image = [obj normalImage];
        //添加图像
        CALayer* imageLayer = [CALayer layer];
        imageLayer.contents = (id)image.CGImage;
        imageLayer.frame = CGRectMake(0.0f, 0.0f, image.size.width, image.
        size.height);                                //设置框架
        imageLayer.anchorPoint = CGPointMake(0.5f, 0.5f);    //设置锚点
        imageLayer.position = CGPointMake(CGRectGetMidX(self.bounds),
        CGRectGetMidY(self.bounds));
        imageLayer.transform = CATransform3DMakeScale(0.01f, 0.01f, 1.0f);
        imageLayer.opacity = 0.6f;                           //设置透明度
        [self.layer addSublayer:imageLayer];
        [_imagesLayers addObject:imageLayer];
    }];
}
```

addLeaves 方法实现为菜单项添加背景的功能。程序代码如下：

```objectivec
- (void)addLeaves {
    [_leavesLayers makeObjectsPerformSelector:@selector(removeFromSuperlayer)];
    [_leavesLayers removeAllObjects];                    //删除数组中的所有对象
    //遍历所有的图像
    [_items enumerateObjectsUsingBlock:^(id obj, NSUInteger idx, BOOL *stop)
{
        UIImage *image = [UIImage imageNamed:@"1.png"];      //实例化对象
        //创建并设置图层
        CALayer* layer = [CALayer layer];
        layer.contents = (id)image.CGImage;
        layer.frame = CGRectMake(0.0f, 0.0f, image.size.width, image.size.height);
                //设置框架
        layer.anchorPoint = CGPointMake(0.0f, 0.5f);          //设置锚点
        layer.position = CGPointMake(CGRectGetMidX(self.bounds), CGRectGet
        MidY(self.bounds));
        layer.transform = CATransform3DMakeScale(0.15f, 0.15f, 1.0f);
        //添加图层
        [self.layer addSublayer:layer];                      //添加图层对象
        [_leavesLayers addObject:layer];
    }];
}
```

实现打开或许隐藏菜单，需要使用 toggleWithAnimation:、setOn:、wheelButtonAction:、expand、fold 和 static inline CGFloat DegreesToRadians(CGFloat inValue)方法实现。setOn:方法设置用来控制菜单的按钮的状态。程序代码如下：

```objectivec
- (void)setOn:(BOOL)on animated:(BOOL)animated {
    _on = on;
    //判断_on
    if (_on) {
        UIImage* image = [UIImage imageNamed:@"3.png"];
        [_wheelButton setImage:image forState:UIControlStateNormal];
            //设置图像
        [self expand];
        _tapGestureRecognizer.enabled = YES;
    } else {
        UIImage* image = [UIImage imageNamed:@"2.png"];      //设置图像
        [_wheelButton setImage:image forState:UIControlStateNormal];
        [self fold];
        _tapGestureRecognizer.enabled = NO;
    }
}
```

expand 方法实现菜单展开的动画。程序代码如下：

```objectivec
- (void)expand {
    CGFloat angle_ = DegreesToRadians(kApertureAngle);
    [CATransaction begin];
    //遍历
    for (NSInteger i=0; i<[_items count]; i++) {
        CALayer* layer = nil;
        CGFloat angle = - angle_ + angle_ * i;
        //实现菜单项背景的动画效果
        CATransform3D transform = CATransform3DConcat(CATransform3DMake
        Scale(1.0f, 1.0f, 1.0f),
        CATransform3DMakeRotation(angle, 0.0f, 0.0f, 1.0f));      //实例化对象
```

```
CABasicAnimation* leafAnimation = [CABasicAnimation animation
WithKeyPath:@"transform"];
[leafAnimation setTimingFunction:[CAMediaTimingFunction functionWithName:
 kCAMediaTimingFunctionEaseOut]];                        //设置动画种类
[leafAnimation setToValue:[NSValue valueWithCATransform3D: transform]];
[leafAnimation setFillMode:kCAFillModeForwards];
[leafAnimation setRemovedOnCompletion: NO];
[leafAnimation setDuration:0.6f];                        //设置持续时间
layer = [_leavesLayers objectAtIndex:i];
[layer addAnimation:leafAnimation forKey:@"expand"];   //添加动画
//缩放图像的动画效果
CABasicAnimation* scaleImageAnimation = [CABasicAnimation animation
WithKeyPath: @"transform"];
[scaleImageAnimation setTimingFunction:[CAMediaTimingFunction
functionWithName:
 kCAMediaTimingFunctionEaseOut]];                        //设置动画种类
[scaleImageAnimation setToValue:[NSValue valueWithCATransform3D:
 CATransform3DMakeScale(1.0f, 1.0f, 1.0f)]];             //设置开始值
[scaleImageAnimation setFillMode:kCAFillModeForwards];
[scaleImageAnimation setRemovedOnCompletion: NO];
//图像位置移动的动画效果
CGPoint point = CGPointMake(0.65*97.0f * cos(angle) + CGRectGetMidX
(self.bounds), 0.65*97.0f * sin(angle) + CGRectGetMidY(self.bounds));
CABasicAnimation* positionImageAnimation = [CABasicAnimation
animationWithKeyPath:@"position"];
[positionImageAnimation setTimingFunction:[CAMediaTimingFunction
functionWithName:
 kCAMediaTimingFunctionEaseOut]];                        //设置动画种类
[positionImageAnimation setToValue:[NSValue valueWithCGPoint:point]];
[positionImageAnimation setFillMode:kCAFillModeForwards];
[positionImageAnimation setRemovedOnCompletion: NO];   //是否在当前层完成动画
//实现图像的缩放和位置移动的动画
CAAnimationGroup* group = [CAAnimationGroup animation];
[group setAnimations:[NSArray arrayWithObjects:scaleImageAnimation,
 positionImageAnimation, nil]];                          //设置动画
[group setFillMode:kCAFillModeForwards];
[group setRemovedOnCompletion: NO];                      //是否在当前层完成动画
[group setDuration:0.3f];
[group setBeginTime:CACurrentMediaTime () + 0.27f];
layer = [_imagesLayers objectAtIndex:i];
[layer addAnimation:group forKey:@"show"];              //添加动画
    }
    [CATransaction commit];
}
```

要实现菜单项的选择,需要使用 tapAction:方法。程序代码如下:

```
- (void)tapAction:(UITapGestureRecognizer*)tapGestureRecognizer {
   CGFloat step = DegreesToRadians(kApertureAngle);
   CGPointpointA=CGPointMake(CGRectGetMidX(self.bounds),CGRectGetMidY(self.bounds));
   //遍历所有的图像
   [_itemsenumerateObjectsUsingBlock:^(MenuItem* obj,NSUIntegeridx,BOOL*stop){
```

```
    //为变量赋值
    CGFloat width = 105.0f;
    CGFloat angle = step * (CGFloat)idx - (step * 1.5);
    CGPoint pointB = CGPointMake(pointA.x + cos(angle) * width, pointA.y
    + sin(angle) * width);
    CGPoint pointC = CGPointMake(pointA.x + cos(angle + step) * width,
    pointA.y + sin(angle + step) * width);
    UIBezierPath* bezierPath = [UIBezierPath bezierPath];
                                                    //实例化对象bezierPath
    [bezierPath moveToPoint: pointA];               //设置开始点
    [bezierPath addLineToPoint:pointB];             //设置结束点
    [bezierPath addLineToPoint:pointC];             //设置结束点
    [bezierPath closePath];
    CGPoint point = [tapGestureRecognizer locationInView:self];
                                                    //获取轻拍的位置
    //判断
    if (CGPathContainsPoint(bezierPath.CGPath, NULL, point, NO)) {
        [obj.target performSelector:obj.action withObject:self];
    }
}];
}
```

（9）打开 ViewController.h 文件，编写代码，实现头文件和对象的声明。程序代码如下：

```
#import <UIKit/UIKit.h>
#import "MenuItem.h"
#import "Menu.h"
@interface ViewController : UIViewController{
    Menu *rosette;              //声明对象
}
@end
```

（10）打开 ViewController.m 文件，编写代码。在此文件中，代码分为了两个部分：创建并设置菜单对象以及选择菜单项后实现的响应。在 viewDidLoad 方法中实现创建并设置菜单对象。程序代码如下：

```
- (void)viewDidLoad
{
    [super viewDidLoad];
    // Do any additional setup after loading the view, typically from a nib.
    //实例化对象
    UIImage* twitterImage = [UIImage imageNamed:@"4.png"];
    UIImage* facebookImage = [UIImage imageNamed:@"5.png"];
    UIImage* mailImage = [UIImage imageNamed:@"6.png"];
    //创建菜单项对象
    MenuItem* twitterItem = [[MenuItem alloc] initWithNormalImage:
    twitterImage target:self action:@selector(twitterAction:)];
    MenuItem* facebookItem = [[MenuItem alloc] initWithNormalImage:
    facebookImage target:selfaction:@selector(facebookAction:)];
    MenuItem* mailItem = [[MenuItem alloc] initWithNormalImage:mailImage
```

```
    target:self
     action:@selector(mailAction:)];
    //创建并设置菜单对象
    rosette = [[Menu alloc] initWithItems: [NSArray arrayWithObjects:
    twitterItem, facebookItem, mailItem, nil]];
    [rosette setCenter:CGPointMake(100.0f, 100.0f)];
    [self.view addSubview:rosette];
}
```

实现选择菜单项后的响应，需要使用 twitterAction:、facebookAction:和 mailAction:方法。程序代码如下：

```
- (void)twitterAction:(id)sender {
//设置背景颜色
    self.view.backgroundColor=[UIColor  colorWithHue:0.3  saturation:0.3
brightness:1.0 alpha:1.0];
}
- (void)facebookAction:(id)sender {
//设置背景颜色
    self.view.backgroundColor=[UIColor  colorWithHue:0.6  saturation:0.3
brightness:1.0 alpha:1.0];
}
- (void)mailAction:(id)sender {
//设置背景颜色
    self.view.backgroundColor=[UIColor  colorWithHue:0.1  saturation:0.3
brightness:1.0 alpha:1.0];
}
```

【代码解析】

本实例关键功能是菜单的选择。下面就是这个知识点的详细讲解。

菜单的选择，要在 tapAction:方法中实现，首先要创建关于 UIBezierPath 的对象，并绘制图形。代码如下：

```
UIBezierPath* bezierPath = [UIBezierPath bezierPath];
[bezierPath moveToPoint: pointA];
[bezierPath addLineToPoint:pointB];
[bezierPath addLineToPoint:pointC];
[bezierPath closePath];
```

然后，使用 locationInView:方法实现获取当前触摸的点，其语法形式如下：

```
- (CGPoint)locationInView:(UIView *)view;
```

其中，(UIView *)view 表示视图对象，表示触摸发生在此视图中。在本代码中，使用了 locationInView:方法实现获取当前触摸的点。代码如下：

```
CGPoint point = [tapGestureRecognizer locationInView:self];
```

其中，self 表示发生触摸的视图对象。最后，使用 CGPathContainsPoint 方法判断触摸的点是否在路径中。其语法形式如下：

```
bool CGPathContainsPoint (
   CGPathRef path,
```

```
    const CGAffineTransform *m,
    CGPoint point,
    bool eoFill
);
```

其中,参数说明如下。

- ❑ CGPathRef path 表示路径。
- ❑ const CGAffineTransform *m 表示仿射变换。
- ❑ CGPoint point 表示给定的点。
- ❑ bool eoFill 表示一个布尔值。

在本代码中,就使用了 CGPathContainsPoint 方法来判断触摸的点是否在绘制的路径中,如果在,就执行此图形路径中的方法;如果不在就不执行。代码如下:

```
if (CGPathContainsPoint(bezierPath.CGPath, NULL, point, NO)) {
    [obj.target performSelector:obj.action withObject:self];
}
```

其中,参数说明如下。

- ❑ bezierPath.CGPath 表示路径。
- ❑ NULL 表示仿射变换。
- ❑ point 表示给定的点。
- ❑ NO 表示一个布尔值。

第 11 章　提醒对话框

在 iPhone 或者 iPad 中，有一些内容是必须让用户知道的，是非常重要的。为了突出它的重要性，苹果公司专门为开发人员提供了警告视图和动作表单，一般将它们统称为提醒对话框。本章主要讲解这类对话框的相关实例。

实例 75　具有文本框的警告视图

【实例描述】

有一些时候，警告视图会被用来作为一个登录窗口。登录窗口中会显示一文本框。本实例实现的功能是在警告视图中显示文本框，作为一个登录窗口使用。运行效果如图 11.1 所示。

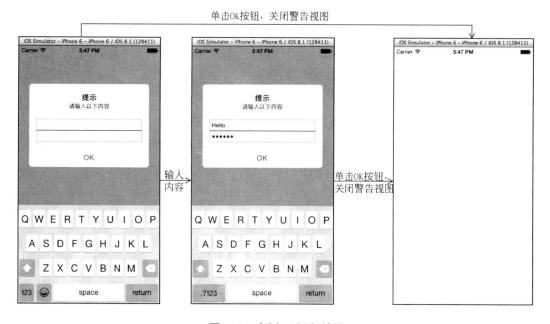

图 11.1　实例 75 运行结果

注意：在文本框中显示的两个文本框，上面的是不加密的，下面的是加密的。

【实现过程】

（1）创建一个项目，命名为"具有文本框的警告视图"。

(2）打开 ViewController.m 文件，编写代码，实现具有文本框警告视图的显示。程序代码如下：

```
- (void)viewDidLoad
{
    [super viewDidLoad];
    // Do any additional setup after loading the view, typically from a nib.
    UIAlertView *alert= [[UIAlertView alloc]initWithTitle:@"提示" message:
    @"请输入以下内容" delegate:nil cancelButtonTitle:@"OK" otherButton
    Titles:nil];                              //创建警告视图
    alert.alertViewStyle=UIAlertViewStyleLoginAndPasswordInput;
                                              //设置警告视图的风格
    [alert show];
}
```

【代码解析】

本实例关键功能是在警告视图中显示文本框。下面就是这个知识点的详细讲解。

UIAlertView 的 alertViewStyle 方法可以实现为用户显示一种格式的警告视图。其语法形式如下：

`@property(nonatomic, assign) UIAlertViewStyle alertViewStyle;`

其中，警告视图的格式如表 11-1 所示。

表 11-1 警告视图的格式

格 式	说 明
UIAlertViewStyleDefault	默认的警告视图，说明只弹出信息和按钮
UIAlertViewStyleSecureTextInpu	有一个加密的文本框
UIAlertViewStylePlainTextInput	有一个不加密的文本框
UIAlertViewStyleLoginAndPasswordInput	有两个文本框，其中一个是加密的文本框，另一个是不加密的文本框

在本代码中使用了 alertViewStyle 的方法实现了在警告视图中显示文本框的功能。代码如下：

`alert.alertViewStyle=UIAlertViewStyleLoginAndPasswordInput;`

实例 76　全屏的警告视图

【实例描述】

在 iPhone 或者 iPad 的应用程序中，为了引起用户的极度关注，有时会使用一些全屏模式的警告视图。那么这些警告视图是怎么实现的呢？本实例就实现了这么一个全屏模式的警告视图。当用户单击"警告视图"按钮，全屏模式的警告视图就会显示；当用户单击警告视图中的 Store 按钮，全屏模式的警告视图就会隐藏。运行效果如图 11.2 所示。

【实现过程】

（1）创建一个项目，命名为"全屏的警告视图"。

（2）创建一个基于 UIButton 类的分类 Block。

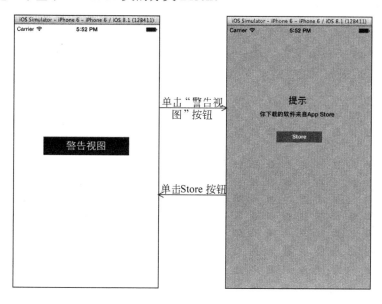

图 11.2　实例 76 运行效果

（3）打开 Block.h 文件，编写代码，实现头文件、宏定义、属性以及方法的声明。程序代码如下：

```
#import <UIKit/UIKit.h>
#import <objc/runtime.h>                                        //头文件
#define kUIButtonBlockTouchUpInside @"TouchInside"              //宏定义
@interface UIButton (Block)
@property (nonatomic, strong) NSMutableDictionary *actions;     //属性
- (void) setAction:(NSString*)action withBlock:(void(^)())block; //方法
@end
```

（4）打开 Block.m 文件，编写代码，实现对动作的设置。使用的方法如表 11-2 所示。

表 11-2　Block.m 文件中方法总结

方　法	功　能
setAction:withBlock:	设置动作
setActions:	进行关联
actions	获取相关联的对象
doTouchUpInside:	实现按下的动作

这里需要讲解几个重要的方法（其他方法请读者参考源代码）。其中，setAction:withBlock:方法实现使用块对动作进行设置。程序代码如下：

```
- (void) setAction:(NSString*)action withBlock:(void(^)())block {
  if ([self actions] == nil) {
    [self setActions:[[NSMutableDictionary alloc] init]];
  }
  [[self actions] setObject:block forKey:action];              //设置对象
  if ([kUIButtonBlockTouchUpInside isEqualToString:action]) {
    [self addTarget:self action:@selector(doTouchUpInside:) forControl
      Events:UIControlEventTouchUpInside];                     //添加动作
```

        }
}

（5）创建一个基于 UIView 类的 AlertView 类。

（6）打开 AlertView.h 文件，编写代码，实现头文件、宏定义、对象、实例变量以及方法的声明。程序代码如下：

```objective-c
#import <UIKit/UIKit.h>
#import "UIButton+Block.h"                              //头文件
//宏定义
#define kMessageWidth 200
#define kTitleMaxHeight 300
……
#define BTN_FONT [UIFont fontWithName:@"HelveticaNeue-Bold" size:12]
#define BTN_CENTERED CGRectMake(kCenteredX - kBtnWidth/2, kBtnY, kBtnWidth, kBtnHeight)
@interface AlertView : UIView{
    //对象
    UIViewController *parentViewController;
    UIView *overlay;
    UILabel *title;                                     //声明标签对象
    UILabel *message;
    NSMutableArray *subviews;
    UIButton * butt;
    BOOL isShown;                                       //实例变量
}
//方法
- (id)initInViewController:(UIViewController *)viewController;
-   (void)showAlertWithText:(NSString  *)alertText  titleText:(NSString
*)titleText   buttonText:(NSString   *)buttonText   onTap:(void
(^)(void))block;;
- (void)removeAlert;                                    //移除警告视图的方法
@end
```

（7）打开 AlertView.m 文件，编写代码，实现全屏的警告视图。使用的方法如表 11-3 所示。

表 11-3　AlertView.m文件中方法总结

方　　法	功　　能
initInViewController:	初始化视图控制器
showAlertWithText:titleText:buttonText:onTap:	显示警告视图
removeAlert	移除警告视图

其中，initInViewController:方法实现视图控制器的初始化。程序代码如下：

```objective-c
- (id)initInViewController:(UIViewController *)viewController {
    self = [super initWithFrame:viewController.view.bounds];
    if (self) {
        parentViewController = viewController;
        //创建并设置空白视图对象
        overlay = [[UIView alloc] initWithFrame:self.frame];
        [overlay setBackgroundColor:[UIColor colorWithHue:0.1 saturation:
        0.3 brightness:0.8 alpha:1.0]];
        subviews = [[NSMutableArray alloc] init];
        isShown = NO;
    }
```

```
    return self;
}
```

showAlertWithText:titleText:buttonText:onTap:方法实现警告视图的显示。程序代码如下：

```
-   (void)showAlertWithText:(NSString   *)alertText   titleText:(NSString
*)titleText buttonText:(NSString *)buttonText onTap:(void (^) (void))block;
{
    isShown = YES;
    UIFont *font = [UIFont fontWithName:@"HelveticaNeue-Medium" size:13];
    //创建并设置标签对象
    message = [[UILabel alloc] init];
    NSMutableParagraphStyle *style1=[[NSMutableParagraphStyle default
ParagraphStyle]mutableCopy];
    [style1 setLineBreakMode:title.lineBreakMode];         //设置换行模式
    NSDictionary *dis1=[NSDictionary dictionaryWithObjectsAndKeys: font,NS
    Font AttributeName,style1,
    NSParagraphStyleAttributeName,nil];
    CGSize textSize =[alertText sizeWithAttributes:dis1];
    CGFloat y = 380/2 - textSize.height/2;
    [message setFrame:CGRectMake(kX, y, kMessageWidth, textSize.height)];
    //设置框架
    [message setFont:font];                                //设置字体
    [message setBackgroundColor:[UIColor clearColor]];
    [message setTextColor:[UIColor blackColor]];
    [message setText:alertText];                           //设置文本内容
    [message setTextAlignment:NSTextAlignmentCenter];
    [message setNumberOfLines:99];
    //创建并设置标签对象
    title = [[UILabel alloc] init];
    NSMutableParagraphStyle *style2=[[NSMutableParagraphStyle default
ParagraphStyle]mutableCopy];
    [style2 setLineBreakMode:title.lineBreakMode];         //设置换行模式
    NSDictionary *dis2=[NSDictionary dictionaryWithObjects AndKeys:font,NS
    FontAttributeName,style2,NSParagraphStyleAttributeName,nil];
    CGSize titleSize=[titleText sizeWithAttributes:dis2];  //获取字体的尺寸
    y = kTitleMaxHeight/2 - titleSize.height/2;
    [title setFrame:CGRectMake(kX, y, kTitleWidth, titleSize.height+20)];
    UIFont * titleFont = [UIFont fontWithName:@"HelveticaNeue-Medium" size:20];
    [title setFont:titleFont];                             //设置字体
    [title setBackgroundColor:[UIColor clearColor]];
    [title setTextColor:[UIColor blackColor]];
    [title setText:titleText];
    [title setTextAlignment:NSTextAlignmentCenter];        //设置对齐方式
[title setNumberOfLines:1];
//创建并设置按钮对象
    butt = [UIButton buttonWithType:UIButtonTypeCustom];
    [butt setBackgroundColor:[UIColor redColor]];          //设置背景颜色
    [butt setTitleColor:[UIColor whiteColor] forState:UIControl StateNormal];
    [butt setTitleColor:[UIColor darkGrayColor] forState:UIControlStateHighlighted];
    [butt.layer setCornerRadius:2];                        //设置圆角半径
    [[butt titleLabel] setFont:BTN_FONT];
    [butt setTitle:buttonText forState:UIControlStateNormal];
    if (block != nil)
    {
        [butt setAction:kUIButtonBlockTouchUpInside withBlock:block];
    //设置动作
    }
```

```
    butt.frame = BTN_CENTERED;
    //实现添加
    [parentViewController.view addSubview:overlay];
    [parentViewController.view addSubview:message];
    [parentViewController.view addSubview:title];
    [parentViewController.view addSubview:butt];
    [subviews addObject:overlay];
    [subviews addObject:message];
    [subviews addObject:title];
    [subviews addObject:butt];
}
```

removeAlert 方法实现移除警告视图。程序代码如下：

```
- (void)removeAlert {
    for (UIView *view in subviews) {
        [view removeFromSuperview];                    //移除指定视图
    }
}
```

（8）打开 ViewController.h 文件，编写代码，实现头文件、插座变量以及动作的声明。程序代码如下：

```
#import <UIKit/UIKit.h>
#import "AlertView.h"
@interface ViewController : UIViewController{
    IBOutlet UIButton *button;                         //插座变量
}
- (IBAction)show:(id)sender;
@end
```

（9）打开 Main.storyboard 文件，添加按钮到设计界面上。将按钮的 Title 设置为"警告视图"，将 Font 设置为 System 22.0，将 Text Color 设置为白色，将 Background 设置为黑色。将此按钮和插座变量 button 以及动作 show:关联。

（10）打开 ViewController.m 文件，编写代码，实现全屏警告视图的显示。程序代码如下：

```
- (IBAction)show:(id)sender {
    AlertView * alert = [[AlertView alloc] initInViewController:self];
    [alert showAlertWithText:@"你下载的软件来自App Store" titleText:@"提示"
    buttonText:@"Store"onTap:^{
        [alert removeAlert];                           //移除警告视图
    } ];
}
```

【代码解析】

本实例关键功能是在全屏警告视图中单击按钮后实现退出。下面就是这个知识点的详细讲解。

如果想实现单击按钮后退出警告视图，需要调用 removeAlert 方法。在程序中的代码如下：

```
[alert showAlertWithText:@"你下载的软件来自 App Store" titleText:@"提示"
buttonText:@"Store"onTap:^{
    [alert removeAlert];
} ];
```

在 removeAlert 方法中，将所有在 subviews 中的视图删除。代码如下：

```
for (UIView *view in subviews) {
    [view removeFromSuperview];
}
```

实例 77 具有进度条的警告视图

【实例描述】

用户下载软件时，为了重点强调下载的进度，常常会在警告视图中显示一个具有加载进度的进度条。本实例实现的功能就是在警告视图中添加一个进度条。

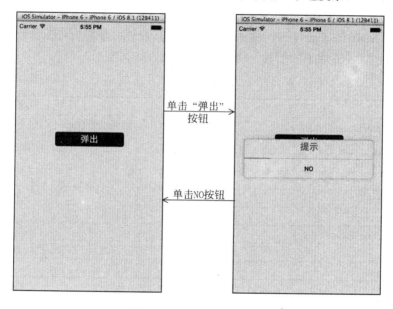

图 11.3 实例 77 运行效果

【实现过程】

（1）创建一个项目，命名为"具有进度条的警告视图"。
（2）添加图片 1.png 到创建项目的 Supporting Files 文件夹中。
（3）创建一个基于 UIView 类的 AlertView 类。
（4）打开 AlertView.h 文件，编写代码，实现宏定义、对象、实例变量、属性以及方法的声明。程序代码如下：

```
#import <UIKit/UIKit.h>
//宏定义
#define kILAlertViewTitleColor [UIColor redColor]
......
#define kOverlayMaxAlpha 0.2
#define kCloseButtonTag 100
#define kSecondButtonTag 101
#define IS_LANDSCAPE  UIInterfaceOrientationIsLandscape([UIApplication sharedApplication].statusBarOrientation)
@interface AlertView : UIView{
    //对象
    UIImageView *_bkgImg;
```

```
    UILabel *_titleLabel;                        //声明标签对象
    UIButton *_closeButton;
    NSTimer *a;                                  //声明定时器对象
    UIWindow *_window;
    UIView *_overlay;
    UIProgressView *_pro;
    float i;                                     //实例变量
}
//属性
@property (strong, nonatomic) NSString *title;
@property (strong, nonatomic) NSString *message;
+ (AlertView *)showWithTitle:(NSString *)title closeButtonTitle:(NSString
*)closeTitle;                        //方法
@end
```

（5）打开 AlertView.m 文件，编写代码，实现具有进度条警告视图的绘制、警告视图的显示与隐藏，以及进度条的加载。使用的方法如表 11-4 所示。

表 11-4　AlertView.m文件中方法总结

方　　法	功　　能
initWithTitle: closeButtonTitle:	初始化警告视图
layoutSubviews	布局
centerWithOrientation	设置中心位置
buttonTapped:	轻拍按钮
showAlertAnimated:	显示动画
dismissAlertAnimated:	隐藏动画
showWithTitle:closeButtonTitle:	显示警告视图
load	实现进度条中进度的加载
pro:	实现加载以及判断

这里需要讲解几个重要的方法（其他方法请读者参考源代码）。其中，具有进度条警告视图的绘制需要使用 initWithTitle:closeButtonTitle:、layoutSubviews 和 centerWithOrientation 方法实现。initWithTitle: closeButtonTitle:方法实现对警告视图的初始化。程序代码如下：

```
- (AlertView *)initWithTitle:(NSString *)title closeButtonTitle:(NSString
*)closeTitle{
    self = [super init];
    if (self) {
        self.title = title;
        if (!_window) {
            UIView *winParent = [[UIApplication sharedApplication].windows
             objectAtIndex:0];
            _window = [winParent.subviews objectAtIndex:0];
        }
        //创建并设置图像视图对象，作为背景
        UIImage *resizableImg = [[UIImage imageNamed: @"1. png"] resizable
        ImageWithCapInsets:UIEdgeInsetsMake(10, 10, 10, 10)];
        _bkgImg = [[UIImageView alloc] initWithImage:resizableImg];
        [self addSubview:_bkgImg];                   //添加视图对象
        _bkgImg.layer.shadowColor = [UIColor blackColor].CGColor;  //设置阴影颜色
        _bkgImg.layer.shadowOffset = CGSizeMake(0, 0);
        _bkgImg.layer.shadowRadius = 5.0f;
```

```
        _bkgImg.layer.shadowOpacity = 0.2;                    //设置透明度
        _bkgImg.layer.shouldRasterize = YES;
        self.autoresizingMask=UIViewAutoresizingFlexibleLeftMargin|UIView
AutoresizingFlexibleRightMargin|UIViewAutoresizingFlexibleTop
Margin| UIViewAutoresizingFlexibleBottomMargin;          //设置自适应
        //创建并设置标签对象,用来显示警告视图的标题
        _titleLabel = [[UILabel alloc] init];
        _titleLabel.text = self.title;                        //设置文本内容
        _titleLabel.numberOfLines = 2;
        _titleLabel.backgroundColor = [UIColor clearColor];
        _titleLabel.textAlignment = NSTextAlignmentCenter;    //设置对齐方式
        _titleLabel.font = kILAlertViewTitleFont;
        _titleLabel.textColor = kILAlertViewTitleColor;
        [self addSubview:_titleLabel];
        //创建并设置进度条对象
        _pro=[[UIProgressView alloc]init];
        [self addSubview:_pro];                               //添加视图对象
        //创建并设置按钮对象
        _closeButton = [UIButton buttonWithType:UIButtonTypeCustom];
        [_closeButton setTitle:closeTitle forState:UIControlStateNormal];//设置标题
        [_closeButton setTitleColor:kILAlertViewCloseButtonColor forState:UIControl
StateNormal];
        [_closeButton setTitleColor:kILAlertViewCloseButtonColorSelected
forState:UIControlStateHighlighted];
        _closeButton.titleLabel.font = kILAlertViewButtonFont;  //设置字体
        _closeButton.tag = kCloseButtonTag;
        [_closeButton addTarget:self action:@selector(buttonTapped:)
forControlEvents:UIControlEventTouchUpInside];
        [self addSubview:_closeButton];                       //添加视图对象
    }
    return self;
}
```

layoutSubviews 方法实现对创建的子视图进行布局。程序代码如下：

```
- (void)layoutSubviews
{
    self.frame                                                                =
CGRectMake(floorf((_window.frame.size.width-kMaxWidth)/2),floorf((_wind
ow.frame.size.height-kMaxHeight)/2),kMaxWidth,kMaxHeight);
    CGSize mainSize = self.frame.size;
    CGFloat yOffset = kBuffer;
    //判断
    if (_title != nil) {
        _titleLabel.frame = CGRectMake(0, yOffset,mainSize.width,20);//设置框架
    }
    _pro.frame = CGRectMake(0,40, mainSize.width,20);         //设置框架
    yOffset = CGRectGetMaxY(_pro.frame) + kBuffer;
    CGFloat buttonXOffset = floorf((mainSize.width-kOneButtonMaxWidth)/2);
    //设置框架
    _closeButton.frame =CGRectMake(buttonXOffset,yOffset,kOneButtonMax Width,
kButtonHeight);
    yOffset = CGRectGetMaxY(_closeButton.frame) + kBuffer;
    CGRect rect = self.frame;
    rect.size.height = yOffset;
    self.frame = rect;
    _bkgImg.frame = CGRectMake(0, 0, self.frame.size.width, self.frame.size.
height);                                                      //设置框架
    [self centerWithOrientation];
```

}
```

警告视图的显示与隐藏需要使用 showAlertAnimated:、dismissAlertAnimated: 和 showWithTitle: closeButtonTitle:方法实现。showAlertAnimated:方法实现显示警告视图时出现的动画效果。程序代码如下:

```
- (void)showAlertAnimated:(BOOL)animated {
    [_window addSubview:self];
    if (animated) {
        self.alpha = 0;
        [UIView animateWithDuration:0.1 animations:^{self.alpha = 1.0;}];
        self.layer.transform = CATransform3DMakeScale(0.5, 0.5, 1.0);      //缩放
        CAKeyframeAnimation *bounceAnimation = [CAKeyframeAnimation animation
        WithKeyPath:@"transform.scale"];
        bounceAnimation.values = [NSArray arrayWithObjects:[NSNumber number
        WithFloat:0.5],[NSNumber numberWithFloat:1.2],[NSNumber numberWith
        Float:0.9],[NSNumber numberWithFloat:1.0], nil];          //设置值
        bounceAnimation.duration = kAnimateInDuration;   //设置动画所需时间
        bounceAnimation.removedOnCompletion = NO;
        [self.layer addAnimation:bounceAnimation forKey:@"bounce"];   //添加动画
        self.layer.transform = CATransform3DIdentity;
    }
}
```

showWithTitle:closeButtonTitle:方法实现警告视图的显示。程序代码如下:

```
+ (AlertView *)showWithTitle:(NSString *)title closeButtonTitle:(NSString
*)closeTitle{
    AlertView *alert = [[AlertView alloc] initWithTitle:title close
    ButtonTitle:closeTitle];                        //实例化对象 alert
    [alert showAlertAnimated:YES];
    [alert performSelector:@selector(load) withObject:self afterDelay:0];
    return alert;
}
```

如果要实现警告视图中进度条的加载,需要使用 load 和 pro:方法。程序代码如下:

```
-(void)load{
    i=0;
    [_pro setProgress:i animated:YES];
    a=[NSTimer scheduledTimerWithTimeInterval:0.2 target:self selector:
    @selector(pro:) userInfo:nil repeats:YES];              //创建定时器
}
-(void)pro:(id)timer{
    i+=1.0f;
    [_pro setProgress:(i/20.0) animated:YES];                //设置进度
    if (_pro.progress==1.0) {
        [a invalidate];
        [self dismissAlertAnimated:YES];
    }
}
```

(6) 打开 ViewController.h 文件,编写代码,实现头文件、对象、插座变量以及方法的声明。程序代码如下:

```
#import <UIKit/UIKit.h>
#import "AlertView.h"
@interface ViewController : UIViewController{
    AlertView *alert;
```
```

```
    IBOutlet UIButton *button;                                           //声明插座变量
}
- (IBAction)show:(id)sender;
@end
```

(7) 打开 Main.storyboard 文件，对 ViewController 视图控制器的设计界面进行设计，效果如图 11.4 所示。

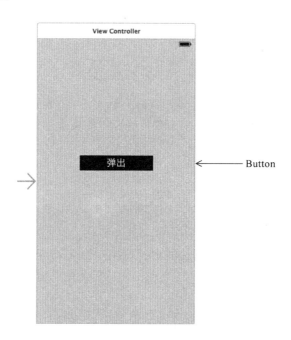

图 11.4　实例 77View Controller 视图控制器的设计界面

需要添加的视图、控件以及对它们的设置如表 11-5 所示。

表 11-5　实例 77 视图、控件设置

视图、控件	属 性 设 置	其　他
设计界面	Background：浅灰色	
Button	Title：弹出 Font：System 19.0 Text Color：白色 Background：黑色	与插座变量 button 关联 与动作 show:关联

(8) 打开 ViewController.m 文件，编写代码，实现头文件、对象、插座变量以及方法的声明。程序代码如下：

```
- (void)viewDidLoad
{
    button.layer.cornerRadius=5;                                         //设置圆角半径
    [super viewDidLoad];
    // Do any additional setup after loading the view, typically from a nib.
}
- (IBAction)show:(id)sender{
    [AlertView showWithTitle:@"提示" closeButtonTitle:@"NO"];   //显示具有进
       度条的警告视图
}
```

【代码解析】

本实例的关键功能是单击警告视图中的按钮退出警告视图，以及警告视图中进度条的进度加载。下面依次讲解这两个知识点。

1. 单击警告视图中的按钮退出警告视图

单击警告视图中的按钮退出警告视图的实现，需要为警告视图中的按钮使用 addTarget:forControlEvents:方法添加一个动作，在此程序中代码如下：

```
[_closeButton addTarget:self action:@selector(buttonTapped:) forControlEvents:UIControlEventTouchUpInside];
```

这样当此按钮被按下后，就会执行 buttonTapped:动作，在 buttonTapped:动作中实现警告视图的隐藏动动画，这时就可以实现单击警告视图中的按钮退出警告视图。代码如下：

```
[self dismissAlertAnimated:YES];
```

2. 警告视图中进度条的进度加载

警告视图中进度条的进度加载，本实例是采用定时器实现的。在定时器中每隔 0.2 秒就会调用一次 pro:方法，代码如下：

```
a=[NSTimer scheduledTimerWithTimeInterval:0.2 target:self selector:@selector(pro:) userInfo:nil repeats:YES];
```

实例 78　具有列表的警告视图

【实例描述】

本实例的功能是实现一个具有列表的警告视图，用户可以在此警告视图中进行选择。运行效果如图 11.5 所示。

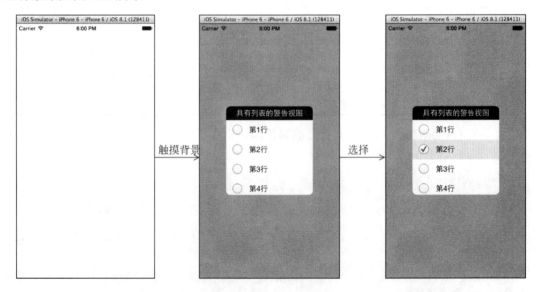

图 11.5　实例 78 运行效果

【实现过程】

当用户选择了列表中的某一行后，会在此行前面的圆圈中打对勾。具体的实现步骤如下。

（1）创建一个项目，命名为"具有表格的警告视图"。

（2）添加图像 1.png 和 2.png 到创建项目的 Supporting Files 文件夹中。

（3）创建一个基于 UIView 类的 AlertView 类。

（4）打开 AlertView.h 文件，编写代码，实现新的类型的定义、协议、属性和方法的声明。程序代码如下：

```objc
#import <UIKit/UIKit.h>
typedef void (^AlerViewButtonBlock)();                    //定义新类型
@class AlertView;
//协议
@protocol AlertViewDatasource <NSObject>
//获取行数的方法
- (NSInteger)popoverListView:(AlertView *)tableView numberOfRowsInSection:(NSInteger)section;
//获取行中内容的方法
- (UITableViewCell *)popoverListView:(AlertView *)tableView cellForRowAtIndex Path:(NSIndexPath *)indexPath;
@end
@protocol AlertViewDelegate <NSObject>
//实现选择的方法
- (void)popoverListView:(AlertView *)tableView didSelectRowAt IndexPath:(NSIndexPath *)indexPath;
//取消选择的方法
- (void)popoverListView:(AlertView *)tableView didDeselectRow AtIndex Path:(NSIndexPath *)indexPat;
@end
@interface AlertView : UIView<UITableViewDataSource,UITableViewDelegate>
//属性
@property (nonatomic, assign) id <AlertViewDelegate>delegate;
@property (nonatomic, retain) id <AlertViewDatasource>datasource;
@property (nonatomic, retain) UILabel *titleName;         //标签
@property (nonatomic, retain) UITableView *mainPopoverListView;//表视图
@property (nonatomic, retain) UIControl *controlForDismiss;
//方法
- (void)initTheInterface;                                 //对界面进行初始化的方法
- (void)animatedIn;
- (void)animatedOut;
- (void)show;                                             //实现显示警告视图的方法
- (void)dismiss;
- (id)dequeueReusablePopoverCellWithIdentifier:(NSString *)identifier;
- (UITableViewCell *)popoverCellForRowAtIndexPath:(NSIndexPath *)indexPath;
- (NSIndexPath *)indexPathForSelectedRow;
@end
```

（5）打开 AlertView.m 文件，编写代码，实现具有列表的警告视图的绘制、显示与隐

藏,以及警告视图的列表选择响应。使用的方法如表 11-6 所示。

表 11-6 AlertView.m 文件中方法总结

方　法	功　能
initWithFrame:	初始化具有列表的警告视图
initTheInterface	初始化界面
indexPathForSelectedRow	获取选中的列表元素
tableView:numberOfRowsInSection:	获取行数
tableView:cellForRowAtIndexPath:	获取某一行的数据
tableView:didDeselectRowAtIndexPath:	取消行的响应
tableView:didSelectRowAtIndexPath:	实现行的响应
animatedIn	显示动画
animatedOut	隐藏动画
show	显示具有列表的警告视图
dismiss	让显示的具有列表的警告视图消失
dequeueReusablePopoverCellWithIdentifier:	列表 cell 的重用
popoverCellForRowAtIndexPath:	根据索引路径获取表单元
touchForDismissSelf:	触摸_controlForDismiss 对象使具有列表的警告视图隐藏

这里需要讲解几个重要的方法(其他方法请读者参考源代码)。其中,要实现具有列表的警告视图的绘制,需要使用 initWithFrame:、initTheInterface、tableView: numberOfRowsInSection:、tableView:cellForRowAtIndexPath: 和 dequeueReusablePopover Cell WithIdentifier:方法。initTheInterface 方法实现对界面进行初始化。程序代码如下:

```
- (void)initTheInterface
{
  //对图层进行设置
  self.layer.borderColor = [[UIColor lightGrayColor] CGColor];
  self.layer.borderWidth = 1.0f;
  self.layer.cornerRadius = 10.0f;                              //设置圆角半径
self.clipsToBounds = TRUE;
//创建并设置标签对象,用来显示警告视图的标题
  _titleName = [[UILabel alloc] initWithFrame:CGRectZero];
  self.titleName.font = [UIFont systemFontOfSize:17.0f];        //设置字体
  self.titleName.backgroundColor = [UIColor blackColor];
  self.titleName.textAlignment = NSTextAlignmentCenter;
  self.titleName.textColor = [UIColor whiteColor];              //设置文本的颜色
  CGFloat xWidth = self.bounds.size.width;
  self.titleName.lineBreakMode = NSLineBreakByTruncatingTail;   //设置换行模式
  self.titleName.frame = CGRectMake(0, 0, xWidth, 32.0f);       //设置框架
  [self addSubview:self.titleName];
  //创建并设置表视图对象
  CGRect tableFrame = CGRectMake(0, 32.0f, xWidth, self.bounds. size.
  height-32.0f);
  _mainPopoverListView = [[UITableView alloc] initWithFrame:tableFrame
  style:UITableViewStylePlain];
  self.mainPopoverListView.dataSource = self;
  self.mainPopoverListView.delegate = self;                     //设置委托
  [self addSubview:self.mainPopoverListView];
  //创建并设置控件对象
  _controlForDismiss = [[UIControl alloc] initWithFrame:[UIScreen
  mainScreen].bounds];
```

```
    _controlForDismiss.backgroundColor = [UIColor colorWithRed:.16 green:.17
    blue:.21 alpha:.5];//背景颜色
    [_controlForDismiss addTarget:self action:@selector (touchForDismiss
    Self:) forControlEvents:UIControlEventTouchUpInside];         //添加动作
}
```

tableView:numberOfRowsInSection:方法实现表视图中行数的获取。程序代码如下：

```
- (NSInteger)tableView:(UITableView *)tableView numberOfRowsInSection:
(NSInteger)section
{
    //判断 popoverListView:numberOfRowsInSection:方法是否实现
    if (self.datasource && [self.datasource respondsToSelector: @selector
    (popoverListView:
    numberOfRowsInSection:)])
    {
        return [self.datasource popoverListView:self numberOfRows InSection:
        section];
    }
    return 0;
}
```

tableView:cellForRowAtIndexPath:方法实现获取表视图中某一行的数据。程序代码如下：

```
- (UITableViewCell *)tableView:(UITableView *)tableView cellForRowAtIndex
Path:(NSIndexPath *)indexPath
{
    if (self.datasource && [self.datasource respondsToSelector: @selector
    (popoverListView:
cellForRowAtIndexPath:)])
    {
        return [self.datasource popoverListView:self cellForRowAtIndex Path:
        indexPath];
    }
    return nil;
}
```

如果要实现显示和隐藏具有列表的警告视图，需要使用 animatedIn、animatedOut、show、dismiss 和 touchForDismissSelf:方法。其中，animatedIn 方法实现显示具有列表的警告视图的动画效果。程序代码如下：

```
- (void)animatedIn
{
    self.transform = CGAffineTransformMakeScale(1.3, 1.3);
    self.alpha = 0;
    [UIView animateWithDuration:.35 animations:^{
        self.alpha = 1;                                          //设置透明度
        self.transform = CGAffineTransformMakeScale(1, 1);       //缩放
    }];
}
```

animatedOut 方法实现当该视图隐藏时的动画效果。程序代码如下：

```
- (void)animatedOut
{
    [UIView animateWithDuration:.35 animations:^{
        self.transform = CGAffineTransformMakeScale(1.3, 1.3);
        self.alpha = 0.0;                                        //设置透明度
    } completion:^(BOOL finished) {
```

```
        if (finished) {
            if (self.controlForDismiss)
            {
                [self.controlForDismiss removeFromSuperview];      //移除
            }
            [self removeFromSuperview];                            //移除
        }
    }];
}
```

show 方法实现具有列表的警告视图的显示。程序代码如下：

```
- (void)show
{
    UIWindow *keywindow = [[UIApplication sharedApplication] keyWindow];
    if (self.controlForDismiss)
    {
        [keywindow addSubview:self.controlForDismiss];             //添加视图对象
    }
    [keywindow addSubview:self];
    self.center = CGPointMake(keywindow.bounds.size.width/2.0f, keywindow.bounds.size.height/2.0f);
    [self animatedIn];
}
```

如果想要实现列表的选择响应，需要使用 indexPathForSelectedRow、tableView: didSelectRowAtIndexPath: 、popoverCellForRowAtIndexPath: 和 tableView:did DeselectRow AtIndex Path:方法。其中，tableView:didSelectRowAtIndexPath:方法实现响应选择的行。程序代码如下：

```
- (void)tableView:(UITableView *)tableView didSelectRow AtIndexPath:(NSIndexPath *)indexPath
{
    if (self.delegate && [self.delegate respondsToSelector: @selector
    (popoverListView:
didSelectRowAtIndexPath:)])
    {
        [self.delegate popoverListView:self didSelectRowAtIndexPath: indexPath];
    }
}
```

tableView:didDeselectRowAtIndexPath:方法实现取消选择行的响应，此方法用于第二次或者多次选择列表中的行。程序代码如下：

```
- (void)tableView:(UITableView *)tableView
didDeselectRowAtIndexPath:(NSIndexPath *)indexPath
{
if (self.delegate && [self.delegate respondsToSelector:@selector (popoverListView:
didDeselectRowAtIndexPath:)])
    {
        [self.delegate popoverListView:self didDeselectRowAtIndexPath: indexPath];
    }
}
```

（6）打开 ViewController.h 文件，编写代码，实现头文件、遵守协议以及属性的声明。程序代码如下：

```
#import <UIKit/UIKit.h>
```

```
#import "AlertView.h"                                           //头文件
@interface ViewController : UIViewController<AlertViewDatasource, Alert
ViewDelegate>
@property (nonatomic, retain) NSIndexPath *selectedIndexPath;   //属性
@end
```

（7）打开 ViewController.m 文件，编写代码，该代码实现具有列表的警告视图的显示，为列表的每一行添加图片（此图片是圆形）和内容，以及选择列表中的行后，在此行的前面的圆圈中打对勾（表示此行已选中）。使用的方法如表 11-7 所示。

表 11-7　ViewController.m文件中方法总结

方　　法	功　　能
touchesBegan:withEvent:	触摸实现具有列表的警告视图的显示
popoverListView:numberOfRowsInSection:	获取行数
popoverListView:cellForRowAtIndexPath:	获取某一行的数据
popoverListView:didDeselectRowAtIndexPath:	实现取消响应的行
popoverListView:didSelectRowAtIndexPath:	实现选择行的响应

其中，如果想要通过触摸实现警告视图的显示，需要使用 touchesBegan:withEvent:方法。程序代码如下：

```
- (void)touchesBegan:(NSSet *)touches withEvent:(UIEvent *)event
{
    AlertView *listView = [[AlertView alloc] initWithFrame:CGRectMake(0, 0,
    200, 200)];                                          //实例化警告视图
    listView.titleName.text = @"具有列表的警告视图";
    listView.datasource = self;
    listView.delegate = self;                            //设置委托
    [listView show];
}
```

如果想要实现为列表中的每一行添加图片，需要使用 popoverList View:numberOfRowsInSection:和 popoverListView:cellForRowAtIndexPath:方法。程序代码如下：

```
//获取行数
-           (NSInteger)popoverListView:(AlertView        *)tableView
numberOfRowsInSection:(NSInteger)section
{
    return 25;
}
//获取某一行的数据
- (UITableViewCell *)popoverListView:(AlertView *)tableView cellForRow
AtIndex Path:(NSIndexPath *)indexPath
{
    static NSString *identifier = @"identifier";
    UITableViewCell *cell = [tableView dequeueReusablePopoverCell With
    Identifier:identifier];
    //判断 cell 是否为空
    if (nil == cell)
    {
        cell = [[UITableViewCell alloc] initWithStyle:UITableViewCellStyle
        Default reuseIdentifier:identifier];
    }
    if ( self.selectedIndexPath && NSOrderedSame == [self.selectedIndexPath
    compare:indexPath])
```

```
    {
        cell.imageView.image = [UIImage imageNamed:@"2.png"];    //设置图像
    }
    else
    {
        cell.imageView.image = [UIImage imageNamed:@"1.png"];    //设置图像
    }
    cell.textLabel.text = [NSString stringWithFormat:@"第%d行", index
    Path.row+1];                                                 //文本内容的设置
    return cell;
}
```

实现在选择行的圆圈中打对勾的功能，需要使用 popoverListView:didSelectRowAtIndexPath:和 popoverListView:didDeselectRowAtIndexPath:方法。其中，popoverListView: didSelectRowAtIndexPath:方法实现在选择行的圆圈中打对勾，程序代码如下：

```
- (void)popoverListView:(AlertView *)tableView didSelectRowAtIndexPath:
(NSIndexPath *)indexPath
{
    self.selectedIndexPath = indexPath;
    UITableViewCell *cell = [tableView popoverCellForRowAtIndexPath:
    indexPath];           //获取选择的表单元
    cell.imageView.image = [UIImage imageNamed:@"2.png"];
}
```

popoverListView:didDeselectRowAtIndexPath:方法实现将选择行中的对勾取消，此方法使用在多次选择中。程序代码如下：

```
- (void)popoverListView:(AlertView *)tableView didDeselectRowAtIndexPath:
(NSIndexPath *)indexPath
{
    UITableViewCell *cell = [tableView popoverCellForRowAtIndexPath:
    indexPath];           //实例化对象
    cell.imageView.image = [UIImage imageNamed:@"1.png"];
}
```

【代码解析】

本实例的功能是在选择的行中添加对勾。下面就是这个知识点的详细讲解。

要在选择的行中打对勾，需要使用 UITabViewDelegate 的 tableView:didSelectRowAtIndexPath:方法，在此方法中，它会调用 popoverListView:didSelectRowAtIndexPath:方法。在 popoverListView:didSelect RowAtIndexPath:方法中实现为选中的行添加对勾的功能。代码如下：

```
self.selectedIndexPath = indexPath;
UITableViewCell *cell = [tableView popoverCellForRow AtIndexPath:
indexPath];
cell.imageView.image = [UIImage imageNamed:@"2.png"];
```

实例79　坠落的警告视图

【实例描述】

iOS 7 提供的警告视图只有一种动画效果，为了使警告视图更能引起观众的注意，本实例实现了有一个具有坠落效果的自定义的警告视图。当用户单击"弹出"按钮，会从界面顶部掉下一个警告视图；当用户单击警告视图中的 Canel 按钮后，警告视图会从停止的位置向界面的底部掉落，直到消失。运行效果如图 11.6 所示。

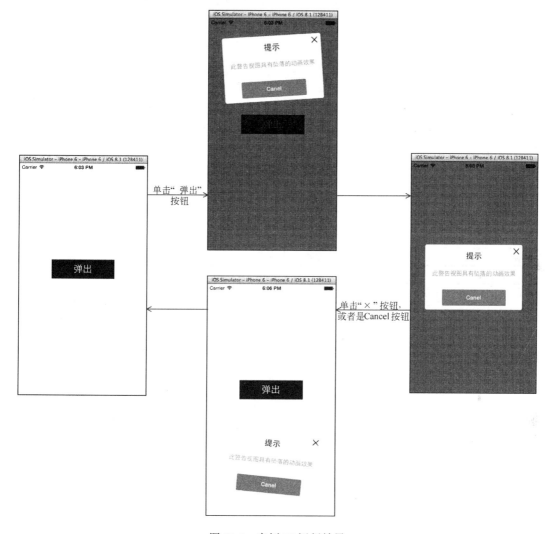

图 11.6　实例 79 运行效果

【程序代码】

（1）创建一个项目，命名为"坠落的警告视图"。
（2）添加图片 1.png 和 2.png 到创建项目的 Supporting Files 文件夹中。
（3）创建一个基于 UIView 类的 AlertView 类。
（4）打开 AlertView.h 文件，编写代码，实现宏定义以及属性以及方法的声明。程序代码如下：

```
#import <UIKit/UIKit.h>
//宏定义
#define kAlertWidth 245.0f
#define kAlertHeight 160.0f
……
```

```
#define kButtonBottomOffset 10.0f
@interface AlertView : UIView
//属性
@property (nonatomic, strong) UILabel *alertTitleLabel;
@property (nonatomic, strong) UILabel *alertContentLabel;
@property (nonatomic, strong) UIButton *Btn;
@property (nonatomic, strong) UIView *backImageView;
//方法
- (void)show;
- (id)initWithTitle:(NSString *)title contentText:(NSString *)content ButtonTitle:(NSString *)Title;
@end
```

（5）打开 AlertView.m 文件，编写代码，实现坠落的警告视图。使用的方法如表 11-8 所示。

表 11-8　AlertView.m文件中方法总结

方　　法	功　　能
initWithTitle:contentText:ButtonTitle:	功能初始化警告视图
BtnClicked:	单击按钮，实现关闭警告视图的功能
show	显示警告视图
dismissAlert	隐藏警告视图
removeFromSuperview	移除
willMoveToSuperview:	在视图移动前调用，将父视图改变为指定的父视图

要实现坠落的警告视图，需要分为两部分：绘制警告视图和警告视图的显示与隐藏。这里需要讲解几个重要的方法（其他方法请读者参考源代码）。其中，要实现警告视图的绘制，需要使用 initWithTitle:contentText:ButtonTitle:方法。程序代码如下：

```
- (id)initWithTitle:(NSString *)title contentText:(NSString *)content ButtonTitle:(NSString *)Title{
    if (self = [super init]) {
        self.layer.cornerRadius = 5.0;
        self.backgroundColor = [UIColor whiteColor];            //设置背景颜色
        //创建并设置标签对象，用来显示警告视图的标题
        self.alertTitleLabel = [[UILabel alloc] initWithFrame:CGRectMake(0, kTitleYOffset, kAlertWidth, kTitleHeight)];
        self.alertTitleLabel.font = [UIFont boldSystemFontOfSize:20.0f];
                                                                //设置字体
        self.alertTitleLabel.textColor = [UIColor colorWithRed:56.0/255.0 green:64.0/255.0 blue:71.0/255.0 alpha:1];
        [self addSubview:self.alertTitleLabel];                 //添加视图对象
        CGFloat contentLabelWidth = kAlertWidth - 16;
        //创建并设置标签栏，用来显示警告视图的信息
        self.alertContentLabel = [[UILabel alloc] initWithFrame:CGRectMake((kAlertWidth - contentLabelWidth) * 0.5, CGRectGetMaxY(self.alertTitleLabel. frame), contentLabelWidth, 60)];
        self.alertContentLabel.numberOfLines = 0;               //设置行数
        self.alertContentLabel.textAlignment = self.alertTitleLabel.textAlignment = NSTextAlignmentCenter;
        self.alertContentLabel.textColor = [UIColor colorWithRed:127.0/255.0 green:127.0/255.0 blue:127.0/255.0 alpha:1];   //设置文本颜色
        self.alertContentLabel.font = [UIFont systemFontOfSize:15.0f];
        [self addSubview:self.alertContentLabel];               //添加视图对象
        //创建并设置按钮对象，作为警告视图的取消按钮
```

```
        CGRect BtnFrame;
        BtnFrame = CGRectMake((kAlertWidth - kSingleButtonWidth) * 0.5,
        kAlertHeight - kButtonBottomOffset - kButtonHeight, kSingle
        ButtonWidth, kButtonHeight);
        self.Btn = [UIButton buttonWithType:UIButtonTypeCustom];
        self.Btn.frame = BtnFrame;                              //设置框架
    }
    [self.Btn setBackgroundColor:[UIColor colorWithRed:227.0/255.0 green:
    100.0/255.0 blue:83.0/255.0 alpha:1]];
    [self.Btn setTitle:Title forState:UIControlStateNormal];
    self.Btn.titleLabel.font=[UIFont boldSystemFontOfSize:14]; //设置字体
    [self.Btn setTitleColor:[UIColor whiteColor] forState: UIControlState
    Normal];
    [self.Btn addTarget:self action:@selector(BtnClicked:) forControl
    Events:UIControlEventTouchUpInside];
    self.Btn.layer.cornerRadius = 3.0;                  //设置圆角半径
    [self addSubview:self.Btn];
    self.alertTitleLabel.text = title;
    self.alertContentLabel.text = content;              //设置文本
    //创建并设置按钮对象
    UIButton *xButton = [UIButton buttonWithType:UIButtonTypeCustom];
    [xButton setImage:[UIImage imageNamed:@"1.png"] forState:UIControl
    StateNormal];
    [xButton setImage:[UIImage imageNamed:@"2.png"] forState:UIControl
    StateHighlighted];
    xButton.frame = CGRectMake(kAlertWidth - 32, 0, 32, 32);  //设置框架
    [self addSubview:xButton];
    [xButton addTarget:self action:@selector(dismissAlert) forControl
    Events: UIControlEventTouchUpInside];
    return self;
}
```

如果要实现警告视图的显示与隐藏功能，需要使用 BtnClicked:、show、dismissAlert、removeFromSuperview 和 willMoveToSuperview:方法。其中，show 方法实现显示警告视图。程序代码如下：

```
- (void)show
{
    UIWindow *shareWindow = [UIApplication sharedApplication].keyWindow;
    self.frame = CGRectMake((CGRectGetWidth(shareWindow.bounds) - kAlert
    Width) * 0.5, - kAlertHeight - 30, kAlertWidth, kAlertHeight);
                                //设置框架
    [shareWindow addSubview:self];
}
```

removeFromSuperview 方法实现以坠落的动画效果删除视图，实现隐藏功能。程序代码如下：

```
- (void)removeFromSuperview
{
    [self.backImageView removeFromSuperview];
    self.backImageView = nil;
    UIWindow *shareWindow = [UIApplication sharedApplication].keyWindow;
    CGRect afterFrame = CGRectMake((CGRectGetWidth(shareWindow.bounds) -
    kAlertWidth) * 0.5, CGRectGetHeight(shareWindow.bounds), kAlertWidth,
    kAlertHeight);
    //坠落的动画效果
    [UIView animateWithDuration:0.35f delay:0.0 options:UIView Animation
```

```objc
    OptionCurveEaseOut animations:^{
        self.frame = afterFrame;                                       //设置框架
        self.transform = CGAffineTransformMakeRotation(M_1_PI / 1.5);//旋转
    } completion:^(BOOL finished) {
        [super removeFromSuperview];                                   //移除
    }];
}
```

willMoveToSuperview:方法在视图移动前调用,将父视图改变为指定的父视图。实现当警告视图隐藏时,以坠落的动画形式隐藏警告视图。程序代码如下:

```objc
- (void)willMoveToSuperview:(UIView *)newSuperview
{
    if (newSuperview == nil) {
        return;
    }
    UIWindow *shareWindow = [UIApplication sharedApplication].keyWindow;
    if (!self.backImageView) {
        self.backImageView = [[UIView alloc] initWithFrame:share Window.
        bounds];                                    //实例化
    }
    self.backImageView.backgroundColor = [UIColor blackColor];   //设置背景颜色
    self.backImageView.alpha = 0.6f;
    [shareWindow addSubview:self.backImageView];                 //添加视图对象
    self.transform = CGAffineTransformMakeRotation(-M_1_PI / 2);
    CGRect afterFrame = CGRectMake((CGRectGetWidth(shareWindow.bounds) -
    kAlertWidth) * 0.5, (CGRectGetHeight(shareWindow.bounds) - kAlertHeight)
    * 0.5, kAlertWidth, kAlertHeight);
    //坠落的动画效果
    [UIView animateWithDuration:0.35f delay:0.0 options: UIView Animation
    OptionCurveEaseIn animations:^{
        self.transform = CGAffineTransformMakeRotation(0);       //设置改变
        self.frame = afterFrame;                                 //设置框架
    } completion:^(BOOL finished) {
    }];
    [super willMoveToSuperview:newSuperview];
}
```

(6) 打开 ViewController.h 文件,编写代码,实现头文件以及动作的声明。程序代码如下:

```objc
#import <UIKit/UIKit.h>
#import "AlertView.h"
@interface ViewController : UIViewController
- (IBAction)show:(id)sender;
@end
```

(7) 打开 Main.storyboard 文件,在视图库中拖动 Button 按钮到界面中,将按钮的背景设置为黑色,将标题设置为"弹出",并将按钮和动作 show:关联。

(8) 打开 ViewController.m 文件,编写代码,实现单击按钮后,以坠落的动画效果隐藏警告视图。程序代码如下:

```objc
- (IBAction)show:(id)sender {
    AlertView *alert = [[AlertView alloc] initWithTitle:@"提示" content
    Text:@"此警告视图具有坠落的动画效果" ButtonTitle:@"Canel" ];  //创建警告视图
    [alert show];
}
```

【代码解析】

本实例的功能是警告视图的坠落动画效果。下面就是这个知识点的详细讲解。

本实例警告视图的坠落动画效果是通过 animateWithDuration:delay:options:animations:completion:方法实现的。在此方法中，对警告视图的旋转角度以及框架进行了重新设置，以显示警告视图的坠落效果为例，代码如下：

```
[UIView    animateWithDuration:0.35f    delay:0.0    options:UIViewAnimation
OptionCurveEaseIn animations:^{
    self.transform = CGAffineTransformMakeRotation(0);
    self.frame = afterFrame;
} completion:^(BOOL finished) {
}];
```

实例 80　弹出视图，模糊界面背景

【实例描述】

为了不让用户分心，起到更好的提醒功能，在弹出视图后，可以将界面除弹出视图以外的东西进行模糊处理。本实例就实现了这么一个功能。当用户单击"警告视图"按钮，就会弹出一个警告视图，并且将界面的背景变模糊；当用户想要还原背景，可以单击"×"按钮，将弹出的警告视图关闭，此时界面的背景就会恢复原来的样子。运行效果如图 11.7 所示。

图 11.7　实例 80 运行效果

【实现过程】

（1）创建一个项目，命名为"弹出视图，模糊背景"。

（2）添加图片 1.png 和 2.png 到创建项目的 Supporting Files 文件夹中。

（3）创建一个基于 UIView 类的分类 ViewSize。

（4）打开 UIView+ViewSize.h 文件，编写代码，实现属性的声明。程序代码如下：

```
#import <UIKit/UIKit.h>
@interface UIView (ViewSize)
//属性
@property (nonatomic) CGFloat left;
@property (nonatomic) CGFloat top;
……
@property (nonatomic) CGSize size;
@end
```

（5）打开 UIView+ViewSize.m 文件，编写代码，实现对视图尺寸的设置和获取。使用的方法如表 11-9 所示。

表 11-9　UIView+ViewSize.m文件中方法总结

方　　法	功　　能
left	获取视图左边位置的 x 值
setLeft:	设置视图左边位置的 x 值
top	获取视图顶部位置的 y 值
setTop:	设置视图顶部位置的 y 值
right	获取视图右边位置的 x 值
setRight:	设置视图右边位置的 x 值
bottom	获取视图底部位置的 y 值
setBottom:	设置视图底部位置的 y 值
width	获取视图的宽度
setWidth:	设置视图的宽度
height	获取视图的高度
setHeight:	设置视图的高度
origin	获取视图的位置
setOrigin:	设置视图的位置
size	获取视图的尺寸
setSize:	设置视图的尺寸

（6）创建一个基于 UILabel 类的分类 LabelSize。

（7）打开 UILabel+LabelSize.h 文件，编写代码，实现方法的声明。程序代码如下：

```
#import <UIKit/UIKit.h>
@interface UILabel (LabelSize)
- (void)autoHeight ;
@end
```

（8）打开 UILabel+LabelSize.m 文件，编写代码，为声明的方法进行定义，实现自动调整标签大小的功能。程序代码如下：

```
- (void)autoHeight {
    CGRect frame = self.frame;
    CGSize maxSize = CGSizeMake(frame.size.width, 9999);
    NSMutableParagraphStyle           *style=[[NSMutableParagraphStyle defaultParagraphStyle]mutableCopy];
    [style setLineBreakMode:self.lineBreakMode];                    //设置换行模式
NSDictionary                                           *dis=[NSDictionary dictionaryWithObjectsAndKeys:self.font,NSFontAttributeName,style,NSParagraphStyleAttributeName,nil];
    CGSize expectedSize =[self.text boundingRectWithSize:maxSize options:NSStringDrawingUsesLineFragmentOrigin  attributes:dis context:nil].size;
    frame.size.height = expectedSize.height;                        //
```

第11章 提醒对话框

```
获取高度
    [self setFrame:frame];
}
```

（9）创建一个基于 UIImage 类的分类 Blur。

（10）打开 UIImage+Blur.h 文件，编写代码，实现头文件和方法的声明。程序代码如下：

```
#import <UIKit/UIKit.h>
#import <Accelerate/Accelerate.h>
@interface UIImage (Blur)
-(UIImage *)boxblurImageWithBlur:(CGFloat)blur;
@end
```

（11）打开 UIImage+Blur.m 文件，编写代码，为声明的方法进行初始化。实现使用模糊度模糊图像。程序代码如下：

```
-(UIImage *)boxblurImageWithBlur:(CGFloat)blur {
    //模糊度
    if (blur < 0.f || blur > 1.f) {
        blur = 0.5f;
    }
    int boxSize = (int)(blur * 50);
    boxSize = boxSize - (boxSize % 2) + 1;
    CGImageRef img = self.CGImage;
    vImage_Buffer inBuffer, outBuffer;               //图像缓存，输入缓存，输出缓存
    vImage_Error error;
    void *pixelBuffer;                               //像素缓存
    CGDataProviderRef inProvider = CGImageGetDataProvider(img);
    CFDataRef inBitmapData = CGDataProviderCopyData(inProvider);
    inBuffer.width = CGImageGetWidth(img);           //设置数据
    inBuffer.height = CGImageGetHeight(img);         //设置宽
    inBuffer.rowBytes = CGImageGetBytesPerRow(img);  //设置高度
    inBuffer.data = (void*)CFDataGetBytePtr(inBitmapData);   //设置字节
    pixelBuffer = malloc(CGImageGetBytesPerRow(img) * CGImageGetHeight(img));
    if(pixelBuffer == NULL)
        NSLog(@"No pixelbuffer");
    //设置数据、宽度、高度和字节
    outBuffer.data = pixelBuffer;
    outBuffer.width = CGImageGetWidth(img);
    outBuffer.height = CGImageGetHeight(img);        //获取高度
    outBuffer.rowBytes = CGImageGetBytesPerRow(img);
    error = vImageBoxConvolve_ARGB8888(&inBuffer, &outBuffer, NULL, 0, 0,boxSize, boxSize, NULL, kvImageEdgeExtend);
    //判断错误
    if (error) {
        NSLog(@"error from convolution %ld", error);  //输出
    }
    //创建颜色空间
    CGColorSpaceRef colorSpace = CGColorSpaceCreateDeviceRGB();
    //用图片创建上下文
    CGContextRef ctx = CGBitmapContextCreate(outBuffer.data, outBuffer.width,outBuffer.height,8,outBuffer.rowBytes,colorSpace, kCGImageAlphaNoneSkipLast);
    CGImageRef imageRef = CGBitmapContextCreateImage (ctx);  //创建位图图形上下文
    UIImage *returnImage = [UIImage imageWithCGImage:imageRef];
    //清空
    CGContextRelease(ctx);                                    //释放图形上下文
    CGColorSpaceRelease(colorSpace);
```

```
    free(pixelBuffer);
    CFRelease(inBitmapData);
    CGColorSpaceRelease(colorSpace);                    //释放色彩空间
    CGImageRelease(imageRef);
    return returnImage;
}
```

（12）创建一个基于 UIView 类的分类 Screenshot.。

（13）打开 UIView+ Screenshot..h 文件，编写代码，实现方法的声明。程序代码如下：

```
#import <UIKit/UIKit.h>
@interface UIView (Screenshot)
- (UIImage*)screenshot;
@end
```

（14）打开 UIView+ Screenshot.m 文件，编写代码，为声明的方法进行定义，实现屏幕截图。程序代码如下：

```
- (UIImage*)screenshot {
    UIGraphicsBeginImageContext(self.bounds.size);//创建一个基于位图的上下文
    [self.layer renderInContext:UIGraphicsGetCurrentContext()];
                                //视图绘制到图像上下文中
    UIImage *image = UIGraphicsGetImageFromCurrentImageContext();
                    //从上下文获取新的图像
    UIGraphicsEndImageContext();
    NSData *imageData = UIImageJPEGRepresentation(image, 1);
    image = [UIImage imageWithData:imageData];
    return image;
}
```

（15）创建一个基于 UIImageView 类的 BlurView 类。

（16）打开 BlurView.h 文件，编写代码，实现头文件、对象以及方法的声明。程序代码如下：

```
#import <UIKit/UIKit.h>
//头文件
#import "UIImage+Blur.h"
#import "UIView+Screenshot.h"
@interface BlurView : UIImageView{
    UIView *_coverView;                             //对象
}
- (id)initWithCoverView:(UIView*)view;              //方法
@end
```

（17）打开 BlurView.m 文件，编写代码，为声明的方法进行定义，实现模糊图像的初始化。程序代码如下：

```
- (id)initWithCoverView:(UIView *)view {
    if (self = [super initWithFrame:CGRectMake(0, 0, view.bounds.size.
    width,view.bounds.size.height)]) {
        CGRect frame;
        frame= view.frame;
        _coverView = view;
        UIImage *blur = [_coverView screenshot];
        self.image = [blur boxblurImageWithBlur:0.2];           //设置图像
    }
    return self;
}
```

（18）创建一个基于 UIButton 类的 CloseButton 类。

（19）打开 CloseButton.m 文件，编写代码，实现关闭按钮的绘制。在此文件中，代码分为了两个部分：初始化和绘制。其中，init 方法实现了关闭按钮的初始化。程序代码如下：

```
- (id)init{
    //判断
    if(!(self = [super initWithFrame:(CGRect){0, 0, 32, 32}])){
        return nil;
    }
    static UIImage *closeButtonImage;
    static dispatch_once_t once;
    dispatch_once(&once, ^{
        closeButtonImage = [self closeButtonImage];          //实例化对象
    });
    [self setBackgroundImage:closeButtonImage forState:UIControl StateNormal];
                                                              //设置背景图像
    return self;
}
```

closeButtonImage 方法实现了关闭按钮的绘制。程序代码如下：

```
- (UIImage *)closeButtonImage{
    UIGraphicsBeginImageContextWithOptions(self.bounds.size, NO, 0);
    CGColorSpaceRef colorSpace = CGColorSpaceCreateDeviceRGB();
CGContextRef context = UIGraphicsGetCurrentContext();
//颜色
    UIColor *topGradient = [UIColor colorWithRed:0.21 green:0.21 blue:0.21 alpha:0.9];
UIColor *bottomGradient = [UIColor colorWithRed:0.03 green:0.03 blue:0.03 alpha:0.9];
//渐变
    NSArray         *gradientColors        =         @[(id)topGradient.CGColor,
(id)bottomGradient.CGColor];
    CGFloat gradientLocations[] = {0, 1};
CGGradientRef gradient = CGGradientCreateWithColors(colorSpace, (__bridge CFArrayRef)gradientColors, gradientLocations);
//阴影
    CGColorRef shadow = [UIColor blackColor].CGColor;
    CGSize shadowOffset = CGSizeMake(0, 1);
    CGFloat shadowBlurRadius = 3;
    CGColorRef shadow2 = [UIColor blackColor].CGColor;
    CGSize shadow2Offset = CGSizeMake(0, 1);
CGFloat shadow2BlurRadius = 0;
//椭圆的绘制
    UIBezierPath             *ovalPath             =             [UIBezierPath
bezierPathWithOvalInRect:CGRectMake(4, 3, 24, 24)];
    CGContextSaveGState(context);
    [ovalPath addClip];
    CGContextDrawLinearGradient(context, gradient, CGPointMake(16, 3),
CGPointMake(16, 27), 0);
    CGContextRestoreGState(context);
    CGContextSaveGState(context);                             //保存
    CGContextSetShadowWithColor(context, shadowOffset, shadowBlurRadius,
shadow);
    [[UIColor whiteColor] setStroke];
    ovalPath.lineWidth = 2;                                   //设置线宽
    [ovalPath stroke];
CGContextRestoreGState(context);
```

```
//关闭按钮中图标的绘制
    UIBezierPath *bezierPath = [UIBezierPath bezierPath];
    [bezierPath moveToPoint:CGPointMake(22.36, 11.46)];     //设置开始点
    [bezierPath addLineToPoint:CGPointMake(18.83, 15)];
    [bezierPath addLineToPoint:CGPointMake(22.36, 18.54)];
    [bezierPath addLineToPoint:CGPointMake(19.54, 21.36)];
    [bezierPath addLineToPoint:CGPointMake(16, 17.83)];
    [bezierPath addLineToPoint:CGPointMake(12.46, 21.36)];  //设置结束点
    [bezierPath addLineToPoint:CGPointMake(9.64, 18.54)];
    [bezierPath addLineToPoint:CGPointMake(13.17, 15)];
    [bezierPath addLineToPoint:CGPointMake(9.64, 11.46)];
    [bezierPath addLineToPoint:CGPointMake(12.46, 8.64)];
    [bezierPath addLineToPoint:CGPointMake(16, 12.17)];
    [bezierPath addLineToPoint:CGPointMake(19.54, 8.64)];
    [bezierPath addLineToPoint:CGPointMake(22.36, 11.46)];
    [bezierPath closePath];                                 //关闭路径
    CGContextSaveGState(context);
    CGContextSetShadowWithColor(context, shadow2Offset, shadow2BlurRadius, shadow2);
    [[UIColor whiteColor] setFill];
    [bezierPath fill];                                      //填充路径
    CGContextRestoreGState(context);
    CGGradientRelease(gradient);
    CGColorSpaceRelease(colorSpace);                        //释放色彩空间
    UIImage *image = UIGraphicsGetImageFromCurrentImageContext();
    UIGraphicsEndImageContext();
    return image;
}
```

（20）创建一个基于 UIView 类的 AlertView 类。

（21）打开 AlertView.h 文件，编写代码，实现头文件、定义新类型、对象、实例变量、属性以及方法的声明。程序代码如下：

```
#import <UIKit/UIKit.h>
//头文件
#import "CloseButton.h"
#import "BlurView.h"
#import "UILabel+LabelSize.h"
#import "UIView+ViewSize.h"
typedef void (^RNBlurCompletion)(void);                     //新类型的
定义
@interface AlertView : UIView{
    //对象
    UIViewController *_controller;
    UIView *_contentView;
    CloseButton *_dismissButton;
    BlurView *_blurView;
    RNBlurCompletion _completion;                           //实例变量
}
//属性
@property (nonatomic) BOOL isVisible;
@property (assign) CGFloat animationDuration;
@property (assign) CGFloat animationDelay;
```

```
@property (assign) UIViewAnimationOptions animationOptions;
//方法
- (id)initWithViewController:(UIViewController*)viewController view:(UIView*)view;
-   (id)initWithViewController:(UIViewController*)viewController  title:
(NSString*)title message:(NSString*)message;
- (void)show;                                              //显示
-   (void)showWithDuration:(CGFloat)duration  delay:(NSTimeInterval)delay
options:
(UIViewAnimationOptions)options completion:(void (^)(void))completion;
- (void)hide;                                              //隐藏
-   (void)hideWithDuration:(CGFloat)duration  delay:(NSTimeInterval)delay
options:(
UIViewAnimationOptions)options completion:(void (^)(void))completion;
@end
```

（22）打开 AlertView.m 文件，编写代码，实现警告视图的绘制，以及弹出警告视图时，界面背景变得模糊。使用的方法如表 11-10 所示。

表 11-10　AlertView.m文件中方法总结

方　　法	功　　能
generateModalViewWithTitle:message:	创建视图
initWithFrame:	使用框架进行初始化
initWithViewController:view:	使用视图控制器和视图进行初始化
initWithViewController:title:	使用视图控件器和标题进行初始化
willMoveToSuperview:	在视图移动前调用，将父视图改变为指定的父视图
show	显示
showWithDuration:delay:options:completion:	使用持续时间、延时、选项和完成处理进行显示
delayedShow	延迟显示
hide	隐藏
hideWithDuration:delay:options:completion:	使用持续时间、延时、选项和完成处理进行隐藏

这里需要讲解几个重要的方法（其他方法请读者参考源代码）。如果要实现警告视图的绘制，需要使用 generateModalViewWithTitle:message:、initWithFrame:、initWithViewController:view: 和 initWithView Controller: title: 方法。其中 generateModal ViewWith Title:message:方法用于创建警告视图，程序代码如下：

```
+ (UIView*)generateModalViewWithTitle:(NSString*)title message:(NSString*)
message {
    CGFloat defaultWidth = 280.f;
    CGRect frame = CGRectMake(0, 0, defaultWidth, 0);
    CGFloat padding = 10.f;
    //创建并设置空白视图对象，作为警告视图的背景
    UIView *view = [[UIView alloc] initWithFrame:frame];
    UIColor *whiteColor = [UIColor colorWithRed:0.816 green:0.788 blue:0.788
    alpha:1.000];
    view.backgroundColor = [UIColor colorWithWhite:0.1 alpha:0.8f];
    view.layer.borderColor = whiteColor.CGColor;             //设置边框颜色
    view.layer.borderWidth = 2.f;
    view.layer.cornerRadius = 10.f;                          //设置圆角半径
    //创建并设置标签对象，用来显示警告视图的标题
```

```
    UILabel *titleLabel = [[UILabel alloc] initWithFrame:CGRectMake(padding,
0, defaultWidth - padding * 2.f, 0)];
    titleLabel.text = title;
    titleLabel.font = [UIFont fontWithName:@"HelveticaNeue-Bold" size:17.f];
    titleLabel.textColor = [UIColor whiteColor];              //设置文本颜色
    titleLabel.shadowColor = [UIColor blackColor];
    titleLabel.shadowOffset = CGSizeMake(0, -1);
    titleLabel.textAlignment = NSTextAlignmentCenter;         //设置对齐方式
    titleLabel.backgroundColor = [UIColor clearColor];
    [titleLabel autoHeight];
    titleLabel.top = padding;
    [view addSubview:titleLabel];                             //添加对象
    //创建并设置标签对象,用来显示警告视图的信息
    UILabel *messageLabel = [[UILabel alloc] initWithFrame:CGRectMake
    (padding, 0, defaultWidth - padding * 2.f, 0)];
    messageLabel.text = message;
    messageLabel.numberOfLines = 0;                           //设置行数
    messageLabel.font = [UIFont fontWithName:@"HelveticaNeue" size:17.f];
    messageLabel.textColor = titleLabel.textColor;
    messageLabel.shadowOffset = titleLabel.shadowOffset;      //设置阴影偏移量
    messageLabel.shadowColor = titleLabel.shadowColor;
    messageLabel.textAlignment = NSTextAlignmentCenter;
    messageLabel.backgroundColor = [UIColor clearColor];      //设置背景颜色
    [messageLabel autoHeight];
    messageLabel.top = titleLabel.bottom + padding;
    [view addSubview:messageLabel];                           //添加视图对象
    view.height = messageLabel.bottom + padding;
    return view;
}
```

initWithFrame:方法实现使用框架进行初始化。程序代码如下:

```
- (id)initWithFrame:(CGRect)frame {
if (self = [super initWithFrame:frame]) {
    //创建并设置关闭按钮
        _dismissButton = [[CloseButton alloc] init];
        _dismissButton.center = CGPointZero;                  //设置中心位置
        [_dismissButton        addTarget:self       action:@selector(hide)
forControlEvents:UIControlEventTouchUpInside];
        self.alpha = 0.f;                                     //设置透明度
        self.backgroundColor = [UIColor clearColor];
        self.autoresizingMask = (UIViewAutoresizingFlexibleWidth |UIView
        Autoresizing FlexibleHeight |UIViewAutoresizingFlexibleLeftMargin
        |UIViewAutoresizingFlexibleTopMargin);
    }
    return self;
}
```

initWithViewController:view:方法实现初始化视图控制器,并对视图进行初始化。程序代码如下:

```
- (id)initWithViewController:(UIViewController*)viewController view: (UI
View*)view {
```

```
    if (self = [self initWithFrame:CGRectMake(0, 0, viewController.view.
    width, viewController.view.height)]) {
        [self addSubview:view];
        _contentView = view;
        //设置中心位置
        _contentView.center = CGPointMake(CGRectGetMidX(self.frame), CGRect
        GetMidY(self.frame));
        _controller = viewController;
        _contentView.clipsToBounds = YES;
        _contentView.layer.masksToBounds = YES;
        _dismissButton.center = CGPointMake(view.left, view.top);  //设置中心位置
        [self addSubview:_dismissButton];
    }
    return self;
}
```

如果想要在弹出警告视图时，使界面背景模糊，需要使用 willMoveToSuperview:、show、showWithDuration:delay:options:completion:、delayedShow、delayedShow、hide 和 hideWithDuration:delay: options:completion:方法。willMoveToSuperview:方法在视图移动前调用，将父视图改变为指定的父视图。程序代码如下：

```
- (void)willMoveToSuperview:(UIView *)newSuperview {
    [super willMoveToSuperview:newSuperview];
    if (newSuperview) {
        self.center = CGPointMake(CGRectGetMidX(newSuperview.frame), CGRect
        GetMidY(newSuperview.frame));                //设置中心位置
    }
}
```

showWithDuration:delay:options:completion:方法实现警告视图以及背景模糊的显示。程序代码如下：

```
- (void)showWithDuration:(CGFloat)duration  delay:(NSTimeInterval)delay
options:(
UIViewAnimationOptions)options
 completion:(void (^)(void))completion {
    self.animationDuration = duration;               //设置动画执行所需时间
    self.animationDelay = delay;
    self.animationOptions = options;
    _completion = [completion copy];
    [self performSelector:@selector(delayedShow) withObject:nil afterDelay:
    kRNBlurDefaultDelay];
}
```

delayedShow 方法实现延时警告视图以及背景模糊的显示。程序代码如下：

```
- (void)delayedShow {
    if (! self.isVisible) {
        if (! self.superview) {
            [_controller.view addSubview:self];        //显示警告视图
            self.top = 0;
        }
        _blurView = [[BlurView alloc] initWithCoverView:_controller.view];
        _blurView.alpha = 0.f;                         //设置透明度
        [_controller.view insertSubview:_blurView belowSubview:self];
        self.transform = CGAffineTransformScale(CGAffineTransformIdentity,
```

```
    0.4, 0.4);                                         //缩放
    //动画效果
    [UIView animateWithDuration:self.animationDuration animations:^{
        _blurView.alpha = 1.f;                          //设置透明度
        self.alpha = 1.f;
        self.transform = CGAffineTransformScale(CGAffineTransf ormIden
        tity, 1.f, 1.f);                                //缩放
    } completion:^(BOOL finished) {
        if (finished) {
            [[NSNotificationCenter defaultCenter] postNotificationName:
             kRNBlurDidShowNotification object:nil];
            self.isVisible = YES;                       //可见
            if (_completion) {
                _completion();
            }
        }
    }];
}
```

hideWithDuration:delay:options:completion:方法实现警告视图以及背景模糊的隐藏。程序代码如下：

```
- (void)hideWithDuration:(CGFloat)duration delay:(NSTimeInterval)delay
options:
(UIViewAnimationOptions)options completion:(void (^)(void))completion {
    if (self.isVisible) {
    //动画效果
        [UIView animateWithDuration:duration delay:delay options:op
        tionsanimations:^{
                        self.alpha = 0.f;
                        _blurView.alpha = 0.f;          //设置透明度
        }
        completion:^(BOOL finished){
            if (finished) {
                [_blurView removeFromSuperview];        //移除
                _blurView = nil;
                [self removeFromSuperview];             //移除
                [[NSNotificationCenter defaultCenter] post
                NotificationName:
kRNBlurDidHidewNotification object:nil];
                self.isVisible = NO;
                if (completion) {
                    completion();
                }
            }
        }];
    }
}
```

（23）打开 ViewController.h 文件，编写代码，实现头文件和动作的声明。程序代码如下：

```
#import <UIKit/UIKit.h>
#import "AlertView.h"                                   //头文件
@interface ViewController : UIViewController
- (IBAction)show:(id)sender;                            //方法
@end
```

（24）打开 Main.storyboard 文件，对 View Controller 视图控制器的设计界面进行设计，

效果如图 11.8 所示。

图 11.8 实例 80View Controller 视图控制器的设计界面

需要添加的视图、控件以及对它们的设置如表 11-11 所示。

表 11-11 实例 80 视图、控件设置

视图、控件	属 性 设 置	其 他
Image View	Image：1.jpg	
Button	Title：警告视图 Text Color：黑色 Background：2.png	与动作 show:关联

（25）打开 ViewController.m 文件，编写代码。实现单击按钮后，弹出警告视图，并且界面的背景会模糊。程序代码如下：

```
- (IBAction)show:(id)sender {
    BOOL useCustomView = NO;
    AlertView *modal;
    if (useCustomView) {
        UIView *view = [[UIView alloc] initWithFrame:CGRectMake(0, 0, 100, 100)];
        view.backgroundColor = [UIColor redColor];
        view.layer.cornerRadius = 5.f;                          //设置圆角半径
        view.layer.borderColor = [UIColor blackColor].CGColor;
        view.layer.borderWidth = 5.f;                           //设置边框宽度
        modal = [[AlertView alloc] initWithViewController:self view:view];
    }
    else {
        modal = [[AlertView alloc] initWithViewController:self title:@"提
            示" message:@"弹出警告视图背景就会变模糊"];
    }
    [modal show];                                               //显示
}
```

【代码解析】

由于本实例中的代码和方法非常的多，为了方便读者的阅读，笔者绘制了一些执行流

程图,如图 11.9~11.10 所示。其中,单击运行按钮后一直到弹出按钮,并将界面的背景变模糊,使用了 showA:、initWithViewController:title:、generateModalViewWithTitle:message:、autoHeight、setTop:、autoHeight、bottom、setHeight:、initWithViewController:view:、width、height、initWithFrame:、init、closeButtonImage、left、top、show、showWithDuration:delay:options:completion:、delayedShow、willMoveToSuperview:、initWithCoverView:、screenshot 和 boxblurImageWithBlur:方法共同实现。它们的执行流程如图 11.9 所示。

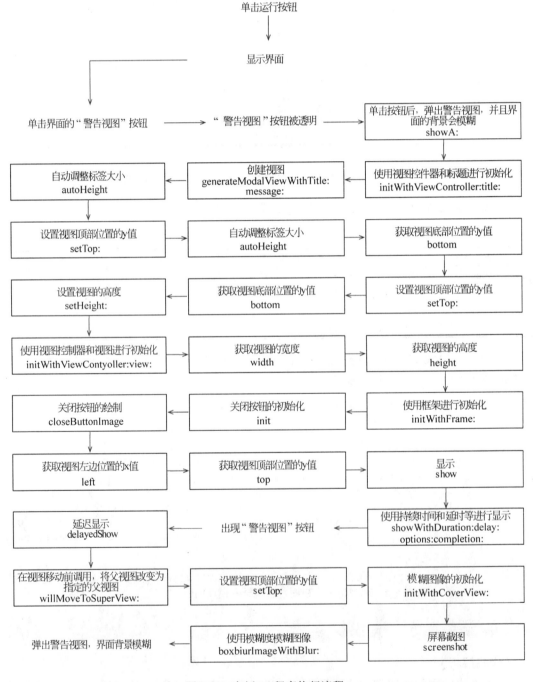

图 11.9　实例 80 程序执行流程 1

如果想要实现关闭警告视图的功能，需要单击警告视图上的按钮。使用了 hide、hideWithDuration:delay:options:completion:和 willMoveToSuperview:方法共同实现。它们的执行流程如图 11.10 所示。

图 11.10　实例 80 程序执行流程 2

第 12 章 文本处理

在 iOS 中，文本是展现信息最简洁明了的方式。iOS 为文本信息的处理提供了很多组件，如常见的文本框、标签和文本视图。本章将详细讲解文本处理的各种经典应用。

实例 81 具有多个颜色的标签

【实例描述】
很多的人都会有这样的问题：怎么在一个标签中显示多个格式呢？对于这个问题，本实例就会为大家解决。本实例实现在一个标签中显示多个颜色的功能，读者可以根据这个功能实现自己想要的格式效果。运行效果如图 12.1 所示。

图 12.1 实例 81 运行效果

【实现过程】
（1）创建一个项目，命名为"具有多个颜色的标签"。
（2）创建一个基于 UILabel 类的 Label 类。

(3) 打开 Label.h 文件，编写代码，实现头文件、对象、属性以及方法的声明。程序代码如下：

```
#import <UIKit/UIKit.h>
#import <CoreText/CoreText.h>
#import <QuartzCore/QuartzCore.h>
@interface Label : UILabel{
    NSMutableAttributedString *_attString;         //声明对象
}
@property (nonatomic,retain)NSMutableAttributedString *attString;
- (void)setColor:(UIColor *)color fromIndex:(NSInteger)location length:
(NSInteger)length;//设置某段字的颜色
- (void)setFont:(UIFont *)font fromIndex:(NSInteger)location length:
(NSInteger)length;
- (void)setStyle:(CTUnderlineStyle)style  fromIndex:(NSInteger)location
length:(NSInteger)length;
@end
```

(4) 打开 Label.m 文件，编写代码，实现具有多个颜色的标签效果。使用的方法如表 12-1 所示。

表 12-1 Label.m文件

方　　法	功　　能
drawRect:	创建并设置文本图层
setText:	设置文本
setColor:fromIndex:length:	设置某段字的颜色
setFont:fromIndex:length:	设置某段字的字体
setStyle:fromIndex:length:	设置某段字的风格

其中，drawRect:方法实现对文本图层的创建与设置。程序代码如下：

```
- (void)drawRect:(CGRect)rect{
    CATextLayer *textLayer = [CATextLayer layer];       //创建对象
    textLayer.string = _attString;                       //设置文本
    textLayer.frame = CGRectMake(0, 0, self.frame.size.width, self.frame.
    size.height);                                        //设置框架
    [self.layer addSublayer:textLayer];
}
```

setText:方法实现对文本的设置。程序代码如下：

```
- (void)setText:(NSString *)text{
    [super setText:text];                                //设置文本
    if (text == nil) {
        self.attString = nil;
    }else{
        self.attString = [[NSMutableAttributedString alloc] initWithString:
        text] ;                                          //实例化对象
    }
}
```

setColor:fromIndex:length:方法为某段字的颜色进行设置。程序代码如下：

```
- (void)setColor:(UIColor *)color fromIndex:(NSInteger)location length:
(NSInteger)length{
    if (location < 0||location>self.text.length-1||length+location>self.
```

```
    text.length) {
        return;
    }
}
//设置字体属性
    [_attString addAttribute:(NSString *)kCTForegroundColorAttributeName
                    value:(id)color.CGColor
                    range:NSMakeRange(location, length)];
}
```

setFont:fromIndex:length:方法为某段字的字体进行设置。程序代码如下:

```
- (void)setFont:(UIFont *)font fromIndex:(NSInteger)location length:
(NSInteger)length{
    if (location < 0||location>self.text.length-1||length+location>self.
    text.length) {
        return;
    }
}
//设置字体属性
    [_attString addAttribute:(NSString *)kCTFontAttributeName
                value:(id)CFBridgingRelease(CTFontCreateWithName
                ((CFStringRef)font.fontName,
                                        font.pointSize,
                                        NULL))
                range:NSMakeRange(location, length)];
}
```

setStyle: fromIndex:length:方法实现对某段字的风格进行设置。程序代码如下:

```
- (void)setStyle:(CTUnderlineStyle)style  fromIndex:(NSInteger)location
length:(NSInteger)
    length{
    if                         (location                         <
0||location>self.text.length-1||length+location>self.text.length) {
        return;
    }
}
//设置字体属性
    [_attString addAttribute:(NSString *)kCTUnderlineStyleAttributeName
                    value:(id)[NSNumber numberWithInt:style]
                    range:NSMakeRange(location, length)];
}
```

(5) 打开 Main.storyboard 文件，将设计界面的颜色设置为浅灰色。

(6) 打开 ViewController.m 文件，编写代码，实现具有多个颜色的标签的创建以及设置，并显示。程序代码如下:

```
- (void)viewDidLoad
{
    [super viewDidLoad];
    // Do any additional setup after loading the view, typically from a nib.
    Label *label = [[Label alloc] initWithFrame:CGRectMake(30, 100, 250,
    80)];
    label.text = @"this is Label";
    [self.view addSubview:label];
    //设置颜色
    [label setColor:[UIColor redColor] fromIndex:0 length:4];
    [label setColor:[UIColor yellowColor] fromIndex:6 length:2];
```

```
    //设置字体
    [label setFont:[UIFont boldSystemFontOfSize:60] fromIndex:0 length:4];
    [label setFont:[UIFont boldSystemFontOfSize:30] fromIndex:6 length:2];
    [label setFont:[UIFont boldSystemFontOfSize:45] fromIndex:10
    length:5];
    [label setStyle:kCTUnderlineStyleDouble fromIndex:0 length:4];
                                                        //设置风格
}
```

【代码解析】

本实例关键功能是对颜色、字体和风格的设置。下面就是这个知识点的详细讲解。

如果想要设置颜色、字体和风格，需要使用 NSMutableAttributedString 的 addAttribute:value:range:方法。其语法形式如下：

```
- (void)addAttribute:(NSString *)name value:(id)value range:(NSRange)
aRange;
```

其中，(NSString *)name 表示属性名称；(id)value 表示属性值；(NSRange)aRange 表示用于属性/值对的字符范围。属性名称的种类很多，如表 12-2 所示。

表 12-2 属性名称

方　　法	功　　能
kCTCharacterShapeAttributeName	字体形状属性
kCTFontAttributeName	字体属性
kCTKernAttributeName	字符间隔属性
kCTLigatureAttributeName	设置是否使用连字属性
kCTForegroundColorAttributeName	字体颜色属性
kCTForegroundColorFromContextAttributeName	上下文的字体颜色属性
kCTParagraphStyleAttributeName	段落样式属性
kCTStrokeWidthAttributeName	笔画线条宽度
kCTStrokeColorAttributeName	笔画的颜色属性
kCTSuperscriptAttributeName	设置字体的上下标属性
kCTUnderlineColorAttributeName	字体下划线颜色属性
kCTUnderlineStyleAttributeName	字体下划线样式属性
kCTVerticalFormsAttributeName	文字的字形方向属性
kCTGlyphInfoAttributeName	字体信息属性
kCTRunDelegateAttributeName	CTRun 委托属性

在本实例中就是使用了 addAttribute:value:range:方法对颜色、字体和风格进行了设置。以设置字体颜色为例，代码如下：

```
[_attString addAttribute:(NSString *)kCTForegroundColorAttributeName
            value:(id)color.CGColor
            range: NSMakeRange(location, length)];
```

其中，(NSString *)kCTForegroundColorAttributeName 表示属性名称，它是字体颜色的属性；(id)color.CGColor 表示与属性名关联的属性值；NSMakeRange(location, length)表示用于属性/值对的字符范围。

实例 82 发光的标签

【代码解析】
本实例实现的功能是让标签中的文本具有发光的效果。其中，用于用户输入的是文本框。运行效果如图 12.2 所示

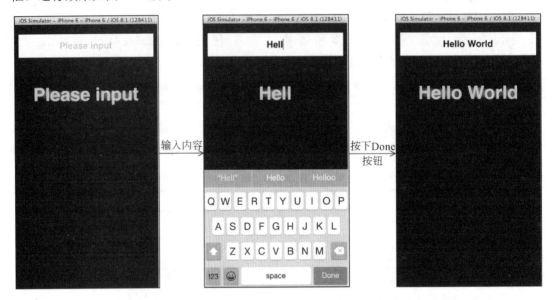

图 12.2 实例 82 运行效果

【实现过程】
（1）创建一个项目，命名为"发光的标签"。
（2）创建一个基于 UILabel 类的 GlowLabel。
（3）打开 GlowLabel.h 文件，编写代码，实现属性的声明。程序代码如下：

```
#import <UIKit/UIKit.h>
@interface GlowLabel : UILabel
//属性
@property (nonatomic, assign) CGFloat glowSize;
@property (nonatomic) UIColor *glowColor;
@end
```

（4）打开 GlowLabel.m 文件，编写代码，实现会发光的标签效果。使用的方法如表 12-3 所示。

表 12-3 GlowLabel.m 文件中方法总结

方法	功能
initWithFrame:	初始化会发光的标签
setup	设置
draw:	绘制发光的标签
drawTextInRect:	将文本绘制到指定的矩形中

这里需要讲解几个重要的方法（其他方法请读者参考源代码）。其中，setup 方法对一些属性进行设置。程序代码如下：

```objc
- (void)setup
{
    //为变量赋值
    self.glowSize = 0.0f;
    self.glowColor = [UIColor clearColor];
}
```

draw:方法实现绘制发光的标签。程序代码如下：

```objc
- (void)draw:(CGRect)rect
{
    CGContextRef ctx = UIGraphicsGetCurrentContext();        //创建图形上下文
    UIGraphicsBeginImageContextWithOptions(rect.size, NO, 0.0);
    [super drawTextInRect:rect];                              //绘制
    UIImage *textImage = UIGraphicsGetImageFromCurrentImageContext();
    UIGraphicsEndImageContext();
    CGContextSaveGState(ctx);                                 //保存
    if (_glowSize > 0) {
        CGContextSetShadowWithColor(ctx, CGSizeZero, _glowSize, _glowColor.CGColor);//设置颜色
    }
    [textImage drawAtPoint:rect.origin];
    CGContextRestoreGState(ctx);
}
```

（5）打开 Main.storyboard 文件，将设计界面的颜色设置为黑色。

（6）打开 ViewController.h 文件，编写代码，实现头文件、宏定义以及属性的声明。程序代码如下：

```objc
#import <UIKit/UIKit.h>
#import "GlowLabel.h"                                        //头文件
//宏定义
#define UIColorFromRGB(rgbValue) [UIColor colorWithRed:((float)(((rgbValue) & 0xFF0000) >> 16))/255.0 green:((float)(((rgbValue) & 0xFF00) >> 8))/255.0 blue:((float)((rgbValue) & 0xFF))/255.0 alpha:1.0]
#define kInitialText @"Please input"
@interface ViewController : UIViewController<UITextFieldDelegate>
//属性
@property (nonatomic) GlowLabel *label;
@property (nonatomic) UITextField *field;
@end
```

（7）打开 ViewController.m 文件，编写代码，实现发光标签的显示。当在文本框中输入内容后，标签中的内容就会变为文本框中的内容。使用的方法如表 12-4 所示。

表 12-4　ViewController.m文件中方法总结

方　　法	功　　能
viewDidLoad	视图加载后调用，实现初始化
setupLabel	创建标签
setupTextField	创建文本框
textFieldShouldReturn:	当用户在键盘上按下 Done 按钮调用，实现键盘的关闭，改变标签的内容

方法	功能
textField:shouldChangeCharactersInRange:replacementString:	每次用户通过键盘输入字符时,字符显示在文本框之前调用,实现标签中内容的同步改变
textFieldShouldBeginEditing:	让文本框成为第一响应

其中,setupLabel 方法实现对发光标签的创建以及设置。程序代码如下:

```
- (void)setupLabel
{
    CGRect frame = CGRectMake(10, 100, 300, 100);
    GlowLabel *v = [[GlowLabel alloc] initWithFrame:frame];
    v.text = kInitialText;
    v.textAlignment = NSTextAlignmentCenter;           //设置对齐方式
    v.clipsToBounds = YES;
    v.backgroundColor = [UIColor clearColor];
    v.font = [UIFont fontWithName:@"Helvetica-Bold" size:40];  //设置字体
    v.alpha = 1.0;
    v.glowSize = 20;
    v.textColor = UIColorFromRGB(0x00ffff);            //设置文本内容
    v.glowColor = [UIColor whiteColor];
    self.label = v;
    [self.view addSubview:v];
}
```

setupTextField 方法实现文本框的创建以及设置。程序代码如下:

```
- (void)setupTextField
{
    CGRect frame = CGRectMake(10, 20, 300, 50);
    UITextField *v = [[UITextField alloc] initWithFrame:frame];
    v.delegate = self;                                  //设置委托
    v.textColor = UIColorFromRGB(0x000000);
    v.placeholder = @"Please input";
    v.returnKeyType = UIReturnKeyDone;                  //设置回车键
    v.font = [UIFont fontWithName:@"Helvetica-Bold" size:20];
    v.backgroundColor = UIColor.whiteColor;             //设置背景颜色
    v.textAlignment = NSTextAlignmentCenter;
    v.contentVerticalAlignment = UIControlContentVerticalAlignmentCenter;
    self.field = v;
    [self.view addSubview:v];
}
```

textFieldShouldReturn:方法,当用户按下键盘上的 Done 按钮后调用,实现键盘的关闭以及将标签中的内容设置为文本框中的内容。程序代码如下:

```
- (BOOL)textFieldShouldReturn:(UITextField *)textField
{
    [self.field resignFirstResponder];
    self.label.text = self.field.text;                  //设置文本内容
    [self.label setNeedsDisplay];
    return YES;
}
```

textField:shouldChangeCharactersInRange:replacementString:方法在每次用户通过键盘输入字符时,字符显示在文本框之前调用,实现标签中内容的同步。程序代码如下:

```
- (BOOL)textField:(UITextField *)textField shouldChangeCharactersInRange:
(NSRange)range
replacementString:(NSString *)string
{
    //设置文本内容
    self.label.text = [self.field.text stringByReplacingCharactersInRange:
    range withString:string];
    [self.label setNeedsDisplay];
    return YES;
}
```

【代码解析】

本实例的关键功能是发光效果和标签显示内容的同步改变。下面依次讲解这两个知识点。

1. 发光效果

发光效果，需要使用 CGContext 的 CGContextSetShadowWithColor 函数实现。它的功能是使用指定颜色绘制阴影，其中语法形式如下：

```
void CGContextSetShadowWithColor (
  CGContextRef context,
  CGSize offset,
  CGFloat blur,
  CGColorRef color
);
```

其中，参数说明如下。

- CGContextRef context 表示上下文。
- CGSize offset 表示阴影的位置。
- CGFloat blur 表示阴影的模糊值，数值越大越模糊。
- CGColorRef color 表示颜色。

在本实例中就使用 CGContextSetShadow 方法实现了标签的发光效果，代码如下：

```
CGContextSetShadowWithColor(ctx, CGSizeZero, _glowSize, _glowColor.
CGColor);
```

其中，参数说明如下。

- ctx 表示上下文。
- CGSizeZero 表示阴影的位置。
- _glowSize 表示阴影的模糊值，数值越大越模糊。
- _glowColor.CGColor 表示颜色。

2. 标签的同步改变

当用户在文本框中输入一个字符时就会在标签中显示一个字符，即标签的同步改变，需要在 textField:shouldChangeCharactersInRange:replacementString:方法中实现。它使用的是 NSString 的 stringByReplacingCharactersInRange:withString:方法。其语法形式如下：

```
- (NSString *)stringByReplacingCharactersInRange:(NSRange)range withString:
(NSString *)replacement;
```

其中，(NSRange)range 表示字符的范围；(NSString *)replacement 表示字符串，它是在该范围内的替换字符串。在本实例中，使用了 stringByReplacingCharactersInRange:withString: 方法实现在文本框中输入一个字符时就会在标签中显示一个字符的效果，代码如下：

```
self.label.text = [self.field.text stringByReplacingCharactersInRange:
range withString:string];
```

其中，range 表示字符的范围；string 表示在该范围内的替换字符串。

实例 83　循环渐变的标签

【实例描述】
　　本实例的关键功能是显示一个循环渐变的标签。每隔几秒就会以动画的形式显示标签的下一个内容。运行效果如图 12.3 所示。

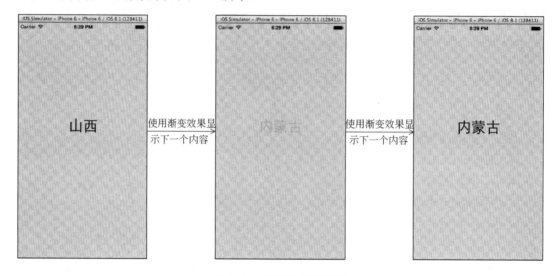

图 12.3　实例 83 运行效果

【实现过程】
（1）创建一个项目，命名为"循环渐变的标签"。
（2）创建一个基于 UILabel 类的 Label 类。
（3）打开 Label.h 文件，编写代码，实现属性和方法的声明。程序代码如下：

```
#import <UIKit/UIKit.h>
@interface Label : UILabel
//属性
@property(nonatomic,retain) NSArray *wordList;
@property(nonatomic,assign) double duration;
//初始化循环渐变的标签的方法
- (void)animateWithWords:(NSArray *)words forDuration:(double)time;
@end
```

（4）打开 Label.m 文件，编写代码，实现循环渐变的标签。使用的方法如表 12-5 所示。

第 12 章 文本处理

表 12-5 Label.m文件中方法总结

方　　法	功　　能
animateWithWords:forDuration:	初始化循环渐变的标签
_startAnimations:	开始动画
_animate:	动作效果

其中，animateWithWords:forDuration:方法实现使用文本内容和持续时间对循环渐变标签的初始化。程序代码如下：

```objc
- (void)animateWithWords:(NSArray *)words forDuration:(double)time {
    self.duration = time;                              //设置持续时间
    if(self.wordList){
        self.wordList = nil;
    }
    self.wordList = [[NSArray alloc] initWithArray:words];
    self.text = [self.wordList objectAtIndex:0];       //设置文本内容
    [NSThread detachNewThreadSelector:@selector(_startAnimations:) toTarget:self withObject:self.wordList];        //创建并启动一个线程
}
```

_startAnimations:方法实现触发动画开始的功能。程序代码如下：

```objc
- (void) _startAnimations:(NSArray *)images {
    for (uint i = 1; i < [images count]; i++) {
        sleep(self.duration);
        [self performSelectorOnMainThread:@selector(_animate:)withObject:[NSNumber numberWithInt:i] waitUntilDone:YES];      //在主线程上执行_animate:方法
        sleep(self.duration);
        if (i == [images count] - 1) {
            i = -1;
        }
    }
}
```

_animate:方法实现渐变的效果。程序代码如下：

```objc
- (void) _animate:(NSNumber*)num
{
    [UIView animateWithDuration:self.duration/2 animations:^{
        self.alpha = 0.0;                              //设置透明度
    } completion:^(BOOL finished) {
        [UIView animateWithDuration:self.duration/2 animations:^{
            self.alpha = 1.0;                          //设置透明度
            self.text = [self.wordList objectAtIndex:[num intValue]];
            //设置文本内容
        } completion:^(BOOL finished) {
        }];
    }];
}
```

（5）打开 ViewController.h 文件，编写代码，实现头文件以及插座变量的声明。程序代码如下：

```objc
#import <UIKit/UIKit.h>
#import "Label.h"                                      //头文件
```

```
@interface ViewController : UIViewController{
    IBOutlet Label *label;                              // Label 插座变量
}
@end
```

（6）打开 Main.storyboard 文件，对 View Controller 视图控制器的设计界面进行设计，效果如图 12.4 所示。

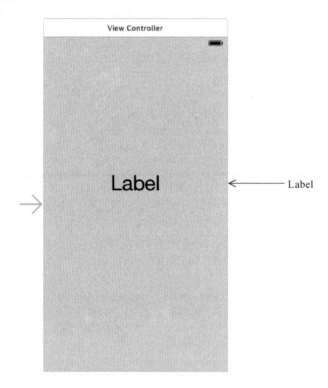

图 12.4 实例 83View Controller 视图控制器的设计界面

需要添加的视图、控件以及对它们的设置如表 12-6 所示。

表 12-6 实例 83 视图、控件设置

视图、控件	属 性 设 置	其 他
设计界面	Background：黑色	
Label	Font：System 36.0 Alignment：居中	Class：Label 与插座变量 label 关联

（7）打开 ViewController.m 文件，编写代码，实现循环渐变标签的创建。程序代码如下：

```
- (void)viewDidLoad
{
    [super viewDidLoad];
    // Do any additional setup after loading the view, typically from a nib.
    [label animateWithWords:@[@"山西",@"河南",@"河北",@"内蒙古",@"西藏",@"浙江"] forDuration:1.0f];
}
```

【代码解析】

本实例关键功能是渐变动画的实现。下面就是这个知识点的详细讲解。

要实现渐变动画效果，需要在_animate:方法中使用 animateWithDuration:animations:completion:方法，代码如下：

```
[UIView animateWithDuration:self.duration/2 animations:^{
    self.alpha = 0.0;
} completion:^(BOOL finished) {
    [UIView animateWithDuration:self.duration/2 animations:^{
        self.alpha = 1.0;
        self.text = [self.wordList objectAtIndex:[num intValue]];
    } completion:^(BOOL finished) {
    }];
}];
```

然后，通过使用 performSelectorOnMainThread:withObject:waitUntilDone:方法实现调用。其语法形式如下：

```
- (void)performSelectorOnMainThread:(SEL)aSelector  withObject:(id)arg waitUntilDone:(BOOL)wait;
```

其中，(SEL)aSelector 表示要调用的方法，是一个选择器；withObject:表示传递的参数；(BOOL)wait 表示一个布尔值，用来指定是否当前线程阻塞，直到指定选择器在主线程中执行。在此代码中，就使用了performSelectorOnMainThread:withObject:waitUntilDone:方法实现了_animate:方法的调用，代码如下：

```
[self performSelectorOnMainThread:@selector(_animate:)withObject:[NSNumber numberWithInt:i] waitUntilDone:YES];
```

实例 84　滚动的标签

【实现描述】

在标签中，显示文字的多少取决于标签的框架大小。当框架大小小于文字的数目时，会将显示不出的文字使用"…"代替。为了防止这种情况的发生。本实例实现的功能是将标签改为一个可以滚动的效果，这样，不管显示多少内容，都不会出现"…"。运行效果如图 12.5 所示。

【实现过程】

（1）创建一个项目，命名为"滚动的标签"。
（2）创建一个基于 UIScrollView 类的 RollLabel 类。
（3）打开 RollLabel.h 文件，编写代码，实现宏定义和方法的声明。程序代码如下：

```
#import <UIKit/UIKit.h>
#define kConstrainedSize CGSizeMake(10000,40)
@interface RollLabel : UIScrollView
+ (void)rollLabelTitle:(NSString *)title color:(UIColor *)color font:(UIFont *)font superView:(UIView *)superView fram:(CGRect)rect;
                //绘制滚动的标签
@end
```

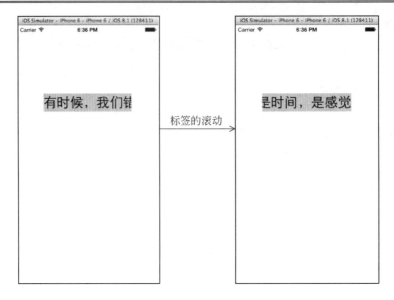

图 12.5 实例 84 运行结果

（4）打开 RollLabel.m 文件，编写代码，在此文件中代码分为两部分：滚动标签的初始化以及绘制。其中，初始化需要使用 initWithFrame:Withsize:方法实现。程序代码：

```objc
- (id)initWithFrame:(CGRect)frame Withsize:(CGSize)size
{
    self = [super initWithFrame:frame];
    if (self) {
        self.showsVerticalScrollIndicator   = NO;
        self.showsHorizontalScrollIndicator = NO;          //水平滚动条
        self.contentSize = size;                            //滚动大小
        self.backgroundColor = [UIColor colorWithHue:0.1 saturation:0.6
        brightness:1.0 alpha:1.0];
    }
    return self;
}
```

绘制滚动的标签，需要使用 rollLabelTitle:color:font:superView:fram:方法实现。程序代码如下：

```objc
+ (void)rollLabelTitle:(NSString *)title color:(UIColor *)color font:
(UIFont *)font superView:(UIView *)superView fram:(CGRect)rect
{
    NSMutableParagraphStyle *style=[[NSMutableParagraphStyle
    defaultParagraphStyle]mutableCopy];
    [style setLineBreakMode:NSLineBreakByWordWrapping];    //设置换行模式
    NSDictionary *dis=[NSDictionary dictionaryWithObjectsAndKeys:font,
    NSFontAttributeName, style, NSParagraphStyleAttributeName, nil];
    //设置标签的大小和滚动视图对象的大小
    CGSize size=[title boundingRectWithSize:kConstrainedSize options:
    NSStringDrawingUsesLineFragmentOrigin attributes:dis context:nil].size;
CGRect frame = CGRectMake(0, 0, size.width, rect.size.height);
    //实例化对象
    RollLabel *roll = [[RollLabel alloc]initWithFrame:rect Withsize:size];
    UILabel *label = [[UILabel alloc]initWithFrame:frame];
    label.text = title;                                    //设置文本内容
    label.font = font;
```

```
    label.textColor = color;
    [roll addSubview:label];
    [superView addSubview:roll];
}
```

（5）打开 ViewController.h 文件，编写代码，实现 RollLabel.h 头文件的声明。

（6）打开 ViewController.m 文件，编写代码，实现滚动标签的创建。程序代码如下：

```
- (void)viewDidLoad
{
    [super viewDidLoad];
    // Do any additional setup after loading the view, typically from a nib.
    [RollLabel rollLabelTitle:@"有时候，我们错过的不是时间，是感觉"
color:[UIColor blackColor] font:[UIFont systemFontOfSize:30] superView:
self.view fram:CGRectMake(60, 150, 200, 40)];
}
```

【代码解析】

本实例关键功能是滚动的实现。下面就是这个知识点的详细讲解。

要实现滚动效果，需要使用 UIScrollView 的 contentSize 属性。在本代码中就使用了 contentSize 属性实现了标签的滚动效果。代码如下：

```
self.contentSize = size;
```

实例 85 具有光晕效果的标签

【实现描述】

由于在标签中显示的内容是很单一的，所以为了去除这种很乏味的效果，本实例在标签中加入了光晕效果。运行效果如图 12.6 所示。

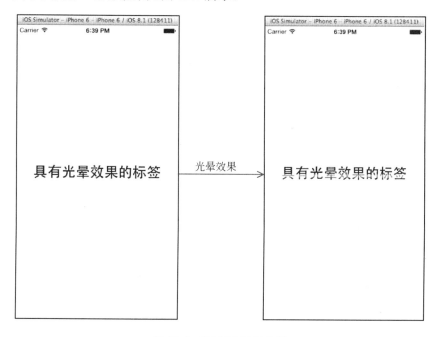

图 12.6 实例 85 运行效果

【实现过程】
(1) 创建一个项目,命名为"具有光晕效果的标签"。
(2) 创建一个基于 UIImage 类的分类 LabelAdditions。
(3) 打开 UIImage+LabelAdditions.h 文件,编写代码,实现方法的声明。程序代码如下:

```
#import <UIKit/UIKit.h>
@interface UIImage (LabelAdditions)
+ (UIImage *)imageWithView:(UIView *)view;              //图像的获取
@end
```

(4) 打开 UIImage+LabelAdditions.m 文件,编写代码,对声明的方法进行定义,实现截屏并获取。程序代码如下:

```
+ (UIImage *)imageWithView:(UIView *)view
{
    UIGraphicsBeginImageContextWithOptions(view.bounds.size, NO, 0.0);
    [view.layer renderInContext:UIGraphicsGetCurrentContext()];
                    //视图绘制到图像上下文中
    UIImage * img = UIGraphicsGetImageFromCurrentImageContext();
                    //从上下文获取新的图像
    UIGraphicsEndImageContext();
    return img;
}
```

(5) 创建一个基于 UIView 类的 Label 类。
(6) 打开 Label.h 文件,编写代码,实现头文件、定义数据结构、实例变量、对象、属性以及方法的声明。程序代码如下:

```
#import <UIKit/UIKit.h>
#import "UIImage+LabelAdditions.h"                      //头文件
//数据类型的定义
typedef enum Direction
{
    EffectDirectionLeftToRight,
    EffectDirectionRightToLeft,
    EffectDirectionTopToBottom,
    EffectDirectionBottomToTop,
    EffectDirectionTopLeftToBottomRight,
    EffectDirectionBottomRightToTopLeft,
    EffectDirectionBottomLeftToTopRight,
    EffectDirectionTopRightToBottomLeft
}Direction;
@interface Label : UIView{
    UILabel *_effectLabel;                              //对象
    CGImageRef _alphaImage;                             //实例变量
    CALayer *_textLayer;                                //对象
}
//属性
@property (strong, nonatomic) UIFont *font;
@property (strong, nonatomic) UIColor *textColor;
@property (strong, nonatomic) UIColor *effectColor;
@property (strong, nonatomic) NSString *text;
@property (assign, nonatomic) Direction direction;
- (void)performEffectAnimation;                         //执行动画效果的方法
@end
```

（7）打开 Label.m 文件，编写代码，实现具有光晕效果的标签。使用的方法如表 12-7 所示。

表 12-7　Label.m文件中方法总结

方　　法	功　　能
initWithFrame:	初始化具有光晕效果的标签
initialize	设置初始值
layoutSubviews	布局
layerClass	获取 CAGradientLayer 类
setFrame:	设置框架
setFont	设置字体
setText:	设置文本
setTextColor:	设置文本的颜色
updateLabel	更新
performEffectAnimation	执行动画效果
animationForStage:	状态的动画
colorsForStage:	状态的颜色

这里需要讲解几个重要的方法（其他方法请读者参考源代码）。其中 initialize 方法实现对一些默认值的进行初始化设置。程序代码如下：

```
- (void)initialize
{
    _textColor = [UIColor whiteColor];
    _effectColor = [UIColor redColor];
    _effectLabel = [[UILabel alloc] initWithFrame:self.bounds];
    _effectLabel.textAlignment = NSTextAlignmentLeft;   //设置对齐方式
    _effectLabel.numberOfLines = 1;
    _effectLabel.backgroundColor = [UIColor clearColor];
    CAGradientLayer *gradientLayer = (CAGradientLayer *)self.layer;
    gradientLayer.backgroundColor = [[UIColor clearColor] CGColor];
                                                         //设置背景颜色
}
```

updateLabel 方法实现对标签的更新。程序代码如下：

```
- (void)updateLabel
{
    CAGradientLayer *gradientLayer = (CAGradientLayer *)self.layer;
    gradientLayer.colors = [self colorsForStage:0];
    [_effectLabel sizeToFit];
    self.frame = CGRectMake(self.frame.origin.x, self.frame.origin.y,
    _effectLabel.frame.size.width, _effectLabel.frame.size.height);
                                                         //设置框架
    _alphaImage = [[UIImage imageWithView:_effectLabel] CGImage];
    _textLayer = [CALayer layer];
    _textLayer.contents = (__bridge id)_alphaImage;
    [self.layer setMask:_textLayer];
    [self setNeedsLayout];
}
```

performEffectAnimation 方法实现动画的执行。程序代码如下：

```
- (void)performEffectAnimation
{
```

```objc
    CAGradientLayer *gradientLayer = (CAGradientLayer *)self.layer;
    gradientLayer.colors = [self colorsForStage:0];
    switch (_direction)
{
    //设置方法
        case EffectDirectionLeftToRight:
            gradientLayer.startPoint = CGPointMake(0.0, 0.5);//设置开始点
            gradientLayer.endPoint = CGPointMake(1.0, 0.5);   //设置结束点
            break;
        case EffectDirectionRightToLeft:
            gradientLayer.startPoint = CGPointMake(1.0, 0.5);
            gradientLayer.endPoint = CGPointMake(0.0, 0.5);
            break;
        case EffectDirectionTopToBottom:
            gradientLayer.startPoint = CGPointMake(0.5, 0.0);
            gradientLayer.endPoint = CGPointMake(0.5, 1.0);
            break;
        case EffectDirectionBottomToTop:
            gradientLayer.startPoint = CGPointMake(0.5, 1.0); //设置开始点
            gradientLayer.endPoint = CGPointMake(0.5, 0.0);    //设置结束点
            break;
        case EffectDirectionBottomLeftToTopRight:
            gradientLayer.startPoint = CGPointMake(0.0, 1.0);
            gradientLayer.endPoint = CGPointMake(1.0, 0.0);
            break;
        case EffectDirectionBottomRightToTopLeft:
            gradientLayer.startPoint = CGPointMake(1.0, 1.0);
            gradientLayer.endPoint = CGPointMake(0.0, 0.0);
            break;
        case EffectDirectionTopLeftToBottomRight:
            gradientLayer.startPoint = CGPointMake(0.0, 0.0); //设置开始点
            gradientLayer.endPoint = CGPointMake(1.0, 1.0);    //设置结束点
            break;
        case EffectDirectionTopRightToBottomLeft:
            gradientLayer.startPoint = CGPointMake(1.0, 0.0);
            gradientLayer.endPoint = CGPointMake(0.0, 1.0);
            break;
}
    //实例化对象
    CABasicAnimation *animation0 = [self animationForStage:0];
    CABasicAnimation *animation1 = [self animationForStage:1];
    CABasicAnimation *animation2 = [self animationForStage:2];
    CABasicAnimation *animation3 = [self animationForStage:3];
    CABasicAnimation *animation4 = [self animationForStage:4];
    CABasicAnimation *animation5 = [self animationForStage:5];
    CABasicAnimation *animation6 = [self animationForStage:6];
    CABasicAnimation *animation7 = [self animationForStage:7];
    CABasicAnimation *animation8 = [self animationForStage:8];
    //创建并设置动画组对象
    CAAnimationGroup *group = [CAAnimationGroup animation];
    group.duration = animation8.beginTime + animation8.duration;
                                                                //设置持续时间
    group.removedOnCompletion = NO;
    group.fillMode = kCAFillModeForwards;
    [group setAnimations:@[animation0, animation1, animation2, animation3, animation4, animation5, animation6, animation7, animation8]]; //设置动画
    [gradientLayer addAnimation:group forKey:@"animationOpacity"];
}
```

animationForStage:方法为每一个状态实现动画效果,程序代码如下:

```
- (CABasicAnimation *)animationForStage:(NSUInteger)stage
{
    CGFloat duration = 0.3;
    CGFloat inset = 0.1;
    CABasicAnimation *animation = [CABasicAnimation animationWithKeyPath:
    @"colors"];
    animation.fromValue = [self colorsForStage:stage];        //设置开始值
    animation.toValue = [self colorsForStage:stage + 1];      //设置结束值
    animation.beginTime = stage * (duration - inset);
    animation.duration = duration;
    animation.repeatCount = 1;                                //设置循环次数
    animation.removedOnCompletion = NO;
    animation.fillMode = kCAFillModeForwards;
    animation.timingFunction = [CAMediaTimingFunction functionWithName:
    kCAMediaTimingFunctionLinear];
    return animation;
}
```

colorsForStage:方法为每一个状态设置颜色组。程序代码如下:

```
- (NSArray *)colorsForStage:(NSUInteger)stage
{
    NSMutableArray *array = [NSMutableArray arrayWithCapacity:8]; //实例化对象
    //变量
    for (int i = 0; i < 9; i++)
    {
        [array addObject:stage != 0 && stage == i ? (id)[_effectColor CGColor] :
        (id)[_textColor CGColor]];
    }
    return [NSArray arrayWithArray:array];
}
```

(8) 打开 ViewController.h 文件,编写代码,实现头文件的声明。

(9) 打开 ViewController.m 文件,编写代码,实现对标签的初始化设置。程序代码如下:

```
- (void)viewDidLoad
{
    [super viewDidLoad];
    // Do any additional setup after loading the view, typically from a nib.
    Label *label = [[Label alloc] initWithFrame:CGRectMake(0, 0, 300, 160)];
    [self.view addSubview:label];                             //添加标签对象
    label.autoresizingMask = UIViewAutoresizingFlexibleTopMargin |
    UIViewAutoresizingFlexibleBottomMargin | UIViewAutoresizingFlexible-
    LeftMargin | UIViewAutoresizingFlexibleRightMargin;       //设置自适应
    label.font = [UIFont boldSystemFontOfSize:28];
    label.text = @"具有光晕效果的标签";
    label.textColor = [UIColor blackColor];                   //设置文本颜色
    label.effectColor = [UIColor yellowColor];
    label.direction = EffectDirectionTopLeftToBottomRight;
    label.center = self.view.center;                          //设置中心位置
    for (int i = 0; i < 8; i++)
    {
        int64_t delayInSeconds = 3 * i;
        dispatch_time_t popTime = dispatch_time(DISPATCH_TIME_NOW, delay-
        InSeconds * NSEC_PER_SEC);
```

```
    //延迟执行动画效果
    dispatch_after(popTime, dispatch_get_main_queue(), ^(void){
        label.direction = i;
        [label performEffectAnimation];
    });
  }
}
```

【代码解析】

本实例关键功能是动画效果的同时执行。下面就是这个知识点的详细讲解。

如果想要让动画效果同时执行,需要使用 CAAnimationGroup 的 animations 属性实现。其语法形式如下:

```
@property(copy) NSArray *animations;
```

在此代码中就使用了 animations 属性实现了光晕的动画效果。代码如下:

```
[group setAnimations:@[animation0, animation1, animation2, animation3,
animation4, animation5, animation6, animation7, animation8]];
```

实例 86 标签云

【实例描述】

在很多的网页中,经常可以看到在一个区域内会有很多的内容标签在移动,一般称之为标签云。本实例就为读者实现一个标签云的效果,当用户单击标签云中的内容标签时,会在特定的标签中显示单击的内容标签中的内容。其中,用于显示指定的内容标签中所包含的内容的是标签。运行效果如图 12.7 所示。

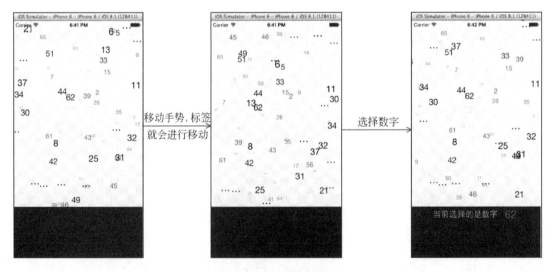

图 12.7 实例 86 运行效果

【实现过程】

当用户单击标签云中的内容标签时,会在特定的标签中显示单击的内容标签中的内容。具体的实现步骤如下。

第 12 章 文本处理

（1）创建一个项目，命名为"标签云"。
（2）创建一个基于 UIButton 类的 Button 类。
（3）打开 Button.h 文件，编写代码，实现属性的声明。程序代码如下：

```
#import <UIKit/UIKit.h>
@interface Button : UIButton
//属性
@property (nonatomic) CGFloat distanceX;
@property (nonatomic) CGFloat distanceY;
@property (nonatomic) CGFloat acceleration;
@end
```

（4）在 Button.m 文件中将声明的属性进行定义。
（5）创建一个基于 UIView 类的 LabelClound 类。
（6）打开 LabelClound.h 文件，编写代码，实现头文件、协议、对象、实例变量、属性以及方法的声明。程序代码如下：

```
#import <UIKit/UIKit.h>
#import "Button.h"                                          //头文件
//协议
@protocol LabelCloudDelegate
- (void)didSelectedNodeButton:(Button *)button;
@end
@interface LabelClound : UIView{
NSTimer *animationTimer;                                    //对象
//实例变量
    CGPoint oldtouchPoint;
    CGFloat speedMAX;
    CGFloat speedMIN;
}
//属性
@property (nonatomic) CGFloat top;
@property (nonatomic) CGFloat bottom;
@property (nonatomic) CGFloat left;
@property (nonatomic) CGFloat right;
@property (nonatomic, assign) id<LabelCloudDelegate>delegate;
@property(nonatomic) Button *cloudButton;
//方法
- (id)initWithFrame:(CGRect)frame andNodeCount:(NSInteger)nodeCount;
- (void)animationUpdate;
- (CGFloat)limitSpeedbettowinMINandMAX:(CGFloat)speedValue;
@end
```

（7）打开 LabelClound.m 文件，编写代码，实现标签栏的效果。使用的方法如表 12-8 所示。

表 12-8 LabelClound.m 文件中方法总结

方　　法	功　　能
initWithFrame:andNodeCount:	初始化标签云的视图
touchesBegan:withEvent:	触摸开始
touchesMoved:withEvent:	手指在屏幕上移动
touchesEnded:withEvent:	触摸结束
animationUpdate	实现移动的动画效果
limitSpeedbettowinMINandMAX:	限制并获取速度

这里需要讲解几个重要的方法（其他方法请读者参考源代码）。其中，initWithFrame:方法实现初始化标签云的视图。程序代码如下：

```objc
- (id)initWithFrame:(CGRect)frame andNodeCount:(NSInteger)nodeCount
{
    self = [super initWithFrame:frame];
    if (self) {
        CGSize nodeSize = CGSizeMake(25, 25);
        speedMAX     = 2.0f;                              //最大移动速度
        speedMIN     = -2.0f;                             //最小移动速度
        //为变量赋值
        self.left    = frame.origin.x - nodeSize.width;
        self.right   = frame.origin.x + frame.size.width;
        self.top     = frame.origin.y - nodeSize.height;
        self.bottom  = frame.origin.y + frame.size.height;
        oldtouchPoint = CGPointMake(-1000, -1000);
        //添加内容标签
        for (NSInteger i = 0;i < nodeCount; i ++) {
            CGFloat x = arc4random()%(int)(self.right  - nodeSize.width);
              //为 x 赋值
            CGFloat y = arc4random()%(int)(self.bottom - nodeSize.height);
              //创建并设置
            cloudButton = [[Button alloc] initWithFrame:CGRectMake(x,y,
            nodeSize.width,nodeSize.height)];
            cloudButton.tag = i;                          //设置 tag 值
            cloudButton.userInteractionEnabled = NO;
            [cloudButton setTitle:[NSString stringWithFormat:@"%d",i]
            forState:UIControlStateNormal];
            [cloudButton addTarget:delegate action:@selector(didSelectedNodeButton:) forControlEvents:UIControlEventTouchUpInside];
                                                          //添加动作

            float fontSize = arc4random()%20+6;
            cloudButton.titleLabel.font = [UIFont systemFontOfSize:fontSize];
            cloudButton.alpha = fontSize/25;              //设置透明度
            [cloudButton setTitleColor:[UIColor blackColor] forState:
            UIControlStateNormal];
            [cloudButton setTitleColor:[UIColor redColor] forState:
            UIControlStateHighlighted];
            [self addSubview:cloudButton];                //添加视图对象
        }
        animationTimer = [NSTimer scheduledTimerWithTimeInterval:1.0/60.0
target:self selector:@selector(animationUpdate) userInfo:nil repeats:YES];
                                                          //创建定时器
        [animationTimer setFireDate:[NSDate distantPast]];
    }
    return self;
}
```

touchesMoved:withEvent:方法实现手指在屏幕上移动。程序代码如下：

```objc
- (void)touchesMoved:(NSSet *)touches withEvent:(UIEvent *)event
{
    UITouch *theTouch = [touches anyObject];
    CGPoint currentPoint = [theTouch  locationInView:self];//获取触摸的位置
    if (oldtouchPoint.x != -1000 && oldtouchPoint.y != -1000) {
        for (int i=0; i<self.subviews.count; i++) {
            Button *cloud = [self.subviews objectAtIndex:i];
              //设置属性
```

```
            cloud.distanceX = [self limitSpeedbettowinMINandMAX:
            (currentPoint.x - oldtouchPoint.x)];
            cloud.distanceY = [self limitSpeedbettowinMINandMAX:
            (currentPoint.y - oldtouchPoint.y)];
            CGFloat distance = sqrt(cloud.distanceX*cloud.distanceX+
cloud.distanceY*cloud.distanceY);
            //判断 distance 是否不为 0
            if (distance!=0) {
                if (0.001 <= cloud.acceleration) {
                    cloud.acceleration = (arc4random()%100)/200.0f;
                }
            }
        }
    }
}
```

animationUpdate 方法实现内容标签的移动动画效果。程序代码如下：

```
- (void)animationUpdate
{
    //遍历
    for (int i=0; i<self.subviews.count; i++) {
        Button *cloud = [self.subviews objectAtIndex:i];
        cloud.center = CGPointMake(cloud.center.x + cloud.distanceX*cloud.
        acceleration,cloud.center.y + cloud.distanceY*cloud.acceleration);
                                                //设置中心位置
        //中心位置放入判断
        if (cloud.center.x < self.left) {
            cloud.center = CGPointMake(self.right,self.bottom - cloud.
            center.y);                          //设置中心位置
        }
        if (cloud.center.x > self.right) {
            cloud.center = CGPointMake(self.left, self.bottom - cloud.
            center.y);                          //设置中心位置
        }
        if (cloud.center.y < self.top){
            cloud.center = CGPointMake(self.right - cloud.center.x, self.
            bottom);                            //设置中心位置
        }
        if (cloud.center.y > self.bottom) {
            cloud.center = CGPointMake(self.right - cloud.center.x, self.
            top);                               //设置中心位置
        }
        cloud.acceleration -= 0.001;
        if (cloud.acceleration < 0.001) {
            cloud.acceleration = 0.001;         //设置加速度
        }
    }
}
```

limitSpeedbettowinMINandMAX:方法实现获取并限制速度。程序代码如下：

```
- (CGFloat)limitSpeedbettowinMINandMAX:(CGFloat)speedValue
{
    //判断 speedValue 是否大于 speedMAX
    if (speedValue > speedMAX) {
        speedValue = speedMAX;
    }
    //判断 speedValue 是否小于 speedMIN
```

```
    if (speedValue < speedMIN) {
        speedValue = speedMIN;
    }
    return speedValue;
}
```

（8）打开 ViewController.h 文件，编写代码，实现头文件、插座变量和对象的声明。程序代码如下：

```
#import <UIKit/UIKit.h>
#import "LabelClound.h"
@interface ViewController : UIViewController<LabelCloudDelegate>{
    IBOutlet UIView *vv;                    //空白视图的插座变量
    IBOutlet UILabel *label;                //标签的插座变量
    IBOutlet UILabel *label1;               //标签的插座变量
    LabelClound *cloud;                     //LabelClound 的对象
}
@end
```

（9）打开 Main.storyboard 文件，对 View Controller 视图控制器的设计界面进行设计，效果如图 12.8 所示。

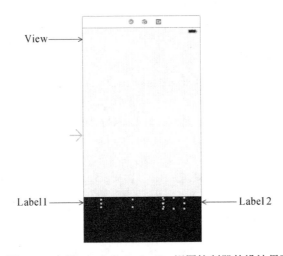

图 12.8　实例 86 View Controller 视图控制器的设计界面

需要添加的视图、控件以及对它们的设置如表 12-9 所示。

表 12-9　实例 86 视图、控件设置

视图、控件	属 性 设 置	其 他
设计界面	Background：黑色	
View	Background：浅灰色	与插座变量 vv 关联
Label1	Text：（空） Color：绿色 Font：System 20.0 Alignment：居中	与插座变量 label 关联
Label2	Text：（空） Color：红色 Font：System 26.0 Alignment：居中	与插座变量 label1 关联

(10)打开 ViewController.m 文件,编写代码,实现标签云对象的创建以及选择。程序代码如下:

```
- (void)viewDidLoad
{
    [super viewDidLoad];
    // Do any additional setup after loading the view, typically from a nib.
    cloud = [[LabelClound alloc] initWithFrame:vv.bounds andNodeCount:66];
                                                            //创建
    [self.view addSubview:cloud];                           //添加
}
//实现选择,并显示选择的数字
- (void)didSelectedNodeButton:(Button *)button
{
    NSString *str=[NSString stringWithFormat:@"%i",button.tag];
    label.text=@"当前选择的是数字";                          //设置标签中文本的内容
    labell.text=str;
}
```

【代码解析】

本实例关键功能是标签的选择以及移动。下面依次讲解这两个知识点。

1. 标签的选择

在本实例中标签云中的每一个标签都是使用按钮实现的,如果想要实现按下任意标签使它和其他的标签不同,需要使用 setTitle:forState:方法。其语法形式如下:

```
- (void)setTitleColor:(UIColor *)color forState:(UIControlState)state;
```

其中,(UIColor *)color 表示按钮的颜色;(UIControlState)state 表示使用指定颜色的状态。在此代码中就使用了 setTitle:forState:方法实现了按下按钮后为红色,按钮弹起后还原为原来的颜色。代码如下:

```
[cloudButton setTitleColor:[UIColor redColor] forState:
UIControlStateHighlighted];
```

如果想要将选择的标签显示在指定的标签控件中,需要使用 addTarget:action:forControlEvents:方法调用 didSelectedNodeButton:方法实现。代码如下:

```
[cloudButton addTarget:delegate action:@selector(didSelectedNodeButton:)
forControlEvents:
UIControlEventTouchUpInside];
```

在 didSelectedNodeButton:方法中实现将选择的标签显示在指定的标签中。

2. 移动

标签云中标签的移动,是通过使用定时器定时更新 animationUpdate 方法实现的,代码如下:

```
animationTimer = [NSTimer scheduledTimerWithTimeInterval:1.0/60.0
target:self selector:@selector
(animationUpdate) userInfo:nil repeats:YES];
```

实例 87 自动计算文本长度

【实例描述】

本实例主要实现的功能是在文本框中输入文本,自动计算文本框中的文本长度,并显示在标签控件中。单击"发表"按钮,就会发表。其中,用于文本输入的是一个文本框,用于显示文本长度的是一个标签。运行效果如图12.9所示。

图12.9 实例87运行效果

【实现过程】

当用户向文本框中输入文本后,会在标签中显示自动更新的文本长度。具体的实现步骤如下。

(1)创建一个项目,命名为"文本长度计算器"。

(2)打开 ViewController.h 文件,编写代码,实现插座变量和动作的声明。程序代码如下:

```
#import <UIKit/UIKit.h>
@interface ViewController : UIViewController{
    IBOutlet UITextField *tf;                           //文本框的插座变量
    IBOutlet UILabel *label;                            //标签的插座变量
    IBOutlet UIActivityIndicatorView *activity;         //指示器的插座变量
}
- (IBAction)result:(id)sender;
@end
```

(3)对文字进行输入,要提供一个输入文字的界面。打开 Main.storyboard 文件,对 View Controller 视图控制器的设计界面进行设计,效果如图12.10所示。

需要添加的视图、控件以及对它们的设置如表12-10所示。

(4)打开 ViewController.m 文件,编写代码。在此文件中,代码分为5个部分:文本的自动计算、单击按钮后的动作、加载动画、显示警告视图以及响应警告视图。添加的aa:

方法用来实现当文本框中的值发生变化后，自动更新标签中的内容。程序代码如下：

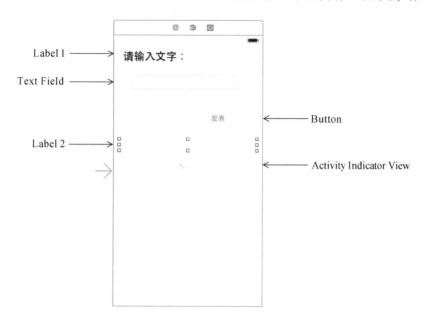

图 12.10　实例 87 View Controller 视图控制器的设计界面

表 12-10　实例 87 视图、控件设置

视图、控件	属 性 设 置	其　　他
Label1	Text：请输入文字 Color：黑色 Font：黑体-繁 中等 23.0	
Text Field		与插座变量 tf 关联
Button	Title：发表	与动作 result:关联
Label2	Text：（空）	与插座变量 label 关联
Activity Indicator View		与插座变量 activity 关联

```
-(void)aa{
int count=tf.text.length;                    //获取文本框内的文本长度
//判断文本框内的文本长度是否超出30
    if (count>=30) {
        UIAlertView *alert=[[UIAlertView alloc]initWithTitle:@"提示"
        message:@"你输入的文字包括空格、标点,不能超出30个" delegate:self
        cancelButtonTitle:@"重新输入" otherButtonTitles: nil];
        [alert show];                        //显示警告视图
    }else{
        NSString *result=[NSString stringWithFormat:@"你还可以输入%d个文字
        ",30-count];
        label.text=result;                   //设置文本内容
        [self performSelector:@selector(aa) withObject:self afterDelay:0];
    }
}
```

在 result:动作方法中，当单击"发表"按钮后，实现键盘的关闭以及指示器的加载。

程序代码如下：

```
- (IBAction)result:(id)sender {
    [tf resignFirstResponder];                                    //关闭键盘
    [self performSelector:@selector(start) withObject:self afterDelay:0.5];
}
```

添加的 star:方法实现指示器的加载动画。程序代码如下：

```
//实现指示器加载
-(void)start{
    [activity setHidden:NO];
    [activity startAnimating];                                    //开始动画
    [self performSelector:@selector(stop) withObject:self afterDelay:1];
}
```

添加的 stop:方法实现当过一段时间后加载完成，出现一个警告视图，表示发表成功。程序代码如下：

```
-(void)stop{
    UIAlertView *alert=[[UIAlertView alloc]initWithTitle:@"成功发布"
      message:nil delegate:self cancelButtonTitle:@"知道了" otherButtonTitles:
      nil];                                                       //创建警告视图
    [alert show];
}
```

添加的 alertView: clickedButtonAtIndex:方法实现当用户单击警告视图中的按钮后，退出警告视图并且将文本框以及标签清空。程序代码如下：

```
-(void)alertView:(UIAlertView *)alertView clickedButtonAtIndex:
(NSInteger)buttonIndex{
NSString *str=[alertView buttonTitleAtIndex:buttonIndex];
//判断 str 是否为"知道了"
    if ([str isEqualToString:@"知道了"]){
        [activity setHidden:YES];                                 //隐藏
        label.text=nil;
        tf.text=nil;
    }else{
        tf.text=nil;                                              //设置文本内容
        [self performSelector:@selector(aa) withObject:self afterDelay:0];
    }
}
```

【代码解析】

本实例关键功能是文本长度的获取和长度值的自动更新显示。下面依次讲解这两个知识点。

1．获取文本长度

要获取文本长度，可以使用 NSString 的 length:方法。其语法形式如下。

```
- (NSUInteger)length;
```

其中，该方法的返回值类型为整型。在代码中，使用了 length:方法进行文本框中文本长度的计算，代码如下：

```
int count=tf.text.length;
```

2. 自动更新文本

自动更新文本是通过调用选择器来实现自身的循环。在代码中，使用了 performSelector: withObject:afterDelay:方法表示自动更新文本，代码如下：

```
[self performSelector:@selector(aa) withObject:self afterDelay:0];
```

实例 88　仿 QQ 登录

【实例描述】

本实例实现的功能仿 QQ 登录。账号为 Love，密码为 123456。当输入正确的账号和密码时，会进入 QQ 登录以后的界面；当输入的账号或者密码不正确，就会进入提示界面。其中，用于用户输入密码与账号的是文本框，运行效果如图 12.11 所示。

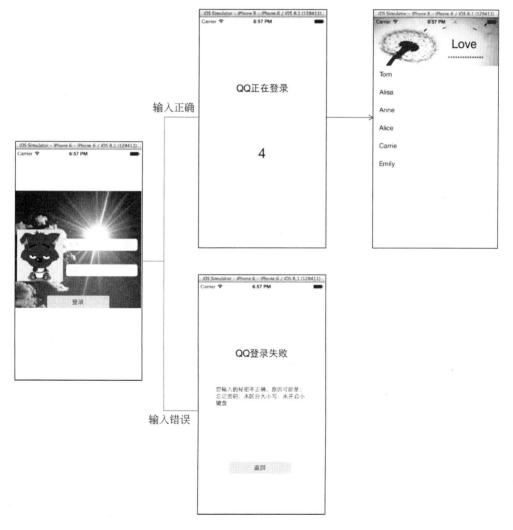

图 12.11　实例 88 运行效果

【实现过程】

当用户在文本框中输入账号和密码后，会进行判断输入的密码和账号是否为指定的内容，如果是则进入登录页；否则，进入登录失败页。具体的实现步骤是：创建一个项目，命名为"qq 登录"。然后添加图片 1.jpg、2.jpg 和 3.jpg 到创建项目的 Supporting Files 文件夹中。下面开始制作各个界面。

1. 制作主界面

在打开此应用程序时，首先看到的是主界面，为用户提供账号和密码的输入。打开 Main.storyboard 文件，对 View Controller 视图控制器的设计界面进行设计，效果如图 12.12 所示。

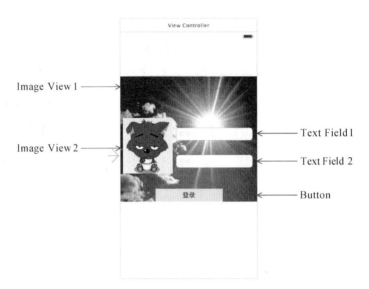

图 12.12　实例 88View Controller 视图控制器设计界面的效果

需要添加的视图、控件以及对它们的设置如表 12-11 所示。

表 12-11　实例 88 视图、控件设置 1

视图、控件	属 性 设 置	其　　他
Image View1	Image：1.jpg	
Image View2	Image：2.jpg	
Text Field1	Placeholder：账号	与 ViewController.h 文件进行动作 close:的声明和关联 将 Did End On Exit 与 close:关联
Text Field2	Placeholder：密码 选择 Secure 复选框	与 ViewController.h 文件进行动作 close:的关联 将 Did End On Exit 与 close:关联
Button	Title：登录	与 ViewController.h 文件进行动作 aa:的声明和关联

2. 制作登录界面

当账号和密码输入成功后，单击"登录"按钮，进入 QQ 登录界面。创建一个基于 UIViewController 类的 aViewController 类。回到 Main.storyboard 文件，从视图库中拖动 View Controller 视图控制器到画布中。将 Class 设置为 aViewController。对 A View Controller 视

图控制器的设计界面进行设计，效果如图 12.13 所示。

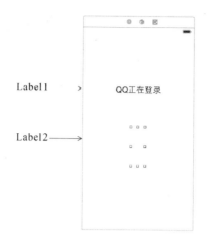

图 12.13　A View Controller 视图控制器的设计界面效果

需要添加的视图、控制以及对它们的设置如表 12-12 所示。

表 12-12　实例 88 视图、控件设置 2

视图、控件	属性设置	其　　他
A View Controller 视图控制器		Storyboard ID：First 选择 Use Storyboard ID 复选框
Label1	Text：QQ 正在登录 Font：System 24.0	
Label2	Text：（空） Font：System 34.0	

创建一个基于 UIViewController 类的 cViewController 类。回到 Main.storyboard 文件，从视图库中拖动 View Controller 视图控制器到画布中。将 Class 设置为 cViewController。对 C View Controller 视图控制器的设计界面进行设计，效果如图 12.14 所示。

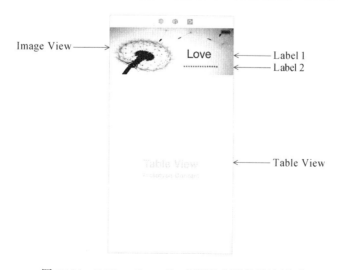

图 12.14　C View Controller 视图控制器的设计界面

需要添加的视图、控件以及对它们的设置如表 12-13 所示。

表 12-13　实例 88 视图、控件设置 3

视图、控件	属 性 设 置	其　　他
C View Controller 视图控制器		Storyboard ID：Third 选择 Use Storyboard ID 复选框
Image View	Image：3.jpg	调整大小
Label1	Text：Love Font：System 30.0	
Label2	Text：***************	
Table View	右击，在出现的菜单中，选择 dataSource，将其和 dock 工作区的 C View Controller 图标进行关联	将 dataSource 和 dock 工作区的 C View Controller 图标进行关联

3. 制作登录失败界面

当账号和密码输入失败后，进入 QQ 登录失败界面。创建一个基于 UIViewController 类的 bViewController 类。回到 Main.storyboard 文件，从视图库中拖动 View Controller 视图控制器到画布中。将 Class 设置为 bViewController。对 B View Controller 视图控制器的设计界面进行设计，效果如图 12.15 所示。

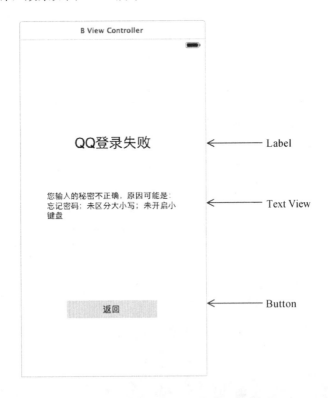

图 12.15　B View Controller 视图控制器的设计界面

需要添加的视图、控件以及对它们的设置如表 12-14 所示。

表 12-14　实例 88 视图、控件设置 4

视图、控件	属　性　设　置	其　　他
B View Controller 视图控制器		Storyboard ID：Second 选择 Use Storyboard ID 复选框
Label	Text：QQ 登录失败 Font：System 24.0	
Text View	Text：您输入的密码不正确，原因可能是： 忘记密码；未分区大小写；未开启小键盘	
Button	Title：返回	与 View Controller 视图控制器的设计界面关联

这时画布的总体效果如图 12.16 所示。

图 12.16　画布效果

打开 ViewController.h 文件，编写代码，实现头文件、插座变量以及动作的声明。程序代码如下：

```
#import <UIKit/UIKit.h>
#import "aViewController.h"
#import "bViewController.h"
@interface ViewController : UIViewController{
    IBOutlet UITextField *t1;                    //文本框的插座变量
    IBOutlet UITextField *t2;                    //文本框的插座变量
    aViewController *aa;
    bViewController *bb;
}
- (IBAction)aa:(id)sender;
- (IBAction)close:(id)sender;                    //关闭键盘
@end
```

打开 ViewController.m 文件，编写代码。在此文件中，代码分为两个部分：实现账号与密码的判断以及键盘的关闭。在 aa:动作方法中，实现文本内容的判断，在文本框中输入密码和账号后，单击"登录"按钮，方法 aa:就会触发，对文本框中的文本内容进行判断，如果文本是要求输入的文本，就会进入登录页；如果文本不是要求输入的文本，就会进入登录失败页。程序代码如下：

```objc
- (IBAction)aa:(id)sender {
    //判断 t1 的文本内容是否为 Love, t2 的文本内容是否为 123456
    if ([t1.text isEqualToString:@"Love"]&&[t2.text isEqualToString:
@"123456"]) {
        aa=[self.storyboard instantiateViewControllerWithIdentifier:
        @"First"];
        [self.view addSubview:aa.view];                    //添加视图对象
    }else{
        bb=[self.storyboard instantiateViewControllerWithIdentifier:
        @"Second"];
        [self.view addSubview:bb.view];
    }
}
```

在 close:动作方法中，实现对键盘的关闭。程序代码如下：

```objc
- (IBAction)close:(id)sender {
    [t1 resignFirstResponder];
    [t2 resignFirstResponder];                             //关闭键盘
}
```

打开 aViewController.h 文件，编写代码，实现头文件、插座变量以及实例变量、对象的声明。程序代码如下：

```objc
#import <UIKit/UIKit.h>
#import "cViewController.h"
@interface aViewController : UIViewController{
    int i;
    IBOutlet UILabel *label;                               //声明操作变量
    NSTimer *timer;
    cViewController *cc;
}
@end
```

打开 aViewController.m 文件，编写代码，实现登录页的倒计时功能。当时间一到，QQ 就成功登录了。程序代码如下：

```objc
- (void)viewDidLoad
{
    i=5;
    timer=[NSTimer        scheduledTimerWithTimeInterval:1        target:self
selector:@selector(aa) userInfo:nil repeats:YES];          //创建时间定时器
    [super viewDidLoad];
    // Do any additional setup after loading the view.
}
//实现倒计时
-(void)aa{
    i--;
    label.text=[NSString stringWithFormat:@"%i",i];
    if (i==0) {
        [timer invalidate];                                //让定时器失效
        cc=[self.storyboard instantiateViewControllerWithIdentifier:
        @"Third"];
        [self.view addSubview:cc.view];
    }
}
```

打开 cViewController.m 文件，编写代码，实现在表视图中添加内容。程序代码如下：

```objc
- (void)viewDidLoad
{
    //创建数组
    arr=[NSArray arrayWithObjects:@"Tom",@"Alisa",@"Anne",@"Alice",@"Carrie",
    @"Emily", nil];
    [super viewDidLoad];
    // Do any additional setup after loading the view.
}
//设置节数
- (NSInteger)numberOfSectionsInTableView:(UITableView *)tableView {
    return 1;
}
//设置行数
- (NSInteger)tableView:(UITableView *)tableView numberOfRowsInSection:
(NSInteger)section {
    return [arr count];
}
//插入表单元
- (UITableViewCell *)tableView:(UITableView *)tableView cellForRowAtIndexPath:
(NSIndexPath *)indexPath {
    static NSString *CellIdentifier = @"Cell";
    UITableViewCell *cell = [tableView dequeueReusableCellWithIdentifier:
    CellIdentifier];
    //判断cell是否为空
    if (cell == nil) {
        cell = [[UITableViewCell alloc] initWithStyle:UITableViewCell
        StyleDefault reuseIdentifier:CellIdentifier];
    }
    cell.textLabel.text = [arr objectAtIndex:indexPath.row]; //设置文本内容
    return cell;
}
```

将声明的插座变量和设计界面中的视图、控件进行关联，如表12-15所示。

表 12-15　插座变量的关联

插座变量	关联的视图、控件
t1	与 text Field1 文本框关联
t2	与 text.Field2 文本框关联
label	与 A View Controller 设计界面上的 label2 标签关联

【代码解析】

本实例关键功能是文本的判断。下面讲解这个知识点。

要判断文本，可以使用 NSString 的 isEqualToString:方法。其语法形式如下：

```objc
-(BOOL)isEqualToString:(NSString *)aString;
```

其中，(NSString *)aString 表示比较的字符串。该方法的返回值类型为 BOOL。如果返回的值为 1，说明两个字符串相等；如果返回的值为 0，说明两个字符串不相等。在代码中使用了 isEqualToString:方法进行了两个文本框中文本的判断，代码如下：

```objc
if ([t1.text isEqualToString:@"Love"]&&[t2.text isEqualToString:@"123456"]){
……
}
```

其中，t1.text 和@"Love"表示比较的两个字符串。

实例 89　阅读浏览器

【实例描述】

本实例实现的功能是一个阅读浏览器。当单击界面上显示的书名后，就会打开对应的图书。当看完此书后，可以单击"返回主菜单"按钮，回到主界面，用户可以重新选择需要阅读的图书。其中，所有的目录名称使用的是标签，对相应内容的显示使用的是文本视图。运行效果如图 12.17 所示。

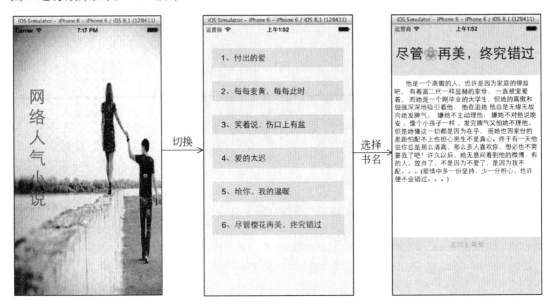

图 12.17　实例 89 运行效果

【实现过程】

当用户打开应用程序后，会进入一个主界面，随后，主界面会被小说的目录覆盖，并且单击目录上的任何一个标题，都会进入相应的界面内容中。具体的实现步骤如下。

（1）创建一个项目，命名为"阅读器"。

（2）添加图片 11.jpg 到创建项目的 Supporting Files 文件夹中。

（3）打开 ViewController.h 文件，编写代码，实现插座变量的声明。程序代码如下：

```
#import <UIKit/UIKit.h>
@interface ViewController : UIViewController{
    IBOutlet UIView *vv;                          //空白视图的插座变量
}
@end
```

（4）打开此阅读器，进入阅读器的主界面。打开 Main.storyboard 文件夹，对 ViewController 视图控制器的设计界面进行设计，效果如图 12.18 所示。

第 12 章　文本处理

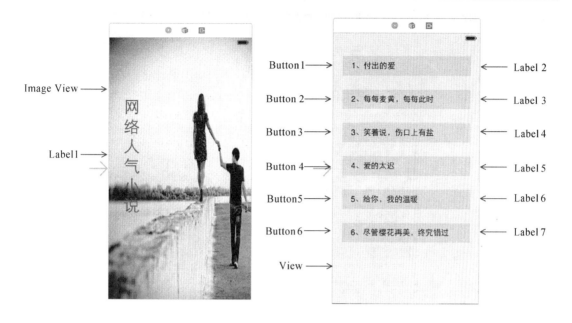

图 12.18　实例 89View Controller 视图控制器的设计界面

需要添加的视图、控件以及对它们的设置如表 12-16 所示。

表 12-16　实例 89 视图、控件设置 1

视图、控件	属性设置	其他
Image View	Image：11.jpg	
Label1	Text：网络人气小说 Color：蓝色 Font：System 36.0 Alignment：居中 Lines：6	
View	Background：浅蓝色	将其覆盖整个设计界面 与插座变量 vv 关联
Label2	Text："　　1、付出的爱"	
Label3	Text："　　2、每每麦黄，每每此时"	
Label4	Text："　　3、笑着说，伤口上有盐"	
Label5	Text："　　4、爱的太迟"	
Label6	Text："　　5、给你，我的温暖"	
Label7	Text："　　6、尽管樱花再美，终究错过"	
Button1	Title：（空）	将其放在 label2 上
Button2	Title：（空）	将其放在 label3 上
Button3	Title：（空）	将其放在 label4 上
Button4	Title：（空）	将其放在 label5 上
Button5	Title：（空）	将其放在 label6 上
Button6	Title：（空）	将其放在 label7 上

（5）在主界面中，选择某一标签，就会进入相应的内容界面。在从视图库中拖动 View

Controller 视图控制器到画布中,对其设计界面进行设计,效果如图 12.19 所示。

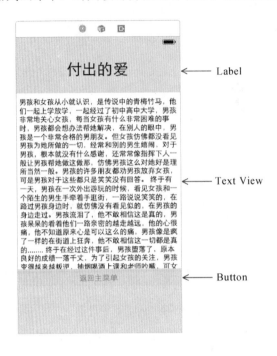

图 12.19　新增 View Controller 视图控制器的设计界面效果

需要添加的视图、控件以及对它们的设置如表 12-17 所示。

表 12-17　实例 89 视图、控件设置 2

视图、控件	属性设置	其他
设计界面	Background:浅绿色	
Label	Text:付出的爱 Font:System 30.0 Alignment:居中对齐	
Text View	Text:男孩和女孩从小就认识,是传说中的青梅竹马,他们一起上学放学,一起经过了初中高中大学,男孩非常地关心女孩,每当女孩有什么非常困难的事时,男孩都会想办法帮她解决,在别人的眼中,男孩是一个非常合格的男朋友。……,是男孩存在的位置.......	
Button	Title:返回主菜单 Background:青色	

剩余的设计界面请参考源代码。

(6)将视图和视图控制器直接进行关联,这时的画布效果如图 12.20 所示。

具体的关联如表 12-18 所示。

(7)将声明的插座变量和设计界面的 View 视图进行关联。

(8)回到 Main.storyboard 文件,将初始视图控制器的设计界面上的 View 视图,向下移动,直到此设计界面上看不到的位置。

(9)打开 ViewController.m 文件,编写代码,实现视图的切换。程序代码如下:

第 12 章 文本处理

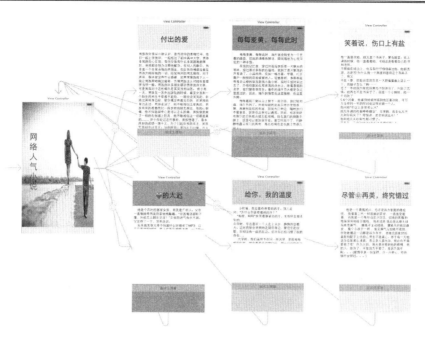

图 12.20 画布效果

表 12-18 关联

进行关联的视图控制器		被关联的视图控制器
初始的视图控制器	Button1	标签内容为"付出的爱"的视图控制器的设计界面
	Button2	标签内容为"每每麦黄,每每此时"的视图控制器的设计界面
	Button3	标签内容为"笑着说,伤口上有盐"的视图控制器的设计界面
	Button4	标签内容为"♥的太迟"的视图控制器的设计界面
	Button5	标签内容为"给你,我的温暖"的视图控制器的设计界面
	Button6	标签内容为"尽管樱花再美,终究错过"的视图控制器的设计界面
标签内容为"付出的爱"的视图控制器		初始的视图控制器的设计界面
标签内容为"每每麦黄,每每此时"的视图控制器		初始的视图控制器的设计界面
标签内容为"笑着说,伤口上有盐"的视图控制器		初始的视图控制器的设计界面
标签内容为"♥的太迟"的视图控制器		初始的视图控制器的设计界面
标签内容为"给你,我的温暖"的视图控制器		初始的视图控制器的设计界面
标签内容为"尽管樱花再美,终究错过"的视图控制器		初始的视图控制器的设计界面

```
- (void)viewDidLoad
{
    [self performSelector:@selector(aa) withObject:self afterDelay:5];
    [super viewDidLoad];
    // Do any additional setup after loading the view, typically from a nib.
}
//实现视图的切换
-(void)aa{
    [UIView beginAnimations:@"" context:nil];
    [UIView setAnimationDuration:3];                              //设置时间
```

• 543 •

```
    vv.frame=CGRectMake(0, 0, 320, 568);
    [UIView commitAnimations];
}
```

【代码分析】

本实例关键功能是 UIView 块动画效果。下面讲解这个知识点。

要实现 UIView 块动画，可以使用 UIView 的 beginAnimations:context:方法，它用来表示块动画的开始。其语法形式如下：

```
+ (void)beginAnimations:(NSString *)animationID context:(void *)context;
```

其中，(NSString *)animationID 表示动画的标识，它是一个字符串；(void *)context 表示传递给动画的消息，一般设置为 nil。一个块动画是有头和尾的，beginAnimations: context:方法，用来表示块动画的开始，使用 commitAnimations:方法表示块动画的结束，其语法形式如下：

```
+ (void)commitAnimations;
```

在代码中使用了 beginAnimations: context:方法和 commitAnimations:方法表示了一个完整的块动画，代码如下：

```
[UIView beginAnimations:@"" context:nil];
[UIView setAnimationDuration:3];
vv.frame=CGRectMake(0, 0, 320, 568);
[UIView commitAnimations];
```

其中，@""表示动画的标识；nil 表示传递给动画的消息。

实例 90 艺术字

【实例描述】

本实例实现的功能是单击 Play 按钮，就会以动画的形式出现 5 种艺术字效果。其中，文字的显示使用的是标签。运行效果如图 12.21 所示。

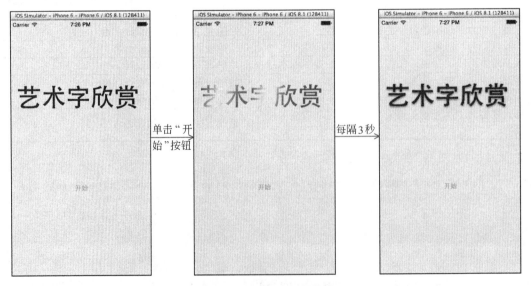

图 12.21 实例 90 运行效果

【实现过程】

（1）创建一个项目，命名为"艺术字"。
（2）创建一个基于 UILabel 类的 label 类。
（3）打开 label.h 文件，编写代码，实现宏定义以及属性的声明。程序代码如下：

```
//宏定义
#ifndef AH_RETAIN
#if __has_feature(objc_arc)
#define AH_RETAIN(x) (x)
#define AH_RELEASE(x) (void)(x)
#define AH_AUTORELEASE(x) (x)
#define AH_SUPER_DEALLOC (void)(0)
#else
#define __AH_WEAK
#define AH_WEAK assign
#define AH_RETAIN(x) [(x) retain]
#define AH_RELEASE(x) [(x) release]
#define AH_AUTORELEASE(x) [(x) autorelease]
#define AH_SUPER_DEALLOC [super dealloc]
#endif
#endif
#import <UIKit/UIKit.h>
@interface label : UILabel
//属性声明
@property (nonatomic, assign) CGFloat shadowBlur;      //用户设置阴影的模糊值
@property (nonatomic, assign) CGSize innerShadowOffset;
@property (nonatomic, strong) UIColor *innerShadowColor;
@property (nonatomic, strong) UIColor *gradientStartColor;
@property (nonatomic, strong) UIColor *gradientEndColor;
                                                       //用于设置渐变的结束颜色
@property (nonatomic, copy) NSArray *gradientColors;
@property (nonatomic, assign) CGPoint gradientStartPoint;
@property (nonatomic, assign) CGPoint gradientEndPoint;
                                                       //用于设置渐变的开始点
@property (nonatomic, assign) NSUInteger oversampling;
@property (nonatomic, assign) UIEdgeInsets textInsets;
@property (nonatomic, assign) NSUInteger minSamples;
@property (nonatomic, assign) NSUInteger maxSamples;
@end
```

（4）打开 label.m 文件，编写代码，实现艺术字的绘制。使用的方法如表 12-19 所示。

表 12-19　label.m 文件中方法总结

方　　法	功　　能
- (void)setDefaults;	默认设置
- (id)initWithFrame:(CGRect)frame;	自定义标签的初始化
- (id)initWithCoder:(NSCoder *)aDecoder;	初始化实例化对象
- (void)setInnerShadowOffset:(CGSize)offset;	设置里面阴影的偏移值
- (void)setInnerShadowColor:(UIColor *)color;	设置里面阴影的颜色
- (UIColor *)gradientStartColor;	渐变的开始颜色
- (void)setGradientStartColor:(UIColor *)color;	设置渐变的开始颜色
- (UIColor *)gradientEndColor;	渐变的结束颜色
- (void)setGradientEndColor:(UIColor *)color;	设置渐变的结束颜色
- (void)setGradientColors:(NSArray *)colors;	设置渐变的多种颜色

续表

方　法	功　能
- (void)setOversampling:(NSUInteger)samples;	取样
- (void)setTextInsets:(UIEdgeInsets)insets;	设置文本的插入
- (void)getComponents:(CGFloat *)rgba forColor:(CGColorRef)color;	得到组件
- (UIColor *)color:(CGColorRef)a blendedWithColor:(CGColorRef)b;	混合颜色
- (void)drawRect:(CGRect)rect;	绘制

这里对几个重要的函数进行讲解，其他函数请读者参考源代码。添加的 setDefaults:方法实现对默认渐变起点、终点等进行设置，代码如下：

```
- (void)setDefaults
{
    gradientStartPoint = CGPointMake(0.5f, 0.0f);           //渐变起点
    gradientEndPoint = CGPointMake(0.5f, 0.75f);            //渐变终点
    minSamples = maxSamples = 1;
    //判断是否实现了 scale 方法
    if ([UIScreen instancesRespondToSelector:@selector(scale)])
    {
        minSamples = [UIScreen mainScreen].scale;
        maxSamples = 32;
    }
    oversampling = minSamples;
}
```

添加 initWithFrame:方法实现标签的初始化，代码如下：

```
- (id)initWithFrame:(CGRect)frame
{
    if ((self = [super initWithFrame:frame]))
    {
        self.backgroundColor = nil;                         //设置背景颜色
        [self setDefaults];
    }
    return self;
}
```

要实现第一种艺术字效果，需要使用添加的 setGradientColors:、color:blendedWithColor:、getComponents:forColor: 和 drawRect:方法。setGradientColors:方法用于设置渐变的多种颜色，代码如下：

```
- (void)setGradientColors:(NSArray *)colors
{
    if (gradientColors != colors)
    {
        AH_RELEASE(gradientColors);
        gradientColors = [colors copy];                     //复制颜色
        [self setNeedsDisplay];
    }
}
```

color:blendedWithColor:方法对多种颜色进行混合，程序代码如下：

```
- (UIColor *)color:(CGColorRef)a blendedWithColor:(CGColorRef)b
{
    CGFloat aRGBA[4];
```

```
    [self getComponents:aRGBA forColor:a];             //得到颜色的组件
    CGFloat bRGBA[4];
    [self getComponents:bRGBA forColor:b];             //获取
    CGFloat source = aRGBA[3];
CGFloat dest = 1.0f - source;
//返回颜色
    return [UIColor colorWithRed:source * aRGBA[0] + dest * bRGBA[0]
                           green:source * aRGBA[1] + dest * bRGBA[1]
                            blue:source * aRGBA[2] + dest * bRGBA[2]
                           alpha:bRGBA[3] + (1.0f - bRGBA[3]) * aRGBA[3]];
}
```

getComponents:方法得到颜色的组件,代码如下:

```
- (void)getComponents:(CGFloat *)rgba forColor:(CGColorRef)color
{
    CGColorSpaceModel model = CGColorSpaceGetModel(CGColorGetColorSpace
    (color));
    const CGFloat *components = CGColorGetComponents(color);    //获取颜色分量
    switch (model)
    {
        case kCGColorSpaceModelMonochrome:
        {
            //设置颜色分量
            rgba[0] = components[0];
            rgba[1] = components[0];
            rgba[2] = components[0];
            rgba[3] = components[1];
            break;
        }
        case kCGColorSpaceModelRGB:
        {
            //设置颜色分量
            rgba[0] = components[0];
            rgba[1] = components[1];
            rgba[2] = components[2];
            rgba[3] = components[3];
            break;
        }
        default:
        {
            rgba[0] = 0.0f;
            rgba[1] = 0.0f;
            rgba[2] = 0.0f;
            rgba[3] = 1.0f;
            break;
        }
    }
}
```

第二种艺术字使用的方法还是 setGradientColors:方法,第三种艺术字使用添加的 setGradientStartColor:、setGradientColors:、setGradientEndColor:、color:blendedWithColor:、getComponents:forColor:和 drawRect:方法。setGradientStartColor:方法设置渐变的开始颜色,程序代码如下:

```
- (void)setGradientStartColor:(UIColor *)color
{
    //判断颜色是否为空
```

```objc
    if (color == nil)
    {
        self.gradientColors = nil;
    }
    else if ([gradientColors count] < 2)//判断 gradientColors 的个数是否小于2
    {
        self.gradientColors = [NSArray arrayWithObjects:color, color, nil];
                                            //创建对象
    }
    else if ([gradientColors objectAtIndex:0] != color)
    {
        NSMutableArray *colors = [gradientColors mutableCopy];
                                            //创建对象
        [colors replaceObjectAtIndex:0 withObject:color];
        self.gradientColors = colors;
                                    //设置渐变颜色
        AH_RELEASE(colors);
    }
}
```

setGradientEndColor:方法设置渐变的结束颜色，程序代码如下：

```objc
- (void)setGradientEndColor:(UIColor *)color
{
    //判断颜色是否为空
    if (color == nil)
    {
        self.gradientColors = nil;
    }
    else if ([gradientColors count] < 2)//判断 gradientColors 的个数是否小于2
    {
        self.gradientColors = [NSArray arrayWithObjects:color, color, nil];
    }
    else if ([gradientColors lastObject] != color)
    {
        NSMutableArray *colors = [gradientColors mutableCopy];  //创建对象
        [colors replaceObjectAtIndex:[colors count] - 1 withObject:color];
        self.gradientColors = colors;                           //设置渐变颜色
        AH_RELEASE(colors);
    }
}
```

剩余两种艺术字的设置代码会在源代码中显示。对于这几种艺术字最后必须要使用drawRect:方法绘制，其语法形式如下：

```objc
- (void)drawRect:(CGRect)rect
{
    if (oversampling > minSamples || (self.backgroundColor && ![self.backgroundColor isEqual:[UIColor clearColor]]))
    {
        UIGraphicsBeginImageContextWithOptions(rect.size, NO, oversampling);
    }
    CGContextRef context = UIGraphicsGetCurrentContext();   //创建图形上下文
    rect = self.bounds;
    rect.origin.x += textInsets.left;
    rect.origin.y += textInsets.top;
    ……
    CGContextRestoreGState(context);
    CGImageRelease(alphaMask);                              //释放 alphaMask
```

```
    }
    if (oversampling)
    {
        UIImage *image = UIGraphicsGetImageFromCurrentImageContext();
                                                        //创建图像上下文
        UIGraphicsEndImageContext();
        [image drawInRect:rect];
    }
}
```

(5)打开 Main.storyboard 文件，对 View Controller 视图控制器的设计界面进行设计，效果如图 12.22 所示。

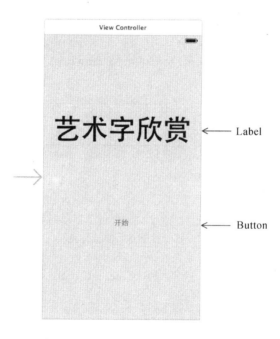

图 12.22　实例 90 View Controller 视图控制器的设计界面效果

需要添加的视图、控件以及对它们的设置如表 12-20 所示。

表 12-20　实例 90 视图、控件设置

视图、控件	属 性 设 置	其　　他
设计界面	Background：浅棕色	
Label	Tex：艺术字欣赏 Color：黑色 Font：Optima Bold 56.0	Class：label
Button	Title：开始	与动作 play:关联

(6)打开 ViewController.m 文件，编写代码，实现艺术字效果的应用。程序代码如下：

```
- (IBAction)play:(id)sender {
    [bb setHidden:YES];
    lab.gradientStartPoint = CGPointMake(0.0f, 0.0f);      //渐变开始点
    lab.gradientEndPoint = CGPointMake(1.0f, 1.0f);        //渐变结束点
    lab.gradientColors = [NSArray arrayWithObjects:         //渐变颜色
                [UIColor blueColor],
```

```
                    [UIColor yellowColor],
                    [UIColor redColor],
                    [UIColor cyanColor],
                    [UIColor blueColor],
                    [UIColor purpleColor],
                    [UIColor orangeColor],
                    nil];
    [self performSelector:@selector(change1) withObject:self afterDelay:3];
}
……
```

以上代码，只是本实例中的部分代码片段，如果想要了解更多，需参考源代码。

【代码分析】

由于本实例中的代码和方法非常的多，为了方便读者的阅读，其中第一种艺术字是由从左到右的蓝、黄、红、青、紫、橘黄渐变色组成的。它是由 setGradientColors:、color:blendedWithColor:、getComponents:forColor:和 drawRect:方法共同实现的，其执行流程如表 12.23 所示。

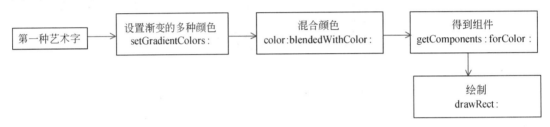

图 12.23　第一种艺术字的执行流程

第二种艺术字是由阴影组成的。它是由 setGradientColors: drawRect:方法实现的，其执行流程如图 12.24 所示。

图 12.24　第二种艺术字的执行流程

第三种艺术字是由橘黄色和青色组成的渐变色组成的。它是由 setGradientStartColor:、setGradientColors:和 setGradientEndColor:、color:blendedWithColor:、getComponents:forColor:和 drawRect:方法共同实现的，其执行流程如图 12.25 所示。

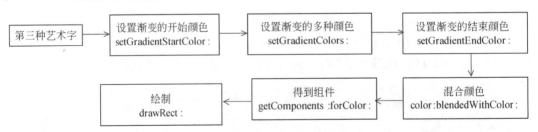

图 12.25　第三种艺术字的执行流程

第 12 章 文本处理

第四种艺术字是由阴影和内阴影组成的。它是由 setGradientStartColor:、setInnerShadowColor:和 setInnerShadowOffset:、drawRect:方法共同实现的,其执行流程如图 12.26 所示。

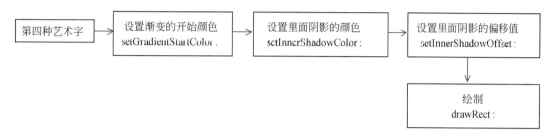

图 12.26　第四种艺术字的执行流程

第五种艺术字是由阴影和红、黄渐变色组成的。它是由 setGradientStartColor:、setInnerShadowOffset:、setGradientStartColor:、setGradientColors:、setGradientEndColor:、setGradientColors:、setOversampling:、getComponents:、color:blendedWithColor:和 drawRect:方法共同实现的,其执行流程如图 12.27 所示。

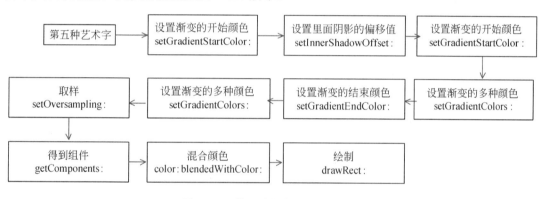

图 12.27　第五种艺术字的执行流程

实例 91　网址管理器

【实例描述】

本实例实现的功能是,使用文本框的键盘类型实现 URL,实现一个网址管理器的功能。当输入网址后,单击"添加到收藏夹"按钮,就会将文本框中的内容添加到收藏夹中。运行效果如图 12.28 所示。

【实现过程】

(1) 创建一个项目,命名为"网址管理器"。

(2) 添加图标 1.jpg 和 2.jpg 到创建项目的 Supporting Files 文件夹中。

(3) 在 ViewController.h 文件中,编写代码,实现插座变量、字符串对象和动作的声明。程序代码如下:

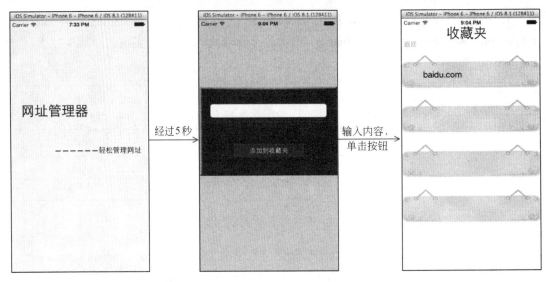

图 12.28　实例 91 运行效果

```
#import <UIKit/UIKit.h>
@interface ViewController : UIViewController{
    IBOutlet UIView *v;
    IBOutlet UIView *v2;                              //声明关于视图的插座变量
    IBOutlet UILabel *labe1;
    IBOutlet UILabel *label2;                         //声明关于标签的插座变量
    IBOutlet UILabel *label3;
    IBOutlet UILabel *label4;
    IBOutlet UILabel *labb;
    IBOutlet UITextField *textfiled;                  //声明关于文本框的插座变量
    NSString *s;
}
- (IBAction)aa:(id)sender;
- (IBAction)close:(id)sender;
- (IBAction)back:(id)sender;
@end
```

（4）单击打开 Main.storyboard 文件，对 View Controller 视图控制器的设计界面进行设计，效果如图 12.29 所示。

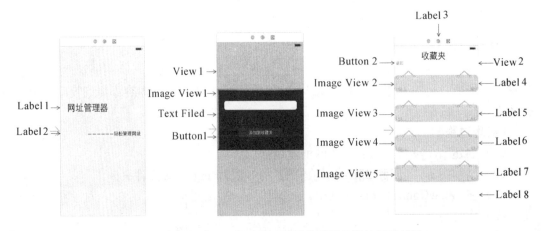

图 12.29　实例 91View Controller 视图控制器的设计界面效果

第 12 章 文本处理

需要添加的视图、控件以及对它们的设置如表 12-21 视图。

表 12-21 实例 91 视图、控件设置

视图、控件	属 性 设 置	其 他
设计界面	Background：浅绿色	
Label1	Text：网址管理器 Font：System 31.0 Alignment：居中	
Label2	Text：————轻松管理网址	
View1	Background：粉色	将其覆盖整个设计界面 与插座变量 v2 关联
Image View1	Image：2.jpg	
Text Field	Keyboard：URL	将 Did End On Exit 与动作 close:关联 与插座变量 textfiled 关联
Button1	Title：添加到收藏夹 Text Color：绿色 Background：深红色	与动作 aa:关联 与动作 close 关联
View2		将其覆盖整个设计界面 与插座变量 v 关联
Label3	Text：收藏夹 Font：System 31.0	
Button2	Title：返回	与动作 back:关联
Image View2	Image：1.jpg	
Image View3	Image：1.jpg	
Image View4	Image：1.jpg	
Image View5	Image：1.jpg	
Label4	Text：（空） Font：System 19.0	与插座变量 label1 关联
Label5	Text：（空） Font：System 19.0	与插座变量 label2 关联
Label6	Text：（空） Font：System 19.0	与插座变量 label3 关联
Label7	Text：（空） Font：System 19.0	与插座变量 label4 关联
Label8	Text：（空） Font：System 19.0	与插座变量 labb 关联

（5）回到 Main.storyboard 文件，将标签控件为"收藏夹"所在的视图向下移动，一直到设计界面中看不到为止。

（6）单击打开 ViewController.m 文件，编写代码。在本文件中，代码分为：显示输入界面、实现在文本框中输入的网址放到收藏夹两部分。如果收藏夹已满，可以将收藏夹中的内容删除。使用的方法如表 12-22 所示。

表 12-22 ViewController.m 文件中方法总结

方 法	功 能
viewDidLoad	视图加载后调用，实现初始化

续表

方　　法	功　　能
ab	显示输入界面
aa:	输入内容并保存
alertView:clickedButtonAtIndex:	实现收藏夹的清空
close:	关闭键盘
back:	返回输入界面

这里需要讲解几个重要的方法（其他方法请读者参考源代码）。其中，aa:方法实现将文本框中输入的内容保存到收藏夹中。程序代码如下：

```objc
- (IBAction)aa:(id)sender {
    s=[NSString stringWithFormat:@"%@",textfiled.text];
    if (label.text.length==0&&label2.text.length==0 &&label3.text.length==0&&label4.text.length==0&&labb.text.length==0) {
        label1.text=s;                                  //设置文本内容
    }else if(label.text.length!=0&&label2.text.length==0 &&label3.text.length==0&&label4.text.length==0
&&labb.text.length==0){
        label2.text=s;                                  //设置文本内容
    }else if (label1.text.length!=0&&label2.text.length!=0&&label3.text.length==0&&label4.text.length==0&&labb.text.length==0 ){
        label3.text=s;                                  //设置文本内容
    }else if (label1.text.length!=0&&label2.text.length!=0&&label3.text.length!=0&&label4.text.length==0&&labb.text.length==0) {
        label4.text=s;                                  //设置文本内容
    }else if(label1.text.length!=0&&label2.text.length!=0&&label3.text.length!=0&&label4.text.length!=0&&labb.text.length==0) {
        UIAlertView *alert=[[UIAlertView alloc]initWithTitle:@"你的收藏夹已满" message:@"是否清空收藏夹" delegate:self cancelButtonTitle:@"NO" otherButtonTitles:@"YES", nil];     //创建警告视图
        [alert show];
    }
    [UIView beginAnimations:@"" context:nil];
    v.frame=CGRectMake(0, 0, 320, 568);                 //设置框架
    [UIView commitAnimations];
}
```

alertView:clickedButtonAtIndex:方法实现的功能是警告视图中按钮的响应。程序代码如下：

```objc
-(void)alertView:(UIAlertView *)alertView clickedButtonAtIndex:(NSInteger)buttonIndex{
NSString *b=[alertView buttonTitleAtIndex:buttonIndex];
//判断 b 是否为 YES
    if([b isEqualToString:@"YES"]){
        label1.text=nil;
        label2.text=nil;
        label3.text=nil;
        label4.text=nil;
        [self aa:s];
    }
}
```

【代码解析】

本实例关键功能是网址格式的输入，下面讲解这个知识点。在文本框属性的第二大属性即输入属性中，有一个 Keyboard 属性，可以对键盘的类型进行设置，在本实例中是要保存输入的网址，所以此属性设置为 URL。

实例 92　拨　号　器

【实例描述】

在 Text Field 文本框控件的 Keyboard 属性中，有一个 Phone Pad 键盘类型，开发者可以使用此键盘类型实现电话的拨号功能。本实例实现的功能就是电话的拨号。运行效果如图 12.30 所示。

图 12.30　实例 92 运行效果

【实现过程】

当用户单击界面的电话按钮，就可以拨打电话。具体操作过程如下。

（1）创建一个项目，命名为"拨号器"。

（2）添加图片 1.jpg 和 2.png 到创建项目的 Supporting Files 文件夹中。

（3）打开 ViewController.h 文件，编写代码，实现插座变量以及动作的声明。程序代码如下：

```
#import <UIKit/UIKit.h>
@interface ViewController : UIViewController{
    IBOutlet UITextField *tf;                    //声明关于文本框的插座变量
}
- (IBAction)aa:(id)sender;
@end
```

（4）单击打开 Main.storyboard 文件，对 View Controller 视图控制器的设计界面进行设计，效果如图 12.31 所示。

需要添加的视图、控件以及对它们的设置如表 12-23 所示。

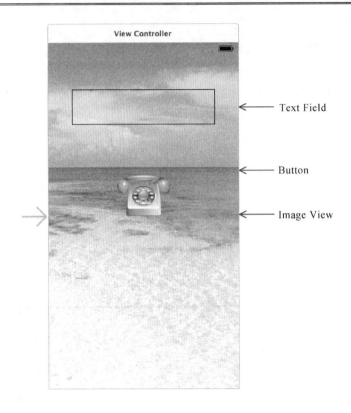

图 12.31　实例 92 设计界面的效果

表 12-23　实例 92 视图、控件设置

视图、控件	属 性 设 置	其 他
Image View	Image：1.jpg	将其覆盖整个设计界面
Text Field	Font：System 33.0 Alignment：居中 Border Style：线框风格 Keyboard：Phone Pad	与插座变量 tf 关联
Button	Title：（空） Image：2.png	与动作 aa:关联

（5）单击打开 ViewController.m 文件，编写代码，实现单击按钮后自动拨号的功能。程序代码如下：

```
- (IBAction)aa:(id)sender {
  NSString *number=[NSString stringWithFormat:@"tel:%@",tf.text];
   //打开内置的拨号器
  [[UIApplication sharedApplication]openURL:[NSURL URLWithString:number]];
}
```

【代码解析】

本实例关键功能是打电话。下面讲解这个知识点。

如果想要实现打电话的功能，可以使用 UIApplication 的 openURL:方法实现。其语法形式如下：

第 12 章 文本处理

```
- (BOOL)openURL:(NSURL *)url;
```

其中，(NSURL *)url 表示一个 URL。在本实例中，使用了 openURL:方法实现了拨号程序打电话的功能。代码如下：

```
[[UIApplication sharedApplication]openURL:[NSURL URLWithString:number]];
```

其中，[NSURL URLWithString:number]表示一个 URL。

实例 93 我的邮箱管理器

【实例描述】

在 Text Field 文本框控件的 Keyboard 属性中，有一个 E-mail 键盘类型，开发者可以使用此键盘类型实现不同邮箱账号的输入。本实例就使用 Keyboard 属性中的 E-mail，来实现在文本框中输入不同的邮箱账号后，进入不同的邮箱登录网页。运行效果如图 12.32 所示。

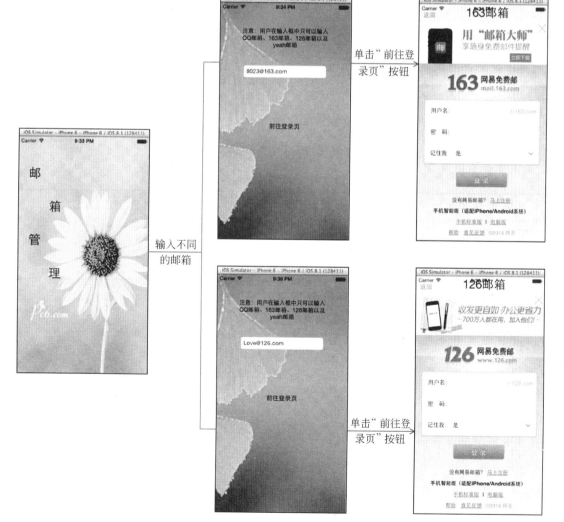

图 12.32 实例 93 运行效果

【实现过程】

（1）创建一个项目，命名为"我的邮箱管理器"。

（2）添加图片 1.jpg 和 2.jpg 到创建项目的 Supporting Files 文件夹中。

（3）单击打开 Main.storyboard 文件，对 View Controller 视图控制器的设计界面进行设计，效果如图 12.33 所示。

图 12.33　实例 93View Controller 视图控制器的设计界面效果

需要添加的视图、控件以及对它们的设置如表 12-24 所示。

表 12-24　实例 93 视图、控件设置 1

视图、控件	属 性 设 置	其　　他
Image View1	Image：1.jpg	
Label1	Text：邮 Font：System 30.0 Alignment：居中	
Label2	Text：箱 Font：System 30.0 Alignment：居中	
Label3	Text：管 Font：System 30.0 Alignment：居中	
Label4	Text：理 Font：System 30.0 Alignment：居中	
Image View2	Image：2.jpg	
Text View	Text：注意：用户在输入框中只可以输入 QQ 邮箱、163 邮箱、126 邮箱以及 yeah 邮箱 Alignment：居中 Background：Default	

续表

视图、控件	属性设置	其他
Text Field	Placeholder：邮箱账号	将此文本框控件和 ViewController.h 文件进行动作 close:的声明和关联 将动作 close:与 Did End On Exit 关联
Button	Title：前往登录页	将此按钮和 ViewController.h 文件进行动作 aa:的声明和关联

（4）创建一个基于 UIViewController 类的 aaViewController 类。

（5）单击打开 aaViewController.h 文件，编写代码，实现插座变量的声明。程序代码如下：

```
#import <UIKit/UIKit.h>
@interface aaViewController : UIViewController{
    IBOutlet UIWebView *web;                //声明关于网页视图的插座变量
}
@end
```

（6）回到 Main.storyboard 文件，从视图库中拖动 View Controller 视图控制器到画布中。将 Class 设置为 aaViewController。这时新增的 View Controller 视图控制器就变为了 Aa View Controller 视图控制器。在 Identity 面板下，将 Storyboard ID 设置为 aa，选择 Use Storyboard ID 复选框。

（7）对 Aa View Controller 视图控制器的设计界面进行设计，效果如图 12.34 所示。

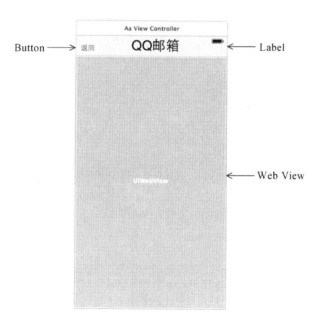

图 12.34 实例 93 Aa View Controller 视图控制器的设计界面效果图

需要添加的视图、控件以及对它们的设置如表 12-25 所示。

表 12-25 实例 93 视图、控件设置 2

视图、控件	属 性 设 置	其 他
设计界面	Background：浅绿色	
Button	Title：返回	将此按钮和 View Controller 视图控件器的界面进行关联
Label	Text：QQ 邮箱 Font：System 29.0 Alignment：居中	
Web View		调整大小 与插座变量 web 关联

（8）单击打开 aaViewController.m 文件，实现加载一个 QQ 邮箱登录的网页。程序代码如下：

```
- (void)viewDidLoad
{
    //加载给定网址的网页
    NSURL *url=[NSURL URLWithString:@"https://mail.qq.com/cgi-bin/
    loginpage"];
    NSURLRequest *request=[NSURLRequest requestWithURL:url];
    [web loadRequest:request];                              //加载
    [super viewDidLoad];
    // Do any additional setup after loading the view.
}
```

（9）创建一个基于 UIViewController 类的 bbViewController 类。

（10）回到 Main.storyboard 文件，从视图库中拖动 View Controller 视图控制器到画布中。将 Class 设置为 bbViewController。这时新增的 View Controller 视图控制器就变为了 Bb View Controller 视图控制器。在 Identity 面板下，将 Storyboard ID 设置为 bb，选择 Use Storyboard ID 复选框。对设计界面的设计请查看源代码。

（11）单击打开 bbViewController.h 文件，编写代码，实现插座变量的声明。程序代码如下：

```
#import <UIKit/UIKit.h>
@interface bbViewController : UIViewController{
    IBOutlet UIWebView *web;                //声明关于网页视图的插座变量
}
@end
```

（12）将声明的插座变量 web 和 Bb View Controller 视图控制器的设计界面中的 Web View 视图进行关联。

（13）单击打开 bbViewController.m 文件，实现加载一个 163 邮箱登录的网页。程序代码如下：

```
- (void)viewDidLoad
{
    NSURL *url=[NSURL URLWithString:@"http://mail.163.com/"];
    NSURLRequest *request=[NSURLRequest requestWithURL:url];
    [web loadRequest:request];                              //加载
    [super viewDidLoad];
    // Do any additional setup after loading the view.
}
@end
```

第 12 章 文本处理

（14）创建一个基于 UIViewController 类的 ccViewController 类。将 Class 设置为 ccViewController。这时新增的 View Controller 视图控制器就变为了 Cc View Controller 视图控制器。在 Identity 面板下，将 Storyboard ID 设置为 cc，选择 Use Storyboard ID 复选框。关于设计界面的设计，请参考源代码。

（15）单击打开 ccViewController.h 文件，编写代码，实现插座变量的声明。程序代码如下：

```
#import <UIKit/UIKit.h>
@interface ccViewController : UIViewController{
    IBOutlet UIWebView *web;                    //声明关于网页视图的插座变量
}
@end
```

（16）将声明的插座变量 web 和 Cc View Controller 视图控制器的设计界面中的 Web View 视图进行关联。

（17）单击打开 ccViewController.m 文件，实现加载一个 163 邮箱登录的网页。程序代码如下：

```
- (void)viewDidLoad
{
    NSURL *url=[NSURL URLWithString:@"http://mail.126.com/"];
    NSURLRequest *request=[NSURLRequest requestWithURL:url];
    [web loadRequest:request];                  //加载
    [super viewDidLoad];
    // Do any additional setup after loading the view.
}
```

（18）创建一个基于 UIViewController 类的 ddViewController 类。

（19）回到 Main.storyboard 文件，从视图库中拖动 View Controller 视图控制器到画布中。将 Class 设置为 ddViewController。这时新增的 View Controller 视图控制器就变为了 Dd View Controller 视图控制器。在 Identity 面板下，将 Storyboard ID 设置为 dd，选择 Use Storyboard ID 复选框。对设计界面的设计请参考源代码。

（20）单击打开 DdViewController.h 文件，编写代码，实现插座变量的声明。程序代码如下：

```
#import <UIKit/UIKit.h>
@interface ddViewController : UIViewController{
    IBOutlet UIWebView *web;                    //声明关于网页视图的插座变量
}
@end
```

（21）将声明的插座变量 web 和 Dd View Controller 视图控制器的设计界面中的 Web View 视图进行关联。

（22）单击打开 ddViewController.m 文件，实现加载一个 yeah 邮箱登录的网页。程序代码如下：

```
- (void)viewDidLoad
{
    NSURL *url=[NSURL URLWithString:@"http://www.yeah.net/"];
    NSURLRequest *request=[NSURLRequest requestWithURL:url];
    [web loadRequest:request];                  //加载
    [super viewDidLoad];
```

```
    // Do any additional setup after loading the view.
}
```

（23）单击打开 ViewController.h 文件，编写代码，实现头文件以及插座变量等的声明。程序代码如下：

```
#import <UIKit/UIKit.h>
#import "aaViewController.h"
#import "bbViewController.h"
#import "ccViewController.h"
#import "ddViewController.h"
@interface ViewController : UIViewController{
    IBOutlet UITextField *tf;
    IBOutlet UIImageView *imagev;              //声明关于图像视图的插座变量
    IBOutlet UIButton *button;
    IBOutlet UITextView *tv;                   //声明关于文本框的插座变量
    aaViewController *av;
    bbViewController *bv;
    ccViewController *cv;
    ddViewController *dv;                      //声明视图控制器对象
}
- (IBAction)aa:(id)sender;
- (IBAction)close:(id)sender;
@end
```

（24）将声明的插座变量和 View Controller 视图控制器的设计界面上的视图和控件进行关联，如表 12-26 所示。

表 12-26 插座变量关联

插 座 变 量	关联的视图、控件
tf	与 TextField 关联
imagev	与 Image View2 关联
button	与 Button 关联
tv	与 Text View 关联

（25）打开 ViewController.m 文件，编写代码。实现输入网址后，显示相应的登录界面。使用的方法如表 12-27 所示。

表 12-27 ViewController.m文件中方法总结

方　　法	功　　能
viewDidLoad	视图加载后调用，实现初始化
show	显示输入界面
aa:	单击按钮后进入相应的登录界面
close:	关闭键盘

这里需要讲解几个重要的方法（其他方法请读者参考源代码）。其中，show 方法实现显示输入界面。程序代码如下：

```
-(void)show{
    [imagev setHidden:NO];
    [tf setHidden:NO];                                   //隐藏文本框
    [button setHidden:NO];
    [tv setHidden:NO];
}
```

aa:方法实现单击按钮后，进入到相应的邮箱登录界面。程序代码如下：

```
- (IBAction)aa:(id)sender {
   NSString *a=[NSString stringWithFormat:@"%@",tf.text];
   //判断字符串 a 是否以 "@qq.com" 结尾
   if([a hasSuffix:@"@qq.com"]==YES){
       av=[self.storyboard instantiateViewControllerWithIdentifier:
       @"aa"];
       [self.view addSubview:av.view];
   }else if([a hasSuffix:@"@163.com"]==YES){
                                      //判断字符串 a 是否以 "@163.com" 结尾
       bv=[self.storyboard instantiateViewControllerWithIdentifier:
       @"bb"];
       [self.view addSubview:bv.view];
   }else if([a hasSuffix:@"@126.com"]==YES){
                                      //判断字符串 a 是否以 "@126.com" 结尾
       cv=[self.storyboard instantiateViewControllerWithIdentifier:
       @"cc"];
       [self.view addSubview:cv.view];
   }else if([a hasSuffix:@"@yeah.net"]){
       dv=[self.storyboard instantiateViewControllerWithIdentifier:
       @"dd"];
       [self.view addSubview:dv.view];       //添加视图对象
   }
}
```

【代码解析】

本实例关键功能是对输入的邮箱进行辨别。下面讲解这个知识点。

如果想要辨别输入的邮箱，需要使用 NSString 的 hasSuffix:方法实现。它的功能是判断字符串是否以某一字符结束。其语法形式如下：

```
- (BOOL)hasSuffix:(NSString *)aString;
```

其中，(NSString *)aString 表示字符串。在此代码中，使用了 hasSuffix:方法实现了在输入邮箱后，对邮箱的辨别，从而进入相应的邮箱。代码如下：

```
NSString *a=[NSString stringWithFormat:@"%@",tf.text];
    if([a hasSuffix:@"@qq.com"]==YES){
        av=[self.storyboard instantiateViewControllerWithIdentifier:@"aa"];
        [self.view addSubview:av.view];
    }
    ……
    else if([a hasSuffix:@"@yeah.net"]){
        dv=[self.storyboard instantiateViewControllerWithIdentifier:@"dd"];
        [self.view addSubview:dv.view];
}
```

实例 94　数 字 天 才

【实例描述】

本实例实现的功能是，在一定的时间限制内对屏幕上出现的数字进行记忆，当屏幕上的数字消失后，在文本框中填写出刚才记忆的数字，若填写正确，进入下一关；若填写失败，可以重新玩本关。运行效果如图 12.35 所示。

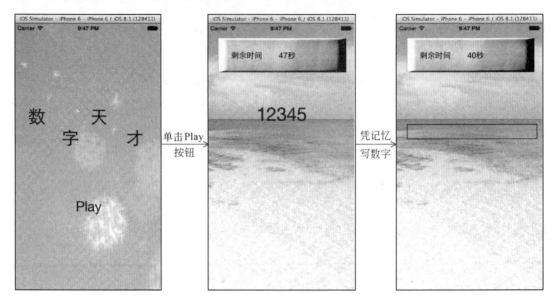

图 12.35 实例 94 运行效果

【实现过程】

（1）创建一个项目，命名为"数字天才"。

（2）添加图片 1.jpg、2.jpg、3.jpg、4.jpg、5.png、6.png 和 7.png 到创建项目的 Supporting Files 文件夹中。

（3）单击打开 Main.storyboard 文件，对 View Controller 视图控制器的设计界面进行设计，效果如图 12.36 所示。

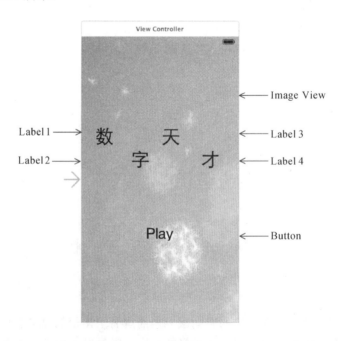

图 12.36 实例 94 View Controller 视图控制器的设计界面效果

需要添加的视图、控件以及对它们的设置如表 12-28 所示。

表 12-28　实例 94 视图、控件设置 1

视图、控件	属性设置	其他
Image View	Image：1.jpg	
Label1	Text：数 Font：System 38.0 Alignment：居中	
Label2	Text：字 Font：System 38.0 Alignment：居中	
Label3	Text：天 Font：System 38.0 Alignment：居中	
Label4	Text：才 Font：System 38.0 Alignment：居中	
Button	Title：Play 将 Font 属性设置为 System 30.0	

（4）创建一个基于 UIViewController 类的 aaViewController 类。

（5）回到 Main.storyboard 文件，从视图库中拖动 View Controller 视图控制器到画布中。将 Class 设置为 aaViewController。这时新增的 View Controller 视图控制器就变为了 Aa View Controller 视图控制器。

（6）在 aaViewController.h 文件中编写代码，实现插座变量、对象、实例变量以及动作的声明。程序代码如下：

```
#import <UIKit/UIKit.h>
@interface aaViewController : UIViewController{
    //插座变量
    IBOutlet UILabel *label;
    IBOutlet UITextField *tf;
    IBOutlet UILabel *aa;
    IBOutlet UIButton *button;              //声明关于按钮的插座变量
IBOutlet UIButton *butt;
    NSTimer *timer;                         //对象
    int i;                                  //实例变量
}
//方法
- (IBAction)close:(id)sender;
- (IBAction)back:(id)sender;
@end
```

（7）回到 Main.storyboard 文件中，将 Play 按钮和 Aa View Controller 视图控制器的设计界面进行关联。

（8）对 Aa View Controller 视图控制器的设计界面进行设计，效果如图 12.37 所示。

需要添加的视图、控件以及对它们的设置如表 12-29 所示。

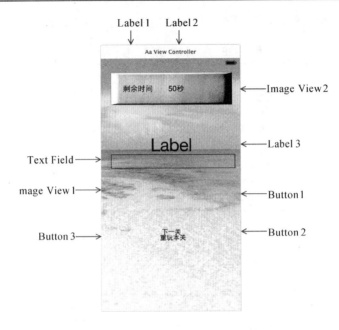

图 12.37　实例 94 Aa View Controller 视图控制器的设计界面效果

表 12-29　实例 94 视图、控件设置 2

视图、控件	属性设置	其他
Image View1	Image：2.jpg	
Button1		覆盖整个设计界面 将此按钮和 aaViewController.h 文件进行动作 close:的声明和关联
Image View2	Image：6.jpg	
Label1	Text：剩余时间	
Label2	Text：50 秒	与插座变量 aa 关联
Label3	Font：System 40.0 Alignment：居中对齐	调整大小 与插座变量 label 关联
Text Field	Font：System 22.0 Alignment：居中 Border Style：线框风格 Keyboard：Number Pad	与插座变量 tf 关联
Button2	Title：下一关 Text Color：黑色	与插座变量 button 关联
Button3	Title：重玩本关 Text Color：黑色	与插座变量 butt 关联 与动作 back:关联

（9）单击打开 aaViewController.m 文件，编写代码，实现在 7 秒内记忆数字，7 秒后写出记忆的数字。使用的方法如表 12-30 所示。

表 12-30　aaViewController.m 文件中方法总结

方　　法	功　　能
viewDidLoad	视图加载后调用，实现初始化
aa	实现倒计时

续表

方　　法	功　　能
ab	实现数字的输入
bb	判断数字的输入
alertView:clickedButtonAtIndex:	实现警告视图的响应
close:	关闭键盘
back:	实现重玩本关

这里需要讲解几个重要的方法（其他方法请读者参考源代码）。其中，viewDidLoad方法实现记忆数字的随机生成。程序代码如下：

```
- (void)viewDidLoad
{
    tf.text=@"";
    [label setHidden:NO];                                    //不隐藏标签
    [butt setHidden:YES];
    [tf setHidden:YES];
    [button setHidden:YES];                                  //隐藏按钮
    i=50;
    int j=arc4random()%4;
    switch (j) {
        case 0:
            label.text=@"12345";                             //设置文本内容
            break;
        case 1:
            label.text=@"12365";                             //设置文本内容
            break;
        case 2:
            label.text=@"231456";                            //设置文本内容
            break;
        default:
            label.text=@"125643";                            //设置文本内容
            break;
    }
    [self performSelector:@selector(ab) withObject:self afterDelay:7];
    timer=[NSTimer scheduledTimerWithTimeInterval:1 target:self selector:
@selector(aa) userInfo:nil repeats:YES];                     //创建时间定时器
    [super viewDidLoad];
    // Do any additional setup after loading the view.
}
```

bb方法实现输入字符串和需要记忆字符串的比较。程序代码如下：

```
-(void)bb{
    NSString *a=[NSString stringWithFormat:@"%@",label.text];
    NSString *b=[NSString stringWithFormat:@"%@",tf.text];
    if([a isEqualToString:b]){
        [button setHidden:NO];                               //不隐藏按钮
    }else{
        UIAlertView *alert=[[UIAlertView alloc]initWithTitle:@"数字白痴"
message:@"你的数字敏感度几乎为0，是否重新挑战" delegate:self
cancelButtonTitle:@"NO" otherButtonTitles:@"YES", nil];
        [alert show];                                        //显示警告视图
    }
}
```

• 567 •

alertView:clickedButtonAtIndex:方法实现警告视图的响应。程序代码如下：

```
-(void)alertView:(UIAlertView *)alertView clickedButtonAtIndex:(NSInteger)
buttonIndex{
    NSString *abb=[alertView buttonTitleAtIndex:buttonIndex];
    if([abb isEqualToString:@"YES"]){          //判断字符串 abb 是否为 YES
        [butt setHidden:NO];
    }
}
```

在本实例中，数字天才游戏分为了 3 关，每一关的难易程度是不一样的，关数越往后，记忆的数字越多，而且记忆的时间也会越来越少。由于篇幅的限制，更多内容请读者参考源代码。

【代码解析】

本实例的关键功能是字符串的比较，和倒计时的实现。下面依次讲解这两个知识点。

1. 字符串的比较

要实现字符串的比较，可以使用 isEqualToString:方法。在此代码中，使用了 isEqualToString:方法对两个从文本标签和文本框中获取的字符串进行比较。代码如下：

```
NSString *a=[NSString stringWithFormat:@"%@",label.text];
NSString *b=[NSString stringWithFormat:@"%@",tf.text];
if([a isEqualToString:b]){
    ……
}else{
    ……
}
```

2. 倒计时的实现

在本实例中，倒计时的实现使用的是定时器。首先，需要创建定时器，代码如下：

```
timer=[NSTimer scheduledTimerWithTimeInterval:1 target:self selector:
@selector(aa) userInfo:nil repeats:YES];
```

此定时器会每隔 1 秒调用一次 aa 方法，然后在 aa 方法中实现倒计时功能，代码如下：

```
i--;
aa.text=[NSString stringWithFormat:@"%i 秒",i];
if(i==0){
    [timer invalidate];
}
```

实例 95　九　宫　格

【实例描述】

本实例实现的功能是对九宫格的挑战。首先是 3*3 的九宫格，如果在一定的时间内输入了正确的答案，单击"下一个"按钮后，会进入 5*5 的九宫格中进行挑战；如果失败，就会出现"重玩本关"的按钮，实现重新开始玩本关。运行效果如图 12.38 所示。

第 12 章 文本处理

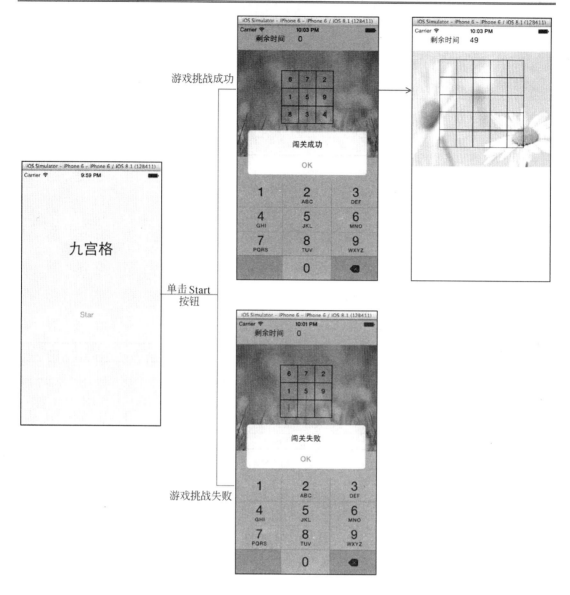

图 12.38 实例 95 运行效果

【实现过程】

（1）创建一个项目，命名为"九宫格"。

（2）添加图片 1.jpg 和 2.jpg 到创建项目的 Supporting Files 文件夹中。

（3）单击打开 Main.storyboard 文件，对 View Controller 视图控制器的设计界面进行设计，效果如图 12.39 所示。

需要添加的视图、控件以及对它们的设置如表 12-31 所示。

（4）创建一个基于 UIViewController 类的 aaViewController 类。

（5）单击打开 aaViewController.h 文件，编写代码，实现插座变量、实例变量以及动作的声明。程序代码如下：

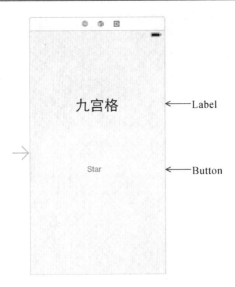

图 12.39　实例 95 View Controller 视图控制器设计界面的效果

表 12-31　实例 95 视图、控件设置 1

视图、控件	属 性 设 置	其　他
设计界面	Background：浅绿色	
Label	Text：九宫格 Font：System 34.0 Alignmen：居中	
Button	Title：Star Font：System 18.0	

```
#import <UIKit/UIKit.h>
@interface aaViewController : UIViewController {
    IBOutlet UITextField *t1;
    IBOutlet UITextField *t2;
    IBOutlet UITextField *t3;                //声明关于文本框的插座变量
    IBOutlet UITextField *t4;
    IBOutlet UITextField *t5;
    IBOutlet UITextField *t6;
    IBOutlet UITextField *t7;
    IBOutlet UITextField *t8;                //声明关于文本框的插座变量
    IBOutlet UITextField *t9;
    NSTimer *timer;
    IBOutlet UILabel *label;
    IBOutlet UIButton *button;               //声明关于按钮的插座变量
    IBOutlet UIButton *butt;
    //用来保存字符串转换为整型的数据
    int a1;
    int a2;
    ……
    int a8;
    int a9;
    //用来保存和
    int sum1;
    int sum2;
    ……
```

```
    int sum9;
    int i;
}
- (IBAction)Close:(id)sender;                    //关闭键盘
- (IBAction)back:(id)sender;
@end
```

（6）回到 Main.storyboard 文件，从视图库中拖动 View Controller 视图控制器到画布中。Class 设置为 aaViewController。这时新增的 View Controller 视图控制器就变为了 Aa View Controller 视图控制器。

（7）对 Aa View Controller 视图控制器的设计界面进行设计，其效果如图 12.40 所示。

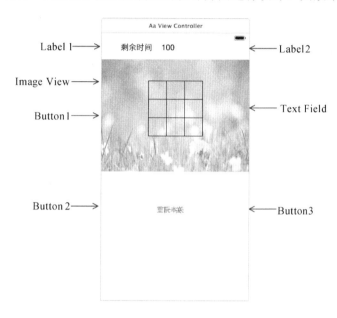

图 12.40　实例 95 Aa View Controller 视图控制器的设计界面效果

需要添加的视图、控件以及对它们的设置如表 12-32 所示。

表 12-32　实例 95 视图、控件设置 2

视图、控件	属 性 设 置	其　　他
Label1	Text：剩余时间	
Label2	Text：100	与插座变量 label 关联
Image View	Image：1.jpg	
Button1	Title：（空）	将大小覆盖整个设计界面 将此按钮和 aaViewController.h 文件进行动作 Close:的声明和关联
9 个 Text Field	Font：System 14.0 Alignment：居中 Border Style：线框风格 Keyboard：Number Pad	将大小设置为（40,40） 与插座变量 t1、t2、t3、t4、t5、t6、t7、t8 和 t9 关联
Button2	Title：下一关	与插座变量 button 关联
Button3	Title：重玩本关	与插座变量 butt 关联 与动作 back 关联

(8) 将 Star 按钮和 Aa View Controller 视图控制器的设计界面关联；将"下一关"按钮和 Bb View Controller 关联，对于此视图控制器的设计界面的设计请参考源代码。

(9) 单击打开 aaViewController.m 文件，编写代码，实现九宫格的功能。使用的方法如表 12-33 所示。

表 12-33　aaViewController.m文件中方法总结

方　　法	功　　能
viewDidLoad	视图加载后调用，实现初始化
time	倒计时
aa	实现判断
Close:	关闭键盘
back:	重新开始

这里需要讲解几个重要的方法（其他方法请读者参考源代码）。其中，viewDidLoad 实现了初始化的设置。程序代码如下：

```
- (void)viewDidLoad
{
    //将 t1～t9 的文字内容设置为空
    t1.text=@"";
    t2.text=@"";
    ……
    t9.text=@"";
    [button setHidden:YES];
    [butt setHidden:YES];
    //将 sum1～sum8 设置为 0
    sum1=0;
    sum2=0;
    ……
    sum7=0;
    sum8=0;
    i=100;
    timer=[NSTimer scheduledTimerWithTimeInterval:1 target:self selector:
@selector(time) userInfo:nil repeats:YES];              //创建时间定时器
    [self performSelector:@selector(aa) withObject:nil afterDelay:100];
    [super viewDidLoad];
    // Do any additional setup after loading the view.
}
```

aa 方法实现对输入的文本进行判断。程序代码如下：

```
-(void)aa{
    NSString *a=[NSString stringWithFormat:@"%@",t1.text];
    a1=[a intValue];                                    //将字符串 a 转换为整型
    NSString *ab=[NSString stringWithFormat:@"%@",t2.text];
    a2=[ab intValue];
    NSString *ac=[NSString stringWithFormat:@"%@",t3.text];
    a3=[ac intValue];                                   //将字符串 ac 转换为整型
    NSString *ad=[NSString stringWithFormat:@"%@",t4.text];
    a4=[ad intValue];
    NSString *ae=[NSString stringWithFormat:@"%@",t5.text];
    a5=[ae intValue];
    NSString *af=[NSString stringWithFormat:@"%@",t6.text];
```

```
        a6=[af intValue];                              //将字符串 af 转换为整型
        NSString *ag=[NSString stringWithFormat:@"%@",t7.text];
        a7=[ag intValue];
        NSString *ah=[NSString stringWithFormat:@"%@",t8.text];
        a8=[ah intValue];                              //将字符串 ah 转换为整型
        NSString *aj=[NSString stringWithFormat:@"%@",t9.text];
        a9=[aj intValue];
        //实现求和
        sum1=a1+a2+a3;
        sum2=a4+a5+a6;
        sum3=a7+a8+a9;
        sum4=a1+a4+a7;
        sum5=a2+a5+a8;
        sum6=a3+a6+a9;
        sum7=a1+a5+a9;
        sum8=a3+a5+a7;
        if((sum1==15)&&(sum2==15)&&(sum3=15)&&(sum4==15)&&(sum5==15)&&(sum6
==15)&&(sum7==15)&&
(sum8=15)){
            UIAlertView *alert=[[UIAlertView alloc]initWithTitle:@"闯关成功"
message:nil delegate:nil cancelButtonTitle:@"OK" otherButtonTitles: nil];
                //创建警告视图
            [alert show];
            [button setHidden:NO];
        }else{
            UIAlertView *alert=[[UIAlertView alloc]initWithTitle:@"闯关失败"
message:nil delegate:nil cancelButtonTitle:@"OK" otherButtonTitles: nil];
                //创建警告视图
            [alert show];
            [butt setHidden:NO];
        }
}
```

【代码解析】

本实例关键功能是字符串的转换。下面就是这个知识点的详细讲解。

如果想要实现字符串的转换功能,可以使用 NSString 的 intValue 方法,它的功能是将字符串转换为整型。其语法形式如下:

```
- (int)intValue;
```

字符串除了可以转换为整型外,还可以转换为其他形式,方法如表 12-34 所示。

表 12-34 字符串的转换方法

方法	功能
doubleValue	将字符串转换为双精度型
floatValue	将字符串转换为单精度型
integerValue	将字符串转换为整型
longLongValue	将字符串转换为长整型
boolValue	将字符串转换为布尔型

在此代码中,文本框输入的文本都是字符串,如果想要实现相加后结果的判断,必须

要将获取的字符串使用 intValue 方法进行转换，代码如下：

```
NSString *a=[NSString stringWithFormat:@"%@",t1.text];
a1=[a intValue];
```

实例 96　单位换算器

【实例描述】

单位换算往往是日常生活中必不可少的。本实例实现的功能实现制作一个单位换算器，让那些还在烦恼单位换算的人不再烦恼。当在任意的一个输入框中输入内容后，按下键盘上的 return 键，就会显示这三个单位之间换算后的结果。其中，用于用户输入的是文本框。运行效果如图 12.41 所示。

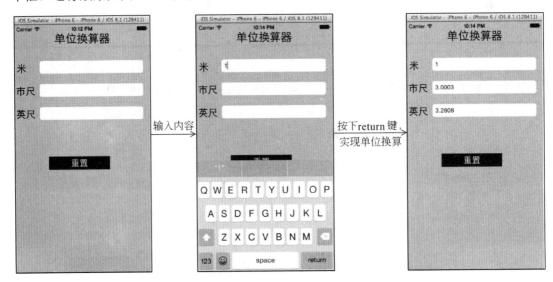

图 12.41　实例 96 运行效果

【实现过程】

（1）创建一个项目，命名为"单位换算器"。

（2）打开 ViewController.h 文件，编写代码，实现遵守协议、插座变量以及动作的声明。程序代码如下：

```
#import <UIKit/UIKit.h>
@interface ViewController : UIViewController<UITextFieldDelegate>{
    IBOutlet UITextField *meter;
    IBOutlet UITextField *chi;                //声明关于文本框的插座变量
    IBOutlet UITextField *feet;
}
- (IBAction)reset:(id)sender;
@end
```

（3）打开 Main.storyboard 文件，对 View Controller 视图控制器的设计界面进行设计，效果如图 12.42 所示。

第 12 章 文本处理

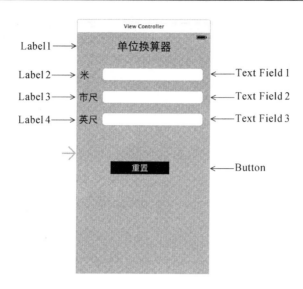

图 12.42 实例 96 View Controller 视图控制器的设计界面

需要添加的视图、控件以及对它们的设置如表 12-35 所示。

表 12-35 实例 96 视图、控件设置

视图、控件	属 性 设 置	其 他
设计界面	Background：浅灰色	
Label1	Text：单位换算器 Font：System 26.0 Alignmen：居中	
Label2	Text：米 Font：System 23.0 Alignmen：居中	
Label3	Text：市尺 Font：System 23.0 Alignmen：居中	
Label4	Text：英尺 Font：System 23.0 Alignmen：居中	
Text Field1		与插座变量 meter 关联
Text Field2		与插座变量 chi 关联
Text Field3		与插座变量 feet 关联
Button	Title：重置 Font：System 20.0 Text Color：白色 Background：黑色	与动作 reset 关联

（4）打开 ViewController.m 文件，编写代码，实现单位之间的换算。使用的方法如表 12-36 所示。

表 12-36　ViewController.m文件中方法总结

方　　法	功　　能
viewDidLoad	视图加载后调用，实现初始化
textFieldShouldReturn:	当用户在键盘上按下 return 键时调用，实现键盘的关闭，以及单位之间的换算
reset:	使文本框的内容为空

这里需要讲解几个重要的方法（其他方法请读者参考源代码）。其中，textFieldShouldReturn:方法实现当用户在键盘上按下 return 键时调用，实现键盘的关闭，以及单位之间的换算。程序代码如下：

```objc
- (BOOL)textFieldShouldReturn:(UITextField *)textField
{
    if (textField==meter) {
        [meter resignFirstResponder];                              //关闭键盘
        double c=[meter.text doubleValue]*3.0003;
        double f=[meter.text doubleValue]*3.2808;
        chi.text=[[NSString alloc]initWithFormat:@"%.4f",c];  //设置文本内容
        feet.text=[[NSString alloc]initWithFormat:@"%.4f",f];
    }
    if (textField==chi) {
        [chi resignFirstResponder];                                //关闭键盘
        double m=[chi.text doubleValue]/3.0003;
        double f=m*3.2808;
        meter.text=[[NSString alloc]initWithFormat:@"%.4f",m];
        feet.text=[[NSString alloc]initWithFormat:@"%.4f",f];  //设置文本内容
    }
    if (textField==feet) {
        [feet resignFirstResponder];                               //关闭键盘
        double m=[feet.text doubleValue]/3.2808;
        double c=m*3.0003;
        meter.text=[[NSString alloc]initWithFormat:@"%.4f",m];
        chi.text=[[NSString alloc]initWithFormat:@"%.4f",c];   //设置文本内容
    }
    return YES;
}
```

【代码解析】

本实例关键功能是按下 return 键后的响应。下面就是这个知识点的详细讲解。

当用户在键盘上按下 return 键后，首先会退出键盘，其次是进行单位换算，要实现这些功能，要调用 UITextFieldDelegate 的 textFieldShouldReturn:方法。它的功能是当按下 return 键后，判断是否结束文本编辑。其语法实现如下：

```objc
- (BOOL)textFieldShouldReturn:(UITextField *)textField;
```

在本代码中，就是重写了 textFieldShouldReturn:方法，实现了退出键盘以及单位换算功能。代码如下：

```objc
- (BOOL)textFieldShouldReturn:(UITextField *)textField
{
    if (textField==meter) {
        [meter resignFirstResponder];                              //关闭键盘
        double c=[meter.text doubleValue]*3.0003;
        double f=[meter.text doubleValue]*3.2808;
        chi.text=[[NSString alloc]initWithFormat:@"%.4f",c];
```

```
        feet.text=[[NSString alloc]initWithFormat:@"%.4f",f];
    }
    ……
    return YES;
}
```

实例 97　计算器

【实例描述】

本实例的功能是实现一个支持两位数运算的计算器。当用户输入操作数和运算符后，单击"="按钮，就会显示最后的结果，其中，用于显示输入和最后结果的是标签。运行效果如图 12.43 所示。

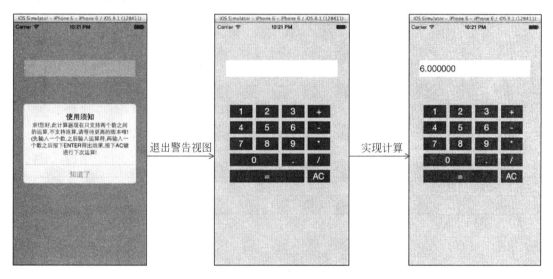

图 12.43　实例 97 运行效果

【实现过程】

（1）创建一个项目，命名为"计算器"。

（2）打开 ViewController.h 文件，编写代码，实现插座变量、实例变量、对象、属性以及动作的声明。程序代码如下：

```
#import <UIKit/UIKit.h>
@interface ViewController : UIViewController{
    double nSum;                                      //存放运算符左边的数
    IBOutlet UILabel *label;
    NSMutableString *string;                          //存放字符串
    double number;                                    //存放运算符右边的数
}
@property double nSum;
@property double number;
@property(strong,nonatomic) NSMutableString *string;
- (IBAction)Press:(id)sender;
@end
```

（3）打开 Main.storyboard 文件，对 View Controller 视图控制器的设计界面进行设计，效果如图 12.44 所示。

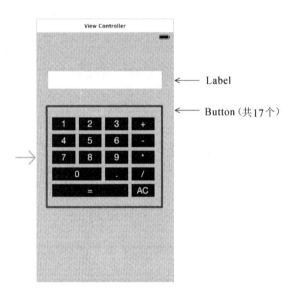

图 12.44　实例 97 View Controller 视图控制器的设计界面

需要添加的视图、控件以及对它们的设置如表 12-37 所示。

表 12-37　实例 97 视图、控件设置

视图、控件	属　性　设　置	其　　他
设计界面	Background：浅灰色	
Label	Text：（空） Font：System 21.0 Background：白色	与插座变量 label 关联
Button（共 17 个）	Text 的设置请参考设计界面 Font：System 20.0 Text Color：白色 Background：黑色 将标题为 0～9 的按钮的 tag 值设置为 0～9 将标题为.的按钮的 tag 值设置为 10 将标题为+-*/的按钮的 tag 值设置为 11～14 将标题为 AC 的按钮的 tag 值设置为 15 将标题为=的按钮的 tag 值设置为 16	与动作 Press 关联

（4）打开 ViewController.m 文件，编写代码，实现计算器的功能。首先在 viewDidLoad 方法中实现警告视图的显示以及创建可变字符串。程序代码如下：

```
- (void)viewDidLoad
{
    [super viewDidLoad];
    // Do any additional setup after loading the view, typically from a nib.
    string=[NSMutableString stringWithCapacity:50];
    //创建警告视图
    UIAlertView *alert=[[UIAlertView alloc] initWithTitle:@"使用须知
"message:@"亲！您好,此计算器现在只支持两个数之间的运算,不支持连算,请等待更高的版本
```

哦！(先输入一个数,之后输入运算符,再输入一个数之后按下 ENTER 得出结果,按下 AC 键进行下次运算!"delegate:nil cancelButtonTitle:@"知道了"otherButtonTitles: nil);
    [alert show];   //显示警告视图
}

然后在 Press 方法中实现计算器的计算功能。程序代码如下：

```objc
- (IBAction)Press:(id)sender {
  UIButton*button=(UIButton*)sender;
  [string appendString:button.titleLabel.text];
  label.text=[NSString stringWithFormat:@"%@",string];    //设置文本内容
  if(11<=button.tag&&button.tag<=14) {                    //判断接收到运算符
      number=[string doubleValue];                        //取左操作数
      [string setString:@""];                             //字符串清零
      [string appendFormat:@"%@",button.titleLabel.text];
  }
  if(16==button.tag) {
      if([string hasPrefix:@"+"]) {                       //字符串以加号开头
          nSum=number+[string doubleValue];
          label.text=[NSString stringWithFormat:@"%f",nSum];
          nSum=0;
      }else if([string hasPrefix:@"-"]) {                 //字符串以减号开头
          nSum=number+[string doubleValue];
          label.text=[NSString stringWithFormat:@"%f",nSum];
          nSum=0;
      }else if([string hasPrefix:@"*"]) {                 //字符串以乘号开头
          NSRange range;
          range=[string rangeOfString:@"*"];
          [string deleteCharactersInRange:range];
          nSum=number*[string doubleValue];
          label.text=[NSString stringWithFormat:@"%f",nSum];
          nSum=0;
      }else if([string hasPrefix:@"/"]) {                 //字符串以除号开头
          NSRange range;
          range=[string rangeOfString:@"/"];
          [string deleteCharactersInRange:range];
          nSum=number/[string doubleValue];
          label.text=[NSString stringWithFormat:@"%f",nSum];
          nSum=0;
      }
  }
  if(15==button.tag) {                                    //当按下 AC 建时,所有数据清零
      [string setString:@""];
      nSum=0;
      number=0;
      label.text=@"0";
  }
}
```

【代码解析】

本实例关键功能是计算器的计算。下面就是这个知识点的详细讲解。

在本实例中，计算器的计算功能主要是通过 tag 实现的。首先，为每一个按钮设置 tag 值。其次再计算输入的字符串，在计算字符串时，需要对运算符运行判断，根据运算符的不同进行相应的计算。程序代码如下：

```objc
if(16==button.tag) {
  if([string hasPrefix:@"+"]) {                           //字符串以加号开头
```

```
            nSum=number+[string doubleValue];
            label.text=[NSString stringWithFormat:@"%f",nSum];
            nSum=0;
        }else if([string hasPrefix:@"-"]) {              //字符串以减号开头
            nSum=number+[string doubleValue];
            label.text=[NSString stringWithFormat:@"%f",nSum];
            nSum=0;
        }else if([string hasPrefix:@"*"]) {              //字符串以乘号开头
            NSRange range;
            range=[string rangeOfString:@"*"];
            [string deleteCharactersInRange:range];
            nSum=number*[string doubleValue];
            label.text=[NSString stringWithFormat:@"%f",nSum];
            nSum=0;
        }else if([string hasPrefix:@"/"]) {              //字符串以除号开头
            NSRange range;
            range=[string rangeOfString:@"/"];
            [string deleteCharactersInRange:range];
            nSum=number/[string doubleValue];
            label.text=[NSString stringWithFormat:@"%f",nSum];
            nSum=0;
        }
    }
```

实例 98　表情键盘

【实例描述】

在一些常用的交流软件中（如 QQ，微信），经常会看到表情键盘的出现，为用户发送的信息添加了不少风趣。那么这样一个表情键盘是如何实现的呢？本实例就为读者来实现这样一个表情键盘。当用户选择键盘中的任意键盘，会将选择的表情添加到文本视图中。运行效果如图 12.45 所示。

图 12.45　实例 98 运行效果

第 12 章 文本处理

【实现过程】
（1）创建一个项目，命名为"表情键盘"。
（2）添加图片 1.png～29.png 到创建项目的 Supporting Files 文件夹中。
（3）创建一个基于 CALayer 类的 ViewLayer 类。
（4）打开 ViewLayer.h 文件，编写代码，实现实例变量和属性的声明。程序代码如下：

```
#import <QuartzCore/QuartzCore.h>
@interface ViewLayer : CALayer{
    CGImageRef _keytopImage;;                              //声明变量
}
@property (nonatomic, retain) UIImage* emoji;
@end
```

（5）打开 ViewLayer.m 文件，编写代码，实现在表情键盘中选择表情后的状态，即显示一个指示器。程序代码如下：

```
- (void)drawInContext:(CGContextRef)context
{
    //从后台返回需要重新获取图片
    _keytopImage = [[UIImage imageNamed:@"29.png"] CGImage];
                                                          //将 UIImage 转换为 CGIMage
    UIGraphicsBeginImageContext(CGSizeMake(82, 111));
                                                          //创建一个基于位图的上下文
    CGContextTranslateCTM(context, 0.0, 111);
    CGContextScaleCTM(context, 1.0, -1.0);
    CGContextDrawImage(context, CGRectMake(0, 0, 82, 111), _keytopImage);
                                                          //绘制图像
    UIGraphicsEndImageContext();
    UIGraphicsBeginImageContext(CGSizeMake(55, 55));  //创建一个基于位图的上下文
    CGContextDrawImage(context, CGRectMake((82 - 55) / 2 , 45, 55, 55),
    [_emoji CGImage]);
    UIGraphicsEndImageContext();
}
```

（6）创建一个基于 UIView 类的 View 类。
（7）打开 View.h 文件，编写代码，实现头文件、协议、对象、实例变量以及属性的声明。程序代码如下：

```
#import <UIKit/UIKit.h>
//头文件
#import "ViewLayer.h"
@class View;
//协议
@protocol ViewDelegate<NSObject>
@optional
- (void)didTouchEmojiView:(View*)emojiView touchedEmoji:(NSString*)string;
@end
@interface View : UIView{
    //对象
    NSArray *_emojiArray;
    NSArray *_symbolArray;
    ViewLayer *_emojiPadLayer;
    NSInteger _touchedIndex;                              //实例变量
}
@property (assign, nonatomic) id<ViewDelegate> delegate;
                                                          //ViewDelegate 协议的属性
@end
```

(8) 打开 View.m 文件，编写代码，实现表情键盘的绘制以及键盘中键的选择。使用的方法如表 12-38 所示。

表 12-38 View.m文件中方法总结

方　　法	功　　能
initWithFrame:	表情键盘的初始化
drawRect:	绘制表情键盘
indexWithEvent:	使用事件获取索引
updateWithIndex:	使用索引更新指示器
touchesBegan:withEvent:	开始触摸
touchesMoved:withEvent:	移动触摸
touchesEnded:withEvent:	结束触摸

这里需要讲解几个重要的方法（其他方法请读者参考源代码）。实现表情键盘的绘制，需要使用 initWithFrame:和 drawRect:方法。其中，initWithFrame:方法使用框架实现表情键盘的初始化。程序代码如下：

```
- (id)initWithFrame:(CGRect)frame
{
    self = [super initWithFrame:frame];
if (self) {
        //创建数组对象_emojiArray
        _emojiArray = [NSArray arrayWithObjects:
                    [UIImage imageNamed:@"1.png"],
                    [UIImage imageNamed:@"2.png"],
                    [UIImage imageNamed:@"3.png"],
                    ……
                    [UIImage imageNamed:@"24.png"],
                    [UIImage imageNamed:@"25.png"],
                    [UIImage imageNamed:@"26.png"],
                    [UIImage imageNamed:@"27.png"],
                    [UIImage imageNamed:@"28.png"],
                    nil];
        //创建数组对象_symbolArray
        _symbolArray = [NSArray arrayWithObjects:
                    @"\U0001F604",
                    @"\U0001F60A",
                    @"\U0001F603",
                    ……
                    @"\U0001F622",
                    @"\U0001F62D",
                    @"\U0001F602",
                    @"\U0001F632",
                    @"\U0001F631",
                    nil];

        //创建指示器
        _emojiPadLayer = [ViewLayer layer];
        [self.layer addSublayer:_emojiPadLayer];
        //背景透明
        [self setBackgroundColor:[UIColor clearColor]];
    }
    return self;
}
```

drawRect:方法实现表情键盘的绘制。程序代码如下：

```objc
- (void)drawRect:(CGRect)rect
{
    int index =0;
    //遍历
    for(UIImage *image in _emojiArray) {
        float originX = (self.bounds.size.width / 7) * (index % 7) + ((self.bounds.size.width / 7) - 45 ) / 2;
        float originY = (index / 7) * (self.bounds.size.width / 7) + ((self.bounds.size.width / 7) - 45 ) / 2;
        [image drawInRect:CGRectMake(originX, originY, 45, 45)];    //绘制
        index++;
    }
}
```

如果想要实现键盘中键的选择，需要使用 indexWithEvent:、updateWithIndex:、touchesBegan:withEvent:、touchesMoved:withEvent:和 touchesEnded:withEvent:方法。其中，updateWithIndex:方法通过使用索引对键盘上的指示器进行更新。程序代码如下：

```objc
- (void)updateWithIndex:(NSUInteger)index
{
    //判断 index 是否小于_emojiArray 的个数
    if(index < _emojiArray.count) {
        _touchedIndex = index;
        if (_emojiPadLayer.opacity != 1.0) {
            _emojiPadLayer.opacity = 1.0;                    //设置透明度
        }
        float originX = (self.bounds.size.width / 7) * (index % 7) + ((self.bounds.size.width / 7) - 45) / 2;
        float originY = (index / 7) * (self.bounds.size.width / 7) + ((self.bounds.size.width / 7) - 45 ) / 2;
        [_emojiPadLayer setEmoji:[_emojiArray objectAtIndex:index]];
                                                            //设置指示器上的表情
        [_emojiPadLayer setFrame:CGRectMake(originX - (82 - 45) / 2, originY - (111 - 45), 82, 111)];
        [_emojiPadLayer setNeedsDisplay];
    }
}
```

touchesBegan:withEvent:、touchesMoved:withEvent:和 touchesEnded:withEvent:方法实现触摸的功能。程序代码如下：

```objc
//开始触摸
- (void)touchesBegan:(NSSet *)touches withEvent:(UIEvent *)event
{
    NSUInteger index = [self indexWithEvent:event];
    //判断 index 是否小于_emojiArray 的个数
    if(index < _emojiArray.count) {
        [CATransaction begin];
        [CATransaction setValue:(id)kCFBooleanTrue forKey:
        kCATransactionDisableActions];
        [self updateWithIndex:index];
        [CATransaction commit];
    }
}
//移动触摸
- (void)touchesMoved:(NSSet *)touches withEvent:(UIEvent *)event
```

```
{
    NSUInteger index = [self indexWithEvent:event];
    if (_touchedIndex >=0 && index != _touchedIndex && index < _emojiArray.
    count) {
        [self updateWithIndex:index];                        //更新
    }
}
//结束触摸
- (void)touchesEnded:(NSSet *)touches withEvent:(UIEvent *)event
{
    if (self.delegate && _touchedIndex >= 0) {
        //判断是否实现了didTouchEmojiView:touchedEmoji:方法
        if ([self.delegate respondsToSelector:@selector(didTouchEmojiView:
        touchedEmoji:)]) {
            [self.delegate didTouchEmojiView:self touchedEmoji:
            [_symbolArray objectAtIndex:_touchedIndex]];
        }
    }
    _touchedIndex = -1;
    _emojiPadLayer.opacity = 0.0;                            //设置透明度
    [self setNeedsDisplay];
    [_emojiPadLayer setNeedsDisplay];
}
```

（9）打开 ViewController.h 文件，编写代码，实现头文件、对象、插座变量以及动作的声明。程序代码如下：

```
#import <UIKit/UIKit.h>
#import "View.h"                        //头文件
@interface ViewController : UIViewController<ViewDelegate>{
    View *_emojiView;
    IBOutlet UITextView *tv;            //声明关于文本视图的插座变量
}
- (IBAction)show:(id)sender;
@end
```

（10）打开 Main.storyboard 文件，对 View Controller 视图控制器的设计界面进行设计，效果如图 12.46 所示。

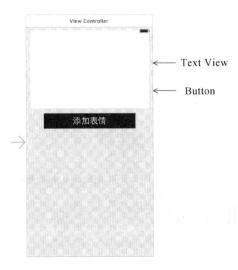

图 12.46　实例 98View Controller 视图控制器的设计界面

第 12 章 文本处理

需要添加的视图、控件以及对它们的设置如表 12-39 所示。

表 12-39 实例 98 视图、控件设置

视图、控件	属性设置	其他
设计界面	Background：浅灰色	
Text View	Text：（空）	与插座变量 tv 关联
Button	Title：添加表情 Font：System 21.0 Text Color：白色 Background：黑色	与动作 show:关联

（11）打开 ViewController.m 文件，编写代码，实现创建表情键盘以及表情键盘的选择。使用的方法如表 12-40 所示。

表 12-40 实例 98ViewController.m文件中方法总结

方法	功能
viewDidLoad	视图加载后调用，实现初始化
show:	显示表情键盘
didTouchEmojiView:touchedEmoji:	将选择的表情放到文本视图中

其中，viewDidLoad 方法在视图加载后调用，实现表情键盘的创建以及对文本视图的设置。程序代码如下：

```
- (void)viewDidLoad
{
    [super viewDidLoad];
    // Do any additional setup after loading the view, typically from a nib.
    //表情键盘对象的创建
    _emojiView = [[View alloc] initWithFrame:CGRectMake(0, self.view.frame.size.height , 320, 216)];
    _emojiView.backgroundColor=[UIColor colorWithHue:0.6 saturation:0.5 brightness:1.0 alpha:1.0];
    _emojiView.delegate = self;
    [self.view addSubview:_emojiView];
    //设置文本视图对象
    [tv setEditable:YES];
    [tv setFont:[UIFont systemFontOfSize:30]];
    [tv setTextColor:[UIColor blackColor]];            //设置文本颜色
    [tv setContentInset:UIEdgeInsetsMake(0, 4, 0, 4)];
    [tv.layer setCornerRadius:6];                      //设置圆角半径
    [tv.layer setMasksToBounds:YES];
}
```

show:方法实现单击界面的按钮后表情键盘的显示，以及普通键盘的退出。程序代码如下：

```
- (IBAction)show:(id)sender {
    [tv resignFirstResponder];
    [UIView beginAnimations:@"" context:nil];
    [UIView setAnimationDuration:0.5];                 //设置持续时间
    _emojiView.frame=CGRectMake(0, self.view.frame.size.height - 216, 320, 216);//设置框架
    [UIView commitAnimations];
}
```

didTouchEmojiView:touchedEmoji:方法实现选择键盘上的表情后,将表情显示在文本视图中。程序代码如下:

```
- (void)didTouchEmojiView:(View*)emojiView touchedEmoji:(NSString*)str
{
    tv.text = [NSString stringWithFormat:@"%@%@", tv.text, str];
                                        //设置文本内容
}
```

【代码解析】

由于本实例中的代码和方法非常多,为了方便读者阅读,笔者绘制了一些执行流程图,如图 12.47 和图 12.48 所示。其中,单击运行按钮在界面显示表情键盘,需要使用 viewDidLoad、initWithFrame:、drawRect:和 show:方法共同实现。它们的执行流程如图 12.47 所示。

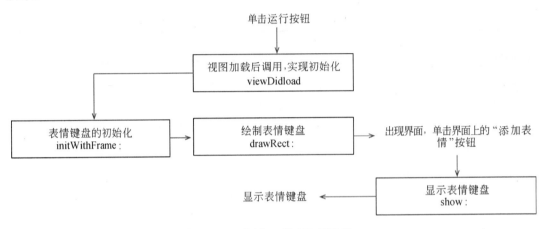

图 12.47 实例 98 程序执行流程 1

如果想要实现表情键盘中任意键的选择,需要使用 touchesBegan:withEvent:、indexWithEvent:、updateWithIndex:、touchesEnded:withEvent:、didTouchEmojiView:touched-Emoji:、drawRect:和 drawInContext:方法共同实现。它们的执行流程如图 12.48 所示。

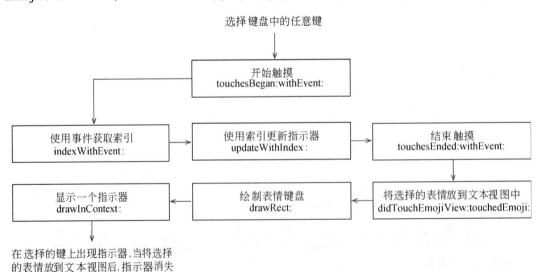

图 12.48 实例 98 程序执行流程 2

第 13 章 表

对于大量数据的显示以及处理，通常需要使用表视图，例如可以在通讯录中看到它的使用。表是排列数据非常好的一种方法。它的利用贯穿了所有的交流、研究和数据分析。本章将详细讲解关于表的各种经典应用。

实例 99　邮编查询

【实例描述】
　　列表的显示形式是多种多样的，在本实例中就实现了列表居左显示了主标签（textLabel），居右显示了附属标签（detailTextLabel）的功能。运行效果如图 13.1 所示。

图 13.1　实例 99 运行效果

【实现过程】
　　在表视图的左边显示地名，右边显示邮编。这些文本都是使用表单元格自带的标签实现的。具体的实现步骤如下。
　　（1）创建一个项目，命名为"山西邮编查询"。

（2）添加图片 1.png 到创建项目的 Supporting Files 文件夹中。
（3）打开 Main.storyboard 文件，对 View Controller 视图控制器的设计界面进行设计，效果如图 13.2 所示。

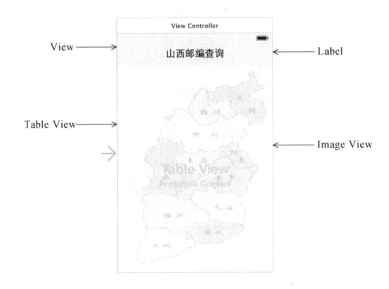

图 13.2 实例 99View Controller 视图控制器的设计界面

需要添加的视图、控件以及对它们的设置如表 13-1 所示。

表 13-1 实例 99 视图、控件设置

视图、控件	属 性 设 置	其 他
View	Background：浅灰色	
Label	Text：山西邮编查询 Font：System 20.0 Alignment：居中	
Image View	Image：1.jpg	
Table View	Alpha：0.8	将其放置到 Image View 上 将 dataSource 与 View Controller 关联 将 delegate 与 View Controller 关联

（4）打开 ViewController.h 文件，编写代码，实现对象的声明。程序代码如下：

```
#import <UIKit/UIKit.h>
@interface ViewController : UIViewController{
    NSArray *city;
    NSArray *code;                          //声明对象
}
@end
```

（5）打开 ViewController.m 文件，编写代码，实现邮编的添加。程序代码如下：

```
- (void)viewDidLoad
{
    city=[[NSArray alloc]initWithObjects:@"太原",@"大同", @"阳泉",@"长治",@"晋城",@"朔州",@"晋中",@"运城",@"忻州",@"临汾",@"吕梁",nil];      //实例化对象
    code=[[NSArray alloc]initWithObjects:@"030000",@"037000",@"045000",
```

```
@"046000",@"048000",@"036000",@"030600",@"044000",@"034000",@"041000",@
"033000", nil];
    [super viewDidLoad];
    // Do any additional setup after loading the view, typically from a nib.
}
//获取块数
- (NSInteger)numberOfSectionsInTableView:(UITableView *)tableView {
    return 1;
}
//获取行数
- (NSInteger)tableView:(UITableView *)tableView numberOfRowsInSection:
(NSInteger)section {
    return [city count];
}
//获取某一行的数据
- (UITableViewCell *)tableView:(UITableView *)tableView cellForRowAtIndexPath:
(NSIndexPath *)indexPath
{
    static NSString *CellIdentifier = @"Cell";
    UITableViewCell *cell = [tableView dequeueReusableCellWithIdentifier:
    CellIdentifier];
    //判断 cell 是否为空
    if (cell==nil)
    {
        cell = [[UITableViewCell alloc] initWithStyle:UITableViewCellStyle
        Value1 reuseIdentifier:CellIdentifier];
    }
    cell.textLabel.text=[city objectAtIndex:indexPath.row];
    //设置文本内容
    cell.detailTextLabel.text=[code objectAtIndex:indexPath.row];
    return cell;
}
```

【代码解析】

本实例关键功能是附属标签的显示以及表单元格的风格。下面依次讲解这两个知识点。

1. 附属标签的显示

在表视图中常常会看到只有一个标签，称之为主标签。其实它还可以有一个附属标签，它的显示需要使用 UITabelViewCell 的 detailTextLabel 属性实现。其语法形式如下：

```
@property(nonatomic, readonly, retain) UILabel *detailTextLabel;
```

在本例中，就使用了 detailTextLabel 属性实现了邮编的添加。代码如下：

```
cell.detailTextLabel.text=[code objectAtIndex:indexPath.row];
```

2. 表单元格的风格

如果想要实现表单元格风格的改变，可以在创建表单元格时，使用 initWithStyle:reuseIdentifier:方法实现。其语法形式如下：

```
- (id)initWithStyle:(UITableViewCellStyle)style reuseIdentifier:(NSString
*)reuseIdentifier;
```

其中，(UITableViewCellStyle)style 表示单元格的风格；(NSString *)reuseIdentifier 是一个字符串，用来标识该单元格对象。其中，单元格的风格有 4 种，如表 13-2 所示。

表 13-2 单元格的风格

风　格	功　能
UITableViewCellStyleDefault	该风格提供了一个简单的左对齐的主标签（textLabel）和一个可选图像视图
UITableViewCellStyleSubtitle	该风格添加了对附属标签（detailTextLabel）的支持，该标签将显示在主标签的下方，字符相对比较小
UITableViewCellStyleValue1	该风格居左显示了主标签（textLabel），居右显示了附属标签（detailTextLabel）
UITableViewCellStyleValue2	该风格居左显示一个蓝色的主标签（textLabel），在右边显示一个小型的黑色附属标签（detailTextLabel）

实例 100　水 平 列 表

【实例描述】

在大多数情况下，列表都是以垂直样子排列的，它的显示占据了很大的屏幕。为了节省屏幕，可以将垂直的列表转换为水平列表。本实例的功能就是实现这么一个水平列表。运行效果如图 13.3 所示。

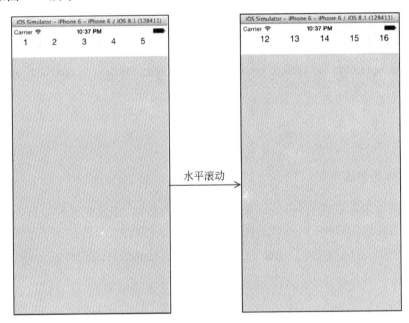

图 13.3　实例 100 运行效果

【实现过程】

（1）创建一个项目，命名为"水平列表"。

（2）打开 Main.storyboard 文件，将设计界面的背景颜色设置为墨绿色。

（3）打开 ViewController.h 文件，编写代码，声明此类遵守的协议。

（4）打开 ViewController.m 文件，编写代码，实现水平列表的绘制和选择任意行的响应。使用的方法如表 13-3 所示。

第13章 表

表 13-3 ViewController.m文件中方法总结

方　　法	功　　能
viewDidLoad	视图加载后调用，实现初始化
tableView:numberOfRowsInSection:	获取行数
tableView:cellForRowAtIndexPath:	获取某一行的数据
tableView:didSelectRowAtIndexPath:	选择行的响应

水平列表的绘制需要使用 viewDidLoad、tableView:numberOfRowsInSection: 和 tableView:cellForRowAtIndexPath:方法。其中，viewDidLoad 方法实现表的创建。程序代码如下：

```
- (void)viewDidLoad
{
    [super viewDidLoad];
    // Do any additional setup after loading the view, typically from a nib.
    UITableView *table = [[UITableView alloc] initWithFrame:CGRectMake(0, 60, 60, 480)];
    table.backgroundColor = [UIColor whiteColor];          //设置背景颜色
    table.transform = CGAffineTransformMakeRotation(M_PI/-2);  //旋转
    table.showsVerticalScrollIndicator = NO;
    table.frame = CGRectMake(0, 0, 320, 60);
    table.rowHeight = 60.0;
    table.delegate = self;                                 //设置委托
    table.dataSource = self;
    [self.view addSubview:table];
}
```

tableView:numberOfRowsInSection: 和 tableView:cellForRowAtIndexPath:方法实现对列表的数据填充。程序代码如下：

```
//获取行数
- (NSInteger)tableView:(UITableView *)tableView numberOfRowsInSection:(NSInteger)section{
    return 16;
}
//获取某一行的内容
- (UITableViewCell *)tableView:(UITableView *)tableView cellForRowAtIndexPath:(NSIndexPath *)indexPath{
    static NSString *CellIdentifier = @"Cell";
    UITableViewCell *cell = [tableView dequeueReusableCellWithIdentifier:CellIdentifier];
    //判断cell是否为空
    if (cell == nil) {
        cell = [[UITableViewCell alloc] initWithStyle:UITableViewCellStyleDefault reuseIdentifier:CellIdentifier] ;
        cell.selectionStyle = UITableViewCellSelectionStyleGray;
                                                //设置选择风格
        cell.textLabel.textAlignment = NSTextAlignmentCenter;
                                                //设置对齐方式
        cell.textLabel.transform = CGAffineTransformMakeRotation(M_PI/2);
    }
    cell.textLabel.text = [NSString stringWithFormat:@"%d",indexPath.row+1];
                                                //设置文本内容
    return cell;
}
```

如果想要实现对选择行的响应，需要使用 tableView:didSelectRowAtIndexPath:方法。

程序代码如下：

```
- (void)tableView:(UITableView *)tableView didSelectRowAtIndexPath:
(NSIndexPath *)indexPath{
    NSString *str=[NSString stringWithFormat:@"你选择的是第%i行",indexPath.
    row +1];
    UIAlertView *alert=[[UIAlertView alloc]initWithTitle:str message:nil
delegate:nil cancelButtonTitle:@"Cancel" otherButtonTitles: nil];
                                        //创建警告视图
    [alert show];
}
```

【代码解析】

本实例关键功能是水平列表的实现。下面就是这个知识点的详细讲解。

如果想要实现水平列表。首先，需要将创建的表视图对象使用 transform 属性进行旋转，随后再对其框架进行重新设置。代码如下：

```
table.transform = CGAffineTransformMakeRotation(M_PI/-2);
......
table.frame = CGRectMake(0, 0, 320, 60);
```

最后，就是改变表单元格中的内容。同样也是使用 transform 属性将表单元格中的标签进行旋转。代码如下：

```
cell.textLabel.transform = CGAffineTransformMakeRotation(M_PI/2);
```

实例 101　表的自动调整

【实例描述】

在表视图中，表单元格中的内容有多有少，那么该如何根据这些内容的多少来自动调整表的行高呢？本实例就为各位开发者解决这一烦恼。运行效果如图 13.4 所示。

图 13.4　实例 101 运行效果

【实现过程】

（1）创建一个项目，命名为"表的自动调整"。

(2)打开 ViewController.h 文件，编写代码，实现宏定义、插座变量和对象的声明。程序代码如下：

```
#import <UIKit/UIKit.h>
//宏定义
#define FONT_SIZE 14.0f
#define CELL_CONTENT_WIDTH 320.0f
#define CELL_CONTENT_MARGIN 10.0f
@interface ViewController : UIViewController{
    IBOutlet UITableView *dataTableView;         //插座变量
    NSMutableArray *items;                        //对象
}
@end
```

(3)打开 Main.storyboard 文件，在设计界面添加 Table View 表视图。将此表视图和插座变量 dataTableView 关联。将此表视图的 dataSource 和 delegate 与 View Controller 关联。

(4)打开 ViewController.m 文件，编写代码，实现表的自动调整。使用的方法如表 13-4 所示。

表 13-4 ViewController.m文件中方法总结

方　法	功　　能
viewDidLoad	视图加载后调用，实现初始化
tableView:numberOfRowsInSection:	获取行数
numberOfSectionsInTableView:	获取表视图的块
tableView:heightForRowAtIndexPath:	获取表中每行的高度
tableView:cellForRowAtIndexPath:	获取某一行的数据

这里需要讲解几个重要的方法（其他方法请读者参考源代码）。其中，viewDidLoad 方法中实现表视图中内容的初始化。程序代码如下：

```
- (void)viewDidLoad {
    [super viewDidLoad];
    items = [[NSMutableArray alloc] init];       //创建可变数组对象
    //添加对象
    [items addObject:@"沉默是一种气质，也是一种风度，更是一种品格。如果没有沉默，就没有孕育，就没有震荡，就没有突破。蛾在沉默了一冬之后，终于把飞的梦幻变成现实；海在沉默了一时之后，终于把惊涛的壮观推出了地平线。"];
    ……
    [items addObject:@"承诺不是天上的白云，逍遥，飘逸；承诺不是绿波上的一朵浪花，轻盈，潇洒；承诺不是水面上的一叶浮萍，漂游不定；承诺不是夜幕中的一朵昙花，转瞬即逝。承诺如同珍珠，它的晶莹是蚌的痛苦的代价，也是蚌的荣耀；承诺如同金黄的谷粒，它的饱满是农民辛勤汗水的结晶，也是农民的希望；承诺如同蜂蜜，它的甘甜是蜜蜂勤劳的结晶，也是蜜蜂的骄傲\n承诺如同流星，它的灿烂是陨石悲壮的付出，也是陨石的辉煌；承诺如同清晨绿草尖的露珠，晶亮而短暂。"];
}
```

tableView:heightForRowAtIndexPath:方法获取表视图中每行的高度。程序代码如下：

```
- (CGFloat)tableView:(UITableView *)tableView heightForRowAtIndexPath:
(NSIndexPath *)indexPath;
{
    NSString *text = [items objectAtIndex:[indexPath row]];
    CGSize constraint = CGSizeMake(CELL_CONTENT_WIDTH - (CELL_CONTENT_
    MARGIN * 2), 20000.0f);
```

```
    NSDictionary * attributes = [NSDictionary dictionaryWithObject:[UIFont
systemFontOfSize:FONT_SIZE] forKey:NSFontAttributeName];    //创建字典对象
    NSAttributedString   *attributedText   =   [[NSAttributedString
alloc]initWithString:text attributes:attributes];
    CGRect rect = [attributedText boundingRectWithSize:constraint options:
NSStringDrawingUsesLineFragmentOrigin context:nil];
    CGSize size = rect.size;                          //获取尺寸
    CGFloat height = MAX(size.height, 44.0f);
    return height + (CELL_CONTENT_MARGIN * 2);
}
```

tableView:cellForRowAtIndexPath:方法实现对每行数据的获取。程序代码如下：

```
- (UITableViewCell *)tableView:(UITableView *)tv cellForRowAtIndexPath:
(NSIndexPath *)indexPath
{
    UITableViewCell *cell;
    UILabel *label = nil;
    cell = [tv dequeueReusableCellWithIdentifier:@"Cell"];
    if (cell == nil)
    {
        cell = [[UITableViewCell alloc] initWithStyle:UITableViewCellSty
        leDefault reuseIdentifier:@"Cell"];
        label = [[UILabel alloc] initWithFrame:CGRectZero]; //实例化标签对象
        [label setLineBreakMode:NSLineBreakByWordWrapping];
        [label setMinimumScaleFactor:FONT_SIZE];
        [label setNumberOfLines:0];                        //设置行数
        [label setFont:[UIFont systemFontOfSize:FONT_SIZE]];
        [label setTag:1];
        [[label layer] setBorderWidth:2.0f];                //设置边框宽度
        [[cell contentView] addSubview:label];
    }
    NSString *text = [items objectAtIndex:[indexPath row]];//创建字符串对象
    CGSize constraint = CGSizeMake(CELL_CONTENT_WIDTH - (CELL_CONTENT_
    MARGIN * 2), 20000.0f);
    NSAttributedString *attributedText = [[NSAttributedString alloc]initW
ithString:text attributes:@{NSFontAttributeName:[UIFont systemFontOfSize:
FONT_SIZE] }];
    CGRect rect = [attributedText boundingRectWithSize:constraintoptions:
NSStringDrawingUsesLineFragmentOrigin context:nil];
CGSize size = rect.size;
    //判断label是否为空
    if (!label)
        label = (UILabel*)[cell viewWithTag:1];
    [label setText:text];                                   //设置文本
    [label setFrame:CGRectMake(CELL_CONTENT_MARGIN, CELL_CONTENT_MARGIN,
CELL_CONTENT_WIDTH - (CELL_CONTENT_MARGIN * 2), MAX(size.height, 44.0f))];
    return cell;
}
```

【代码解析】

本实例关键功能是表单元格的行高的调整。下面就是这个知识点的详细讲解。

如果想要实现表单元格行高的调整，需要在 tableView:heightForRowAtIndexPath:方法中进行。它的功能就是对表单元格的行高进行设置。其中，要想获取行高，需要使用到 NSAttributedString 的 boundingRectWithSize:options:context:方法。它的功能是返回文本绘制所占据的矩形空间。其语法形式如下：

```
- (CGRect)boundingRectWithSize:(CGSize)size options:(NSStringDrawingOptions)
```

第 13 章 表

```
options
 context:(NSStringDrawingContext *)context;
```

其中，(CGSize)size 表示宽高限制。宽高限制，用于计算文本绘制时占据的矩形块；(NSStringDrawingOptions)options 表示文本绘制时的附加选项；(NSStringDrawingContext *)context 表示上下文。代码如下：

```
CGRect rect = [attributedText boundingRectWithSize:constraint options:
NSStringDrawingUsesLineFragmentOrigin context:nil];
```

其中，constraint 表示宽高限制；NSStringDrawingUsesLineFragmentOrigin 表示文本绘制时的附加选项；nil 表示没有图形上下文。

实例 102　排　排　看

【实例描述】

在表视图中每一个表单元格都是可以实现编辑功能的。本实例就是采用了表单元格的编辑功能之一，移动实现了一个排排看的小游戏。当用户的顺序排对了，就表示游戏成功完成；否则就失败了。运行效果如图 13.5 所示。

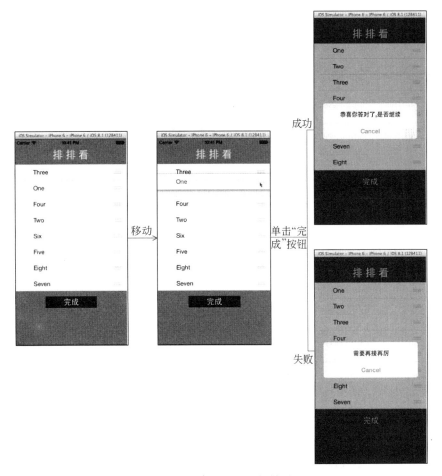

图 13.5　实例 102 运行效果

【实现过程】

当用户单击"完成"按钮后,将表视图中的内容与正确的内容进行比较。如果正确,表示游戏成功;如果错误,表示游戏失败。具体的实现步骤如下。

(1)创建一个项目,命名为"排排看"。

(2)打开 ViewController.h 文件,编写代码,实现插座变量、对象和动作的声明。程序代码如下:

```
#import <UIKit/UIKit.h>
@interface ViewController : UIViewController{
    IBOutlet UITableView *table;              //插座变量
    NSMutableArray *array;                     //对象
}
- (IBAction)finsh:(id)sender;
@end
```

(3)打开 Main.storyboard 文件,对 View Controller 视图控制器的设计界面进行设计,效果如图 13.6 所示。

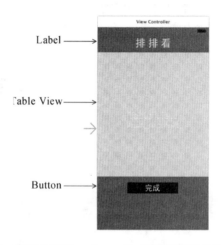

图 13.6 实例 102 View Controller 视图控制器的设计界面

需要添加的视图、控件以及对它们的设置如表 13-5 所示。

表 13-5 实例 102 视图、控件设置

视图、控件	属性设置	其他
设计界面	Background:墨绿色	
Label	Text:排 排 看 Color:白色 Font:System 29.0 Alignment:居中	
Table View		与插座变量 table 关联 将 dataSource 与 View Controller 关联 将 delegate 与 View Controller 关联
Button	Title:完成 Font:System 22.0 Text Color:白色 Background:黑色	与动作 finsh:关联

· 596 ·

（4）打开 ViewController.m 文件，编写代码，实现排排看的游戏功能。在此文件中，代码分为内容的填充、移动以及判断等几个部分。使用的方法如表 13-6 所示。

表 13-6　ViewController.m文件中方法总结

方　　法	功　　能
viewDidLoad	视图加载后调用，实现初始化
numberOfSectionsInTableView:	获取表视图的块
tableView:numberOfRowsInSection:	获取行数
tableView:cellForRowAtIndexPath:	获取某一行的数据
tableView:editingStyleForRowAtIndexPath:	获取每一行应该显示的编辑控件
tableView:moveRowAtIndexPath:toIndexPath:	实现表单元格的移动
finsh:	实现单击"完成"按钮后的判断

其中，表视图的填充需要使用 viewDidLoad、numberOfSectionsInTableView:、tableView:numberOfRowsInSection:和 tableView:cellForRowAtIndexPath:方法。程序代码如下：

```
- (void)viewDidLoad
{
   [super viewDidLoad];
   // Do any additional setup after loading the view, typically from a nib.
   int i=arc4random()%2;
   switch (i) {
      case 0:
         array=[NSMutableArray arrayWithObjects:@"Three",@"One",@"Four",@"Two",@"Six",@"Five",@"Eight",@"Seven", nil];         //创建数组对象
         break;
      case 1:
         array=[NSMutableArray arrayWithObjects:@"Five",@"One",@"Four",@"Two",@"Six",@"Three" ,@"Eight",@"Seven", nil];         //创建数组对象
         break;
      default:
         array=[NSMutableArray arrayWithObjects:@"One",@"Five",@"Four",@"Two",@"Three",@"Six" ,@"Eight",@"Seven", nil];         //创建数组对象
         break;
   }
   table.scrollEnabled=NO;
   [table setEditing:YES];                                    //设置表视图的编辑性
}
//获取表视图的块
- (NSInteger)numberOfSectionsInTableView:(UITableView *)tableView {
   return 1;
}
//获取行数
- (NSInteger)tableView:(UITableView *)tableView numberOfRowsInSection:(NSInteger)section {
   return [array count];
}
//获取某一行的数据
- (UITableViewCell *)tableView:(UITableView *)tableView cellForRowAtIndexPath: (NSIndexPath *)indexPath {
   static NSString *CellIdentifier = @"Cell";
   UITableViewCell *cell = [tableView dequeueReusableCellWithIdentifier:CellIdentifier];
   //判断cell是否为空
```

```
    if (cell == nil) {
        cell = [[UITableViewCell alloc] initWithStyle:UITableViewCell
        StyleDefault reuseIdentifier:CellIdentifier];
    }
    cell.textLabel.text = [array objectAtIndex:indexPath.row];
                                                            //设置文本内容
    return cell;
}
```

如果想要实现移动的功能,即排序功能,需要使用 ableView:editingStyleForRowAtIndexPath: 和 tableView: moveRowAtIndexPath:toIndexPath:方法实现。程序代码如下:

```
//获取每一行应该显示的编辑控件
-(UITableViewCellEditingStyle)tableView:(UITableView *)tableView editingStyle
ForRowAtIndexPath:
(NSIndexPath *)indexPath{
    return UITableViewCellEditingStyleNone;
}
//移动行
-(void)tableView:(UITableView *)tableView moveRowAtIndexPath:(NSIndexPath
*)sourceIndexPath toIndexPath:(NSIndexPath *)destinationIndexPath{
    NSInteger fromRow = [sourceIndexPath row];          //获取需要移动的行
    NSInteger toRow = [destinationIndexPath row];       //获取需要移动的位置
    id object = [array objectAtIndex:fromRow];          //读取需要移动行的数据
    [array removeObjectAtIndex:fromRow];
    [array insertObject:object atIndex:toRow];
}
```

当排序之后,就可以单击"完成"按钮来进行判断,看一看排序是否正确。程序代码如下:

```
- (IBAction)finsh:(id)sender {
    NSArray *aa=[NSArray arrayWithObjects:@"One",@"Two",@"Three",@"Four",
@"Five",@"Six",
@"Seven",@"Eight",nil];
    if ([array isEqualToArray:aa] ) {
        UIAlertView *alert=[[UIAlertView alloc]initWithTitle:@"恭喜你答对了,
是否继续" message:nil delegate:self cancelButtonTitle:@"Cancel" otherButton
Titles: nil];                                           //创建警告视图
        [alert show];
    }else{
        UIAlertView *alert=[[UIAlertView alloc]initWithTitle:@"需要再接再厉"
message:nil delegate:self cancelButtonTitle:@"Cancel" otherButtonTitles:
nil];                                                   //创建警告视图
        [alert show];
    }
}
```

【代码解析】

本实例关键功能是表的编辑模式的设置以及表单元格的移动。下面依次讲解这两个知识点。

1. 表的编辑模式的设置

表的编辑模式的设置,需要使用 UITableViewDelegate 的 tableView:editingStyleForRowAtIndexPath:方法,其语法形式如下:

```
- (UITableViewCellEditingStyle)tableView:(UITableView *)tableView
editingStyleForRowAtIndexPath:
(NSIndexPath *)indexPath;
```

其中，(UITableView *)tableView 表示表视图对象；(NSIndexPath *)indexPath 表示在表视图中索引路径定位的行。该方法的返回值类型为该行的编辑样式。其中，表视图的编辑样式如下所示。

- ❑ UITableViewCellEditingStyleNone：无。
- ❑ UITableViewCellEditingStyleDelete：删除。
- ❑ UITableViewCellEditingStyleInsert：插入。

在此代码中，就使用了 tableView:editingStyleForRowAtIndexPath:方法将表视图中的编辑样式设置为了移动的样式。代码如下：

```
-(UITableViewCellEditingStyle)tableView:(UITableView *)tableView editingStyle
ForRowAtIndexPath:
(NSIndexPath *)indexPath{
    return UITableViewCellEditingStyleNone;
}
```

2. 表单元格的移动

如果想要实现表单元格的移动，需要使用 UITableViewDataSource 的 tableView:moveRowAtIndexPath:toIndexPath:方法。其语法形式如下：

```
- (void)tableView:(UITableView *)tableView moveRowAtIndexPath:(NSIndexPath *)
fromIndexPath toIndexPath:
(NSIndexPath *)toIndexPath;
```

其中，(UITableView *)tableView 表示表视图对象；(NSIndexPath *)fromIndexPath 表示要移动的索引路径定位的行；(NSIndexPath *)toIndexPath 表示要移动的索引路径定位的行的目标。在此代码中，就使用了 tableView:moveRowAtIndexPath:toIndexPath:方法实现了表单元格的移动。代码如下：

```
-(void)tableView:(UITableView *)tableView moveRowAtIndexPath:(NSIndexPath
*)sourceIndexPath toIndexPath:(NSIndexPath *)destinationIndexPath{
    NSInteger fromRow = [sourceIndexPath row];
    NSInteger toRow = [destinationIndexPath row];
    id object = [array objectAtIndex:fromRow];
    [array removeObjectAtIndex:fromRow];
    [array insertObject:object atIndex:toRow];
}
```

实例 103　归　归　类

【实例描述】

本实例采用了表单元格的删除功能实现了一个归归类的小游戏。当用户单击"完成"按钮，对游戏是否成功进行判断。运行效果如图 13.7 所示。

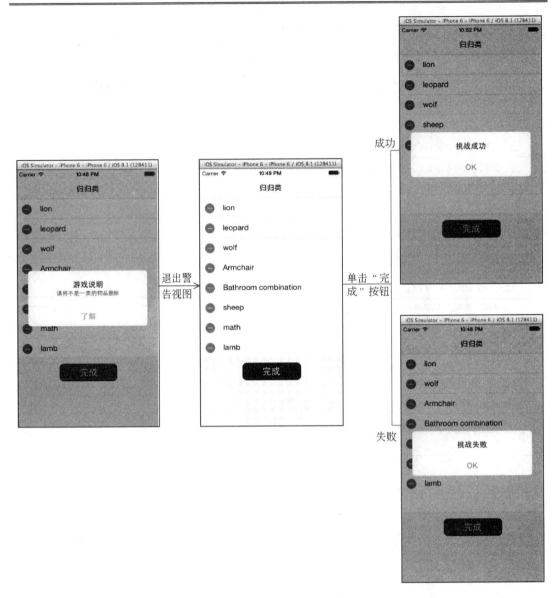

图 13.7　实例 103 运行效果

【实现过程】

当用户单击"完成"按钮后,将表视图中的内容与正确的内容进行比较,如果正确,表示游戏成功;如果错误,表示游戏失败。具体的实现步骤如下。

(1)创建一个项目,命名为"归归类"。

(2)打开 ViewController.h 文件,编写代码,实现对象、插座变量以及动作的声明。程序代码如下:

```
#import <UIKit/UIKit.h>
@interface ViewController : UIViewController{
    NSMutableArray *array;                         //对象
    IBOutlet UITableView *table;                   //表视图的插座变量
    IBOutlet UIButton *button;                     //按钮的插座变量
}
```

第 13 章 表

```
- (IBAction)finsh:(id)sender;                    //动作
@end
```

（3）打开 Main.storyboard 文件，将 Navigation Controller 导航控件器添加到画布中，将与导航控制器关联的根视图设置为 View Controller 视图控制器的视图，将 Is Initial View Controller 复选框选中。这时再对 View Controller 视图控制器的设计界面进行设计，效果如图 13.8 所示。

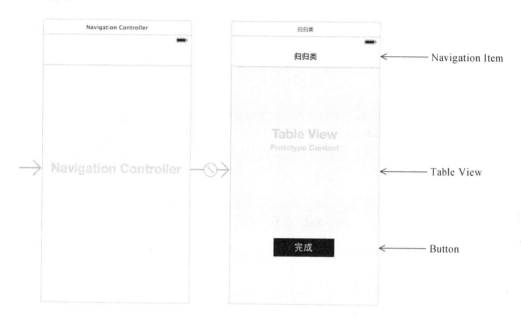

图 13.8　实例 103 View Controller 视图控制器的设计界面效果

需要添加的视图、控件以及对它们的设置如表 13-7 所示。

表 13-7　实例 103 视图、控件设置

视图、控件	属 性 设 置	其　　他
Navigation Item	Title：归归类	
Table View		与插座变量 table 关联 将 dataSource 与 View Controller 关联 将 delegate 与 View Controller 关联
Button	Title：完成 Font：System 20.0 Text Color：白色 Background：黑色	与插座变量 button 关联 与动作 finsh:关联

（4）打开 ViewController.m 文件，编写代码，实现归归类的游戏功能，在此文件中，代码分为内容的填充、删除以及判断等几个部分。使用的方法如表 13-8 所示。

表 13-8　ViewController.m 文件中方法总结

方　　法	功　　能
viewDidLoad	视图加载后调用，实现初始化
tableView:numberOfRowsInSection:	获取行数

方 法	功 能
tableView:cellForRowAtIndexPath:	获取某一行的数据
tableView:commitEditingStyle:forRowAtIndexPath:	实现删除
finsh:	实现单击"完成"按钮后的判断
alertView:clickedButtonAtIndex:	实现警告视图的响应

其中，内容的填充需要使用 viewDidLoad、tableView:numberOfRowsInSection: 和 tableView:cellForRowAtIndexPath:方法实现。程序代码如下：

```
//视图加载后调用，实现初始化
- (void)viewDidLoad
{
    [super viewDidLoad];
    // Do any additional setup after loading the view, typically from a nib.
    UIAlertView *alert=[[UIAlertView alloc]initWithTitle:@"游戏说明"
message:@"请将不是一类的物品删除" delegate:nil cancelButtonTitle:@"连接"
otherButtonTitles: nil];                    //创建警告视图
    [alert show];
    array=[NSMutableArray
arrayWithObjects:@"lion",@"leopard",@"wolf",@"Armchair     ",@"Bathroom
combination",@"sheep ",@"math",@"lamb", nil]; //创建数组
    [table setEditing:YES];
    button.layer.cornerRadius=10;               //设置圆角半径
}
//获取行数
- (NSInteger)tableView:(UITableView *)tableView numberOfRowsInSection:
(NSInteger)section {
    return [array count];
}
//获取某一行的数据
- (UITableViewCell *)tableView:(UITableView *)tableView cellForRowAtIndexPath:
(NSIndexPath *)indexPath {
    static NSString *CellIdentifier = @"Cell";
    UITableViewCell *cell = [tableView dequeueReusableCellWithIdentifier:
    CellIdentifier];
    //判断cell是否为空
    if (cell == nil) {
        cell = [[UITableViewCell alloc] initWithStyle:UITableViewCell
        StyleDefault reuseIdentifier:CellIdentifier];
    }
    cell.textLabel.text = [array objectAtIndex:indexPath.row];
                                                //设置文本内容
    return cell;
}
```

表视图中表单元格的删除，需要使用 tableView:commitEditingStyle:forRowAtIndexPath: 方法实现。程序代码如下：

```
-(void)tableView:(UITableView *)tableView commitEditingStyle:
(UITableViewCellEditingStyle)editingStyle forRowAtIndexPath:(NSIndexPath
*)indexPath{
    if(editingStyle==UITableViewCellEditingStyleDelete){
        [array removeObjectAtIndex:indexPath.row];  //删除选择行的数组元素
        [table deleteRowsAtIndexPaths:[NSArray arrayWithObject:indexPath]
withRowAnimation:UITableViewRowAnimationAutomatic];//删除行
```

单击"完成"按钮后实现的判断,需要使用 finsh:和 alertView:clickedButtonAtIndex:方法实现。其中 finsh:方法实现判断当前没有删除的内容,是否和指定的内容相等。程序代码如下:

```
- (IBAction)finsh:(id)sender {
   NSArray *array2=[NSArray arrayWithObjects:@"lion",@"leopard",@"wolf",
   @"sheep ",@"lamb", nil];
   if ([array isEqualToArray:array2]) {
      //判断,如果相等
      UIAlertView *alert=[[UIAlertView alloc]initWithTitle:@"挑战成功"
message:nil delegate:self cancelButtonTitle:@"OK" otherButtonTitles: nil];
      [alert show];
   }else{
      //判断,如果不相等
      UIAlertView *alert=[[UIAlertView alloc]initWithTitle:@"挑战失败"
message:nil delegate:nil cancelButtonTitle:@"OK" otherButtonTitles: nil];
      [alert show];
   }
}
```

alertView:clickedButtonAtIndex:方法实现对警告视图中被选按钮的响应。程序代码如下:

```
- (void)alertView:(UIAlertView *)alertView clickedButtonAtIndex:(NSInteger)
buttonIndex{
   NSString *b=[alertView buttonTitleAtIndex:buttonIndex];
   if([b isEqualToString:@"OK"]){
       exit(1);                     //退出
   }
}
```

【代码解析】

本实例的关键功能是数组的比较,和表单元格的删除。下面依次讲解这两个知识点。

1. 数组的比较

如果想要实现数组的判断,需要使用 NSArray 的 isEqualToArray:方法。其语法形式如下:

```
-(BOOL)isEqualToArray:(NSArray *)otherArray;
```

其中,(NSArray *)otherArray 表示数组对象。在此代码中,就使用了 isEqualToArray:方法实现了两个数组的比较及判断。代码如下:

```
if ([array isEqualToArray:array2]) {
   ......
}else{
   ......
}
```

其中,array2 表示数组对象。

2. 表单元格的删除

表单元格的删除需要使用 UITableViewDataSource 的 tableView:commitEditingStyle:

forRowAtIndexPath:方法。它的功能是在表视图的指定行提交插入或删除的数据源。其语法形式如下：

```
- (void)tableView:(UITableView *)tableView commitEditingStyle:
(UITableViewCellEditingStyle)editingStyle
 forRowAtIndexPath:(NSIndexPath *)indexPath;
```

其中，(UITableView *)tableView 表示表视图；(UITableViewCellEditingStyle)editingStyle 表示插入或删除行对应的表单元格编辑风格；(NSIndexPath *)indexPath 表示一个索引路径。用来实现删除功能的是 deleteRowsAtIndexPaths:withRowAnimation:方法。其语法形式如下：

```
- (void)deleteRowsAtIndexPaths:(NSArray *)indexPaths withRowAnimation:
(UITableViewRowAnimation)
animation;
```

其中，(NSArray *)indexPaths 用来表示数组对象，确定要删除的行；(UITableViewRowAnimation)animation 表示动画效果。这些动画效果如表 13-9 所示。

表 13-9 动画效果

动　　画	功　　能
UITableViewRowAnimationFade	表单元格淡出
UITableViewRowAnimationRight	表单元格从右侧滑出
UITableViewRowAnimationLeft	表单元格从左侧滑出
UITableViewRowAnimationTop,	表单元格滑动到相邻单元格之上
UITableViewRowAnimationBottom	表单元格滑动到相邻单元格之下
UITableViewRowAnimationNone	无动画效果
UITableViewRowAnimationMiddle	表视图试图保留旧的和新的单元格为中心的空间或者将会占据
UITableViewRowAnimationAutomatic	表视图选择一个合适的动画风格

在此代码中，就是在 tableView:commitEditingStyle:forRowAtIndexPath:方法中使用了 deleteRowsAtIndexPaths:withRowAnimation:方法实现了表单元格的删除功能。代码如下：

```
-(void)tableView:(UITableView *)tableView commitEditingStyle:
(UITableViewCellEditingStyle)editingStyle forRowAtIndexPath:(NSIndexPath
*)indexPath{
    if(editingStyle==UITableViewCellEditingStyleDelete){
        [array removeObjectAtIndex:indexPath.row];   //删除选择行的数组元素
        [table deleteRowsAtIndexPaths:[NSArray arrayWithObject:indexPath]
withRowAnimation:UITableViewRowAnimationAutomatic];//删除行
    }
}
```

其中，[NSArray arrayWithObject:indexPath] 表示数组对象；UITableViewRowAnimationAutomatic 表示动画效果。

实例 104　自定义索引的表

【实例描述】

本实例实现的功能是自制一个具有索引的表，索引可以快速地访问列表中的特定消息。当用户在索引栏上滑动时，索引栏会跟随手指产生阴影效果。运行效果如图 13.9 所示。

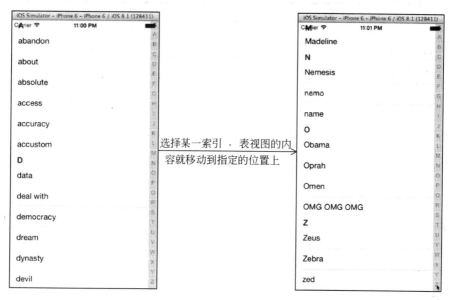

图 13.9　实例 104 运行效果

【实现过程】

（1）创建一个项目，命名为"自定义索引的表"。
（2）创建一个基于 UIView 类的 Index 类。
（3）打开 Index.h 文件，编写代码，实现宏定义、协议、对象、实例变量以及属性的声明。程序代码如下：

```
#import <UIKit/UIKit.h>
@class Index;
#define RGB(r,g,b,a)   [UIColor colorWithRed:(double)r/255.0f green:(double)g/255.0f blue:(double)b/255.0f alpha:a]                     //宏定义
//协议
@protocol TableViewIndexBarDelegate <NSObject>
- (void)tableViewIndexBar:(Index*)indexBar didSelectSectionAtIndex:(NSInteger)index;
@end
@interface Index : UIView{
    //实例变量
    BOOL isLayedOut;
    CGFloat letterHeight;
    //对象
    NSArray *letters;
    CAShapeLayer *shapeLayer;
}
//属性
@property (nonatomic, strong) NSArray *indexes;
@property (nonatomic, weak) id <TableViewIndexBarDelegate> delegate;
@end
```

（4）打开 Index.m 文件，编写代码，实现自定义索引的绘制以及选择。使用的方法如表 13-10 所示。

表 13-10 Index.m 文件中方法总结

方 法	功 能
initWithFrame:	自定义索引的初始化
initWithCoder:	初始化实例化对象
setup	创建自定义索引
setIndexes:	设置索引
layoutSubviews	布局
textLayerWithSize:string:andFrame:	获取文本层
touchesBegan:withEvent:	开始触摸
touchesMoved:withEvent:	移动触摸
sendEventToDelegate:	发送事件给委托
animateLayerAtIndex:	动画效果

这里需要讲解几个重要的方法（其他方法请读者参考源代码）。如果想要绘制自定义的索引，需要使用 initWithFrame:、initWithCoder:、setup、setIndexes:、layoutSubviews 和 textLayerWithSize:string:andFrame:方法。其中，setup 方法实现自定义索引的创建。程序代码如下：

```objc
- (void)setup{
    letters = @[@"#", @"A", @"B", @"C",@"D", @"E", @"F", @"G",@"H", @"I",
 @"J", @"K",@"L", @"M", @"N", @"O",@"P", @"Q", @"R", @"S",@"T", @"U", @"V",
 @"W",@"X", @"Y", @"Z"];
    shapeLayer = [CAShapeLayer layer];
    shapeLayer.lineWidth = 1.0f;                              //设置线宽
    shapeLayer.fillColor = [UIColor clearColor].CGColor;      //设置填充颜色
    shapeLayer.lineJoin = kCALineCapSquare;
    shapeLayer.strokeColor = [RGB(218, 218, 218, 1) CGColor];
    shapeLayer.strokeEnd = 1.0f;
    self.layer.masksToBounds = NO;
}
```

layoutSubviews 方法实现对创建的自定义索引进行布局。程序代码如下：

```objc
- (void) layoutSubviews {
    [super layoutSubviews];
    if (!isLayedOut){
        [self.layer.sublayers makeObjectsPerformSelector:@selector
 (removeFromSuperlayer)];
        shapeLayer.frame = (CGRect) {.origin = CGPointZero, .size = self.
 layer.frame.size};                                  //设置框架
        UIBezierPath *bezierPath = [UIBezierPath bezierPath];
        [bezierPath moveToPoint:CGPointZero];             //设置开始点
        [bezierPath addLineToPoint:CGPointMake(0, self.frame.size.height)];
                                                          //设置结束点
        letterHeight = self.frame.size.height / [letters count];
        CGFloat fontSize = 12;
        if (letterHeight < 14){
            fontSize = 11;
        }
        //对索引的文本层和绘制的路径进行设置
        [letters enumerateObjectsUsingBlock:^(NSString *letter, NSUInteger
 idx, BOOL *stop) {
            CGFloat originY = idx * letterHeight;
            CATextLayer *ctl = [self textLayerWithSize:fontSize string:letter
```

```
andFrame:CGRectMake(0, originY, self.frame.size.width, letterHeight)];
                                        //实例化对象
        [self.layer addSublayer:ctl];
        [bezierPath moveToPoint:CGPointMake(0, originY)];//设置开始点
        [bezierPath addLineToPoint:CGPointMake(ctl.frame.size.width,
        originY)];
    }];
    shapeLayer.path = bezierPath.CGPath;
    [self.layer addSublayer:shapeLayer];           //添加图层对象
    isLayedOut = YES;
    }
}
```

textLayerWithSize:string:andFrame:方法实现对文本层的获取。程序代码如下：

```
- (CATextLayer*)textLayerWithSize:(CGFloat)size string:(NSString*)string
andFrame:(CGRect)frame{
    CATextLayer *tl = [CATextLayer layer];
    [tl setFont:@"ArialMT"];
    [tl setFontSize:size];                           //设置文字的尺寸
    [tl setFrame:frame];
    [tl setAlignmentMode:kCAAlignmentCenter];        //设置对齐方式
    [tl setContentsScale:[[UIScreen mainScreen] scale]];
    [tl setForegroundColor:RGB(50, 100,100, 1).CGColor];
    [tl setString:string];                           //设置字符串
    return tl;
}
```

自定义索引的选择，需要使用 touchesBegan:withEvent:、touchesMoved:withEvent:、sendEventToDelegate: 和 animateLayerAtIndex: 方法 。 其中， touchesBegan:withEvent: 和 touchesMoved:withEvent:方法实现索引的触摸。程序代码如下：

```
//开始触摸
- (void)touchesBegan:(NSSet *)touches withEvent:(UIEvent *)event{
    [super touchesBegan:touches withEvent:event];
    [self sendEventToDelegate:event];
}
//移动触摸
- (void)touchesMoved:(NSSet *)touches withEvent:(UIEvent *)event{
    [super touchesMoved:touches withEvent:event];
    [self sendEventToDelegate:event];
}
```

sendEventToDelegate:方法实现将事件发送给委托的功能。程序代码如下：

```
- (void)sendEventToDelegate:(UIEvent*)event{
    UITouch *touch = [[event allTouches] anyObject];
    CGPoint point = [touch locationInView:self];              //获取触摸点
    NSInteger indx = (NSInteger) floorf(fabs(point.y) / letterHeight);
    indx = indx < [letters count] ? indx : [letters count] - 1;
    [self animateLayerAtIndex:indx];
    __block NSInteger scrollIndex;
NSIndexSet *indexSet = [[NSIndexSet alloc] initWithIndexesInRange:
NSMakeRange(0, indx+1)];
//遍历数组中的每一个元素
    [letters enumerateObjectsAtIndexes:indexSet options:NSEnumerationReverse
usingBlock:、^(NSString *letter, NSUInteger idx, BOOL *stop) {
        scrollIndex = [indexes indexOfObject:letter];
```

```
        *stop = scrollIndex != NSNotFound;
    }];
    [delegate tableViewIndexBar:self didSelectSectionAtIndex:scrollIndex];
                                                            //调用
}
```

animateLayerAtIndex:方法实现自定义索引的动画效果。程序代码如下：

```
- (void)animateLayerAtIndex:(NSInteger)index{
    if ([self.layer.sublayers count] - 1 > index){
        CABasicAnimation *animation = [CABasicAnimation animationWithKeyPath:
        @"backgroundColor"];
        animation.toValue = (id)[RGB(180, 180, 180, 1) CGColor];
                                                            //设置结束值
        animation.duration = 0.5f;                          //设置持续时间
        animation.autoreverses = YES;
        animation.repeatCount = 1;                          //设置循环次数
        animation.timingFunction = [CAMediaTimingFunction functionWithName:
kCAMediaTimingFunctionEaseOut];
        [self.layer.sublayers[index] addAnimation:animation forKey:@"
        myAnimation"];
    }
}
```

（5）打开 ViewController.h 文件，编写代码，实现头文件、插座变量以及对象的声明。程序代码如下：

```
#import <UIKit/UIKit.h>
#import "Index.h"                                           //头文件
@interface ViewController : UIViewController<TableViewIndexBarDelegate>{
    //插座变量
    IBOutlet Index *indexBar;
    IBOutlet UITableView *plainTableView;
    //对象
    NSArray *sections;
    NSArray *rows;
}
@end
```

（6）打开 Main.storyboard 文件，对 View Controller 视图控制器的设计界面进行设计，效果如图 13.10 所示。

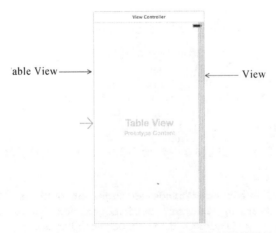

图 13.10　实例 104 View Controller 视图控制器的设计界面

第 13 章 表

需要添加的视图、控件以及对它们的设置如表 13-11 所示。

表 13-11 实例 104 视图、控件设置

视图、控件	属 性 设 置	其 他
View	Background：绿色	Class：Index 与插座变量 indexBar 关联
Table View		与插座变量 plainTableView 关联 将 dataSource 与 View Controller 关联 将 delegate 与 View Controller 关联

（7）打开 ViewController.m 文件，编写代码，实现在表视图上显示绘制的自定义索引。使用的方法如表 13-12 所示。

表 13-12 ViewController.m文件中方法总结

方 法	功 能
viewDidLoad	视图加载后调用，实现初始化
numberOfSectionsInTableView:	获取块数
tableView:numberOfRowsInSection:	获取行数
tableView:titleForHeaderInSection:	获取每一个块的标题
tableView:cellForRowAtIndexPath:	获取某一行的数据
tableViewIndexBar:didSelectSectionAtIndex:	实现选择某一索引的响应

这此文件中代码被分为了：填充数据和实现选择的响应两个部分。其中，实现内容的填充需要使用 viewDidLoad、numberOfSectionsInTableView:、tableView:numberOfRowsInSection:、tableView:titleForHeaderInSection:和 tableView:cellForRowAtIndexPath:方法。程序代码如下：

```
//视图加载后调用，实现初始化
- (void)viewDidLoad
{
    [super viewDidLoad];
    // Do any additional setup after loading the view from its nib.
    sections = @[@"A", @"D", @"F", @"M", @"N", @"O", @"Z"];
    //创建数组
    rows = @[@[@"abandon", @"about", @"absolute", @"access", @"accuracy",
    @"accustom"],
            @[@"data", @"deal with", @"democracy", @"dream", @"dynasty",
            @"devil", @"delectation"],
            @[@"facility", @"fact", @"friends", @"family", @"fatish",
            @"funeral"],
            @[@"Mark", @"Madeline"],
            @[@"Nemesis", @"nemo", @"name"],
            @[@"Obama", @"Oprah", @"Omen", @"OMG OMG OMG"],
            @[@"Zeus", @"Zebra", @"zed"]];
    indexBar.delegate = self;                       //设置委托
}
//获取块数
- (NSInteger)numberOfSectionsInTableView:(UITableView *)tableView{
    [indexBar setIndexes:sections];
    return [sections count];
```

```objc
}
//获取行数
- (NSInteger)tableView:(UITableView *)tableView numberOfRowsInSection:
(NSInteger)section{
    return [rows[section] count];
}
//获取每一个块的标题
- (NSString*)tableView:(UITableView *)tableView titleForHeaderInSection:
(NSInteger)section{
    return sections[section];
}
//获取某一行的数据
- (UITableViewCell*)tableView:(UITableView *)tableView cellForRowAtIndexPath:
(NSIndexPath *)indexPath{
    static NSString *cellId = @"TableViewCellId";
    UITableViewCell *cell = [tableView dequeueReusableCellWithIdentifier:
    cellId];
    //判断cell是否为空
    if (cell == nil){
        cell = [[UITableViewCell alloc] initWithStyle:UITableViewCellStyle
        Default reuseIdentifier:cellId];
    }
    [cell.textLabel setText:rows[indexPath.section][indexPath.row]];
                                                                //设置文本
    return cell;
}
```

tableViewIndexBar:didSelectSectionAtIndex:方法实现对选择索引的响应。程序代码如下：

```objc
- (void)tableViewIndexBar:(Index *)indexBar didSelectSectionAtIndex:
(NSInteger)index{
    if ([plainTableView numberOfSections] > index && index > -1){
        [plainTableView scrollToRowAtIndexPath:[NSIndexPath indexPathForRow:
0 inSection:
index]atScrollPosition:UITableViewScrollPositionTop animated:YES];
    }
}
```

【代码解析】

由于本实例中的代码和方法非常的多，为了方便读者的阅读，笔者绘制了一些执行流程图如图13.11～图13.12所示。其中，单击运行按钮一直到表视图的自定义索引的显示，使用了initWithCoder:、setup、setIndexes:、layoutSubviews、textLayerWithSize:string:andFrame:、viewDidLoad、numberOfSectionsInTableView:、tableView:numberOfRowsInSection:、tableView:titleForHeaderInSection:和tableView:cellForRowAtIndexPath:方法。它们的执行流程如图13.11所示。

如果想要实现对索引的选择，可以使用touchesBegan:withEvent:、sendEventToDelegate:、animateLayerAtIndex:、tableView:cellForRowAtIndexPath:和ableView:titleForHeaderInSection:、tableViewIndexBar:didSelectSectionAtIndex:方法。它们的执行流程如图13.12所示。

第 13 章 表

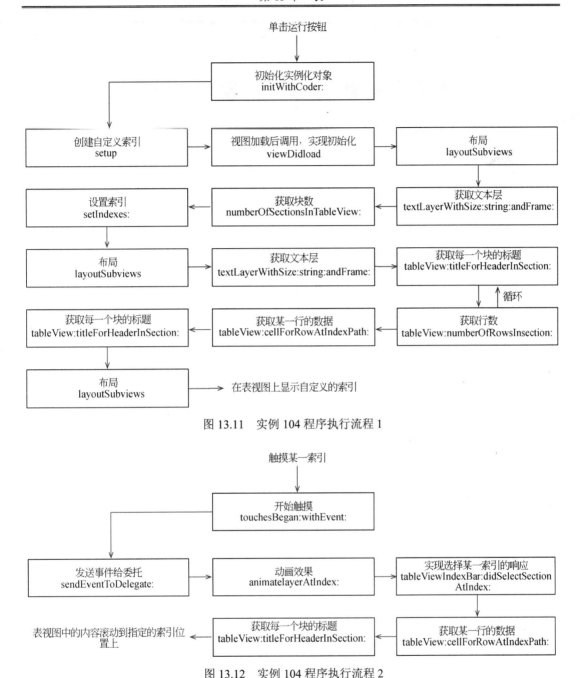

图 13.11 实例 104 程序执行流程 1

图 13.12 实例 104 程序执行流程 2

实例 105 自制的列表单选控件

【实例描述】

在表视图中可以实现列表的单选功能，但是除了表视图外，还有其他控件可以实现列表中的单选功能吗？本实例就要为开发者实现了一个自制的列表单选控件。当用户选择其中的某一项后，选择的行会发生改变。运行效果如图 13.13 所示

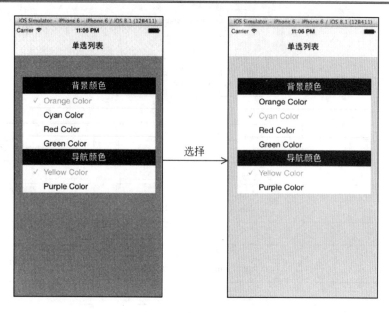

图 13.13　实例 105 运行效果

【实现过程】

（1）创建一个项目，命名为"列表单选控件"。

（2）创建一个基于 UIView 类的 View 类。

（3）打开 View.h 文件，编写代码，实现协议和属性的声明。程序代码如下：

```
#import <UIKit/UIKit.h>
@class View;
//协议
@protocol ViewDelegate
-(void)radioView:(View *)radioView didSelectOption:(int)optionNo fortag:(int)tagNo;
@end
@interface View : UIView<UITableViewDataSource, UITableViewDelegate>
//属性
@property (assign, nonatomic) int optionNo;
@property (assign, nonatomic) int tag;
……
@property (assign) int rowHeight;
@property (assign) id <ViewDelegate> delegate;
@end
```

（4）打开 View.m 文件，编写代码，实现列表单选控件的绘制以及选择功能。使用的方法如表 13-13 所示。

表 13-13　View.m 文件中方法总结

方　　法	功　　能
initWithFrame:	初始化自定义的列表控件
initWithCoder:	初始化实例化对象
baseInit	基本内容的初始化
tableView:heightForRowAtIndexPath:	获取表中每行的高度
numberOfSectionsInTableView:	获取表视图的块

第13章 表

续表

方　　法	功　　能
tableView:numberOfRowsInSection:	获取行数
tableView:cellForRowAtIndexPath:	获取某一行的数据
tableView:didSelectRowAtIndexPath:	实现选择

这里需要讲解几个重要的方法（其他方法请读者参考源代码）。列表单选控件的绘制需要使用 initWithFrame:、initWithCoder:、baseInit、tableView:heightForRowAtIndexPath:、numberOfSectionsInTableView:、tableView:numberOfRowsInSection:和 tableView:cellForRowAtIndexPath:方法。其中，baseInit 方法实现对一些基本内容的初始化。程序代码如下：

```
-(void) baseInit {
    selectedRow = optionNo;
    optionNo = 0;
    editable = NO;
    delegate = nil;                                          //委托
    rowItems = [[NSMutableArray alloc]init];                 //实例化可变数组对象
    //创建并设置表视图对象
    UITableView* tableView = [[UITableView alloc] initWithFrame:self.
    bounds];
    tableView.dataSource = self;
    tableView.delegate = self;                               //设置委托
    tableView.bounces = NO;
    [tableView setTableFooterView:[[UIView alloc] initWithFrame:CGRectZero]];
    [tableView setBackgroundColor:[UIColor colorWithRed:242/255.f green:235/
    255.f blue:235/255.f alpha:1]];
    [self addSubview:tableView];                             //添加视图对象
}
```

tableView:heightForRowAtIndexPath:、numberOfSectionsInTableView:、tableView:numberOfRowsInSection:和 tableView:cellForRowAtIndexPath:方法实现了对行高的设置以及内容的填充。程序代码如下：

```
//获取表中每行的高度
-(CGFloat)tableView:(UITableView *)tableView heightForRowAtIndexPath:
(NSIndexPath *)indexPath {
    return 30;
}
//获取表视图的块
- (NSInteger)numberOfSectionsInTableView:(UITableView *)tableView {
    return 1;
}
//获取行数
- (NSInteger)tableView:(UITableView *)tableView numberOfRowsInSection:
(NSInteger)section {
    return maxRow;
}
//获取某一行的数据
- (UITableViewCell *)tableView:(UITableView *)tableView cellForRowAtIndexPath:
(NSIndexPath *)indexPath {
    static NSString *cellIdentifier = @"Cell";
    UITableViewCell *cell = [tableView dequeueReusableCellWithIdentifier:
    cellIdentifier];
    cell = [[UITableViewCell alloc] initWithStyle:UITableViewCellStyleDefault
```

```
    reuseIdentifier:cellIdentifier];
    [cell setSelectionStyle:UITableViewCellSelectionStyleNone];
    //创建并设置标签对象
    UILabel *labelOne = [[UILabel alloc]initWithFrame:CGRectMake(20, 5, 140, 20)];
    [labelOne setBackgroundColor:[UIColor clearColor]];    //设置背景颜色
    [labelOne setTextColor:[UIColor colorWithRed:5/255.f green:5/255.f
    blue:5/255.f alpha:1]];                                //设置文字颜色
    NSString *textString = [[NSString alloc] init];
    textString = @"\u2001 ";
    if (indexPath.row == optionNo) {
        textString = @"\u2713 ";
        //设置文字颜色
        [labelOne setTextColor:[UIColor colorWithRed:35/255.f green:150/255.
        f blue:200/255.f alpha:1]];
    }
    textString = [textString stringByAppendingString:[rowItems objectAtIndex:
    indexPath.row]];
    labelOne.textAlignment = NSTextAlignmentLeft;
    labelOne.text = textString;                            //设置文本内容
    [cell.contentView addSubview:labelOne];
    return cell;
}
```

列表中选择的实现，需要使用 tableView:didSelectRowAtIndexPath:方法。程序代码如下：

```
- (void)tableView:(UITableView *)tableView didSelectRowAtIndexPath:
(NSIndexPath *)indexPath {
    optionNo = indexPath.row;
    [self.delegate radioView:self didSelectOption:indexPath.row fortag:
    tag];
    [tableView reloadData];
}
```

（5）打开 ViewController.h 文件，编写代码，实现头文件、遵守协议以及插座变量的声明。程序代码如下：

```
#import <UIKit/UIKit.h>
#import "View.h"                      //头文件
@interface ViewController : UIViewController<ViewDelegate>{
    //插座变量
    IBOutlet View *bg;
    IBOutlet View *Nav;
}
@end
```

（6）打开 Main.storyboard 文件，将 Navigation Controller 导航控件器添加到画布中，将与导航控制器关联的根视图设置为 View Controller 视图控制器的视图，将 Is Initial View Controller 复选框选中。这时再对 View Controller 视图控制器的设计界面进行设计，效果如图 13.14 所示。

需要添加的视图、控件以及对它们的设置如表 13-14 所示。

（7）打开 ViewController.m 文件，编写代码，实现对单选列表控件的设置，以及选择某一行后的响应。其中，要实现设置，需要使用 viewDidLoad 方法。程序代码如下：

第 13 章 表

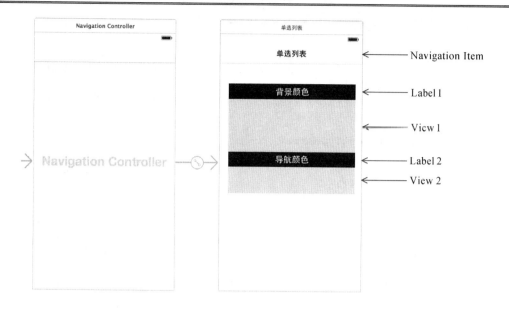

图 13.14 实例 105 View Controller 视图控制器的设计界面

表 13-14 实例 105 视图、控件设置

视图、控件	属 性 设 置	其 他
Navigation Item	Title：单选列表	
Label1	Title：背景颜色 Color：白色 Font：System Bold 19.0 Alignment：居中 Background：黑色	与插座变量 table 关联 将 dataSource 与 View Controller 关联 将 delegate 与 View Controller 关联
View1	Background：浅灰色	Class：View 与插座变量 bg 关联
Label2	Title：导航颜色 Color：白色 Font：System Bold 19.0 Alignment：居中 Background：黑色	
View2	Background：浅灰色	Class：View 与插座变量 Nav 关联

```
- (void)viewDidLoad
{
    [super viewDidLoad];
    // Do any additional setup after loading the view, typically from a nib.
    //第一个列表对象的设置
    NSMutableArray *sortByItemsArray = [[NSMutableArray alloc] init];
    [sortByItemsArray addObject:@"Orange Color"];
    [sortByItemsArray addObject:@"Cyan Color"];
    [sortByItemsArray addObject:@"Red Color"];
    [sortByItemsArray addObject:@"Green Color"];
    bg.rowItems = sortByItemsArray;                              //设置条目
    bg.maxRow = [sortByItemsArray count];
    bg.editable = YES;
```

```
    bg.delegate = self;                                    //设置委托
    bg.tag = 1;
    bg.optionNo = 0;
    self.view.backgroundColor=[UIColor orangeColor];       //设置背景颜色
    //第二个列表对象的设置
    NSMutableArray *sortTypeArray = [[NSMutableArray alloc] init];
    [sortTypeArray addObject:@"Yellow Color"];
    [sortTypeArray addObject:@"Purple Color"];
    Nav.rowItems = sortTypeArray;
    Nav.optionNo = 0;
    Nav.tag = 2;                                           //设置tag值
    Nav.delegate = self;
    Nav.maxRow = [sortTypeArray count];
    Nav.editable = YES;
    self.navigationController.navigationBar.backgroundColor =[UIColor
yellowColor];                                              //设置背景颜色
}
```

选择某一行后的响应,需要使用radioView:didSelectOption:方法实现。程序代码如下:

```
-(void)radioView:(View *)radioView didSelectOption:(int)optionNo fortag:
(int)tagNo{
    if (optionNo==0&&tagNo==1) {
        self.view.backgroundColor=[UIColor orangeColor];   //设置背景颜色
    }else if (optionNo==1&&tagNo==1) {
        self.view.backgroundColor=[UIColor cyanColor];
    }else if (optionNo==2&&tagNo==1) {
        self.view.backgroundColor=[UIColor redColor];      //设置背景颜色
    }else if (optionNo==3&&tagNo==1) {
        self.view.backgroundColor=[UIColor greenColor];
    }else if (optionNo==0&&tagNo==2) {
        self.navigationController.navigationBar.backgroundColor =[UIColor
yellowColor];                                              //设置背景颜色
    }else if (optionNo==1&&tagNo==2) {
         self.navigationController.navigationBar.backgroundColor =[UIColor
purpleColor];
    }
}
```

【代码解析】

本实例关键功能是列表中选择的实现以及响应。下面依次讲解这两个知识点。

1. 列表中的选择

在自定义的列表单选控件中,实现选择,需要使用 UITableDelegate 的 tableView:didSelectRowAtIndexPath:方法,代码如下:

```
- (void)tableView:(UITableView *)tableView didSelectRowAtIndexPath:
(NSIndexPath *)indexPath {
    optionNo = indexPath.row;
    [self.delegate radioView:self didSelectOption:indexPath.row fortag:
tag];
    [tableView reloadData];
}
```

2. 选择行的响应

如果要实现选择任意的行背景或者导航的颜色发生相应的改变,需要在 tableView:

didSelectRowAtIndexPath:方法中调用 radioView:didSelectOption:方法,它的功能就是实现响应。radioView:didSelectOption:方法中实现响应的代码如下:

```
-(void)radioView:(View *)radioView didSelectOption:(int)optionNo fortag:
(int)tagNo{
    if (optionNo==0&&tagNo==1) {
        self.view.backgroundColor=[UIColor orangeColor];
    }
    ……
    else if (optionNo==1&&tagNo==2) {
        self.navigationController.navigationBar.backgroundColor =[UIColor
        purpleColor];
    }
}
```

实例 106 下拉刷新列表

【实例描述】

刷新列表常常用在更新比较繁琐的冗长列表中。当用户进行刷新后,被刷新出来的内容永远都是最新的,它常常在社交更新流、主动收件箱等地方出现。本实例就为各位开发者实现一个下拉刷新的列表。当用户向下拖动时,界面就会开始刷新,刷新停止后,在最前面就会出现一个当前的时间。运行效果如图 13.15 所示。

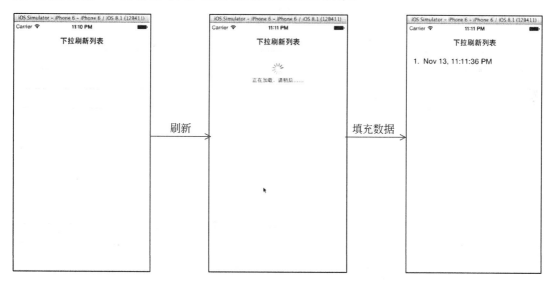

图 13.15 实例 106 运行效果

【实现过程】

(1)创建一个项目,命名为"下拉刷新"。

(2)创建一个基于 UITableViewController 类的 TableViewController 类。

(3)打开 TableViewController.h 文件,编写代码,实现属性的声明。程序代码如下:

```
#import <UIKit/UIKit.h>
@interface TableViewController : UITableViewController
```

```
//属性
@property (nonatomic) int count;
@property (nonatomic,retain) NSMutableArray *countArr;
@end
```

(4)打开 TableViewController.m 文件,编写代码,实现列表的下拉刷新功能。这里分为了两个部分:列表刷新和数据的填充。使用的方法如表 13-15 所示。

表 13-15　TableViewController.m文件中方法总结

方　　法	功　　能
viewDidLoad	视图加载后调用,实现初始化
refreshView:	刷新
handleData	刷新数据
numberOfSectionsInTableView:	获取表视图的块
tableView:numberOfRowsInSection:	获取行数
tableView:cellForRowAtIndexPath:	获取某一行的数据

列表刷新需要使用 viewDidLoad、refreshView:和 handleData 方法实现。其中,viewDidLoad 方法实现了下拉刷新对象的创建。程序代码如下:

```
- (void)viewDidLoad
{
    [super viewDidLoad];
    self.count = 0;
    self.countArr = [[NSMutableArray alloc] initWithCapacity:16] ;
    UIRefreshControl *refresh = [[UIRefreshControl alloc] init];
    refresh.tintColor = [UIColor redColor];                         //设置颜色
    refresh.attributedTitle = [[NSAttributedString alloc] initWithString:
    @"下拉刷新"] ;
    //添加动作
    [refresh addTarget:self action:@selector(refreshView:) forControlEvents:
    UIControlEventValueChanged];
    self.refreshControl = refresh;
}
```

refreshView:方法实现刷新功能。程序代码如下:

```
-(void)refreshView:(UIRefreshControl *)refresh
{
    if (refresh.refreshing) {
        refresh.attributedTitle = [[NSAttributedString alloc]initWithString:
        @"正在加载,请稍后……"] ;
        //在2秒后执行动作handleData
        [self performSelector:@selector(handleData) withObject:nil
        afterDelay:2];
    }
}
```

handleData 方法实现数据的刷新。程序代码如下:

```
-(void)handleData
{
    NSDateFormatter *formatter = [[NSDateFormatter alloc] init];
    [formatter setDateFormat:@"MMM d, h:mm:ss a"];                  //设置格式
    NSString *lastUpdated = [NSString stringWithFormat:@"最后加载的内容为:%@",
    [formatter stringFromDate:[NSDate date]]];                      //创建字符串对象
```

```
    self.refreshControl.attributedTitle = [[NSAttributedString alloc]
    initWithString:lastUpdated];
    self.count++;
    [self.countArr addObject:[NSString stringWithFormat:@"%d. %@",self.count,
    [formatter stringFromDate:[NSDate date]]]];       //添加对象
    [self.refreshControl endRefreshing];
    [self.tableView reloadData];
}
```

numberOfSectionsInTableView:、tableView:numberOfRowsInSection:和 tableView:cellForRowAtIndexPath:方法实现表视图中内容的填充功能。程序代码如下：

```
//获取表视图的块
- (NSInteger)numberOfSectionsInTableView:(UITableView *)tableView
{
    return 1;
}
//获取行数
- (NSInteger)tableView:(UITableView *)tableView numberOfRowsInSection:
(NSInteger)section
{
    return self.countArr.count;
}
//获取某一行的数据
- (UITableViewCell *)tableView:(UITableView *)tableView cellForRowAtIndexPath:
(NSIndexPath *)indexPath
{
    static NSString *CellIdentifier = @"Cell";
    UITableViewCell *cell = [tableView dequeueReusableCellWithIdentifier:
    CellIdentifier];
    //判断 cell 是否为空
    if (cell == nil) {
        cell = [[UITableViewCell alloc] initWithStyle:UITableViewCellStyleDefault
        reuseIdentifier:CellIdentifier] ;
    }
    cell.textLabel.text = [self.countArr objectAtIndex:indexPath.row];
                                                //设置文本内容
    cell.textLabel.font = [UIFont systemFontOfSize:17];
    return cell;
}
```

（5）打开 Main.storyboard 文件，将 Navigation Controller 导航控件器添加到画布中。将 Navigation Item 的 Title 设置为"下拉刷新列表"，将 dataSource、delegate 与 Table View Controller 进行关联。

【代码解析】

本实例关键功能是表视图的刷新功能。下面就是这个知识点的详细讲解。

在 iOS 6 以后的版本中，引入了一个新的 API，就是 UIRefreshControl。它使用在 UITableViewController 表视图控制器中。如果想要实现表视图的下拉刷新功能，就要使用 UITableViewController 的 refreshControl 属性。其语法形式如下：

```
@property (nonatomic,retain) UIRefreshControl *refreshControl;
```

在此代码中就是使用了 refreshControl 属性实现了下拉刷新功能。代码如下：

```
UIRefreshControl *refresh = [[UIRefreshControl alloc] init];
……
```

```
self.refreshControl = refresh;
```

实例 107 背景随动

【实例描述】

本实例主要实现的是背景随意的功能。当用户滑动表视图时，表视图后面的背景也会跟随一起滑动。运行效果如图 13.16 所示。

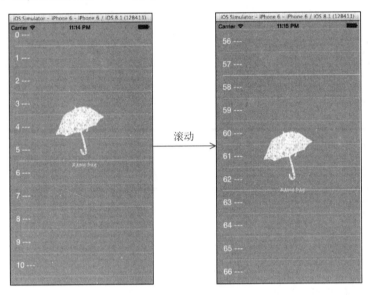

图 13.16 实例 107 运行效果

【实现过程】

（1）创建一个项目，命名为"表的背景随动"。

（2）添加图片 1.jpg 到创建项目的 Supporting Files 文件夹中。

（3）打开 ViewController.h 文件，编写代码，实现宏定义、遵守协议、实例变量以及对象的声明。程序代码如下：

```
#import <UIKit/UIKit.h>
#define bgHeight tableView.frame.size.height
#define beginBgY -tableView.frame.size.height
#define RefreshViewHight 60
@interface ViewController : UIViewController<UITableViewDataSource,
UITableViewDelegate>{
    NSInteger alreadyAddBg;
    BOOL _loadingMore;                              //声明布尔变量
    UITableView *tableView;
    NSMutableArray *dataArr;                        //声明可变数组对象
}
@end
```

（4）打开 ViewController.m 文件，编写代码，实现表的背景随动效果。在此文件中代码分为了界面的初始化和数据的加载两个部分。使用的方法如表 13-16 所示。

表 13-16　ViewController.m文件中方法总结

方　　法	功　　能
viewDidLoad	视图加载后调用，实现初始化
initData	初始化数据
addBgLayer	添加背景层
addLayerWithFrame:	使用框架初始化添加的背景层
tableView:heightForRowAtIndexPath:	获取表中每行的高度
tableView:numberOfRowsInSection:	获取行数
tableView:cellForRowAtIndexPath:	获取某一行的数据
addFooterView:	添加加载视图
isCanLoadData	是否可以加载更多数据
scrollViewDidEndDragging:willDecelerate:	在用户拖动并且停止时调用，显示更多数据
scrollViewDidScroll:	如果拖动的距离大于 footview 的高度，并且没有正在加载数据时调用
loadDataBegin	准备开始加载数据
loadDataing	加载数据中
loadDataEnd	加载数据完毕

这里需要讲解几个重要的方法（其他方法请读者参考源代码）。其中，界面的初始化使用了 viewDidLoad、initData、addBgLayer、addLayerWithFrame:、tableView:heightForRowAtIndexPath:、tableView:numberOfRowsInSection: 和 tableView:cellForRowAtIndexPath: 方法实现。viewDidLoad 方法实现对表视图对象的创建以及设置。程序代码如下：

```
- (void)viewDidLoad{
    [super viewDidLoad];
    tableView = [[UITableView alloc]init];
    [tableView setFrame:self.view.bounds];
    [tableView setBackgroundColor:[UIColor clearColor]];
    [tableView setDelegate:self];                          //设置委托
    [tableView setDataSource:self];
    [self.view addSubview:tableView];                      //添加对象
    alreadyAddBg = 0;
    [self initData];
}
```

initData 方法为表视图中填充的数据进行初始化。程序代码如下：

```
- (void)initData{
    dataArr = [[NSMutableArray alloc]init];                //实例化对象
    for (int i=0; i<20; i++) {
        [dataArr addObject:[NSString stringWithFormat:@"%d ---",i]];
                                                           //添加对象
    }
    [tableView reloadData];
    [self addBgLayer];
}
```

addBgLayer 方法实现为表视图添加背景图层。程序代码如下：

```
- (void)addBgLayer{
    CGFloat contentHeight = tableView.contentSize.height;
    NSInteger needNum = (contentHeight+bgHeight - beginBgY)/bgHeight + 1;
    //判断 needNum 是否大于 alreadyAddBg
```

```
        if (needNum > alreadyAddBg) {
            //遍历
            for(int i = alreadyAddBg; i < needNum; i++) {
                CGRect rect = CGRectMake(0, beginBgY+i*bgHeight, tableView.frame.size.width,tableView.frame.size.height);
                [self addLayerWithFrame:rect];
            }
            alreadyAddBg = needNum;
        }
    }
```

addLayerWithFrame:方法实现背景图层的初始化。程序代码如下：

```
- (void)addLayerWithFrame:(CGRect)frame{
    CALayer *background = [CALayer layer];
    background.zPosition = -1;
    background.frame = frame;                           //设置框架
    background.contents = (__bridge id)([[UIImage imageNamed:@"1.jpg"] CGImage]);
    [tableView.layer addSublayer:background];
}
```

tableView:numberOfRowsInSection:和 tableView:cellForRowAtIndexPath:方法实现了表视图的内容填充置。程序代码如下：

```
- (NSInteger)tableView:(UITableView *)tableView numberOfRowsInSection:(NSInteger)section
{
    return dataArr.count;
}
//获取某一行的数据
- (UITableViewCell *)tableView:(UITableView *)_tableView cellForRowAtIndexPath:(NSIndexPath *)indexPath{
    static NSString * showUserInfoCellIdentifier = @"ShowUserInfoCell";
    UITableViewCell * cell = [_tableView dequeueReusableCellWithIdentifier:showUserInfoCellIdentifier];
    //判断cell是否为空
    if (cell == nil)
    {
        cell = [[UITableViewCell alloc] initWithStyle:UITableViewCellStyleSubtitle reuseIdentifier:showUserInfoCellIdentifier];
        [cell setBackgroundColor:[UIColor clearColor]];      //设置背景颜色
    }
    cell.textLabel.text=[dataArr objectAtIndex:indexPath.row];
    cell.textLabel.textColor=[UIColor whiteColor];           //设置文本颜色
    return cell;
}
```

表视图中数据的加载，需要使用 addFooterView:、isCanLoadData、scrollViewDidEndDragging:willDecelerate:、scrollViewDidScroll:、loadDataBegin、loadDataing 和 loadDataEnd 方法。其中，addFooterView:方法实现添加加载视图的功能。程序代码如下：

```
- (void)addFooterView:(UITableView*)table{
    UIView *footView = [[UIView alloc]initWithFrame:CGRectMake(0, 0, 320, RefreshViewHight)];
    [footView setBackgroundColor:[UIColor clearColor]];
    //创建并设置加载视图对象
    UIActivityIndicatorView *tableFooterActivityIndicator = [[UIActivityIndicatorView alloc] init];
```

```
    [tableFooterActivityIndicator sizeToFit];
    [tableFooterActivityIndicator setCenter:footView.center];
     tableFooterActivityIndicator.color=[UIColor greenColor] ;    //设置颜色
    [tableFooterActivityIndicator startAnimating];               //开始动画
    [footView addSubview:tableFooterActivityIndicator];
    table.tableFooterView = footView;
}
```

scrollViewDidEndDragging.willDecelerate:方法实现用户拖动后显示更多数据。程序代码如下:

```
- (void)scrollViewDidEndDragging:(UIScrollView *)scrollView willDecelerate:
(BOOL)decelerate{
    //下拉到最底部时显示更多数据
    if(!_loadingMore && scrollView.contentOffset.y > ((scrollView.contentSize.
height - scrollView.frame.size.height)))    {
        if ([self isCanLoadData]) {
            [self loadDataBegin];
        }
    }
}
```

loadDataBegin 方法实现准备开始加载数据的功能。程序代码如下:

```
- (void)loadDataBegin {
    if (_loadingMore == NO){
        _loadingMore = YES;
        [self addFooterView:tableView];
        //在 0.2 秒后执行 loadDataing 方法
        [self performSelector:@selector(loadDataing) withObject:nil afterDelay:
0.2];
    }
}
```

loadDataing 方法实现数据正在加载中的功能。程序代码如下:

```
- (void)loadDataing{
NSInteger num = dataArr.count+20;
//遍历
    for (int i=dataArr.count; i<num; i++) {
        [dataArr addObject:[NSString stringWithFormat:@"%d ---",i]];
    }
    [tableView reloadData];
    [self addBgLayer];
    [self loadDataEnd];                              //调用数据加载完毕的方法
}
```

【代码解析】

本实例关键功能是背景随动的实现。下面就是这个知识点的详细讲解。

在本实例中,背景随动主要是通过使用 addBgLayer 方法实现的。在此方法中使用了 CGRectMake 函数实现了矩形框架的获取,从而实现了背景随动。其语法形式如下:

```
CGRect CGRectMake (
  CGFloat x,
  CGFloat y,
  CGFloat width,
  CGFloat height
);
```

其中，参数表示如下。
- CGFloat x 表示 x 的坐标位置。
- CGFloat y 表示 y 的坐标位置。
- CGFloat width 表示宽度。
- CGFloat height 表示高度。

在此代码中，CGRectMake 函数的代码如下：

```
CGRect rect = CGRectMake(0, beginBgY+i*bgHeight, tableView.frame.size.width,tableView.frame.size.height);
```

其中，参数表示如下。
- 0 表示 x 的坐标位置。
- beginBgY+i*bgHeight 表示 y 的坐标位置（背景随动的关键）。
- tableView.frame.size.width 表示宽度。
- tableView.frame.size.height 表示高度。

实例 108　卡片插入式列表

【实例描述】

本实例主要实现的功能是卡片插入式列表，表单元格可以以动态的形式插入。运行效果如图 13.17 所示。

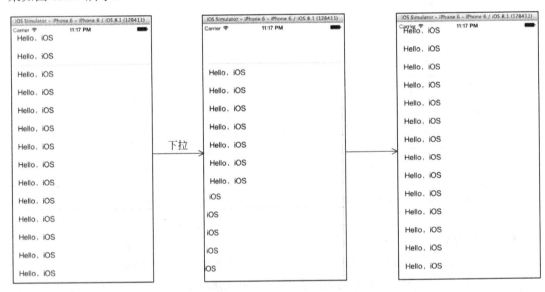

图 13.17　实例 108 运行效果

【实现过程】

（1）创建一个项目，命名为"卡片插入式列表"。

（2）打开 Main.storyboard 文件，将 Table View 表视图添加到 View Controller 视图控制器的设计界面上。将 dataSource、delegate 与 View Controller 关联。

（3）打开 ViewController.m 文件，编写代码，实现卡片插入式列表，代码分为数据的填充以及使用两部分，方法如表 13-17 所示。

表 13-17　ViewController.m文件中方法总结

方　　法	功　　能
tableView:heightForRowAtIndexPath:	获取表中每行的高度
numberOfSectionsInTableView:	获取表视图的块
tableView:numberOfRowsInSection:	获取行数
tableView:cellForRowAtIndexPath:	获取某一行的数据
tableView:willDisplayCell:forRowAtIndexPath:	在表单元格将要显示时调用此方法，实现卡片插入式动画效果

这里需要讲解几个重要的方法（其他方法请读者参考源代码）。其中，代码的数据填充需要使用 numberOfSectionsInTableView:、numberOfSectionsInTableView:、tableView:numberOfRowsInSection:和 tableView:cellForRowAtIndexPath:方法实现。程序代码如下：

```
//获取表视图的块
- (NSInteger)numberOfSectionsInTableView:(UITableView *)tableView
{
    // Return the number of sections.
    return 1;
}
//获取行数
- (NSInteger)tableView:(UITableView *)tableView numberOfRowsInSection:
(NSInteger)section
{
    // Return the number of rows in the section.
    return 20;
}
//获取某一行的数据
- (UITableViewCell *)tableView:(UITableView *)tableView cellForRowAtIndexPath:
(NSIndexPath *)indexPath
{
    static NSString *CellIdentifier = @"RoyCard";
    UITableViewCell *cell = [tableView dequeueReusableCellWithIdentifier:
    CellIdentifier];
    //判断cell是否为空
    if (cell == nil) {
        cell = [[UITableViewCell alloc] initWithStyle:UITableViewCellStyle
        Default reuseIdentifier:CellIdentifier];
    }
    cell.textLabel.text =@"Hello, iOS";
    return cell;
}
```

tableView:willDisplayCell:forRowAtIndexPath:方法在表单元格将要显示时调用,它的功能是实现卡片插入式动画效果。程序代码如下：

```
- (void)tableView:(UITableView *)tableView willDisplayCell:(UITableViewCell *)
cell forRowAtIndexPath:
(NSIndexPath *)indexPath
{
    CGFloat rotationAngleDegrees = 0;
    CGFloat rotationAngleRadians = rotationAngleDegrees * (M_PI/180);
    CGPoint offsetPositioning = CGPointMake(-200, -20);
```

```objc
CATransform3D transform = CATransform3DIdentity;
transform = CATransform3DRotate(transform, rotationAngleRadians, 0.0,
0.0, 1.0);                //旋转
transform = CATransform3DTranslate(transform, offsetPositioning.x,
offsetPositioning.y, 0.0);
UIView *card = [cell contentView];
card.layer.transform = transform;
card.layer.opacity = 0.8;                        //设置透明度
[UIView animateWithDuration:1.0f animations:^{
    card.layer.transform = CATransform3DIdentity;
    card.layer.opacity = 1;
}];
}
```

【代码解析】

本实例关键功能是表单元格在显示时出现的效果。下面就是这个知识点的详细讲解。

要实现表单元格在显示时出现的效果，可以使用 UITableViewDelegate 的 tableView:willDisplayCell:forRowAtIndexPath:方法。其语法形式如下：

```objc
- (void)tableView:(UITableView *)tableView willDisplayCell:(UITableViewCell *)
cell forRowAtIndexPath:
(NSIndexPath *)indexPath;
```

其中，(UITableView *)tableView 表示表视图对象；(UITableViewCell *)cell 表示表视图的单元格对象；(NSIndexPath *)indexPath 表示一个索引路径，用来定位表视图的行。在本代码中，就是使用了 tableView:willDisplayCell:forRowAtIndexPath:方法实现了在表单元格出现时显示卡片插入式效果，代码如下：

```objc
- (void)tableView:(UITableView *)tableView willDisplayCell:(UITableViewCell *)
cell forRowAtIndexPath:
(NSIndexPath *)indexPath
{
   CGFloat rotationAngleDegrees = 0;
   ……
   [UIView animateWithDuration:1.0f animations:^{
       card.layer.transform = CATransform3DIdentity;
       card.layer.opacity = 1;
   }];
}
```

实例 109　嵌套的表

【实例描述】

在 iOS 开发的界面中，经常会看到在一个表视图中又嵌套了另外的一个表视图。本实例就为开发者实现这样的一个功能。运行效果如图 13.18 所示。

【实现过程】

（1）创建一个项目，命名为"嵌套的表"。

（2）创建一个基于 UITableViewCell 类的 Cell 类。

（3）打开 Cell.h 文件，编写代码，实现遵守协议、实例变量、对象以及属性的声明。程序代码如下：

第 13 章 表

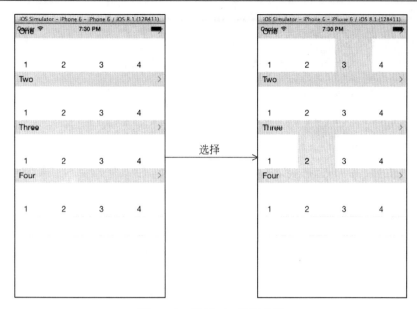

图 13.18 实例 109 运行效果

```
#import <UIKit/UIKit.h>
@interface Cell : UITableViewCell<UITableViewDataSource,UITableViewDelegate>{
NSInteger porsection;                          //实例变量
//对象
    UITableView *table;
    UILabel *lbl;
}
//属性
@property (nonatomic, retain) NSMutableArray* dataArray1;
@property(nonatomic,copy)NSString *lblHead;
@end
```

（4）打开 Cell.m 文件，编写代码，实现将视图放入到表单元格中的功能。使用的方法如表 13-18 所示。

表 13-18 Cell.m 文件中方法总结

方法	功能
initWithStyle:reuseIdentifier:	表单元格的初始化
tableView:numberOfRowsInSection:	获取行数
tableView:cellForRowAtIndexPath:	获取某一行的数据
tableView:heightForRowAtIndexPath:	获取高度

这里需要讲解几个重要的方法（其他方法请读者参考源代码）。其中，initWithStyle: reuseIdentifier:方法对表视图单元格进行初始化。程序代码如下：

```
- (id)initWithStyle:(UITableViewCellStyle)style reuseIdentifier:(NSString
*)reuseIdentifier
{
    self = [super initWithStyle:style reuseIdentifier:reuseIdentifier];
    if (self) {
    //创建并设置图像视图对象
        UIImageView *image=[[UIImageView alloc]initWithFrame:CGRectMake
        (0,0,320,30)];
        image.backgroundColor=[UIColor clearColor];
```

```
        image.image=[UIImage imageNamed:@"1.png"];
        //创建并设置标签对象
        lbl=[[UILabel alloc]initWithFrame:CGRectMake(10,5,200,20)];
        lbl.backgroundColor=[UIColor clearColor];
        [image addSubview:lbl];
        //创建并设置表视图对象
        table = [[UITableView alloc]initWithFrame:CGRectMake(110,-75,110,320)
        style:UITableViewStylePlain];
        table.delegate = self;
        table.dataSource = self;
        table.showsHorizontalScrollIndicator=NO;
        table.showsVerticalScrollIndicator=NO;
        table.transform = CGAffineTransformMakeRotation(M_PI / 2 *3);
                                                                    //进行旋转
        [self addSubview:image];
        [self addSubview:table];
    }
    return self;
}
```

tableView:cellForRowAtIndexPath:方法实现获取某一行的数据。程序代码如下:

```
- (UITableViewCell *)tableView:(UITableView *)tableView cellForRowAtIndexPath:
(NSIndexPath *)indexPath
{
    NSString *CellIdentifier = [NSString stringWithFormat:@"cell%d",
    indexPath.row];
    UITableViewCell *cell = [tableView dequeueReusableCellWithIdentifier:
    CellIdentifier];
    //判断cell是否为空
    if (cell == nil)
    {
        cell = [[UITableViewCell alloc]initWithStyle:UITableViewCellStyleDefault
        reuseIdentifier:CellIdentifier];
        cell.transform = CGAffineTransformMakeRotation(M_PI/2);//进行旋转
        lbl.text=lblHead;
    }
    [[cell textLabel] setText:[dataArray1 objectAtIndex:indexPath.row]];
                                                                    //设置文本
    cell.textLabel.numberOfLines = 0;
    return cell;
}
```

(5) 打开 ViewController.h 文件,编写代码,实现头文件、遵守协议以及对象的声明。程序代码如下:

```
#import <UIKit/UIKit.h>
#import "CustomCell.h"                          //头文件
@interface ViewController : UIViewController<UITableViewDelegate,UITable
ViewDataSource>{
    //对象
    UITableView *skytable;
    NSMutableArray *array;
    NSMutableArray *Title;
}
@end
```

(6) 打开 ViewController.m 文件,编写代码,实现表视图的嵌套功能。使用的方法如表 13-19 所示。

第 13 章 表

表 13-19 ViewController.m 文件中方法总结

方　　法	功　　能
viewDidLoad	视图加载后调用，实现初始化
numberOfSectionsInTableView:	获取块数
tableView:numberOfRowsInSection:	获取行数
tableView:cellForRowAtIndexPath:	获取某一行的数据
tableView:heightForRowAtIndexPath:	获取高度

其中，viewDidLoad 方法实现对表视图对象的创建和设置。程序代码如下：

```
- (void)viewDidLoad
{
   [super viewDidLoad];
   // Do any additional setup after loading the view, typically from a nib.
   skytable = [[UITableView alloc]initWithFrame:CGRectMake(0, 0, 320,460)
   style:UITableViewStylePlain];
   [skytable setDelegate:self];                        //设置委托
   [skytable setDataSource:self];
   array= [[NSMutableArray alloc]initWithObjects:@"1",@"2",@"3",@"4",@"5",
   @"6",@"7",@"8",nil];
   Title= [[NSMutableArray alloc]initWithObjects:@"One",@"Two",@"Three",
   @"Four",nil];
   [self.view addSubview:skytable];                    //添加视图对象
}
```

tableView:cellForRowAtIndexPath:方法实现获取表视图的某一行数据。程序代码如下：

```
-(UITableViewCell *)tableView:(UITableView *)tableView cellForRowAtIndexPath:
(NSIndexPath *)indexPath
{
   static NSString *identity = @"cell";
   CustomCell *cell = (CustomCell *)[tableView dequeueReusableCellWith
   Identifier:identity];
   //判断 cell 是否为空
   if (cell == nil){
      cell = [[CustomCell alloc]initWithStyle:UITableViewCellStyleDefault
      reuseIdentifier:identity];
   }else{
      cell = [[CustomCell alloc]initWithStyle:UITableViewCellStyleDefault
      reuseIdentifier:identity];
   }
   cell.selectionStyle = UITableViewCellSelectionStyleNone;//设置选择风格
   cell.lblHead=[Title objectAtIndex:indexPath.row];
   cell.dataArray1=array;
   return cell;
}
```

【代码解析】

本实例关键功能是表视图嵌套的实现。下面就是这个知识点的详细讲解。

在本实例中表视图的嵌套，首先要对表单元格进行设置，即在创建的 CustomCell 类中设置。在此类中，实现表视图对象创建的代码如下：

```
table = [[UITableView alloc]initWithFrame:CGRectMake(110,-75,110,320) style:
UITableViewStylePlain];
table.delegate = self;
table.dataSource = self;
……
```

```
table.transform = CGAffineTransformMakeRotation(M_PI / 2 *3);
[self addSubview:table];
```

表视图创建好后,实现内容的填充。这时在每一个单元格中就存在一个表视图了。最后,嵌套的实现,需要再创建一个表视图,此时表视图的单元格需要重新创建。代码如下:

```
CustomCell *cell = (CustomCell *)[tableView dequeueReusableCellWithIdentifier:
identity];
if (cell == nil){
    cell = [[CustomCell alloc]initWithStyle:UITableViewCellStyleDefault
    reuseIdentifier:identity];
}else{
    cell = [[CustomCell alloc]initWithStyle:UITableViewCellStyleDefault
    reuseIdentifier:identity];
}
```

实例 110 仿 QQ 聊天

【实例描述】

读者对 QQ 相信都不陌生,本实例实现的功能就是模仿 QQ 中的聊天效果。运行效果如图 13.19 所示。其中用于用户输入的是文本框,用于显示内容的是表视图。开发者可以由此例模仿出短信发送等各种类似的聊天效果。

图 13.19 实例 110 运行效果

第 13 章 表

【实现过程】

当用户单击文本框后，可以在其中输入内容，之后按下键盘上的 return 键，将信息进行发送，并显示在表视图中。当单击文本框左边的话筒按钮，就会转入声音的发送中。具体的实现步骤如下。

（1）创建一个项目，命名为"仿 QQ 聊天"。

（2）添加图片 1.png、2.png、3.png、4.png、5.png、6.png、7.png、8.png、9.png、10.png、11.jpg、12.png、13.png、14.png、15.png、16.jpg、17.jpg 和 18.png 到创建项目的 Supporting Files 文件夹中。

（3）创建一个名为 message.plist 的文件。对文件中的内容进行设置，如表 13-20 所示。

表 13-20　message.plist文件中的内容

Key	Type	Value
Root	Array	
Item0	Dictionary	
content	String	喂，你好！
icon	String	16.jpg
time	String	前天 10:25
type	Number	0
Item1	Dictionary	
content	String	问你个严肃的事
icon	String	16.jpg
time	String	前天 10:25
type	Number	0
Item2	Dictionary	
content	String	说，什么事？
icon	String	17.jpg
time	String	昨天 20:05
type	Number	1
Item3	Dictionary	
content	String	我就想问下：你是不是我最好的朋友？
icon	String	16.jpg
time	String	昨天 20:07
type	Number	0
Item4	Dictionary	
content	String	不借
icon	String	17.jpg
time	String	昨天 20:30
type	Number	0
Item5	Dictionary	
content	String	你就回答"是"或者"不是"。
icon	String	16.jpg
time	String	今天 22:23
type	Number	0

（4）创建一个基于 NSObject 的 Message 文件。

（5）打开 Message.h 文件，编写代码，实现数据结构的定义和属性的声明。程序代码如下：

```
#import <Foundation/Foundation.h>
//数据结构的定义
typedef enum {
    MessageTypeMe = 0, // 自己发的信息
    MessageTypeOther = 1 //别人发的信息
} MessageType;
@interface Message : NSObject
//属性
@property (nonatomic, copy) NSString *icon;
@property (nonatomic, copy) NSString *time;
@property (nonatomic, copy) NSString *content;
@property (nonatomic, assign) MessageType type;
@property (nonatomic, copy) NSDictionary *dict;
@end
```

（6）打开 Message.m 文件，编写代码，实现 setDict:方法的功能，它的功能是对 QQ 的一些属性进行设置。程序代码如下：

```
- (void)setDict:(NSDictionary *)dict{
    //赋值
    _dict = dict;
    self.icon = dict[@"icon"];
    self.time = dict[@"time"];
    self.content = dict[@"content"];
    self.type = [dict[@"type"] intValue];
}
```

（7）创建一个基于 NSObject 的 MessageFrame 文件。

（8）打开 MessageFrame.h 文件，编写代码，实现头文件、宏定义以及属性的声明。程序代码如下：

```
#import <Foundation/Foundation.h>
#import "Message.h"
#define kMargin 10
#define kIconWH 40
......
#define kTimeFont [UIFont systemFontOfSize:12]
#define kContentFont [UIFont systemFontOfSize:16]
@interface MessageFrame : NSObject
@property (nonatomic, assign, readonly) CGRect iconF;
......
@property (nonatomic, strong) Message *message;
@property (nonatomic, assign) BOOL showTime;
@end
```

（9）打开 MessageFrame.m 文件，编写代码，实现 setMessage:方法的功能，它的功能是对发送的 QQ 信息进行设置。程序代码如下：

```
- (void)setMessage:(Message *)message{
    _message = message;
    //获取屏幕宽度
    CGFloat screenW = [UIScreen mainScreen].bounds.size.width;
    //计算时间的位置
```

```
    if (_showTime){
        CGFloat timeY = kMargin;
        CGSize timeSize = [_message.time sizeWithAttributes:@{UIFont
        DescriptorSizeAttribute: @"16"}];
        NSLog(@"----%@", NSStringFromCGSize(timeSize));          //输出
        CGFloat timeX = (screenW - timeSize.width) / 2;
        _timeF = CGRectMake(timeX, timeY, timeSize.width + kTimeMarginW,
timeSize.height + kTimeMarginH);
    }
    //计算头像位置
    CGFloat iconX = kMargin;
    //如果是自己发的信息,头像在右边
    if (_message.type == MessageTypeMe) {
        iconX = screenW - kMargin - kIconWH;
    }
    CGFloat iconY = CGRectGetMaxY(_timeF) + kMargin;
    _iconF = CGRectMake(iconX, iconY, kIconWH, kIconWH);
    //计算内容位置
    CGFloat contentX = CGRectGetMaxX(_iconF) + kMargin;
    CGFloat contentY = iconY;
    NSDictionary *dis=[NSDictionary dictionaryWithObjectsAndKeys:kContentFont,
    NSFontAttributeName, nil];
    CGSize contentSize = [_message.content boundingRectWithSize:CGSizeMake
(kContentW, CGFLOAT_MAX) options:NSStringDrawingUsesLineFragmentOrigin
attributes:dis context:nil].size;
    //判断
    if (_message.type == MessageTypeMe) {
        contentX = iconX - kMargin - contentSize.width - kContentLeft -
        kContentRight;
    }
    _contentF = CGRectMake(contentX, contentY, contentSize.width +
kContentLeft + kContentRight, contentSize.height + kContentTop +
kContentBottom);
    //计算高度
    _cellHeight = MAX(CGRectGetMaxY(_contentF), CGRectGetMaxY(_iconF)) +
    kMargin;
}
```

(10) 创建一个基于 UITableViewCell 的 MessageCell 文件。

(11) 打开 MessageCell.h 文件,编写代码,实现头文件、对象以及属性的声明。程序代码如下:

```
#import <UIKit/UIKit.h>
//头文件
#import "MessageFrame.h"
#import "Message.h"
@interface MessageCell : UITableViewCell{
    //对象
    UIButton     *_timeBtn;
    UIImageView *_iconView;
    UIButton     *_contentBtn;
}
//属性
@property (nonatomic, strong) MessageFrame *messageFrame;
@end
```

(12) 打开 MessageCell.m 文件,编写代码,实现对发送信息进行绘制。其中,initWithStyle:reuseIdentifier:方法实现 QQ 信息框架初始化功能。程序代码如下:

```objc
- (id)initWithStyle:(UITableViewCellStyle)style reuseIdentifier:(NSString
*)reuseIdentifier
{
    self = [super initWithStyle:style reuseIdentifier:reuseIdentifier];
    if (self) {
        self.backgroundColor = [UIColor clearColor];
        //创建时间按钮
        _timeBtn = [[UIButton alloc] init];
        [_timeBtn setTitleColor:[UIColor blackColor] forState:
        UIControlStateNormal];
        _timeBtn.titleLabel.font = kTimeFont;              //设置字体
        _timeBtn.enabled = NO;
        [_timeBtn setBackgroundImage:[UIImage imageNamed:@"18.png"] forState:
        UIControlStateNormal];
        [self.contentView addSubview:_timeBtn];            //添加视图对象
        //创建头像
        _iconView = [[UIImageView alloc] init];
        [self.contentView addSubview:_iconView];
        //创建内容
        _contentBtn = [UIButton buttonWithType:UIButtonTypeCustom];
        [_contentBtn setTitleColor:[UIColor blackColor] forState:
        UIControlStateNormal];
        _contentBtn.titleLabel.font = kContentFont;
        _contentBtn.titleLabel.numberOfLines = 0;          //设置行数
        [self.contentView addSubview:_contentBtn];
    }
    return self;
}
```

setMessageFrame:方法实现对 QQ 信息的框架的设置。程序代码如下：

```objc
- (void)setMessageFrame:(MessageFrame *)messageFrame{
    _messageFrame = messageFrame;
    Message *message = _messageFrame.message;
    //设置时间
    [_timeBtn setTitle:message.time forState:UIControlStateNormal];
    _timeBtn.frame = _messageFrame.timeF;
    //设置头像
    _iconView.image = [UIImage imageNamed:message.icon];
    _iconView.frame = _messageFrame.iconF;
    //设置内容
    [_contentBtn setTitle:message.content forState:UIControlStateNormal];
    _contentBtn.contentEdgeInsets = UIEdgeInsetsMake(kContentTop, kContentLeft,
    kContentBottom, kContentRight);
    _contentBtn.frame = _messageFrame.contentF;
    //判断 message.type 是否等于 MessageTypeMe
    if (message.type == MessageTypeMe) {
        _contentBtn.contentEdgeInsets = UIEdgeInsetsMake(kContentTop,
        kContentRight, kContentBottom, kContentLeft);
    }
    UIImage *normal , *focused;
    //判断 message.type 是否等于 MessageTypeMe
    if (message.type == MessageTypeMe) {
        normal = [UIImage imageNamed:@"13.png"];
        normal = [normal stretchableImageWithLeftCapWidth:normal.size.width
        * 0.5 topCapHeight:normal.size.height * 0.7];
                                   //创建一个内容可拉伸，而边角不拉伸的图片
        focused = [UIImage imageNamed:@"12.png"];
        focused = [focused stretchableImageWithLeftCapWidth:focused.size.
```

```
            width * 0.5 topCapHeight:focused.size.height * 0.7];
                                     //创建一个内容可拉伸,而边角不拉伸的图片
    }else{
        normal = [UIImage imageNamed:@"15.png"];
        normal = [normal stretchableImageWithLeftCapWidth:normal.size.width
        * 0.5 topCapHeight:normal.size.height * 0.7];
                                     //创建一个内容可拉伸,而边角不拉伸的图片
        focused = [UIImage imageNamed:@"14.png"];
        focused = [focused stretchableImageWithLeftCapWidth:focused.size.
        width * 0.5 topCapHeight:focused.size.height * 0.7];
                                     //创建一个内容可拉伸,而边角不拉伸的图片
    }
    [_contentBtn setBackgroundImage:normal forState:UIControlStateNormal];
    [_contentBtn setBackgroundImage:focused forState:UIControlStateHighlighted];
}
```

(13) 打开 ViewController.h 文件,编写代码,实现头文件、插座变量、对象以及方法的声明。程序代码如下:

```
#import <UIKit/UIKit.h>
#import "MessageFrame.h"
#import "Message.h"
#import "MessageCell.h"
@interface ViewController : UIViewController<UITextFieldDelegate>{
    IBOutlet UITableView *tableV;
    IBOutlet UITextField *messageField;       //声明关于文本框的插座变量
    IBOutlet UIButton *speakBtn;
    NSMutableArray *_allMessagesFrame;
}
- (IBAction)speak:(id)sender;
@end
```

(14) 打开 Main.storyboard 文件,对 View Controller 视图控制器的设计界面进行设计,效果如图 13.20 所示。

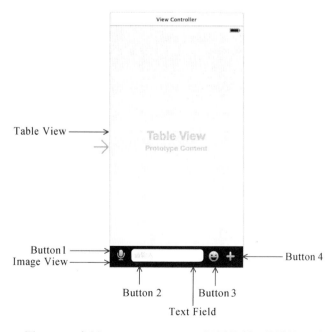

图 13.20　实例 110 View Controller 视图控制器的设计界面

需要添加的视图、控件以及对它们的设置如表 13-21 所示。

表 13-21　实例 110 视图、控件设置

视图、控件	属 性 设 置	其　　他
Table View		与插座变量 table 关联 将 dataSource 与 View Controller 关联 将 delegate 与 View Controller 关联
Image View	Image：10.pngV	
Button1	Title：（空） Background：7.png	与动作 speak:关联
Button2	Title：按住说话 Background：9.png	与插座变量 speakBtn 关联
Button3	Title：（空） Background：3.png	
Button4	Title：（空） Background：5.png	与动作 finsh:关联
Text Field	Font：System 16.0 Placeholder：请输入	与插座变量 messageField 关联

（15）打开 ViewController.m 文件，编写代码，实现仿 QQ 聊天。在此文件中，代码分为：初始化、对键盘的设置、对文本框和数据的设置以及对语音按钮的设置。使用的方法如表 13-22 所示。

表 13-22　ViewController.m 文件中方法总结

方　　法	功　　能
viewDidLoad	视图加载后调用，实现初始化
keyBoardWillShow:	键盘即将显示
keyBoardWillHide	键盘即将退出
textFieldShouldReturn:	单击文本框键盘的回车按钮
addMessageWithContent:time:	给数据源增加内容
tableView:numberOfRowsInSection:	获取行数
tableView: cellForRowAtIndexPath:	获取某一行数据
tableView: heightForRowAtIndexPath:	获取高度
scrollViewWillBeginDragging:	在滚动时，让视图退出第一响应者的状态
speak:	语音按钮单击

其中，viewDidLoad 方法实现的功能就是初始化。程序代码如下：

```
- (void)viewDidLoad
{
    [super viewDidLoad];
    tableV.separatorStyle = UITableViewCellSeparatorStyleNone;
                                                        //设置分割线的风格
    tableV.allowsSelection = NO;
    tableV.backgroundView = [[UIImageView alloc] initWithImage:[UIImage
    imageNamed:@"11.jpg"]];
    NSArray *array = [NSArray arrayWithContentsOfFile:[[NSBundle mainBundle]
    pathForResource:@"messages" ofType:@"plist"]];      //创建数组对象
```

```objc
    _allMessagesFrame = [NSMutableArray array];
    NSString *previousTime = nil;
    //遍历
    for (NSDictionary *dict in array) {
        MessageFrame *messageFrame = [[MessageFrame alloc] init];
        Message *message = [[Message alloc] init];
        message.dict = dict;
        messageFrame.showTime = ![previousTime isEqualToString:message.time];
        messageFrame.message = message;                        //设置信息
        previousTime = message.time;
        [_allMessagesFrame addObject:messageFrame];
    }
    [[NSNotificationCenter defaultCenter] addObserver:self selector:@selector(keyBoardWillShow:) name:UIKeyboardWillShowNotification object:nil];
                                                            //添加观察者
    [[NSNotificationCenter defaultCenter] addObserver:self selector:@selector(keyBoardWillHide:) name:UIKeyboardWillHideNotification object:nil];
    //设置textField输入起始位置
    messageField.leftView = [[UIView alloc] initWithFrame:CGRectMake(0, 0, 10, 0)];
    messageField.leftViewMode = UITextFieldViewModeAlways;
    messageField.delegate = self;                           //设置委托
}
```

keyBoardWillShow:方法实现键盘即将显示的功能。程序代码如下：

```objc
- (void)keyBoardWillShow:(NSNotification *)note{
    CGRect rect = [note.userInfo[UIKeyboardFrameEndUserInfoKey] CGRectValue];
    CGFloat ty = - rect.size.height;
    [UIView animateWithDuration:[note.userInfo[UIKeyboardAnimationDurationUserInfoKey] doubleValue] animations:^{
        self.view.transform = CGAffineTransformMakeTranslation(0, ty);
                                                            //上下移动
    }];
}
```

keyBoardWillHide:方法实现键盘即将关闭的功能。程序代码如下：

```objc
- (void)keyBoardWillHide:(NSNotification *)note{
    [UIView animateWithDuration:[note.userInfo[UIKeyboardAnimationDurationUserInfoKey] doubleValue] animations:^{
        self.view.transform = CGAffineTransformIdentity;    //设置改变
    }];
}
```

textFieldShouldReturn:方法实现单击键盘上的return键的功能。程序代码如下：

```objc
- (BOOL)textFieldShouldReturn:(UITextField *)textField{

    //增加数据源
    NSString *content = textField.text;
```

```
    NSDateFormatter *fmt = [[NSDateFormatter alloc] init];
    NSDate *date = [NSDate date];
    fmt.dateFormat = @"MM-dd";                              //设置格式
    NSString *time = [fmt stringFromDate:date];
    [self addMessageWithContent:content time:time];
    //刷新表格
    [tableV reloadData];                                    //加载数据
    //滚动至当前行
    NSIndexPath *indexPath = [NSIndexPath indexPathForRow:_allMessagesFrame.count - 1 inSection:0];
    [tableV scrollToRowAtIndexPath:indexPath atScrollPosition:UITableViewScrollPositionBottom animated:YES];
    //清空文本框内容
    messageField.text = nil;
    return YES;
}
```

addMessageWithContent:方法实现将文本框中的内容添加到数据源中。程序代码如下：

```
- (void)addMessageWithContent:(NSString *)content time:(NSString *)time{
    MessageFrame *mf = [[MessageFrame alloc] init];
    Message *msg = [[Message alloc] init];
    msg.content = content;
    msg.time = time;                                        //设置时间
    msg.icon = @"16.jpg";
    msg.type = MessageTypeMe;
    mf.message = msg;
    [_allMessagesFrame addObject:mf];
}
```

tableView:cellForRowAtIndexPath:方法实现表视图的内容填充功能。程序代码如下：

```
- (UITableViewCell *)tableView:(UITableView *)tableView cellForRowAtIndexPath:(NSIndexPath *)indexPath
{
    static NSString *CellIdentifier = @"Cell";
    MessageCell *cell = [tableView dequeueReusableCellWithIdentifier:CellIdentifier];
    //判断cell是否为空
    if (cell == nil) {
        cell = [[MessageCell alloc] initWithStyle:UITableViewCellStyleDefault
            reuseIdentifier:CellIdentifier];
    }
    //设置数据
    cell.messageFrame = _allMessagesFrame[indexPath.row];
    return cell;
}
```

speak:方法实现单击语音按钮后的语音说话功能。程序代码如下：

```
- (IBAction)speak:(id)sender {
    if (messageField.hidden) {
```

```objc
        //输入框隐藏,按住说话按钮显示
        messageField.hidden = NO;
        speakBtn.hidden = YES;
        //设置背景图像
        [sender setBackgroundImage:[UIImage imageNamed:@"7.png"] forState:
        UIControlStateNormal];
        [sender setBackgroundImage:[UIImage imageNamed:@"8.png"] forState:
        UIControlStateHighlighted];
        [messageField becomeFirstResponder];
    }else{
        //输入框处于显示状态,按住说话按钮处于隐藏状态
        messageField.hidden = YES;
        speakBtn.hidden = NO;
        //设置背景图像
        [sender setBackgroundImage:[UIImage imageNamed:@"1.png"] forState:
        UIControlStateNormal];
        [sender setBackgroundImage:[UIImage imageNamed:@"2.png"] forState:
        UIControlStateHighlighted];
        [messageField resignFirstResponder];
    }
}
```

【代码解析】

由于本实例中的代码和方法非常的多,为了方便读者的阅读,笔者绘制了一些执行流程图如图 13.21~图 13.22 所示。其中,单击运行按钮一直到 QQ 界面的出现,使用了 viewDidLoad、setDict:、setMessage:、tableView:numberOfRowsInSection:、tableView:heightForRowAtIndexPath:、tableView:cellForRowAtIndexPath:、initWithStyle:reuseIdentifier:和 setMessageFrame:方法。它们的执行流程如图 13.21 所示。

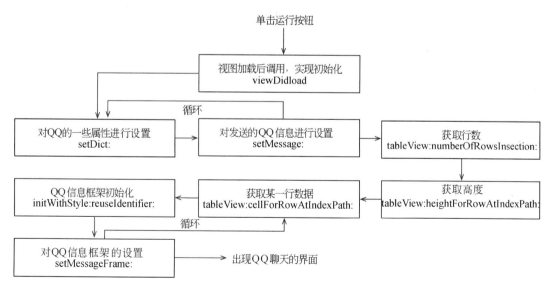

图 13.21 实例 110 程序执行流程 1

要实现信息的发送以及显示，需要使用 keyBoardWillShow:、textFieldShouldReturn:、addMessageWithContent:time: 、setMessage: 、tableView:numberOfRowsInSection: 、tableView:heightForRowAtIndexPath:、tableView:cellForRowAtIndexPath:、setMessageFrame:、tableView:cellForRowAtIndexPath: 、scrollViewWillBeginDragging:和 keyBoardWillHide:方法。它们的执行流程如图 13.22 所示。

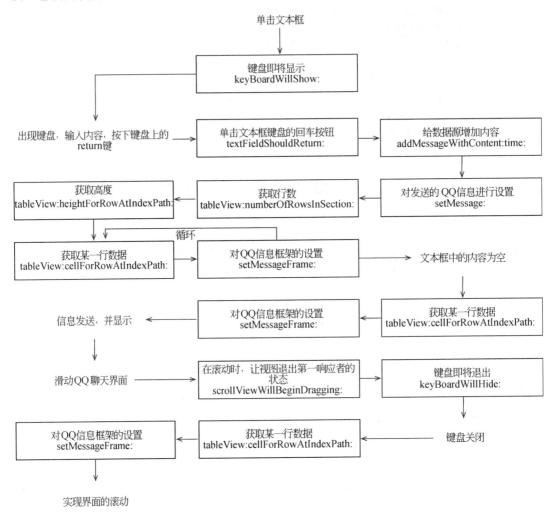

图 13.22　实例 110 程序执行流程 2

实例 111　树形展开列表

【实例描述】
　　本实例主要实现以树形展开的列表，它有点类似于文件浏览器的文件夹树形展开的效果。运行效果如图 13.23 所示。

第 13 章　表

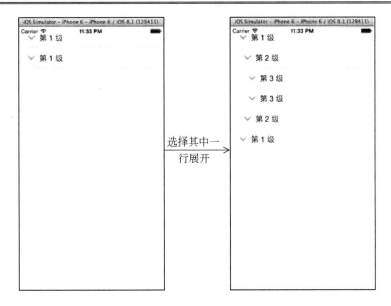

图 13.23　实例 111 运行效果

【实现过程】

（1）创建一个项目，命名为"树形展开列表"。
（2）添加图片 1.png 到创建项目的 Supporting Files 文件夹中。
（3）创建一个基于 UITableViewCell 类的 Cell 类。
（4）打开 Cell.h 文件，编写代码，实现属性的声明。程序代码如下：

```
#import <UIKit/UIKit.h>
#define kLevelOffset    10
@interface Cell : UITableViewCell
//属性声明
@property (nonatomic)                       NSInteger       level;
@property (nonatomic, weak) IBOutlet  UIImageView*    iconView;
@property (nonatomic, weak) IBOutlet  UILabel*        contextLabel;
@end
```

（5）打开 Cell.m 文件，编写代码，实现对表单元格的内容设置。程序代码如下：

```
- (void) setLevel:(NSInteger)level
{
   _level = level;
   CGRect rect = _contextLabel.frame;
   rect.origin.x =46 + kLevelOffset * _level;
   _contextLabel.frame = rect;                                      //设置框架
   rect = _iconView.frame;
   rect.origin.x =20 + kLevelOffset * _level;
   _iconView.frame = rect;
   _contextLabel.text = [NSString stringWithFormat:@"第 %d 级", _level + 1];
       //设置文本内容
}
```

（6）创建一个基于 NSObject 类的 CellMode 类。
（7）打开 CellMode.h 文件，编写代码，实现属性的声明。程序代码如下：

```
#import <Foundation/Foundation.h>
@interface CellModel : NSObject
```

```
//属性
@property (nonatomic) NSInteger    level;
@property (nonatomic) NSInteger    hide;
@end
```

（8）打开 CellMode.m 文件，编写代码，实现字符串的获取。程序代码如下：

```
- (NSString*) description
{
    return [NSString stringWithFormat:@"%@ level = %d hide = %d", [super
    description], _level, _hide];
}
```

（9）打开 ViewController.h 文件，编写代码，实现头文件、宏定义、对象以及插座变量的声明。程序代码如下：

```
#import <UIKit/UIKit.h>
//头文件
#import "CellModel.h"
#import "Cell.h"
#define kMaxDeep 7                                    //宏定义
@interface ViewController : UIViewController{
    NSMutableArray * _tableViewDataSource;            //对象
    IBOutlet UITableView*   tableV;                   //插座变量
}
@end
```

（10）打开 Main.storyboard 文件，对 View Controller 视图控制器的设计界面进行设计，效果如图 13.24 所示。

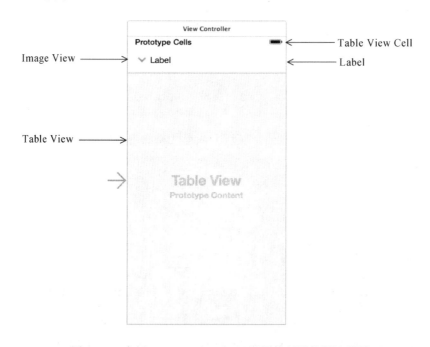

图 13.24　实例 111 View Controller 视图控制器的设计界面

需要添加的视图、控件以及对它们的设置如表 13-23 所示。

第 13 章 表

表 13-23 实例 111 视图、控件设置

视图、控件	属 性 设 置	其 他
Table View	Prototype Cells：1	与插座变量 tableV 关联 将 dataSource 与 View Controller 关联 将 delegate 与 View Controller 关联
Table View Cell	Identifier：default	Class：Cell
Label		与插座变量 contextLabel 关联
Image View	Image：1.png	与插座变量 iconView 关联

（11）打开 ViewController.m 文件，编写代码，实现树形展开列表的功能。在此文件中，代码分为：表视图的内容填充以及选择表单元格后展开的功能。使用的方法如表 13-24 所示。

表 13-24 ViewController.m 文件中方法总结

方 法	功 能
dataFiller:	添加数据
viewDidLoad	视图加载后调用，实现初始化
tableView:numberOfRowsInSection:	获取行数
tableView:cellForRowAtIndexPath:	获取某一行的数据
tableView:didSelectRowAtIndexPath:	响应选择的行

其中，表视图的内容填充需要使用 dataFiller:、viewDidLoad、tableView:numberOfRowsInSection:和 tableView:cellForRowAtIndexPath:方法。dataFiller:方法实现为可变数组添加数据的功能。程序代码如下：

```
- (void)dataFiller:(NSInteger) level
{
    CellModel* model = [[CellModel alloc] init];
    model.level = level;
    model.hide = (1 << (kMaxDeep - level)) - 1;
    [_tableViewDataSource addObject:model];           //添加对象
    //判断 level 是否不等于 kMaxDeep - 1
    if (level != kMaxDeep - 1)
    {
        int t = random() % 3 + 1;
        for (int i=0; i<t; i++)
            [self dataFiller:level + 1];
    }
}
```

tableView:numberOfRowsInSection:方法实现获取表视图的行数。程序代码如下：

```
- (NSInteger) tableView:(UITableView *)tableView numberOfRowsInSection:
(NSInteger)section
{
    int n = 0;
    //遍历
    for (CellModel* model in _tableViewDataSource) {
        if (model.hide == kShowFlag) n ++ ;
    }
    return n;
}
```

tableView:cellForRowAtIndexPath:方法实现获取某一行数据的功能。程序代码如下：

```objc
- (UITableViewCell*) tableView:(UITableView *)tableView cellForRowAtIndexPath:
(NSIndexPath *)indexPath
{
    int t = -1;
    CellModel* m;
    //遍历
    for (CellModel* model in _tableViewDataSource) {
        if (model.hide == kShowFlag) t ++;
        //判断t是否等于indexPath.row
        if (t == indexPath.row) {
            m = model;
            break;
        }
    }
    static NSString* key = @"default";
    Cell* cell = [tableView dequeueReusableCellWithIdentifier:key];
    cell.level = m.level;                                  //设置级别
    return cell;
}
```

选择表单元格后展开的功能需要使用 tableView:didSelectRowAtIndexPath:方法实现。程序代码如下：

```objc
- (void) tableView:(UITableView *)tableView didSelectRowAtIndexPath:
(NSIndexPath *)indexPath
{
    int t = -1, p = 0;
    CellModel* m;
    //遍历
    for (CellModel* model in _tableViewDataSource) {
        if (model.hide == kShowFlag) t ++;
        //判断t是否等于indexPath.row
        if (t == indexPath.row) {
            m = model;
            break;
        }
        p++;
    }
    //判断m.level是否等于6
    if (m.level == 6)
        return;
    p ++;
    if (p == _tableViewDataSource.count)
        return;
    CellModel* nxtModel = _tableViewDataSource[p];
    //判断nxtModel.level是否大于m.level
    if (nxtModel.level > m.level)
    {
        if (nxtModel.hide == kShowFlag){
            NSMutableArray* arr = [NSMutableArray array];     //实例化对象
            while (true) {
                if (nxtModel.hide == kShowFlag) {
                    t ++;
                    NSIndexPath* path = [NSIndexPath indexPathForRow:t
                    inSection:0];                             //实例化对象
                    [arr addObject:path];
                }
```

第 13 章 表

```
            nxtModel.hide ^= 1 << (kMaxDeep - m.level - 1);
            p ++;
            if (p == _tableViewDataSource.count) break;
            nxtModel = _tableViewDataSource[p];
            //判断 nxtModel.level 是否小于等于 m.level
            if (nxtModel.level <= m.level) break;
        }
        //选择行
        [tableView deleteRowsAtIndexPaths:arr
                        withRowAnimation:UITableViewRowAnimationFade];
    }else{
        NSMutableArray* arr = [NSMutableArray array];
        while (true){
            nxtModel.hide ^= 1 << (kMaxDeep - m.level - 1);
            //判断 nxtModel.hide 是否等于 kShowFlag
            if (nxtModel.hide == kShowFlag){
                t ++;
                NSIndexPath* path = [NSIndexPath indexPathForRow:t
                inSection:0];
                [arr addObject:path];                         //添加对象
            }
            p ++;
            if (p == _tableViewDataSource.count) break;
            nxtModel = _tableViewDataSource[p];
            if (nxtModel.level <= m.level) break;
        }
        //插入行
        [tableView insertRowsAtIndexPaths:arr
                        withRowAnimation:UITableViewRowAnimationFade];
    }
}
```

【代码解析】

本实例的关键功能是列表的展开和折叠。下面依次讲解这两个知识点。

1. 列表的展开

如果想要实现列表的展开，需要使用 UITableView 的 insertRowsAtIndexPaths:withRowAnimation:方法，它的功能是根据索引路径数组插入行。其语法形式如下：

```
- (void)insertRowsAtIndexPaths:(NSArray *)indexPaths withRowAnimation:
(UITableViewRowAnimation)animation;
```

其中，(NSArray *)indexPaths 表示目标行所对应的索引路径数组；(UITableViewRowAnimation)animation 表示需要使用的动画效果。在此实例中，使用了 insertRowsAtIndexPaths:withRowAnimation:方法实现了列表的展开效果，代码如下：

```
[tableView insertRowsAtIndexPaths: arr
withRowAnimation:UITableViewRowAnimationFade];
```

其中，arr 表示目标行所对应的索引路径数组；UITableViewRowAnimationFade 表示使用的动画效果。

2. 列表的折叠

在此实例中，使用了 deleteRowsAtIndexPaths:withRowAnimation:实现了列表的折叠效

果,代码如下:

```
[tableView deleteRowsAtIndexPaths:arr
                    withRowAnimation:UITableViewRowAnimationFade];
```

实例 112 圆角表视图

【实例描述】
本实例主要的功能是实现一个带有圆角效果的表视图。当用户选择其中的某一行数据时,会在此行上出现选取标记。运行效果如图 13.25 所示。

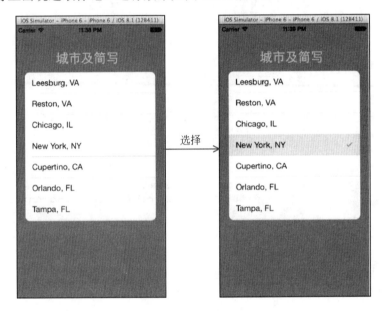

图 13.25 实例 112 运行效果

【实现过程】
(1) 创建一个项目,命名为"圆角表视图"。
(2) 创建一个基于 UIView 类的 Rounded 类。
(3) 打开 Rounded.m 文件,编写代码,绘制一个具有圆角的图形。在此文件中,代码分为了初始化以及绘制两部分。其中,initWithFrame:方法实现初始化的功能。程序代码如下:

```
- (id)initWithFrame:(CGRect)frame {
  if (self = [super initWithFrame:frame]) {
    self.backgroundColor = [UIColor clearColor];          //设置背景颜色
    self.userInteractionEnabled = NO;
  }
  return self;
}
```

圆角图形的绘制需要使用 drawRect:方法实现。程序代码如下:

```
- (void)drawRect:(CGRect)rect {
```

```
    CGContextRef context = UIGraphicsGetCurrentContext();    //创建图形上下文
    CGFloat cornerRadius = 10.0;
    CGFloat minx = CGRectGetMinX(rect);                      //获取最小的 x 值
    CGFloat midx = CGRectGetMidX(rect);
    CGFloat maxx = CGRectGetMaxX(rect);
    CGFloat miny = CGRectGetMinY(rect);                      //获取最小的 y 值
    CGFloat midy = CGRectGetMidY(rect);
    CGFloat maxy = CGRectGetMaxY(rect);
    CGContextMoveToPoint(context, minx, midy);               //设置开始点
    CGContextAddArcToPoint(context, minx, miny, midx, miny, cornerRadius);
    //设置结束点
    CGContextAddArcToPoint(context, maxx, miny, maxx, midy, cornerRadius);
    CGContextAddArcToPoint(context, maxx, maxy, midx, maxy, cornerRadius);
    CGContextAddArcToPoint(context, minx, maxy, minx, midy, cornerRadius);
    CGContextClosePath(context);
    CGContextFillPath(context);                              //填充路径
}
```

（4）创建一个基于 UITableView 类的 RoundedTableView 类。

（5）打开 RoundedTableView.h 文件，编写代码，实现头文件、对象以及方法的声明。程序代码如下：

```
#import <UIKit/UIKit.h>
#import "Rounded.h"
@interface RoundedTableView : UITableView{
    Rounded *mask;                                           //声明对象
}
- (void)adjustMask;
@end
```

（6）打开 RoundedTableView.m 文件，编写代码，实现具有圆角的表视图。使用的方法如表 13-25 所示。

表 13-25　RoundedTableView.m文件中方法总结

方　　法	功　　能
setupView	创建视图
initWithFrame:	初始化具有圆角的表视图
initWithCoder:	初始化实例化对象
adjustMask	调整具有圆角图形的视图
reloadData	加载数据

这里需要讲解几个重要的方法（其他方法请读者参考源代码）。其中，setupView 方法实现创建并设置具有圆角图形的视图。程序代码如下：

```
- (void)setupView {
    mask = [[Rounded alloc] initWithFrame:CGRectZero];
    self.layer.mask = mask.layer;
    self.layer.cornerRadius = 10;                            //设置圆角半径
    self.showsVerticalScrollIndicator = NO;
    self.showsHorizontalScrollIndicator = NO;
}
```

adjustMask 方法实现对具有圆角图形的视图进行调整。程序代码如下：

```
- (void)adjustMask {
    CGRect frame = mask.frame;
```

```
    if(self.contentSize.height > self.frame.size.height) {
     frame.size = self.contentSize;                          //设置尺寸
    } else {
        frame.size = self.frame.size;
    }
    mask.frame = frame;                                      //设置框架
    [mask setNeedsDisplay];
    [self setNeedsDisplay];
}
```

（7）打开 ViewController.h 文件，编写代码，实现头文件、插座变量以及对象的声明。程序代码如下：

```
#import <UIKit/UIKit.h>
#import "RoundedTableView.h"
@interface ViewController : UIViewController{
    IBOutlet RoundedTableView *Rtv;
    NSMutableArray *cities;                                  //声明对象
}
@end
```

（8）打开 Main.storyboard 文件，对 View Controller 视图控制器的设计界面进行设计，效果如图 13.26 所示。

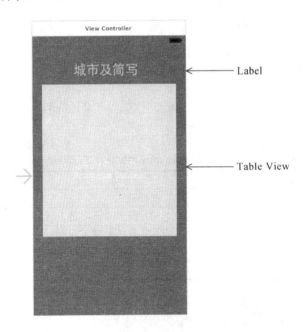

图 13.26　实例 112 View Controller 视图控制器的设计界面

需要添加的视图、控件以及对它们的设置如表 13-26 所示。

表 13-26　实例 112 视图、控件设置

视图、控件	属 性 设 置	其　　他
设计界面	Background：深黑色	与插座变量 table 关联 将 dataSource 与 View Controller 关联 将 delegate 与 View Controller 关联

视图、控件	属性设置	其他
Label	Text：城市及简写 Color：绿色 Font：System 27.0 Alignment：居中	
Table View		Class：RoundedTableView 与插座变量 Rtv 关联 将 dataSource 与 View Controller 关联 将 delegate 与 View Controller 关联

（9）打开 ViewController.m 文件，编写代码，实现表视图中内容的填充以及选择。使用的方法如表 13-27 所示。

表 13-27　ViewController.m文件中方法总结

方　　法	功　　能
viewDidLoad	视图加载后调用，实现初始化
numberOfSectionsInTableView:	获取块数
tableView:numberOfRowsInSection:	获取行数
tableView:cellForRowAtIndexPath:	获取某一行数据
tableView: didSelectRowAtIndexPath:	选择行的响应

其中，viewDidLoad、numberOfSectionsInTableView:、tableView:numberOfRowsInSection: 和 tableView:cellForRowAtIndexPath:方法实现表视图内容的填充。程序代码如下：

```
//视图加载后调用，实现初始化
- (void)viewDidLoad
{
    [super viewDidLoad];
    // Do any additional setup after loading the view, typically from a nib.
    cities = [[NSMutableArray alloc] initWithObjects:@"Leesburg, VA",
@"Reston, VA",@"Chicago, IL", @"New York, NY", @"Cupertino, CA", @"Orlando,
FL",@"Tampa, FL", nil];
    [Rtv reloadData];                              //加载数据
}
//获取块数
- (NSInteger)numberOfSectionsInTableView:(UITableView *)tableView {
    return 1;
}
//获取行数
- (NSInteger)tableView:(UITableView *)tableView numberOfRowsInSection:
(NSInteger)section {
    return cities.count;
}
//获取某一行的数据
- (UITableViewCell *)tableView:(UITableView *)aTableView cellForRowAtIndexPath:
(NSIndexPath *)indexPath {
    static NSString *CellIdentifier = @"Cell";
    UITableViewCell *cell = [Rtv dequeueReusableCellWithIdentifier:
    CellIdentifier];
    //判断 cell 是否为空
    if(cell == nil) {
        cell = [[UITableViewCell alloc] initWithStyle:UITableViewCellStyleDefault
```

```
            reuseIdentifier:CellIdentifier];
    }
    if(indexPath.row < cities.count) {
        cell.textLabel.text = cities[indexPath.row];            //设置文本内容
    }
    return cell;
}
```

如要实现在选择的表视图单元格中添加标记的功能，需要使用 tableView:didSelectRowAtIndexPath:方法。程序代码如下：

```
-(void)tableView:(UITableView *)tableView didSelectRowAtIndexPath:
(NSIndexPath *)indexPath{
    UITableViewCell *cellView = [tableView cellForRowAtIndexPath:
    indexPath];
    if (cellView.accessoryType == UITableViewCellAccessoryNone) {
        cellView.accessoryType=UITableViewCellAccessoryCheckmark;
                                                        //设置附件类型
    }
    else {
        cellView.accessoryType = UITableViewCellAccessoryNone;
                                                        //设置附件类型
        [tableView deselectRowAtIndexPath:indexPath animated:YES];
    }
}
```

【代码解析】

本实例的关键功能是圆角表视图以及选取标记的实现。下面依次讲解这两个知识点。

1．圆角表视图的实现

在本实例中要实现圆角表视图，首先，需要绘制一个圆角矩形，此时需要使用 UIView 的 drawRect:方法，代码如下：

```
- (void)drawRect:(CGRect)rect {
    CGContextRef context = UIGraphicsGetCurrentContext();
    ……
    CGContextAddArcToPoint(context, minx, miny, midx, miny, cornerRadius);
    CGContextAddArcToPoint(context, maxx, miny, maxx, midy, cornerRadius);
    CGContextAddArcToPoint(context, maxx, maxy, midx, maxy, cornerRadius);
    CGContextAddArcToPoint(context, minx, maxy, minx, midy, cornerRadius);
    CGContextClosePath(context);
    CGContextFillPath(context);
}
```

接着再创建一个 RoundedTableView 类。在此类中，使用 mask 属性为此表视图设置一个遮罩层（此遮罩层就是绘制的圆角矩形），从而实现圆角表视图的效果。代码如下：

```
mask = [[Rounded alloc] initWithFrame:CGRectZero];
self.layer.mask = mask.layer;
```

2．选取标记的实现

如果想要实现选取标记的设置，需要使用 UITableViewCell 的 accessoryType 属性。其语法形式如下：

```
@property(nonatomic) UITableViewCellAccessoryType accessoryType;
```

其中，选取标记的类型如表 13-28 所示。

表 13-28　选取标记的类型

类　　型	功　　能
UITableViewCellAccessoryNone	无
UITableViewCellAccessoryDisclosureIndicator	右边有一个小箭头，它距离最右边有十几像素
UITableViewCellAccessoryDetailDisclosureButton	右边有一个蓝色的圆形按钮和一个小箭头
UITableViewCellAccessoryCheckmark	右边有一个对号
UITableViewCellAccessoryDetailButton	右边有一个蓝色的圆形按钮

在此代码中，就使用了 accessoryType 属性将选取标记设置为了对号。代码如下：

cellView.accessoryType=UITableViewCellAccessoryCheckmark;

实例 113　表单元格的自定义折叠

【实例描述】

如果表单元格中的内容很多，为了节省空间，是不可以全部显示的。这时就需要将表单元格进行展开和折叠。展开后，显示表单元格中的所有内容；折叠后，只显示部分内容。本实例的主要功能就是要实现此效果。运行效果如图 13.27 所示。

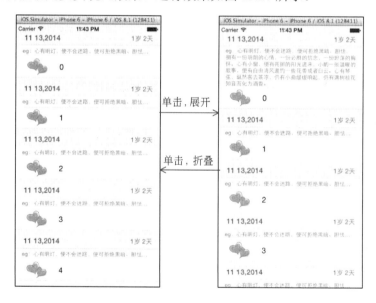

图 13.27　实例 113 运行效果

【实现过程】

（1）创建一个项目，命名为"表单元格的自定义折叠"。

（2）添加图片 1.png 到创建项目的 Supporting Files 文件夹中。

（3）创建一个基于 UITableViewController 类的 CustomTableViewController 类。

（4）打开 CustomTableViewController.h 文件，编写代码，实现宏定义、对象以及实例变量的声明。程序代码如下：

```objc
#import <UIKit/UIKit.h>
//宏定义
#define originalHeight 25.0f
#define isOpen @"90.0f"
@interface CustomTableViewController : UITableViewController{
    NSMutableDictionary *dicClicked;                //对象
//实例变量
    NSInteger count;
    CGFloat mHeight;
    NSInteger sectionIndex;
}
@end
```

(5)打开 CustomTableViewController.m 文件,编写代码,实现表单元格的自定义折叠,在此文件中代码分为了初始化、内容的填充、块标题栏的设置以及折叠的实现。使用的方法如表 13-29 所示。

表 13-29　CustomTableViewController.m文件中方法总结

方　　法	功　　能
viewDidLoad	视图加载后调用,实现初始化
numberOfSectionsInTableView:	获取块数
tableView:numberOfRowsInSection:	获取行数
tableView:cellForRowAtIndexPath:	获取某一行的数据
tableView:heightForHeaderInSection:	获取块标题栏高度
tableView:viewForHeaderInSection:	获取块标题栏的内容
tableView:didSelectRowAtIndexPath:	选择行的响应
tableView:heightForRowAtIndexPath:	获取高度

这里需要讲解几个重要的方法(其他方法请读者参考源代码)。其中,内容的填充需要使用 numberOfSectionsInTableView:、tableView:numberOfRowsInSection:和 tableView:cellForRowAtIndexPath:方法。其语法形式如下:

```objc
//获取块数
- (NSInteger)numberOfSectionsInTableView:(UITableView *)tableView
{
    // Return the number of sections.
    return 50;
}
//获取行数
- (NSInteger)tableView:(UITableView *)tableView numberOfRowsInSection:(NSInteger)section
{
    return 2;
}
//获取某一行的数据
- (UITableViewCell *)tableView:(UITableView *)tableView cellForRowAtIndexPath:(NSIndexPath *)indexPath
{
    static NSString *contentIndentifer = @"Container";
    if (indexPath.row == 0) {
        UITableViewCell *cell = [tableView dequeueReusableCellWithIdentifier:contentIndentifer];
        //判断cell是否为空
        if (cell == nil) {
            cell = [[UITableViewCell alloc] initWithStyle:UITableView
```

```
            CellStyleDefault reuseIdentifier:contentIndentifer];
      }
      //内容的设置
      NSString *statisticsContent = @"eg：心有明灯，便不会迷路，便可拒绝黑暗、
胆怯，拥有一份明朗的心情，一份必胜的信念，一份坦荡的胸怀……心有小窗，便有亮丽的阳光进
来，小酌一些温暖的故事，便有自由清风邀约一些花香或者白云。心有琴弦，纵然客去茶凉，仍有
小曲缓缓响起，仍有满树桂花知音而化为酒香。";
      cell.textLabel.font = [UIFont systemFontOfSize:12.0f];    //设置字体
      cell.textLabel.text = statisticsContent;                  //设置文本内容
      cell.textLabel.textColor = [UIColor brownColor];
      cell.textLabel.opaque = NO;
                        //选中Opaque表示视图后面的任何内容都不应该绘制
      cell.textLabel.numberOfLines = 8;                         //设置行数
      cell.selectionStyle = UITableViewCellSelectionStyleNone;
      return cell;
}
   static NSString *CellIdentifier = @"Cell";
   UITableViewCell *cell = [tableView dequeueReusableCellWithIdentifier:
   CellIdentifier];
   //判断cell是否为空
   if (cell == nil) {
      cell = [[UITableViewCell alloc] initWithStyle:UITableViewCellStyleDefault
      reuseIdentifier:CellIdentifier];
   }
   cell.imageView.image = [UIImage imageNamed:@"1.png"];        //设置图像
   cell.textLabel.text = [NSString stringWithFormat:@"%d",count];
   count++;
   return cell;
}
```

tableView:heightForHeaderInSection:和 tableView:viewForHeaderInSection:方法实现了对块标题栏的设置，其中，tableView:heightForHeaderInSection:方法对块标题栏的高度进行获取。程序代码如下：

```
-(CGFloat)tableView:(UITableView *)tableView heightForHeaderInSection:
(NSInteger)section
{
   if (section == 0)
      return 46;
   else
      return 30.0f;
}
```

tableView:viewForHeaderInSection:方法对块标题栏的内容进行获取。程序代码如下：

```
-(UIView *)tableView:(UITableView *)tableView viewForHeaderInSection:
(NSInteger)section
{
   CGRect headerFrame = CGRectMake(0, 0, 300, 30);
   CGFloat y = 2;
   //判断section是否为0
   if (section == 0) {
      headerFrame = CGRectMake(0, 0, 300, 100);
      y = 18;
   }
   UIView *headerView = [[UIView alloc] initWithFrame:headerFrame];
   //创建并设置标签对象，用来存放日期
   UILabel *dateLabel=[[UILabel alloc] initWithFrame:CGRectMake(20, y, 240,
```

```
24)];
dateLabel.font=[UIFont boldSystemFontOfSize:16.0f];     //设置字体
dateLabel.textColor = [UIColor darkGrayColor];
dateLabel.backgroundColor=[UIColor clearColor];         //设置背景颜色
//创建并设置标签对象,用来存放年龄
UILabel *ageLabel=[[UILabel alloc] initWithFrame:CGRectMake(216, y, 88,
24)];
ageLabel.font=[UIFont systemFontOfSize:14.0];
ageLabel.textAlignment=NSTextAlignmentRight;            //设置对齐方式
ageLabel.textColor = [UIColor darkGrayColor];
ageLabel.backgroundColor=[UIColor clearColor];
NSDateFormatter *dateFormatter = [[NSDateFormatter alloc] init];
dateFormatter.dateFormat = @"MM dd,yyyy";               //设置日期格式
dateLabel.text = [NSString stringWithFormat:@"%@",[dateFormatter
stringFromDate:[NSDate date]]];
ageLabel.text = @"1岁2天";
[headerView addSubview:dateLabel];
[headerView addSubview:ageLabel];                       //添加对象
return headerView;
}
```

选择的表单元格的打开或者折叠,需要使用 tableView:didSelectRowAtIndexPath:和 tableView:heightForRowAtIndexPath:方法实现。其中,tableView:didSelectRowAtIndexPath: 方法实现选择的表单元格的打开或者折叠。程序代码如下:

```
- (void)tableView:(UITableView *)tableView didSelectRowAtIndexPath:
(NSIndexPath *)indexPath
{
    //判断 indexPath.row 是否为 0
    if (indexPath.row == 0) {
        UITableViewCell *targetCell = [tableView cellForRowAtIndexPath:
        indexPath];
        if (targetCell.frame.size.height == originalHeight){
            [dicClicked setObject:isOpen forKey:indexPath];   //设置对象
        }
        else{
            [dicClicked removeObjectForKey:indexPath];        //移除对象
        }
        [self.tableView reloadRowsAtIndexPaths:[NSArray arrayWithObject:
indexPath] withRowAnimation:UITableViewRowAnimationFade];
    }
}
```

tableView:heightForRowAtIndexPath:方法实现高度的获取。程序代码如下:

```
-(CGFloat)tableView:(UITableView *)tableView heightForRowAtIndexPath:
(NSIndexPath *)indexPath
{
    //判断 indexPath.row 是否为 0
    if (indexPath.row == 0) {
        if ([[dicClicked objectForKey:indexPath] isEqualToString: isOpen])
            return [[dicClicked objectForKey:indexPath] floatValue];
        else
            return originalHeight;                     //返回 originalHeight
    }
    else {
        return 45.0f;
    }
}
```

【代码解析】

本实例关键功能是表单元格的自定义折叠。下面就是这个知识点的详细讲解。

在本实例中表单元格的自定义折叠的实现,首先要对表单元格进行高度的设置。其次,单击某一行后,表单元格的展开是通过 tableView:didSelectRowAtIndexPath:方法实现的。在此方法中首先判断选择的表单元格是否为第 0 行的,然后再判断此时表单元格的高度是否与之前的高度一致,如果一致就展开表单元格,如果不一致就折叠。代码如下:

```
- (void)tableView:(UITableView *)tableView didSelectRowAtIndexPath:
(NSIndexPath *)indexPath
{
    if (indexPath.row == 0) {
        UITableViewCell *targetCell = [tableView cellForRowAtIndexPath:
        indexPath];
        if (targetCell.frame.size.height == originalHeight){
            [dicClicked setObject:isOpen forKey:indexPath];
        }
        else{
            [dicClicked removeObjectForKey:indexPath];
        }
        [self.tableView reloadRowsAtIndexPaths:[NSArray arrayWithObject:
indexPath] withRowAnimation:UITableViewRowAnimationFade];
    }
}
```

实例 114　具有搜索功能的表视图

【实例描述】

在列表中保存了一组数据,当此数据很多时,要找到一个内容是相当困难的,那么如何在列表中实现即快速又高效的查找呢?本实例就位各位开发者解决了这一难题,它实现了一个具有索引功能的表视图,用户只需要将想要搜索的内容输入到搜索框中就可以很快地找到相关的内容。此应用一般使用在电话簿等地方。运行效果如图 13.28 所示。

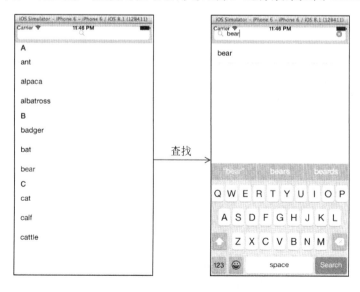

图 13.28　实例 114 运行效果

【实现过程】
(1) 创建一个项目,命名为"具有搜索功能的表视图"。
(2) 创建一个名为 message.plist 的文件。对文件中的内容进行设置,如表 13-30 所示。

表 13-30 message.plist 文件中的内容

Key	Type	Value
Root	Dictionary	
A	Array	
Item0	String	ant
Item1	String	alpaca
Item2	String	albatross
B	Array	
Item0	String	badger
Item1	String	bat
Item2	String	bear
C	Array	
Item0	String	cat
Item1	String	calf
Item2	String	cattle

(3) 打开 ViewController.h 文件,编写代码,实现对象、实例变量、插座变量以及方法的声明。程序代码如下:

```
#import <UIKit/UIKit.h>
@interface ViewController : UIViewController{
   //对象
   NSDictionary *list;
   NSArray *ff;
   NSMutableArray *listOfMovies;
NSMutableArray *searchResult;
//插座变量
   IBOutlet UITableView *tv;
IBOutlet UISearchBar *searchBar;
//实例变量
   BOOL isSearchOn;
   BOOL canSelectRow;
}
- (void) searchMoviesTableView;            //方法
@end
```

(4) 打开 Main.storyboard 文件,对 ViewController 视图控制器的设计界面进行设置,效果如图 13.29 所示。

将 Search bar 搜索控件和插座变量 searchBar 关联;将 Table View 表视图与插座变量 tv 关联。

(5) 打开 ViewController.m 文件,编写代码,在表视图中实现搜索功能。在此文件中代码分为内容填充、块标题栏的设置以及搜索。使用的方法如表 13-31 所示。

第 13 章 表

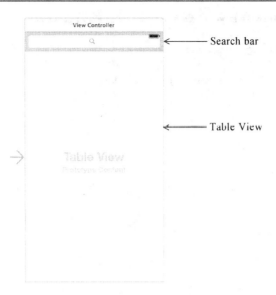

图 13.29 实例 114 View Controller 视图控制器的设计界面

表 13-31 ViewController.m 文件中方法总结

方　　法	功　　能
viewDidLoad	视图加载后调用，实现初始化
searchBar:textDidChange:	当用户在搜索栏中进行输入时调用，实现搜索
searchMoviesTableView	搜索结果
numberOfSectionsInTableView:	获取块数
tableView:numberOfRowsInSection:	获取行数
tableView:cellForRowAtIndexPath:	获取某一行的数据
tableView:titleForHeaderInSection:	获取块标题栏的标题

其中，内容填充需要使用 viewDidLoad、numberOfSectionsInTableView:、tableView:numberOfRowsInSection:和 tableView:cellForRowAtIndexPath:方法实现。viewDidLoad 方法实现初始化功能。程序代码如下：

```
- (void)viewDidLoad
{
    [super viewDidLoad];
    // Do any additional setup after loading the view, typically from a nib.
    NSString *path=[[NSBundle mainBundle]pathForResource:@"1" ofType:@"plist"];
    NSDictionary *dic = [[NSDictionary alloc] initWithContentsOfFile:path];
    list = dic;
    NSArray *array = [[list allKeys] sortedArrayUsingSelector:@selector(compare:)];          //创建数组对象
    ff = array;
    tv.tableHeaderView = searchBar;
    searchBar.autocorrectionType = UITextAutocorrectionTypeYes;
    listOfMovies = [[NSMutableArray alloc] init];
    //创建数组对象
    //遍历
    for (NSString *year in array)
    {
```

```
        NSArray *movies = [list objectForKey:year];
        //遍历
        for (NSString *title in movies)
        {
            [listOfMovies addObject:title];                          //添加对象
        }
    }
    searchResult = [[NSMutableArray alloc] init];
    isSearchOn = NO;
    canSelectRow = YES;
}
```

numberOfSectionsInTableView:方法获取表视图中的块数。程序代码如下:

```
-(NSInteger)numberOfSectionsInTableView:(UITableView *)tableView{
    //判断 isSearchOn
    if (isSearchOn)
        return 1;
    else
        return [ff count];
}
```

tableView:numberOfRowsInSection:方法获取表视图中的行数。程序代码如下:

```
-(NSInteger)tableView:(UITableView  *)tableView  numberOfRowsInSection:
(NSInteger) section{
//判断 isSearchOn
    if (isSearchOn) {
        return [searchResult count];
    } else{
        NSString *year = [ff objectAtIndex:section];
        NSArray *movieSection = [list objectForKey:year];
        return [movieSection count];                                 //返回
    }
}
```

tableView:cellForRowAtIndexPath:方法获取表视图中单元格的内容。程序代码如下:

```
-       (UITableViewCell         *)tableView:(UITableView         *)tableView
cellForRowAtIndexPath: (NSIndexPath *)indexPath {
    static NSString *CellIdentifier = @"Cell";
UITableViewCell *cell = [tableView dequeueReusableCellWithIdentifier:
CellIdentifier];
//判断 cell 是否为空
    if (cell == nil) {
        cell = [[UITableViewCell alloc] initWithStyle:UITableViewCellStyleDefault
        reuseIdentifier:CellIdentifier];
    }
    if (isSearchOn) {
        NSString *cellValue = [searchResult objectAtIndex:indexPath.row];
        cell.textLabel.text = cellValue;                             //设置文本内容
    } else {
        NSString *year = [ff objectAtIndex:[indexPath section]];
        NSArray *movieSection = [list objectForKey:year];
        cell.textLabel.text = [movieSection objectAtIndex:[indexPath row]];
    //设置文本内容
    }
    return cell;
}
```

tableView:titleForHeaderInSection:方法实现获取块标题栏的标题。程序代码如下:

```
- (NSString *)tableView:(UITableView *)tableView titleForHeaderInSection:
(NSInteger)section {
    NSString *year = [ff objectAtIndex:section];
    //判断 isSearchOn
    if (isSearchOn)
        return nil;
    else
        return year;
}
```

搜索的实现需要使用 searchBar:textDidChange:和 searchMoviesTableView 方法。其中，searchBar:textDidChange:方法当用户在搜索栏中进行输入时调用，实现搜索。程序代码如下：

```
- (void)searchBar:(UISearchBar *)searchBar textDidChange:(NSString *)
searchText {
    //判断
    if ([searchText length] > 0) {
        //当文本长度大于 0 时，实现搜索
        isSearchOn = YES;
        canSelectRow = YES;
        tv.scrollEnabled = YES;           //可以滚动
        [self searchMoviesTableView];
    }
    else {
        isSearchOn = NO;
        canSelectRow = NO;
        tv.scrollEnabled = NO;            //不可以滚动
    }
    [tv reloadData];                       //重新加载数据
}
```

searchMoviesTableView 方法实现对结果的搜索。程序代码如下：

```
- (void)searchMoviesTableView {
    [searchResult removeAllObjects];      //移除对象
    //遍历
    for (NSString *str in listOfMovies)
    {
        NSRange titleResultsRange = [str rangeOfString:searchBar.text
options:NSCaseInsensitiveSearch];
        if (titleResultsRange.length > 0)
            [searchResult addObject:str];//添加对象
    }
}
```

【代码解析】

本实例关键功能是在表视图中实现搜索。下面就是这个知识点的详细讲解。

要在搜索框中输入内容时实现搜索功能，首先要告知委托，当搜索框中的内容发生了改变，使用 UISearchBarDelegate 的 searchBar:textDidChange:方法，其语法形式如下：

```
- (void)searchBar:(UISearchBar *)searchBar textDidChange:(NSString
*)searchText;
```

其中，(UISearchBar *)searchBar 表示搜索栏对象；(NSString *)searchText 表示在搜索

框中的当前文本。在此代码中，就使用了 searchBar:textDidChange:方法实现了当在搜索框中输入内容时进行搜索，代码如下：

```
- (void)searchBar:(UISearchBar *)searchBar textDidChange:(NSString *)searchText {
    if ([searchText length] > 0) {
        ……
        [self searchMoviesTableView];
    }
    else {
        ……
    }
    [tv reloadData];
}
```

最后，结果的搜索是在 searchMoviesTableView 方法中实现的。

实例 115 自定义表单元格的动画效果

【实例描述】
本实例实现的功能是自定义表单元格的动画效果。它会以动画的形式显示表单元格，从而将表视图以动画的形式显示出来。这样将会增加表视图的活泼感。运行效果如图 13.30 所示。

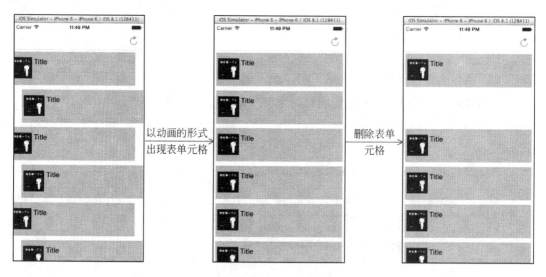

图 13.30 实例 115 运行效果

【实现过程】
（1）创建一个项目，命名为"自定义表单元格的动画效果"。
（2）添加图片 1.jpg 到创建项目的 Supporting Files 文件夹中。
（3）创建一个基于 UIView 类的分类 Position。
（4）打开 UIView+Position.h 文件，编写代码，实现方法的声明。程序代码如下：

```
#import <UIKit/UIKit.h>
```

```
@interface UIView (Position)
- (void) setOrigin:(float)x :(float)y;                    //设置位置
- (void) resetOriginToTopLeft;
- (void) resetOriginToTopRight;
@end
```

（5）打开 UIView+Position.m 文件，编写代码，实现对以上方法的定义（请读者参考源代码）。

（6）创建一个基于 UITableViewCell 类的 AnimatedCell 类。

（7）打开 AnimatedCell.h 文件，编写代码，实现头文件、宏定义、属性以及方法的声明。程序代码如下：

```
#import <UIKit/UIKit.h>
#import "UIView+Position.h"
#define animationDelay 0.0
……
#define sizeTitleWidth 200
#define sizeTitleheight 20
@interface AnimatedCell : UITableViewCell
@property (nonatomic) bool popped;
@property (nonatomic,strong) UIView *atcContentView;
- (void) resetPosition;
- (void) configureCellContentSizeWidth:(float) width height:(float)height;
……
- (void) popCellWithAnimation:(bool)animated;
- (void) popCellWithAnimation:(bool)animated duration:(float)duration;
@end
```

（8）打开 AnimatedCell.m 文件，编写代码，实现表单元格的绘制以及动画效果，使用的方法如表 13-32 所示。

表 13-32　AnimatedCell.m 文件中方法总结

方　　法	功　　能
initWithStyle:reuseIdentifier:	表单元格的初始化
setSelected:animated:	设置选择和动画
resetPosition	重置位置
configureCellContentSizeWidth:height:	设置表单元格的大小和宽度
pushCellWithAnimation:	使用动画效果显示表单元格
pushCellWithAnimation:direction:	使用动画以及方向显示表单元格
pushCellWithAnimation:duration:	使用动画以及时间显示表单元格
pushCellWithAnimation:duration:direction:	设置动画效果的方向，以及实现动画效果
popCellWithAnimation:	使用动画隐藏表单元格
popCellWithAnimation:duration:	使用动画以及时间隐藏表单元格

这里需要讲解几个重要的方法（其他方法请读者参考源代码）。其中，对表单元格的绘制，需要使用 initWithStyle:reuseIdentifier:、setSelected:animated:、resetPosition 和 configureCellContentSizeWidth:height: 方法。initWithStyle:reuseIdentifier: 方法实现对表单元格的初始化。程序代码如下：

```
- (id)initWithStyle:(UITableViewCellStyle)style reuseIdentifier:(NSString *)reuseIdentifier{
    self = [super initWithStyle:style reuseIdentifier:reuseIdentifier];
```

```objc
    if (self) {
        self.backgroundColor = [UIColor clearColor];       //设置背景颜色
        self.selectionStyle = UITableViewCellSelectionStyleNone;
        atcContentView = [[UIView alloc] initWithFrame:CGRectZero];
         //实例化对象
        [atcContentView resetOriginToTopLeft];
        [self addSubview:atcContentView];                 //添加视图对象
        //创建并设置空白视图对象,作为表单元格的背景
        UIView *bgView = [[UIView alloc] initWithFrame:CGRectMake
          (defaultPadding, defaultPadding,sizeCellWidth, sizeCellHeight)];
        bgView.backgroundColor = [UIColor lightGrayColor];
        //创建并设置图像视图对象,作为表单元格的图标
        UIImageView *picture = [[UIImageView alloc] initWithFrame:CGRectMake
  (defaultPadding, defaultPadding+9, sizePicture, sizePicture)];
                                                          //实例化对象
        picture.image = [UIImage imageNamed:@"1.jpg"];    //设置图像
        //创建并设置标签对象,作为表单元格的内容
        UILabel *title = [[UILabel alloc] initWithFrame:CGRectMake(picture.
frame.origin.x    +   picture.frame.size.width    +    defaultPadding,
defaultPadding+9, sizeTitleWidth, sizeTitleheight)];
        title.text = @"Title";                            //设置文本内容
        [self.atcContentView addSubview:bgView];
        [bgView addSubview:title];
        [bgView addSubview:picture];
    }
    return self;
}
```

表单元格的动画效果有两种:显示以及隐藏。其中,显示表单元的动画效果需要使用 pushCellWithAnimation:、pushCellWithAnimation:direction:、pushCellWithAnimation:duration: 和 pushCellWithAnimation:duration:direction:方法。其中,pushCellWithAnimation:duration 方法使用动画效果以及时间来显示表单元格。程序代码如下:

```objc
-(void)pushCellWithAnimation:(bool)animated duration:(float)duration{
    if (!animated)
        duration = 0.0;
    //动画效果
    [UIView animateWithDuration:duration
                     delay:animationDelay
                     options: UIViewAnimationOptionCurveEaseInOut
                     animations:^{
                         atcContentView.center = CGPointMake(0, 0);
                     }completion:nil];
}
```

pushCellWithAnimation:duration:direction:方法通过设置动画方向、动画效果以及时间, 来实现表单元格的显示。程序代码如下:

```objc
-  (void)pushCellWithAnimation:(bool)animated  duration:(float)duration
direction:(NSString *)direction{
    //判断方向,从而进行中心点的设置
    //判断direction是否为"up"
    if ([direction isEqualToString:@"up"]) {
        [atcContentView setCenter:CGPointMake(0, -atcContentView.frame.
          size.height)];
    } else if ([direction isEqualToString:@"down"]) {
```

```
                                              //判断 direction 是否为 "down"
    [atcContentView setCenter:CGPointMake(0, atcContentView.frame.size.
      height)];
} else if ([direction isEqualToString:@"left"]) {
                                              //判断 direction 是否为 "left"
    [atcContentView setCenter:CGPointMake(-atcContentView.frame.size.
      width, 0)];
} else if ([direction isEqualToString:@"right"]) {
                                              //判断 direction 是否为 "right"
    [atcContentView setCenter:CGPointMake(atcContentView.frame.size.
      width, 0)];
} else {
    [atcContentView setCenter:CGPointMake(0, -atcContentView.frame.
      size.height)];                          //设置中心位置
}
[self pushCellWithAnimation:animated duration:duration];
}
```

表单元格的隐藏，需要使用 popCellWithAnimation:和 popCellWithAnimation:duration:方法。其中，popCellWithAnimation:duration:方法通过使用动画以及时间实现表单元格的隐藏。程序代码如下：

```
- (void) popCellWithAnimation:(bool)animated duration:(float)duration{
    popped = YES;
    if (!animated) duration = 0.0;
    //动画效果
    [UIView animateWithDuration:duration
                    delay:animationDelay
                    options: UIViewAnimationOptionCurveEaseInOut
                    animations:^{
                            atcContentView.center = CGPointMake
(atcContentView.frame.size.width, 0);
                    }completion:nil];
}
```

（9）打开 ViewController.h 文件，编写代码，实现头文件、遵守协议、实例变量、插座变量以及动作的声明。程序代码如下：

```
#import <UIKit/UIKit.h>
#import "AnimatedCell.h"
#define tableViewCellHeight 90
@interface   ViewController   :   UIViewController<UITableViewDelegate,
UITableViewDataSource>{
    bool tableAnimated;
    IBOutlet UITableView *table;               //声明关于表视图的插座变量
}
- (IBAction)replayAnimation:(id)sender;
@end
```

（10）打开 Main.storyboard 文件，拖动 Navigation Controller 导航控制器到画布中，将此控制器关联的根视图设置为 View Controller 视图控制器。对 View Controller 视图控制器的设计界面进行设计，效果如图 13.31 所示。

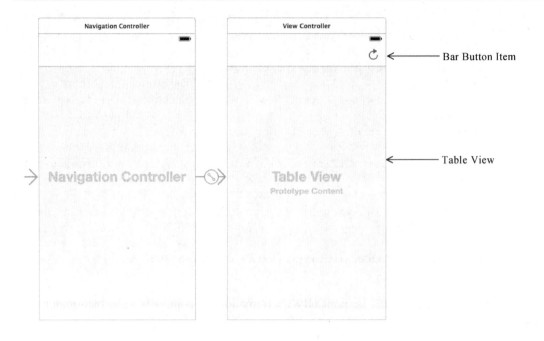

图 13.31　实例 115 View Controller 视图控制器设计界面

需要添加的视图、控件以及对它们的设置如表 13-33 所示。

表 13-33　实例 115 视图、控件设置

视图、控件	属 性 设 置	其 他
Bar Button Item	Identifier：Refresh	与动作 replayAnimation:关联
Table View		与插座变量 table 关联 将 dataSource 与 View Controller 关联 将 delegate 与 View Controller 关联

（11）打开 ViewController.m 文件，编写代码，实现界面、自定义表单元格的设置显示以及行的选择。使用的方法如表 13-34 所示。

表 13-34　ViewController.m 文件中方法总结

方　　法	功　　能
viewDidLoad	视图加载后调用，实现初始化
tableView:numberOfRowsInSection:	获取行数
tableView:cellForRowAtIndexPath:	获取某一行数据
tableView:didSelectRowAtIndexPath:	选择行的响应
startTableViewAnimation:	实现动画效果
replayAnimation	单击导航栏上的按钮后实现重复动画
tableView: heightForRowAtIndexPath:	获取高度

界面、自定义表单元格的设置显示需要 viewDidLoad、tableView:numberOfRowsInSection:、tableView:cellForRowAtIndexPath:、startTableViewAnimation:和 tableView: heightForRowAtIndex-Path:方法。其中，tableView:cellForRowAtIndexPath:方法实现获取某一行，程序代码如下：

```
- (UITableViewCell *)tableView:(UITableView *)tableView cellForRowAtIndexPath:
```

```
(NSIndexPath *)indexPath{
    static NSString *CellIdentifier = @"Cell";
    AnimatedCell *cell = [tableView dequeueReusableCellWithIdentifier:
    CellIdentifier];
    //判断 cell 是否为空
    if (cell == nil) {
        cell = [[AnimatedCell alloc] initWithStyle:UITableViewCellStyleDefault
reuseIdentifier:CellIdentifier];                            //实例化对象
        [cell configureCellContentSizeWidth:tableView.frame.size.width height:
        tableViewCellHeight];
    }
    [cell resetPosition];                                   //重设位置
    if([indexPath row] == ((NSIndexPath*)[[tableView
indexPathsForVisibleRows] lastObject]).row){
        if (!tableAnimated)
            [self startTableViewAnimation:tableView];       //动画
    }
    return cell;
}
```

startTableViewAnimation:方法实现表视图的动画效果。程序代码如下：

```
- (void)startTableViewAnimation:(UITableView *)tableView
{
tableAnimated = YES;
//遍历
for (AnimatedCell *atCell in table.visibleCells) {
    //判断
    if ([table.visibleCells indexOfObject:atCell] % 2 == 0)
        [atCell pushCellWithAnimation:YES direction:@"left"];
    else
        [atCell pushCellWithAnimation:YES direction:@"right"];
}
}
```

选择某一行(某一表单元格)实现的响应，需要使用 tableView:didSelectRowAtIndexPath:方法。程序代码如下：

```
- (void)tableView:(UITableView *)tableView didSelectRowAtIndexPath:
(NSIndexPath *)indexPath
{
    AnimatedCell *selectedCell = (AnimatedCell *)[tableView
    cellForRowAtIndexPath:indexPath];       //实例化对象
    [selectedCell popCellWithAnimation:YES];
}
```

【代码解析】

本实例的关键功能是表单元格的显示和隐藏动画效果的实现。下面依次讲解这两个知识点。

1. 表单元格的显示动画

在本实例中表单元格的显示动画是通过使用 animateWithDuration:delay:options:animations:completion:方法实现的。代码如下：

```
[UIView animateWithDuration:duration
                delay:animationDelay
                options: UIViewAnimationOptionCurveEaseInOut
```

```
                animations:^{
                    atcContentView.center = CGPointMake(0, 0);
                }completion:nil];
```

对于实例中实现的判断动画运动的方向性，需要使用 pushCellWithAnimation:duration: direction:方法。程序代码如下：

```
- (void)pushCellWithAnimation:(bool)animated duration:(float)duration
direction:(NSString *)direction{
    //判断方向，从而进行中心点的设置
    if ([direction isEqualToString:@"up"]) {
        [atcContentView setCenter:CGPointMake(0, -atcContentView.
        frame.size.height)];
    } ……
    else {
        [atcContentView setCenter:CGPointMake(0, -atcContentView.frame.
        size.height)];
    }
    [self pushCellWithAnimation:animated duration:duration];
}
```

2. 表单元格的隐藏动画

在本实例中表单元格的隐藏动画是通过使用 animateWithDuration:delay:options:animations: completion:方法实现的。

实例 116　两个列表的显示

【实例描述】

有的人经常在问：由于手机的屏幕有限，是否可以在一个界面中显示多个不相干的列表？答案是肯定的。本实例就实现了在一个界面中同时显示两个不相干的列表。运行效果如图 13.32 所示。

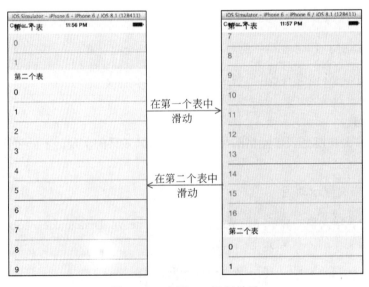

图 13.32　实例 116 运行效果

【实现过程】

（1）创建一个项目，命名为"显示两个列表"。

（2）创建一个基于 UIViewController 类的 SubViewController 类。

（3）打开 SubViewController.h 文件，编写代码，实现宏、数据结构的定义以及属性的声明。程序代码如下：

```objc
#import <UIKit/UIKit.h>
//宏定义
#define heightDefaultCell    44.0
#define heightSectionHeader22.0
#define minimumHeightSubTable  (heightSectionHeader+(heightDefaultCell*2.0))
#define shouldCoverWholeView  0
//数据结构的定义
typedef enum {
    doubleTableStateMaxFront,
    doubleTableStateMinFront,
} DOUBLE_TABLE_STATE;
@interface SubViewController : UIViewController<UITableViewDataSource,
UITableViewDelegate, UIScrollViewDelegate>
//属性
@property (assign, nonatomic) BOOL isBehindSubTable;
@property (assign, nonatomic) CGFloat lastContentOffsetY;
@property (strong, nonatomic) IBOutlet UIView *backgroundView;
@property (strong, nonatomic) IBOutlet UITableView *tableView;
@end
```

（4）打开 Main.storyboard 文件，从视图库中拖动 View Controller 视图控制器到画布中。将 Class 设置为 SubViewController。这时新增的 View Controller 视图控制器就变为了 Sub View Controller 视图控制器。将 View 与插座变量 backgroundView 关联。添加 Table View 表视图控制器到 Sub View Controller 视图控制器的设计界面中，将此表视图与插座变量 backgroundView 关联。

（5）打开 SubViewController.m 文件，编写代码，实现头文件的声明、表视图的创建和设置以及滚动。使用的方法如表 13-35 所示。

表 13-35　SubViewController.m文件中方法总结

方　　法	功　　能
viewDidLoad	视图加载后调用，实现初始化
numberOfSectionsInTableView:	获取块数
tableView:numberOfRowsInSection:	获取行数
tableView:cellForRowAtIndexPath:	获取某一行数据
tableView:titleForHeaderInSection:	获取每一个块的标题
scrollViewWillBeginDragging:	在滚动即将开始时调用

这里需要讲解几个重要的方法（其他方法请读者参考源代码）。其中表视图的创建和设置需要使用 viewDidLoad、numberOfSectionsInTableView:、tableView:numberOfRowsInSection:、tableView:cellForRowAtIndexPath: 和 tableView:titleForHeaderInSection: 方法。其中，viewDidLoad 方法实现对表视图的初始化设置。程序代码如下：

```objc
- (void)viewDidLoad
{
    [super viewDidLoad];
```

```objc
    // Do any additional setup after loading the view.
    //判断表视图是否为空
    if (self.tableView == nil) {
        self.tableView = [[UITableView alloc] initWithFrame:self.view.
        frame style:UITableViewStylePlain];
        self.tableView.autoresizingMask = UIViewAutoresizingFlexibleWidth|
        UIViewAutoresizingFlexibleHeight;          //设置自适应
        self.tableView.separatorColor=[UIColor blackColor];
                                                  //设置分割线的颜色
        [self.view addSubview:self.tableView];
        [self.view bringSubviewToFront:self.tableView];
    }
    //判断表视图是否遵守了 dataSource 和 delegate 委托
    if (self.tableView.dataSource == nil || self.tableView.delegate == nil) {
        [self.tableView setDataSource:self];      //设置数据源
        [self.tableView setDelegate:self];        //设置委托
    }
    self.tableView.backgroundColor = [UIColor clearColor];
    self.tableView.scrollsToTop = NO;
}
```

tableView:cellForRowAtIndexPath:方法实现获取表视图中某一行的数据。程序代码如下:

```objc
- (UITableViewCell *)tableView:(UITableView *)tableView cellForRowAtIndexPath:
(NSIndexPath *)indexPath {
    static NSString *CellIdentifier = @"Cell";
    UITableViewCell *cell = [tableView dequeueReusableCellWithIdentifier:
    CellIdentifier];
    //判断 cell 是否为空
    if (cell == nil) {
        cell = [[UITableViewCell alloc] initWithStyle:
        UITableViewCellStyleDefault reuseIdentifier:CellIdentifier];
        if (self.isBehindSubTable) {
            cell.textLabel.textColor = [UIColor redColor]; //设置文本颜色
            cell.backgroundColor=[UIColor colorWithHue:0.3 saturation:0.3
            brightness:1.0 alpha:1.0];
        }
        else {
            cell.textLabel.textColor = [UIColor blackColor];
            //设置背景颜色
            cell.backgroundColor=[UIColor colorWithHue:0.5 saturation:0.3
            brightness:1.0 alpha:1.0];
        }
    }
    cell.textLabel.text = [NSString stringWithFormat:@"%d", indexPath.row];
                                                  //设置文本内容
    return cell;
}
```

tableView:titleForHeaderInSection:方法获取每一个块的标题。程序代码如下:

```objc
- (NSString *)tableView:(UITableView *)tableView titleForHeaderInSection:
(NSInteger)section {
    NSString *headerTitle = nil;
    //判断
    if (self.isBehindSubTable) {
        headerTitle = @"第一个表";
    } else {
        headerTitle = @"第二个表";
```

```
    }
    return headerTitle;
}
```

scrollViewWillBeginDragging: 方法在滚动即将开始时调用，实现对双表状态的重新设置。程序代码如下：

```
- (void)scrollViewWillBeginDragging:(UIScrollView *)scrollView {
    ViewController *mainController = (ViewController*)self.parentViewController;
    DOUBLE_TABLE_STATE nextDoubleTableState = mainController.lastDoubleTableState;
    BOOL shouldConfigure = NO;
    //判断 self.isBehindSubTable
    if (self.isBehindSubTable) {
        //判断 nextDoubleTableState 是否不等于 doubleTableStateMinFront
        if (nextDoubleTableState != doubleTableStateMinFront) {
            nextDoubleTableState = doubleTableStateMinFront;
            shouldConfigure = YES;
        }
    } else {
        //判断 nextDoubleTableState 是否不等于 doubleTableStateMaxFront
        if (nextDoubleTableState != doubleTableStateMaxFront) {
            nextDoubleTableState = doubleTableStateMaxFront;
            shouldConfigure = YES;
        }
    }
    //判断 shouldConfigure
    if (shouldConfigure) {
        [mainController configureSubTablesForDoubleTableState:
        nextDoubleTableState];
    }
}
```

（6）打开 ViewController.h 文件，编写代码，实现头文件、属性以及方法的声明。程序代码如下：

```
#import <UIKit/UIKit.h>
#import "SubViewController.h"                           //头文件
@interface ViewController : UIViewController
//属性
@property DOUBLE_TABLE_STATE lastDoubleTableState;
@property (strong, nonatomic) SubViewController *behindSubTable;
@property (strong, nonatomic) SubViewController *frontSubTable;
//方法
- (void)configureSubTablesForDoubleTableState:(DOUBLE_TABLE_STATE)
doubleTableState;
- (CGFloat)frontSubTableOriginYforDoubleTableState:(DOUBLE_TABLE_STATE)
doubleTableState;
@end
```

（7）打开 ViewController.m 文件，编写代码，实现两个列表的同时显示。使用的方法如表 13-36 所示。

表 13-36 ViewController.m文件中方法总结

方　法	功　能
viewDidLoad	视图加载后调用，实现初始化
configureSubTablesForDoubleTableState:	设置子表
frontSubTableOriginYforDoubleTableState:	获取子表的 y 值

其中，viewDidLoad 方法实现对列表以及界面的初始化设置。程序代码如下：

```objc
- (void)viewDidLoad
{
    [super viewDidLoad];
    // Do any additional setup after loading the view, typically from a nib.
    //判断 self.behindSubTable 是否为空
    if (self.behindSubTable == nil) {
        self.behindSubTable = [[SubViewController alloc] initWithNibName:
        nil bundle:nil];
        [self addChildViewController:self.behindSubTable];
    }
    //判断 self.frontSubTable 是否为空
    if (self.frontSubTable == nil) {
        self.frontSubTable = [[SubViewController alloc] initWithNibName:
        nil bundle:nil];
        [self addChildViewController:self.frontSubTable];
    }
    self.behindSubTable.isBehindSubTable = YES;
    if ([self.behindSubTable.view.superview isEqual:self.view] == NO) {
        [self.view addSubview:self.behindSubTable.view];    //添加视图对象
    }
    if ([self.frontSubTable.view.superview isEqual:self.view] == NO) {
        [self.view addSubview:self.frontSubTable.view];    //添加视图对象
    }
    [self configureSubTablesForDoubleTableState:doubleTableStateMaxFront];
}
```

configureSubTablesForDoubleTableState:方法实现的功能是对子表进行设置。程序代码如下：

```objc
- (void)configureSubTablesForDoubleTableState:(DOUBLE_TABLE_STATE)
doubleTableState {
    CGRect animatedFrontSubTableFrame = self.frontSubTable.view.frame;
                                                                //获取框架
    animatedFrontSubTableFrame.origin.y = [self frontSubTableOriginYfor
    DoubleTableState:doubleTableState];             //设置 y 的值
    UIEdgeInsets modifiedBehindSubTableInset = self.behindSubTable.
    tableView.contentInset;
    modifiedBehindSubTableInset.bottom  =  (self.view.frame.size.height
    -animatedFrontSubTableFrame.origin.y);
    [self.behindSubTable.tableView setContentInset:modifiedBehindSubTable
    Inset];                                         //设置内容视图的位置
    UIEdgeInsets modifiedFrontSubTableInset = self.frontSubTable.
    tableView.contentInset;
    modifiedFrontSubTableInset.bottom = animatedFrontSubTableFrame.
    origin.y;
    [self.frontSubTable.tableView setContentInset:modifiedFrontSubTableInset];
                                                    //设置内容视图的位置
    self.lastDoubleTableState = doubleTableState;
```

```
        //动画效果
        [UIView animateWithDuration:0.3
                              delay:0
                            options: UIViewAnimationOptionCurveEaseOut
                         animations:^{
                             //设置框架
                             [self.frontSubTable.view setFrame:animatedFront
                             SubTableFrame];
                         } completion:^(BOOL finished) {
                             //判断 self.lastDoubleTableState 是否等于
                             doubleTableStateMaxFront
                             if (self.lastDoubleTableState ==
                             doubleTableStateMaxFront) {
                                 self.behindSubTable.tableView.scrollsToTop = NO;
                                 self.frontSubTable.tableView.scrollsToTop = YES;
                             }
                             else {
                                 self.behindSubTable.tableView.scrollsToTop = YES;
                                 self.frontSubTable.tableView.scrollsToTop = NO;
                             }
                         }];
}
```

frontSubTableOriginYforDoubleTableState:方法获取子表的 y 值。程序代码如下：

```
- (CGFloat)frontSubTableOriginYforDoubleTableState:(DOUBLE_TABLE_STATE)
doubleTableState {

    CGFloat frontSubTableOriginY = minimumHeightSubTable;
    //判断 doubleTableState 是否等于 doubleTableStateMaxFront
    if (doubleTableState == doubleTableStateMaxFront) {
        return frontSubTableOriginY;
    }
    CGFloat maximumVisibleHeight = self.view.frame.size.height
    -minimumHeightSubTable;
    CGFloat actualContentHeight = self.behindSubTable.tableView.contentSize.
    height;
    //判断 actualContentHeight 是否大于 maximumVisibleHeight
    if (actualContentHeight > maximumVisibleHeight) {
        frontSubTableOriginY = maximumVisibleHeight;
    } else {
        frontSubTableOriginY = actualContentHeight;
    }
    return frontSubTableOriginY;                            //返回
}
```

【代码解析】

本实例关键功能是表的移动。下面就是这个知识点的详细讲解。

在运行结果中可以看到，当滚动第一个表中的内容时，第二个表就会向下移动到达指定的位置停止，此功能需要在 scrollViewWillBeginDragging:方法中实现。其中，实现此功能的关键代码如下：

```
if (self.isBehindSubTable) {
    if (nextDoubleTableState != doubleTableStateMinFront) {
        nextDoubleTableState = doubleTableStateMinFront;
        shouldConfigure = YES;
    }
} else {
    if (nextDoubleTableState != doubleTableStateMaxFront) {
```

```
            nextDoubleTableState = doubleTableStateMaxFront;
            shouldConfigure = YES;
        }
    }
    if (shouldConfigure) {
        [mainController
configureSubTablesForDoubleTableState:nextDoubleTableState];
    }
```

在此代码中，对两个表的状态进行了重新的设置。

实例 117　表单元格内容的复制

【实例描述】

在表视图中，表单元格可以进行删除、移动、添加等操作。除此以外，还可以进行其他的操作。本实例主要的功能是实现表单元格内容的复制。它是为了方便用户对表视图中内容的提取而设计。运行效果如图 13.33 所示。

图 13.33　实例 117 运行效果

第 13 章 表

【实现过程】

（1）创建一个项目，命名为"表单元格内容的复制"。

（2）创建一个基于 UITableViewCell 类的 CopyCell 类。

（3）打开 CopyCell.h 文件，编写代码，实现协议、实例变量、属性以及方法的声明。程序代码如下：

```
#import <UIKit/UIKit.h>
@class CopyCell;
//协议
@protocol CopyCellDelegate<NSObject>
@required
- (void) copyableCell:(CopyCell *)cell selectCellAtIndexPath:(NSIndexPath *)indexPath;
- (void) copyableCell:(CopyCell *)cell deselectCellAtIndexPath:(NSIndexPath *)indexPath;
@optional
- (NSString *) copyableCell:(CopyCell *)cell dataForCellAtIndexPath:(NSIndexPath *)indexPath;
@end
@interface CopyCell : UITableViewCell{
    //实例变量
    NSString *data;
    NSIndexPath *indexPath;
}
//属性
@property (nonatomic, retain) NSString *data;
@property (nonatomic, retain) NSIndexPath *indexPath;
@property (nonatomic, assign) id<CopyCellDelegate> delegate;
//方法
- (void) initialize;
- (void) menuWillHide:(NSNotification *)notification;
- (void) menuWillShow:(NSNotification *)notification;
- (void) handleLongPress:(UILongPressGestureRecognizer *)longPressRecognizer;
@end
```

（4）打开 Copy.m 文件，编写代码，实现对表单元格的设置、菜单的显示与复制以及菜单的隐藏功能。使用的方法如表 13-37 所示。

表 13-37　Copy.m 文件中方法总结

方　　法	功　　能
initWithStyle:reuseIdentifier:	表单元格的初始化
initialize	初始化设置
setSelected:animated:	设置选择和动画
canPerformAction:withSender:	告知菜单栏哪些菜单可以显示
copy:	复制
canBecomeFirstResponder	判断对象是否为第一响应者，并返回 YES
becomeFirstResponder	获取第一响应者
touchesEnded:withEvent:	触摸
menuWillHide:	隐藏菜单
menuWillShow:	显示菜单
handleLongPress:	长按

这里需要讲解几个重要的方法（其他方法请读者参考源代码）。表单元格的设置需要使用 initWithStyle:reuseIdentifier:、initialize 和 setSelected:animated:方法。其中，initialize 方法实现对一些默认设置的初始化。程序代码如下：

```
- (void)initialize
{
    self.data = nil;
    self.indexPath = nil;
    self.delegate = nil;
    self.selectionStyle = UITableViewCellSelectionStyleNone;
    //长按手势识别器对象的创建以及设置
    UILongPressGestureRecognizer *recognizer =[[UILongPressGestureRecognizer
    alloc] initWithTarget:self action:@selector(handleLongPress:)];
    [recognizer setMinimumPressDuration:kLongPressMinimumDurationSeconds];
    [self addGestureRecognizer:recognizer];
}
```

菜单的显示需要使用 canPerformAction:withSender:、canBecomeFirstResponder、becomeFirstResponder、menuWillShow:和 handleLongPress:方法。canPerformAction:withSender:方法实现告知菜单栏哪些菜单可以显示。程序代码如下：

```
- (BOOL)canPerformAction:(SEL)action withSender:(id)sender
{
    if (action == @selector(copy:)) {
        return YES;
    }
    return [super canPerformAction:action withSender:sender];    //返回
}
```

menuWillShow:方法实现菜单的显示。程序代码如下：

```
- (void)menuWillShow:(NSNotification *)notification
{
    self.selectionStyle = UITableViewCellSelectionStyleBlue;
    if ((self.delegate != nil) && [self.delegate respondsToSelector:@selector
    (copyableCell:selectCellAtIndexPath:)]) {
        //调用 copyableCell:selectCellAtIndexPath:方法
        [self.delegate copyableCell:self selectCellAtIndexPath:self.
        indexPath]
    }
    //移除通知
    [[NSNotificationCenter defaultCenter] removeObserver:self
name:UIMenuControllerWillShowMenuNotification object:nil];
    //注册通知
    [[NSNotificationCenter defaultCenter] addObserver:self selector:
@selector(menuWillHide:)
name:UIMenuControllerWillHideMenuNotification object:nil];
}
```

handleLongPress:方法实现长按的手势动作，在长按手势动作后，出现菜单。程序代码如下：

```
- (void)handleLongPress:(UILongPressGestureRecognizer *)
longPressRecognizer
{
    //判断 longPressRecognizer.state 是否不等于 UIGestureRecognizerStateBegan
    if (longPressRecognizer.state != UIGestureRecognizerStateBegan) {
```

```
        return;
    }
    if ([self becomeFirstResponder] == NO) {
        return;
    }
    UIMenuController *menu = [UIMenuController sharedMenuController];
                                                                //实例化对象
    [menu setTargetRect:self.bounds inView:self];
    [[NSNotificationCenter defaultCenter] addObserver:self selector:
@selector(menuWillShow:)
name:UIMenuControllerWillShowMenuNotification object:nil];//添加观察者
    [menu setMenuVisible:YES animated:YES];                  //显示
}
```

如果想要实现复制以及隐藏功能,需要使用copy:和menuWillHide:方法。其中,copy:方法实现对表单元格内容的复制。程序代码如下:

```
- (void)copy:(id)sender
{
    if ((self.delegate != nil) && [self.delegate respondsToSelector:
@selector(copyableCell:dataForCellAtIndexPath:)]) {
        //实例化对象
        NSString *dataText = [self.delegate copyableCell:self
        dataForCellAtIndexPath:self.indexPath];
        [UIPasteboard generalPasteboard].string=dataText;    //设置字符串
    }
    else if (self.data != nil) {
        [UIPasteboard generalPasteboard].string=self.data;   //设置字符串
    }
    [self resignFirstResponder];
}
```

menuWillHide:方法实现菜单的隐藏功能。程序代码如下:

```
- (void)menuWillHide:(NSNotification *)notification
{
if ((self.delegate != nil) && [self.delegate respondsToSelector:
@selector(copyableCell:deselectCellAtIndexPath:)]) {
        [self.delegate copyableCell:self deselectCellAtIndexPath:
        self.indexPath];
    }
    self.selectionStyle = UITableViewCellSelectionStyleNone;
        //设置选择块的风格
    [[NSNotificationCenter defaultCenter] removeObserver:self name:
    UIMenuControllerWillHideMenuNotification object:nil];
}
```

(5)打开 ViewController.h 文件,编写代码,实现头文件、遵守协议、插座变量以及对象的声明。程序代码如下:

```
#import <UIKit/UIKit.h>
#import "CopyCell.h"
@interface ViewController : UIViewController<CopyCellDelegate,UITableView
Delegate,UITableViewDataSource>{
    IBOutlet UITableView *demoTableView;
    IBOutlet UITextField *tf;                    //声明关于文本框的插座变量
    NSArray *tableData;
}
@end
```

（6）打开 Main.storyboard 文件，对 View Controller 视图控制器的设计界面进行设计，效果如图 13.34 所示。

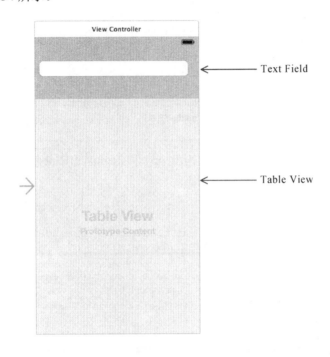

图 13.34 实例 117 View Controller 视图控制器的设计界面

需要添加的视图、控件以及对它们的设置如表 13-38 所示。

表 13-38 实例 117 视图、控件设置

视图、控件	属 性 设 置	其 他
设计界面	Background：浅灰色	
Text Field		与插座变量 tf 关联
Table View		与插座变量 demoTableView 关联 将 dataSource 与 View Controller 关联 将 delegate 与 View Controller 关联

（7）打开 ViewController.m 文件，编写代码，实现表视图内容的填充以及表单元格内容的复制。使用的方法如表 13-39 所示。

表 13-39 ViewController.m文件中方法总结

方 法	功 能
viewDidLoad	视图加载后调用，实现初始化
numberOfSectionsInTableView:	获取块数
tableView:numberOfRowsInSection:	获取行数
tableView:cellForRowAtIndexPath:	获取某一行的数据
copyableCell:selectCellAtIndexPath:	选择某一行
copyableCell:deselectCellAtIndexPath:	取消选择的行
copyableCell:dataForCellAtIndexPath:	获取选择行的内容

其中，表视图内容的填充需要使用 viewDidLoad、numberOfSectionsInTableView:、tableView:numberOfRowsInSection:和 tableView:cellForRowAtIndexPath:方法实现。程序代码如下：

```
//初始化
- (void)viewDidLoad{
    [super viewDidLoad];
    // Do any additional setup after loading the view, typically from a nib.
    tableData = [NSArray arrayWithObjects: @"leopard 豹子", @"bird 鸟",
@"elephant 大象",@"fox 狐狸",@"deer 鹿", @"duck 鸭子",@"orangutan 猩猩",
@"panda 熊猫",@"frog 青蛙",@"chicken 小鸡",@"ant 蚂蚁",nil];        //创建数组
    tf.inputView = [[UIView alloc]initWithFrame:CGRectZero];
}
//获取块数
- (NSInteger)numberOfSectionsInTableView:(UITableView *)tableView{
    return 1;
}
//获取行数
- (NSInteger)tableView:(UITableView *)tableView numberOfRowsInSection:
(NSInteger)section{
    return tableData.count;
}
//获取某一行的数据
- (UITableViewCell *)tableView:(UITableView *)tableView cellForRowAtIndexPath:
(NSIndexPath *)indexPath{
    static NSString *cellId = @"CopyableCell";
    CopyCell *cell = (CopyCell *) [tableView dequeueReusableCellWithIdentifier:
    cellId];
    //判断cell是否为空
    if (cell == nil) {
        cell = [[CopyCell alloc] initWithStyle:UITableViewCellStyleDefault
        reuseIdentifier:cellId] ;
    }
    [cell setIndexPath:indexPath];
    [cell setDelegate:self];                                //设置委托
    cell.selectionStyle = UITableViewCellSelectionStyleNone;
    cell.accessoryType = UITableViewCellAccessoryNone;      //设置附件类型
    cell.textLabel.text = [tableData objectAtIndex:indexPath.row];
    return cell;
}
```

表单元格内容的复制需要使用 copyableCell:selectCellAtIndexPath:和 copyableCell:dataForCellAtIndexPath:方法。程序代码如下：

```
//选择某一行
- (void)copyableCell:(CopyCell *)cell selectCellAtIndexPath:(NSIndexPath
*)indexPath{
    [demoTableView selectRowAtIndexPath:indexPath animated:NO scrollPosition:
    UITableViewScrollPositionNone];                         //选择行
}
//获取选择行的内容
- (NSString *)copyableCell:(CopyCell *)cell dataForCellAtIndexPath:
(NSIndexPath *)indexPath{
    //判断indexPath.row是否小于tableData.count
    if (indexPath.row < tableData.count) {
        return [tableData objectAtIndex:indexPath.row];
```

```
        }
        return @"";
}
```

copyableCell:deselectCellAtIndexPath:方法实现的功能是将选择的行取消。程序代码如下：

```
- (void)copyableCell:(CopyCell *)cell deselectCellAtIndexPath:
(NSIndexPath *)indexPath{
    [demoTableView deselectRowAtIndexPath:indexPath animated:NO];
}
```

【代码解析】

本实例的关键功能是复制的实现以及它的执行流程。下面依次讲解这两个知识点。

1. 复制的实现

所谓复制，就是将内容放置在粘贴板上，在 iOS 中，使用 UIPasteboard 的 string 属性实现，其语法形式如下：

```
@property(nonatomic, copy) NSString *string;
```

在此代码中就是使用 string 方法，实现了表视图内容的复制功能。代码如下：

```
[UIPasteboard generalPasteboard].string=dataText;
```

2. 执行流程

由于本实例中的代码和方法非常的多，为了方便读者的阅读，笔者绘制了一些执行流程图如图 13.35～图 13.36 所示。其中，单击运行按钮一直到表视图的显示，使用了 viewDidLoad、numberOfSectionsInTableView:、tableView:numberOfRowsInSection:、tableView:cellForRowAtIndexPath:、initWithStyle:reuseIdentifier:、initialize 和 setSelected:animated:方法共同实现。它们的执行流程如图 13.35 所示。

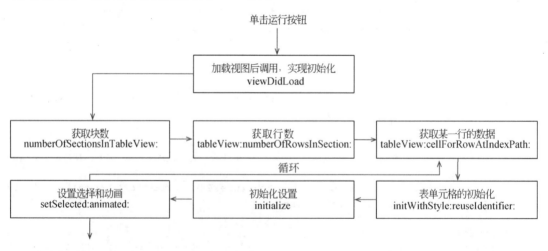

图 13.35　实例 117 程序执行流程 1

复制表单元格中的内容并且粘贴到文本框中，需要使用 handleLongPress:、becomeFirstResponder、canBecomeFirstResponder、canPerformAction:withSender:、menuWillShow:、copyableCell:selectCellAtIndexPath:、setSelected:animated:、copy:、copyableCell:dataForCellAtIndexPath:、menuWillHide 和 copyableCell:deselectCellAtIndexPath: 方法共同实现。它们的执行流程如图 13.36 所示。

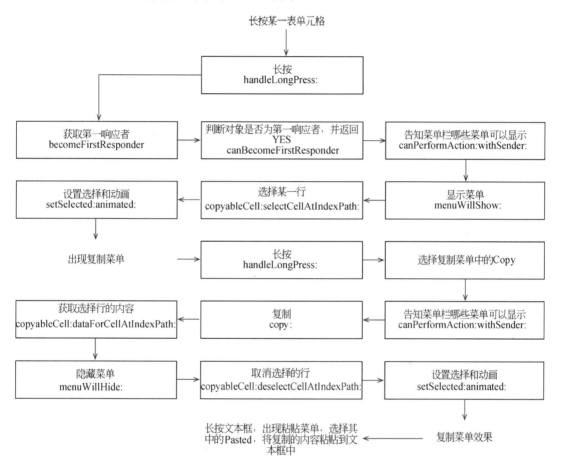

图 13.36　实例 117 程序执行流程 2